Compound Semiconductor Bulk Materials and Characterizations

Volume 2

Compound Semiconductor Bulk Materials and Characterizations

Volume 2

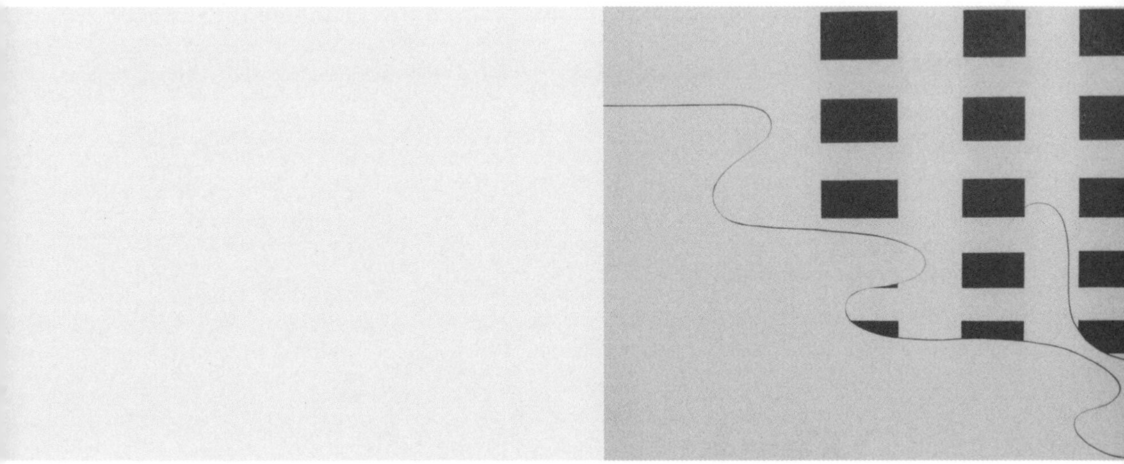

Osamu Oda

Nagoya Institute of Technology

World Scientific

NEW JERSEY • LONDON • SINGAPORE • BEIJING • SHANGHAI • HONG KONG • TAIPEI • CHENNAI

Published by

World Scientific Publishing Co. Pte. Ltd.
5 Toh Tuck Link, Singapore 596224
USA office: 27 Warren Street, Suite 401-402, Hackensack, NJ 07601
UK office: 57 Shelton Street, Covent Garden, London WC2H 9HE

British Library Cataloguing-in-Publication Data
A catalogue record for this book is available from the British Library.

The Author

Dr. Osamu Oda
Special Professor
Nagoya Institute of Technology
Nagoya, Aichi
Japan

Cover
Sakuzo Ohyama
https://sites.google.com/site/galleryohyama/

COMPOUND SEMICONDUCTOR BULK MATERIALS AND CHARACTERIZATIONS
Volume 2

ISBN 978-981-283-505-5

Printed in Singapore by B & Jo Enterprise Pte Ltd

PREFACE

This book is the second volume of a two-volume set entitled "Compound Semiconductor Bulk Materials and Characterization". This second volume deals with materials which could not be described in the first one (referred to herein as Vol. 1). These are III-V mixed crystals, nitrides and other III-V compounds, ZnO, mercury compounds, SiC, lead compounds and chalcopyrite compounds. Some of these materials have been developed quite recently. In the appendix, raw materials necessary for these compound semiconductors including those described in Vol. 1 are explained.

As mentioned in the preface of Vol. 1, compound semiconductor materials have found many applications such as in LEDs and LDs based on GaP, GaAs, and GaN which are used for displays, lighting, data storage, CDs and DVDs and in communications and electronic devices based on GaAs which are used for satellite communication systems, cellular phones and wireless base stations. Compound semiconductors are still developing, especially GaN and SiC for high power electronic devices. In this volume, these recent developments are also reviewed.

Because the world's energy consumption is increasing, energy saving technologies are indispensable to future society. In fact, GaN based white LEDs are being rapidly developed in the world, including emerging countries. The development of GaN and SiC power electronic devices is being strongly pursued in order to apply them in the field of offline power supplies, home electronics, hybrid and electric vehicles and smartgrid systems, especially for integrating solar cells to the power electric supply networks. The fact that CO_2 emissions per day are reaching about half the weight of a person, 37 kg/person/day for instance in the case of Japan, shows the urgent necessity of decreasing our energy consumption. It is also urgent since at present it seems that nuclear power is necessary though evil as proved by the Great East Japan Earthquake. In order to realize a society without nuclear power, it is indispensable to develop sustainable energy sources and power electronics and compound semiconductor materials may be the key technologies for their realization.

In fact, LEDs using GaP, GaAs, GaN are energy saving displays, especially GaN based white LEDs which are replacing incandescent bulbs and phosphorescent lights in the world. Electronic devices based on GaAs are indispensable for mobile devices such as cellular phones, smartphones and tablets. GaN and SiC are now looking at application outlets in energy saving systems. It will therefore give me great pleasure if my books including Vol. 1 serve as an introduction for young students and researchers and engineers who are looking for solutions for energy problems.

I here at the outset would like to express my respect to many researchers who dedicated most of their lives in developing compound semiconductor materials. The reader can trace their persistent efforts reading these books and seeing the references. Most of them worked hard, with enthusiasm, conviction and hope that their efforts would be useful and contribute to society.

On this occasion, I would like to express many thanks to ex-vice president I. Ichikawa and ex-executive manager M. Ohno for their encouragement in our research in NGK where I was engaged in AlN and GaN materials and power devices, Drs. M. Tanaka, U. Hori, M. Miyoshi, T. Shibata who worked with me in the field of nitride materials and Dr. N. Shimizu and Mrs. K. IIda, Y. Imanishi and S. Tange in that of power devices. I am deeply grateful to those who collaborate and discuss with me during my work in NGK, Dr. B. Daudin and E. Monroy at CEA Grenoble, Prof. N. Grandjean at EPFL, Drs. D. Ueda, T. Tanaka and T. Ueda at Panasonic Co., and ex-executive manager S. Yoshida and ex-director H. Iwakuro at Shidengen Electric Manufacturing Co. I would like to express my appreciation to ex-president T. Ishimaru, president S. Kato, Profs. S. Maruse and Y. Inagaki, project manager Mr. O. Takenaka and Dr. S. Noda at Aichi Science and Technology Foundation (ASTF) for their encouragement of my activity in the promotion of science and technology in the Tokai area and Ms. M. Moriya, Y. Kaminani, L. Nakagawa, K. Osa and T. Ito for their assistance in my work and to start-up the International Symposium on advanced Plasma science and its applications for nitrides and nanomaterials (ISPlama) and Nitride Semiconductor Workshop (NSW) which has been sponsored by ASTF. During the activity of ASTF, I could discuss with Profs. M. Hori, H. Amano and H. Toyoda at Nagoya University, Prof. A. Wakahara at Toyohashi University of Technology, Prof. H. Miyake at Mie University, Prof. M. Hiramatsu at Meijou University, Prof. K. Nakamura at Chubu University and Director G.C. Guibert at CEA-MINATEC, ex-director C.D. Dilks at Philadelphire Sience Center, Director J. Duboz at CREHA-CNRS, Prof. Y. Nanishi at Ritsumeikan University, Dr. R. Toba and Mr. H. Kobayashi at Dowa Mining Co., Dr. M. Hata at Sumitomo Chemical Co. and Mr. M. Isemura at Itochu Plastics Inc. and I deeply appreciate their useful discussion and sincere help. I especially would like to express my great thanks to Prof. T. Egawa at Nagoya Institute Technology who accepted me as a special professor in the field of GaN materials and devices and Drs. L. Selvaraj and Y. Sakai for their work with me and Ms. K. Watanabe, Y. Sakabe and N. Imura for their continuous help.

I would like to express my deep thanks to those who are related with my university era, ex-Profs. Y. Masuda and R. Watanabe at Tohoku University where I finished my bachelor and master course in the field of metallurgy, where I have learned the fundamentals of physical chemistry which has been a great help to my scientific understanding of crystal growth phenomena. Ex-director Prof. J. Benard at ENSCP of Paris University who accepted me as a French government scholarship student and ex-Prof. G. Rhead and Prof. F. Marcus, who directed my doctorate work in the field of surface science, and ex-Prof. G.A. Bootsma and Dr. L.J. Hanekamp at Twente University of Technology where I did my post-doctorate work in the field of surface science. Even though crystal growth itself was not my speciality when I started it in Japan Energy Co. (presently JX Nippon Oil & Energy Co.), knowledge at universities has helped me greatly in this field since fundamental science is always in demand even though the field is different and the academic experience was of great help to me in my work in the field of compound semiconductors.

I want to acknowledge Dr. F.S. Skeffington, a physicist in Cambridge, for his careful and through review of the manuscript with interest in this field and B. Wong at World Scientific Co. for his patience awaiting the completion of this book and his understanding of its value. Finally, I wish specially to thank my wife Mikiko and my daughter Tomoko for their encouragement during the writing of this book.

Osamu Oda

Nagoya, Japan
August 2012

CONTENTS

PART 4 OTHER COMPOUND SEMICONDUCTOR MATERIALS....................1

19. III-V MIXED CRYSTALS 3

19.1 INTRODUCTION 3

19.2 PHYSICAL PROPERTIES 3

19.3 CRYSTAL GROWTH 5

19.4 APPLICATIONS 21

20. NITRIDE AND OTHER III-V COMPOUNDS 27

20.1 INTRODUCTION 27

20.2 PHYSICAL PROPERTIES 27

20.3 CRYSTAL GROWTH 29

20.4 CHARACTERIZATION 73

20.5 APPLICATIONS 92

21. ZnO 127

21.1 INTRODUCTION 127

21.2 PHYSICAL PROPERTIES 127

21.3 CRYSTAL GROWTH 127

21.4 CHARACTERIZATION 137

21.5 APPLICATIONS 143

22. MERCURY COMPOUNDS 151

22.1 INTRODUCTION 151

22.2 PHYSICAL PROPERTIES 151

22.3 CRYSTAL GROWTH 153

22.4 CHARACTERIZATION 178

22.5 APPLICATIONS 193

23. SiC 205

23.1 INTRODUCTION 205
23.2 PHYSICAL PROPERTIES 205
23.3 CRYSTAL GROWTH 208
23.4 CHARACTERIZATION 252
23.5 APPLICATIONS 278

24. LEAD COMPOUNDS 315

24.1 INTRODUCTION 315
24.2 PHYSICAL PROPERTIES 315
24.3 CRYSTAL GROWTH 316
24.4 CHARACTERIZATION 332
24.5 APPLICATIONS 340

25. CHALCOPYRITE COMPOUNDS 349

25.1 INTRODUCTION 349
25.2 PHYSICAL PROPERTIES 349
25.3 CRYSTAL GROWTH 352
25.4 CHARACTERIZATION 361
25.5 APPLICATIONS 363

APPENDIX: RAW MATERIALS 369

1. INTRODUCTION 369
2. PURIFICATION PROCESS 369
3. ZINC 372
4. CADMIUM 373
5. MERCURY 375
6. GALLIUM 376
7. INDIUM 377
8. PHOSPHORUS 378
9. ARSENIC 380
10. ANTIMONY 381
11. BISMUTH 382
12. SELENIUM 383

13. TELLURIUM 384
14. BORIC OXIDE 386
15. PYROLYTIC BORON NITRIDE (pBN) 386
16. GRAPHITE 387

INDEX 391

PART 4

OTHER COMPOUND SEMICONDUCTOR MATERIALS

19. III-V Mixed Crystals

20. Nitride and Other III-V Compounds

21. ZnO

22. Mercury Compounds

23. SiC

24. Lead Compounds

25. Chalcophyrite Compounds

19. III-V MIXED CRYSTALS

19.1 INTRODUCTION

As explained in Vol. 1, most III-V crystals are very important materials for a variety of devices. These devices are based on epitaxial ternary and quaternary compound semiconductor layers grown on one of these III-V material substrates. Clearly, however, the composition of these ternary and quaternary layers is constrained since the composition must be made appropriate in such a way that the lattice constant of the epitaxial layers corresponds to that of the substrates. When there is a big difference in these lattice constants, lattice misfit dislocations arise at the interface between the substrate and the epitaxial layer and they propagate in the epitaxial layer and compromise its electro-optical properties. This is the context which informs the preparation of III-V mixed single crystals as substrates for the desired ternary and quaternary epitaxial layers. When crystals can be realized in this way, the fabrication possibility of various devices expands. In fact, this possibility has been shown in the case of infrared $In_{1-x}Ga_xAs$ lasers on $In_{1-x}Ga_xAs$[1] and $In_{1-x}As_xP$[2] substrates.

The difficulty in the growth of III-V crystals stems from constitutional supercooling because of the wide deviation of the solidus and liquidus lines in III-V mixed crystal phase diagrams. Various crystal growth methods have been developed to prevent the constitutional supercooling problem as explained below. These crystal growth methods may be helpful not only for the growth of III-V mixed crystals described in this chapter but also for those of III-nitrides and II-VI materials.

19.2 PHYSICAL PROPERTIES

Phase diagrams for mixed crystals are very important in considering crystal growth. Some of them are shown In Fig. 19.1, Phase diagrams for Ga-P-As[3], Ga-In-P[4-7], Ga-In-As[4,8-10], In-P-As[11-13], Ga-In-Sb[14-16] and quaternary III-V materials[13,17] have been extensively studied. All III-V mixed crystals show isomorphous solid solutions. There are however wide deviations between solidus and liquidus lines and constitutional supercooling gives rise to the difficulty of crystal growth as explained in Sec. 3.6.2 of Vol. 1.

Lattice constants of III-V mixed crystals are predicted by Vegard's law but some deviations are probable. Lattice constants of various III-V mixed crystals are summarized in Fig. 19.2[18]. Lattice constants for Ga-P-As[19,20], Ga-In-P[21], Ga-In-As[22], In-P-As[23], Ga-In-Sb[24,25] have been extensively studied.

Various physical properties of III-V mixed crystals are summarized in Refs. 26 and 27. In Table 19.1, the physical properties of the main III-V ternary crystals are reviewed.[28] In the case of mixed crystals, the miscibility gap where dissolution is not possible must be considered. The miscibility gap has been calculated by Onda and Ito.[29,30] Nishinaga and Hiramatsu[31] have calculated the relationship between the excess free energy of mixing and the miscibility gap. The position and the range of the miscibility

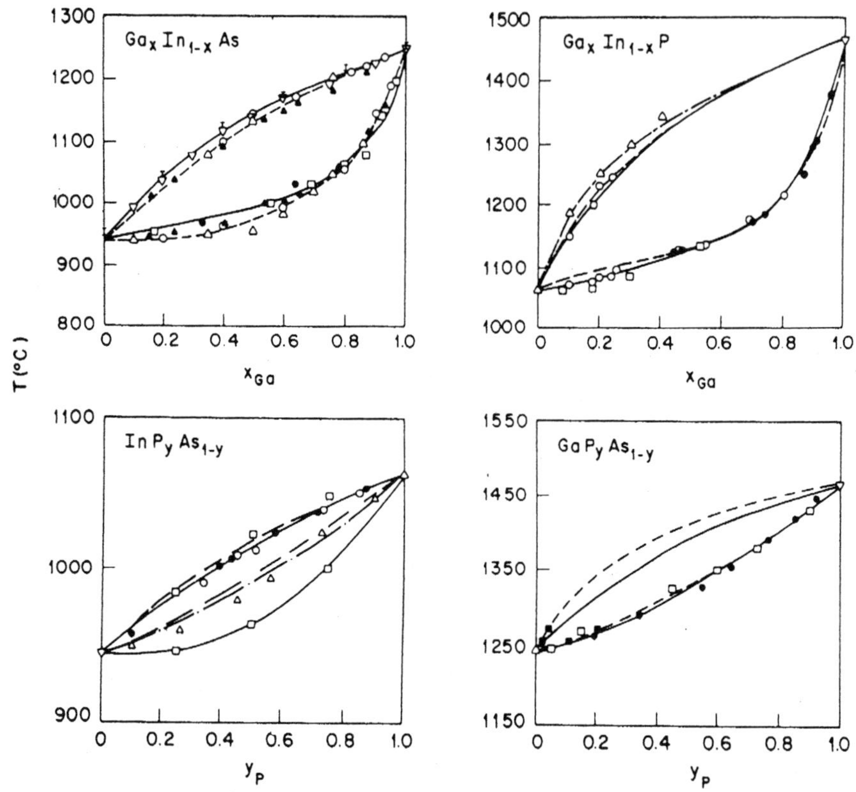

Fig. 19.1 Pseudobinary phase diagrams of various III-V mixed crystals. Solid lines calculated (reprinted from Ref. 13 with permission, copyright 1981 Elsevier).

Table 19.1 Physical properties of ternary III-V crystals

Ternary System	Bandgap(eV)	Mel2Tting Point (°C)	Lattice Constant (Å)
GaP-GaAs	2,261-1.1424	1467-1238	5.447-5.654
GaP-InP	2.261-1.351	1062-1467	5.447-6.869
GaAs-InAs	1.424-0.360	943-1238	5.654-6.058
InP-InAs	1.351-0.360	943-1062	5.869-6.058
GaSb-InSb	0.726-0.172	525-706	6.095-6.497

gap in the phase diagram has been explained by the regular solution approximation and the delta lattice parameter model.[31,32] Stringfellow[33] has examined the criteria for clustering and spinodal decomposition in III-V mixed crystals. Analytical expressions are derived for the temperature above which no clustering or phase separation occurs. Yamazakei et al.[34] have calculated the excess free energy of the mixing of III-V pseudobinary alloys and analyzed the phonon spectra on the basis of a cluster model.

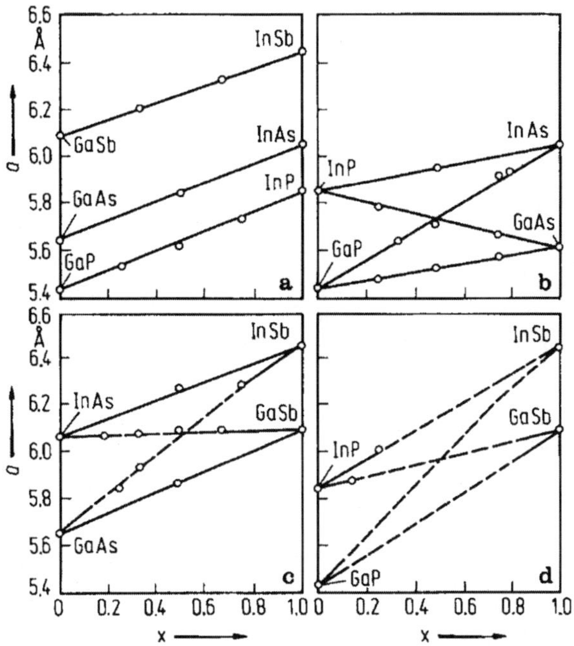

Fig. 19.2 III-V solid solutions. Variation of lattice parameters in various pseudobinary systems. Immiscibility is indicated by broke lines (reprinted from Ref. 18 with permission, copyright 1981 Elsevier).

Jones et al.[35] have used the difference of the bond probability between identical cations (anions) in quasi-chemical and random solid solutions as a quantitative measure of the tendency for clustering to occur in ternary III-V crystals. Onabe[36] has derived an expression for the spinodal curves for quaternary solid solutions.

The thermal conductivity of mixed III-V materials has been theoretically studied as a function of the composition, based on the phenomenological approach of Abeles and a more exact method.[37]

19.3 CRYSTAL GROWTH

The main difficulty in growing mixed III-V crystals is in controlling the composition and in preventing constitutional supercooling as explained in Sec. 3.6.2 of Vol. 1. When a crystal with a certain composition is grown from a mixed melt, the composition of the melt varies greatly during crystal growth because the composition of the melt is different from that of the grown crystal. This changes the crystal composition along the growth axis. The change of the melt composition alters the melting point and supercooling occurs and polycrystals are grown.

To overcome this difficulty, various technologies to keep the melt composition constant have had to be introduced to conventional crystal growth methods. These technologies are reviewed in Refs. 38 and 39. Table 19.2 summarizes the growth of various

Table 19.2 Crystal Growth of Mixed III-V Compounds

Method	Contents	Authors	Year	Ref.
CVT	Growth of $GaAs_{1-x}P_x$ by iodine transport	Pizzarello	1962	130
CVT	Growth of $In_{1-x}Ga_xAs$ by chlorine transport	Minden	1965	41
CVT	Growth of $GaAs_{1-x}P_x$ by iodine transport	Rubenstein	1965	42
THM	Growth of $GaAs_{1-x}P_x$ $(0.5 < x < 0.7)$ from Ga solution	Wolff et al.	1968	129
ZL	Growth of $GaAs_{1-x}P_x$ using B_2O_3 as the encapsulant	Draper et al.		49
CZ, LEC	Growth of $In_{1-x}Ga_xAs$ for $0 \le x \le 0.2$ and $0.8 \le x \le 1.0$	Wagner	1970	70
Bridgman	Solution growth of $In_{1-x}Ga_xP$ by the Bridgman method	Burnham et al.	1970	66
THM	Growth of $In_{1-x}Ga_xP$ from the InP zone	Blum et al.	1970	54, 55
HSG	Horizontal solution growth of $In_{1-x}Ga_xP$	Itoh et al.	1971	67
CZ, LEC	Growth of $InAs_xP_{1-x}$ by the LEC and Gremmelmaier methods	Thompson et al.	1971	122
TDM-CVP	Growth of $In_{1-x}Ga_xP$ from various solution compositions	Okuno et al.	1971	68
CZ	Growth of $InSb_{1-x}Bi_x$ up to $x = 0.026$	Joukoff et al.	1972	135
LEC	Growth of $GaAs_{1-x}P_x$ from Ga-rich solution	Cerrina et al.	1973	126
SSD	Growth of $In_{1-x}Ga_xP$ from In(+Te) solution	Kato et al.	1974	57
LPEE	Growth of $Al_xGa_{1-x}As$ $(0.1 < x < 0.3)$ of thickness 15-58 μm	Daniele et al.	1975	132,133
SSD	Growth of $In_{1-x}Ga_xP$	Gillesen et al.	1977	58
CZ	Growth of $InGa_{1-x}Sb_x$ with $0.9 < x < 1$ with high speed rotation	Bachmann et al.	1980	105
LPEE	Growth of $AlxGa_{1-x}As(0.03 < x < 0.38)$ up to 1.4 mm thick	Daniele et al.	1981	134
LEC	Growth of $In_{1-x}P_xAs$ with pulling rate less than 100 μm/min	Watanabe et al.	1983	125
LPEE	Growth of thick $In_{1-x}Ga_xP$ epilayers on (100) GaAs	Daniel et al.	1983	62
CVT	Growth of $GaAs_{1-x}P_x$ by iodine transport	Takeda et al.	1984	131
CZ	Growth of $In_{1-x}Gax_xSb$ by the solute-feeding CZ method	Sukegawa et al.	1985-2000	46, 106-114
TDM-CVP	Growth of $In_{1-x}Ga_xP$ measuring $4 \times (4\text{-}5) \times 5\ mm^3$	Nishizawa et al.	1986	60
LEC	Growth of $In_{1-x}Ga_xAs$ using B_2O_3-Na_3AlF_6 encapsulant in the solute-transport method	Sigiura et al.	1987	76
LEC	Growth of $GaAs_{1-x}P_x$ of diameter 20 mm and length 20 mm	Hibiya et al.	1987-1988	127, 128
LPEE	Growth of $In_{1-x}Ga_xAs$ of thickness 1.5 mm	Bryskiewicz	1988	101
YO-YO	Growth of $In_{1-x}Ga_xP$ by a modified LPE method	Sukegawa	1988	100
LEC	Growth of $In_{1-x}Ga_xAs$ for $0.01 < x < 0.12$ with 25 mm diameter	Bonner et al.	1989-1990	72-73
THM	Growth of $In_{1-x}Ga_xP$ from In solution	Bischopink et al.	1989	56
CZ	Application of ultrasonic vibration to the growth of $In_{1-x}Ga_xSb$	Tsuruta et al.	1990	102
Bridgman	Growth of $InSb_{1-x}Bi_x$ of thickness 17 mm with rotationary-continuous feeding	Ozawa et al.	1991	136
LEC	Growth of $In_{1-x}Ga_xAs$ with various methods for source feeding to the crucible	Nakajima et al.	1991-1996	77-78
LEC	Growth of $In_{1-x}Ga_xAs$ with $0 < x < 0.10$ of 50 mm diameter	Bonner et al.	1994	74, 75
CZ	Double crucible method for the growth of $InGa_{1-x}Sb_x$	Ashley et al.	1995	103
ZM	Multicomponent zone-melting method for the growth of $In_{1-x}Ga_xAs$	Nakajima et al.	1995-2005	79-83
SG	Growth of $In_{1-x}Ga_xAs$ by the temperature difference method	Ozawa et al.	1996	85
Bridgman	Growth of $In_{1-x}Ga_xAs$ in microgravity	Fujiwara et al.	1996	95
Bridgman	Growth of $In_{1-x}Ga_xAs$ by rotation of the furnace	Ozawa et al.	1997	64
LEC	Growth of $In_{1-x}P_xAs$ $(0 < x < 0.10)$	Bonner	1997	28
SHM	Growth of $In_{1-x}Ga_xSb$ through forced convection	Dutta et al.	1998	115
VB	Growth of $In_{1-x}Ga_xSb$ and $In_{1-x}Ga_xAs$ by thermochemical analysis	Dutta et al.	2000	65

Table 19.2 Crystal Growth of Mixed III-V Compounds (continued)

Method	Contents	Authors	Year	Ref.
TLZ	Growth of $In_{1-x}Ga_xAs$	Kinoshita et al.	2001-2005	52, 87-92
VGF	Growth of $In_{1-x}Ga_xSb$ with GaSb/InSb/GaSb sandwich structure	Kumagawa et al.	2003-2006	117-121
VB	Growth of $In_{1-x}Ga_xSb$ under alternating magnetic field	Miltric et al.	2006	116
CZ	Homogeneous growth of $In_{1-x}Ga_xSb$ using a double crucible method	Tsaur et al.	2007	104

CZ: Czochralski, LEC: Liquid Encapsulated Czochralski, VB: Vertical Bridgmann, VGF: Vertical Gradient Freezing, SHM: Submerged Heater Method, ZM: Zone Melting, ZL: Zone Leveling, TLZ: Travelling Liquidus Zone, SG: Solution Growth, THM: Travelling Heater Method, HSG: Horizontal Solution Growth, SSD: Solute Solution Diffusion, TDM-VCP: the Temperature Difference Method under Controlled Vapor Presure, LPEE: Lquid Phase Electroepitaxy, CVT: Chemical Vapor Transport

mixed crystals by various methods.

In the past, chemical vapor transport[40-42] (Vol. 1, Sec. 2.4.3) was studied because supercooling is not a problem. This method has however the disadvantage that growth is slow and that large crystals can not be grown.

Later, III-V mixed crystals have been grown by the LEC method (Vol. 1, Sec. 2.2.3) used for binary III-V crystals with various modifications to improve the composition homogeneity along the growth axis. A typical modified method is the double crucible method (or solute transfer pulling method).[43-46] The solute feeding method has also been applied to the LEC method.[46] Shindo et al.[47] have proposed a solvent travelling method for mixed oxide single crystal growth but this can not be used for III-V mixed crystals of larger diameter. Shindo[45] therefore developed a travelling plate method as shown in Fig. 19.3.

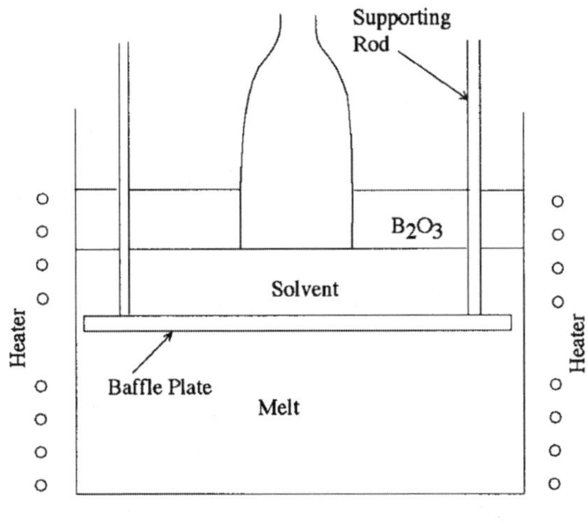

Fig. 19.3 Solute transfer pulling method (from Ref. 45 with permission).

Zone refining methods with various modifications such as temperature-gradient zone melting [48], zone leveling [49,50], solute feeding zone melting[51] and the traveling liquidus-zone (TLZ) method[52] have been explored.

Solution growth methods[53] have been used in the past and modified methods such as the traveling heater method (THM) [54-56], solute solution diffusion (SSD)[57,58], and the temperature difference method under controlled vapor pressure (TDM-CVP)[59,60], have been extensively studied .

Other methods such as vapor-liquid-solid (VLS)[61], the recharging method[54], liquid phase electroepitaxy (LPEE)[62,63], and the rotational Bridgeman method with continuous source feeding[64] have also been studied. Dutta et al.[65] have developed a thermochemistry simulation method for growing compositionally uniform ternary and quaternary alloys.

Further difficulties arise from the fact that a seed crystal of appropriate composition can not be obtained easily. Seed crystals must be prepared by a series of iterative growth cycles from a binary compound.

19.3.1 $In_{1-x}Ga_xP$

Burnham et al.[66] have applied a modified Bridgman solution growth method as shown in Fig. 19.4. Source InP and GaP were dissolved into an In solution at 935°C to saturate the In solution and $In_{1-x}Ga_xP$ crystals were deposited at the lower temperature region (925°C) to obtain crystals of the same composition (x = 0.3). Itoh et al.[67] have applied a horizontal solution growth method as shown in Fig. 19.5. At the high temperature region, a GaP crystal was placed as the source material in the high temperature region and a controlled composition $In_{1-x}Ga_xP$ single crystal was grown in the low temperature region from the solution.

Blum and Chicotak[54,55] have used a modified THM method to grow $In_{0.18}Ga_{0.82}P$ crystals using the InP zone as the solvent as shown in Fig. 19.6 and obtained mixed

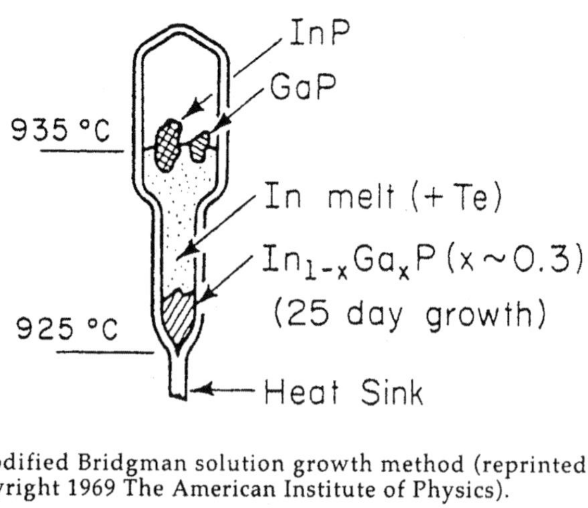

Fig. 19.4 The modified Bridgman solution growth method (reprinted from Ref. 66 with permission, copyright 1969 The American Institute of Physics).

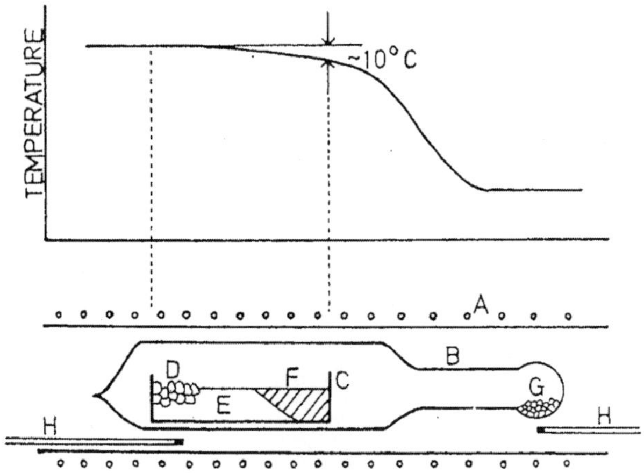

Fig. 19.5 Horizontal solution growth method for $In_{1-x}Ga_xP$ growth and temperature profile. A: furnace, B: quartz ampoule, C: boat, D: GaP source, E: In melt, F: $In_{1-x}Ga_xP$, G: red phosphorous, H: thermocouple (reprinted from Ref. 67 with permission, copyright 1969 The American Institute of Physics).

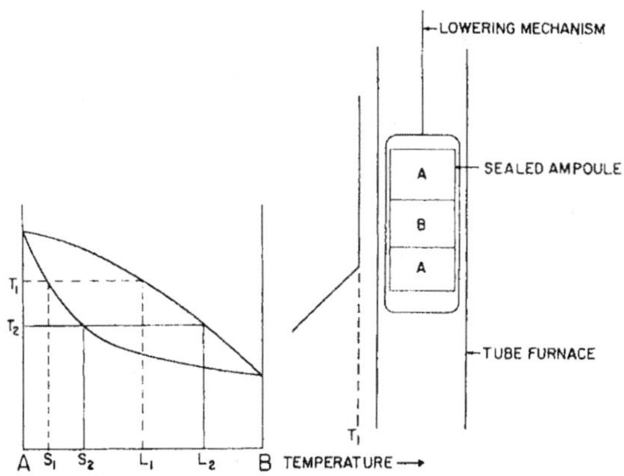

Fig. 19. 6 Explanaton of the THM method with the phase diagrm (reprinted from Ref. 55 with permission, copyright 1970 The Electrochemical Society).

crystals of diameter 15 mm and length 10 mm. By this method, the composition can be kept constant since the composition of the InP zone can be kept stable.

$In_{1-x}Ga_xP$ single crystals have been grown by the THM method, using binary GaP and InP feed materials and an In solution[56] as shown in Fig. 19.7. Stable growth was achieved for $x \geq 0.8$ at a growth temperature of 900°C. Single crystals with a diameter of 15 mm with $0.95 \leq x \leq 0.98$ have been grown by presaturation of the In solution zone and by using (111) GaP seed crystals.

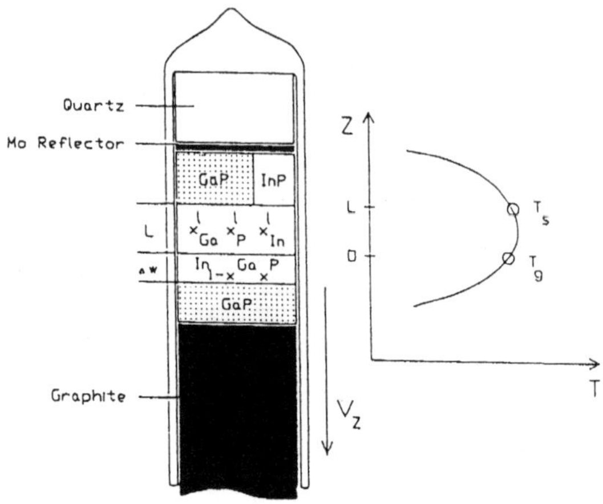

Fig. 19.7 THM method for the growth of $In_{1-x}Ga_xP$. Schematic sketch of the growth ampoule and temperature profile. x^l_{Ga}, x^l_P, x^l_{In} : liquid concentration in the solution zone, x: solid composition, L: zone length, Δw: length of the growing ternary material, T_s: temperature at interface solution/feed, T_g: temperature at interface seed/solution, v: growth velocity, z: growth direction (reprinted from Ref. 56 with permission, copyright 1989 Elsevier).

$In_{1-x}Ga_xP$ single crystals have been grown by the solute solution diffusion (SSD) method, firstly by Kato et al.[57] as shown in Fig. 19.8 and then by Gillesen and Marshall[58] and the method was called the SSD method. The principle of SSD is explained in Sec. 2.3.3 of Vol. 1.

Okuno et al.[59] have grown $In_{1-x}Ga_xP$ single crystals from an In-rich In-Ga-P solution. The zone containing the elements, Ga, In and P was heated to about 1150°C and cooled at a rate of about 100°C/hr. Nishizawa and Yoshida[60] have applied the temperature difference method under controlled vapor pressure (TDM-CVP) (Sec. 2.3.5 of Vol. 1) and have obtained $In_{1-x}Ga_xP$ single crystals measurng 4 × (4-5) × 5 mm^2.

Daniele and Lewis[62,68] used liquid phase electroepitaxy (LPEE), which was first developed for growing GaAs/GaAlAs double hetero stucture lasers[69], to grow thick $In_{1-x}Ga_xP$ layers on (100) GaAs substrates over the composition range $0.30 < x < 0.57$ and demonstated growth of uniform composition with x varying less than 0.02 over a thickness less than 100 nm.

19.3.2 $In_{1-x}Ga_xAs$

Wagner[70] has grown $In_{1-x}Ga_xAs$ single crystals by two methods, the sealed CZ method which is called the Gremmelmeier method[71] as explained in Sec. 2.2.3(1) of Vol.1 and the LEC method in Sec.2.2.3(2) of Vol. 1.

Bonner et al. have grown $In_{1-x}Ga_xAs$ with x varying from 0.01 to 0.05[72] and from 0.02 to 0.12[73] by the LEC method. They were able to grow single crystals 25 mm in

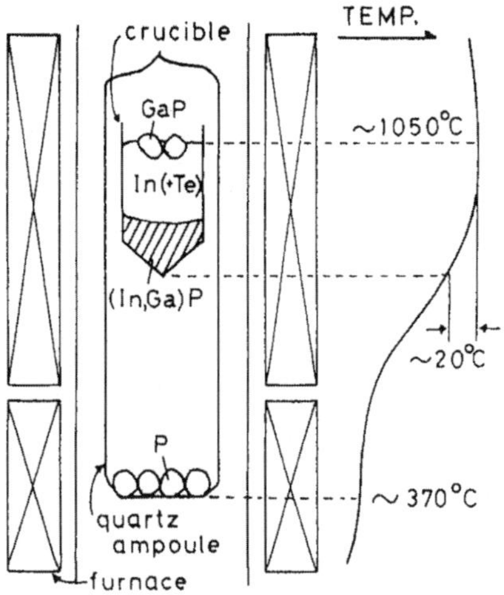

Fig. 19.8 Crystal growth of $In_{1-x}Ga_xP$ by SSD method (reprinted from Ref. 57 with permission, copyright 1974 Japan Society of Applied Physics).

diameter and 50-75 mm in length. To grow crystals with x > 0.05, $In_{1-x}Ga_xAs$ seed crystals nearly lattice matched with the grown crystals were necessary. They further increased the crystal size up to 50 mm in diameter and 1000 g in weight when the composition ranged from x = 0 to 0.10.[74, 75]

Sugiura et al.[76] first applied a solute transport pulling method to the growth of $In_{1-x}Ga_xAs$ bulk crystals, using B_2O_3-Na_3AlF_6 (3.6 mol.%) as the encapsulant of the LEC method. Nakajima et al.[77] have grown $In_{1-x}Ga_xAs$ bulk crystals by the LEC method supplying GaAs source as seen in Fig. 19.9, using <111B> $In_{1-x}Ga_xAs$ ($0.05 \leq x \leq 0.08$) seed crystal. They were able to grow $In_{1-x}Ga_xAs$ single crystals with a uniform composition of x = 0.14.[78]

Kusuniki et al.[79] have applied a two melt reservoir method as shown in Fig. 19.10 to the LEC method and grew $In_{1-x}Ga_xAs$ (x = 0.055 ± 0.004) single crystals. This method is however limited for the growth of large crystals because of the difficulty in continuously supplying the source material. To overcome this drawback, Suzuki et al.[80] developed a multicomponent zone melting method. This method is principally based on the zone melting (leveling) method[48] but the melt composition is $In_{1-x}Ga_xAs$ and from this melt a crystal of $In_yGa_{1-y}As$ can be grown whose composition is in equilibrium with the melt composition based on the phase diagram. By this method, they were successful in growing 13 mm thick $In_{1-x}Ga_xAs$ single crystals with x = 0.14.[81-84]

$In_{1-x}Ga_xAs$ alloys have been grown from solution by the solute-transport method. Ozawa et al.[85] have grown $In_{1-x}Ga_xAs$ bulk crystals on GaAs seed crystals, using the temperature difference method[86] as seen in Fig.19.11. In this method, the temperature difference is created in the solvent and crystals are grown at the lower temperature side.

Ozawa et al.[64] have grown $In_{1-x}Ga_xAs$ single crystals by the rotational Bridgman

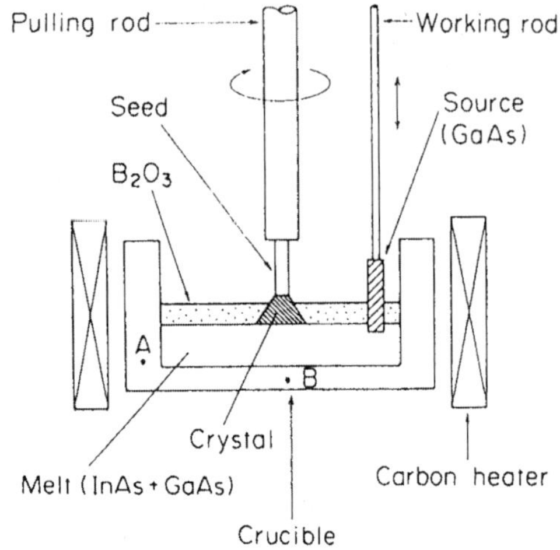

Fig. 19.9 Schematic drawing of the apparatus for a high pressure crystal growth system (reprinted from Ref. 77 with permission, copyright 1991 Elsevier).

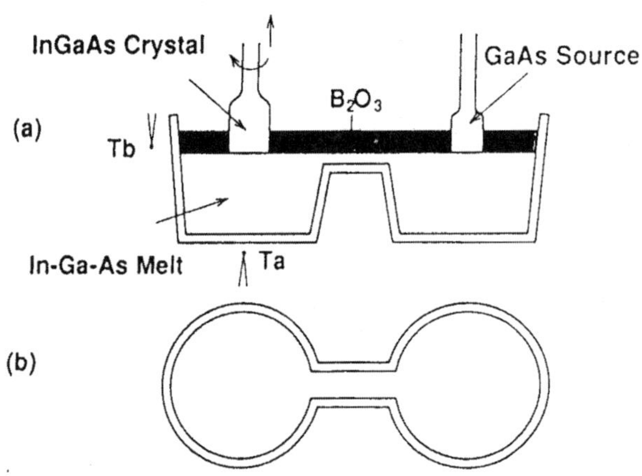

Fig. 19.10 Schematic diagram of the crucible with two melt reservoirs: (a) in cross-section and (b) from the top (reprinted from Ref. 79 with permission, copyright 1991 Elsevier).

method as shown in Fig. 19.12 in order to avoid the constitutional supercooling phenomenon. x ranged from 0.03 to 0.15 and crystals were grown at 800°C with the ampoule rotation rate of 75 rpm.

Kinoshita's group has developed the traveling liquidus zone (TLZ) method (Fig.

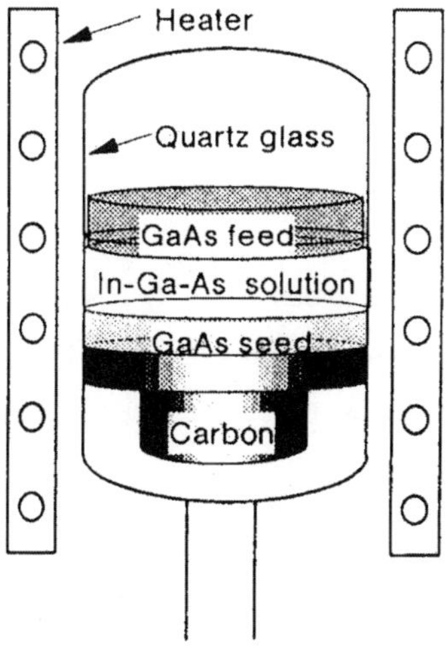

Fig. 19.11 Schematic diagram of apparatus for the temperature difference crystal growth method (reprinted from Ref. 85 with permission, copyright 1997 The Institute of Electronics, Information and Communication Engineers).

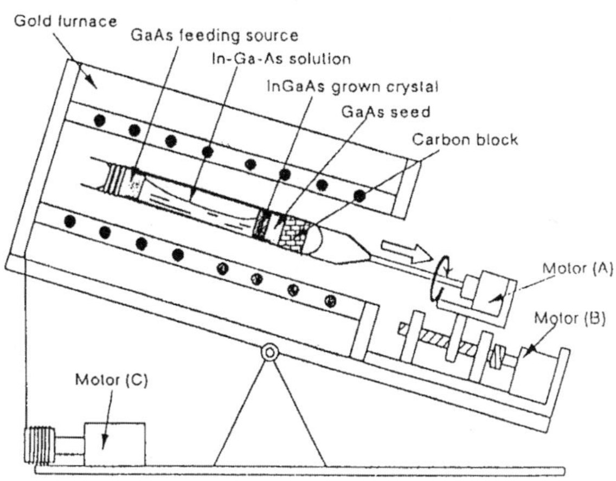

Fig. 19.12 Rotational Bridgman method (reprinted from Ref. 64 with permission, copyright 1996 The Institute of Electronics, Information and Communication Engineers).

19.13) for growing $In_{1-x}Ga_xAs$ alloy crystals.[52, 87-92] This method was invented after the directional solidification study[93] and the multicomponent zone melting method.[94] In this TLZ method, as shown in Fig. 19.13, because of the composition difference of InAs at the feed interface and crystal interface, InAs diffuses from the crystal interface to the feed interface. The melting point at the crystal interface increases so that crystal growth proceeds. Segregated InAs at the crystal interface diffuses to the feed interface and the feed crystal dissolves in the liquidus. The liquidus zone thus travels from the

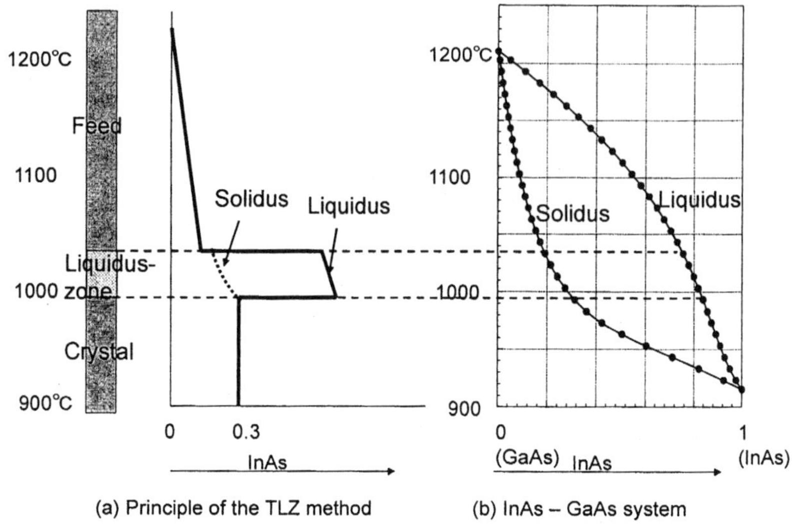

Fig. 19.13 Schematic drawing of the TLZ method: (a) principle of the TLZ method, (b) pseudobinary phase diagram of the InAs-GaAs system (reprinted from Ref. 87 with permission, copyright 2002 The Japanese Association for Crystal Growth).

crystal side to the feed side and crystal growth proceeds. During this crystal growth, the ampoule is moved in such a way that the interface position is fixed so that a crystal of uniform composition can be grown. By this method, they have been able to grow a single crystal of $In_{1-x}Ga_xAs$ of 15 mm diameter and 28 mm length and composition 0.34 ± 0.01.

The effect of weightlessness on the crystal growth of $In_{1-x}Ga_xAs$ by the Bridgman method has been studied as compared to crystal growth on earth[95] and it was concluded that lowered gravity had a considerable effect on heat transfer in the melt facilitating the control of the shape of the solid liquid interface because the heat transfer was governed by heat conduction. Dutta and Miller[65] have applied a thermochemical analysis to the vertical Bridgman method to grow homogeneous $In_{1-x}Ga_xAs$ crystals.

In order to obtain very thick $In_{1-x}Ga_xAs$ epitaxial layers applicable as substrates, various modified methods of LPE such as the source current controlled method[96-99], liquid phase electroepitaxy (LPEE)[63] and the YO-YO solute feeding method[100] have been applied. Bryskiewicz et al.[101] have grown $In_{1-x}Ga_xAs$ bulk crystals by the LPEE method as shown in Fig. 19.14. $In_{1-x}Ga_xAs$ ingots of 14 mm diameter and up to 3 mm thickness have been grown on (100) InP substrates.

19.3.3 $In_{1-x}Ga_xSb$

$In_{1-x}Ga_xSb$ single crystals have been grown by the CZ method using the double crucible method[43] in order to improve composition uniformity along the growth axis. Various designs of double crucible method have been attempted to prevent the difficulties

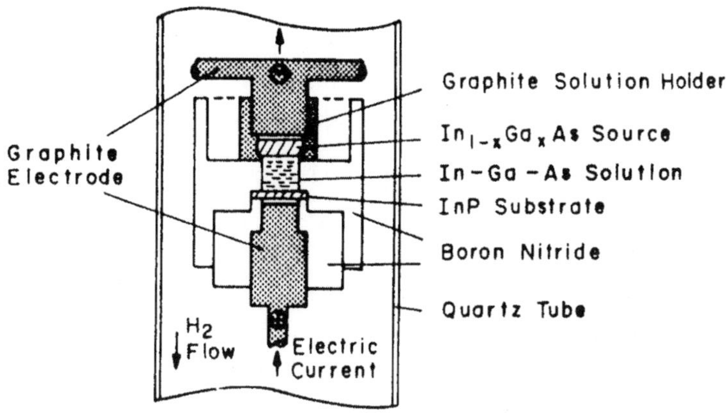

Fig. 19.14 Schematic diagram of the growth cell used for LPEE growth of $In_{1-x}Ga_xAs$ bulk crystals (reprinted from Ref. 101 with permission, copyright 1987 Institute of Physics).

encountered in the method.

The double crucible method has however a drawback in that InSb is back diffused from the floating crucible. Tsuruta et al.[102] have introduced ultrasonic vibrations in the double crucible CZ method. Ashley et al.[103] have improved the method working on the liquidus-solidus temperature difference, friction between the inner and outer crucibles, location of the inner crucible, back diffusion between two crucibles and freezing of the channel between the two crucibles. As a result, [111] single crystals of $In_{1-x}Ga_xSb$ with $0 < x < 0.11$ up to 24 mm in diameter and 50 mm in length and with dislocation densities of 2×10^4 cm^{-2} and compositional variation of better than 10 % have been grown. Tsaur and Kou[104] have modified the forthcoming double crucible method by using a long bottom tube as shown in Fig. 19.15.

Bachman et al.[105] have grown $In_{1-x}Ga_xSb$ single crystals by the CZ method in the range of $0.9 < x < 1$ and obtained single crystals whose EPD on (111)A face was less than 10^4 cm^{-2}. The distribution coefficient of Ga as a function of the Ga liquidus composition has been determined as shown in Fig. 19.16.

Sukegawa et al.[46, 106-114] prepared $In_{1-x}Ga_xSb$ bulk crystals by the solute-feeding Czochralski (SF-CZ) method as seen in Fig. 19.17(a).[107] They grew $In_{0.025}Ga_{0.975}Sb$ single crystals using a GaSb (111)B seed crystal. Various SF-CZ methods have been studied by Watanabe et al.[108-111] Using a crucible with a hole in the solution as seen in Fig. 19.17(b), they have grown a $Ga_{0.97}In_{0.03}Sb$ single crystal using a GaSb seed crystal at 691°C and then using this mexed crytal seed at 672°C, have grown a $Ga_{0.95}In_{0.05}Sb$ single crystal whose diameter is 17 mm and the length is 18 mm.

Dutta and Ostrogorsky[115] have applied the submerged heater method (SHM) (Fig. 12.5 of Vol. 1) and used the submerged baffle to stir the melt by forced convection. Mitric et al.[116] have grown $In_{1-x}Ga_xSb$ bulk crystals by the vertical Bridgman method under alternating magnetic field as shown in Fig. 19.18. Dutta and Miller[65] have ap-

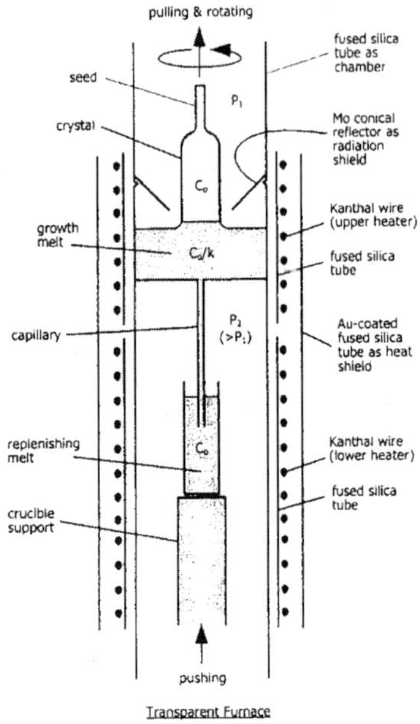

Fig. 19.15 New double-crucible Czochralski growth of an alloy of uniform composition C_0 using pressure difference (P_2-P_1) to cause replenishing through a long capillary tube that suppresses diffusion and mixing between growth melt (dense) and replenishing melt (lighter) (reprinted from Ref. 104 with permission, copyright 2007 Elsevier).

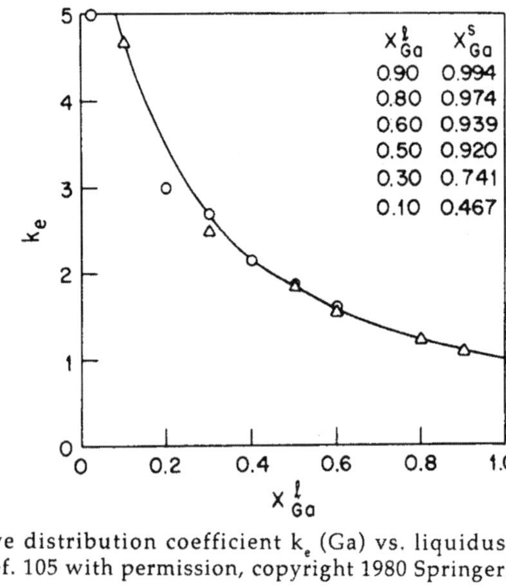

Fig. 19.16 Effective distribution coefficient k_e (Ga) vs. liquidus composition x^l (Ga). (reprinted from Ref. 105 with permission, copyright 1980 Springer).

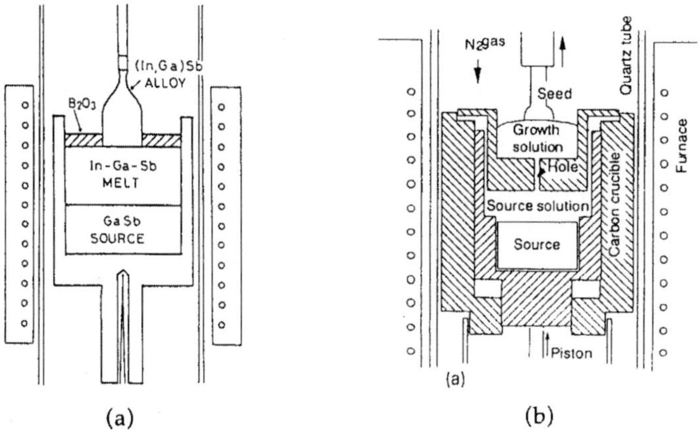

Fig. 19.17 (a) Schematic diagram of the solute feeding CZ method (from Ref. 107 with permission). (b) Modified solute feeding CZ method with a hole (reprinted from Ref. 108 with permission, copyright 1994 Elsevier).

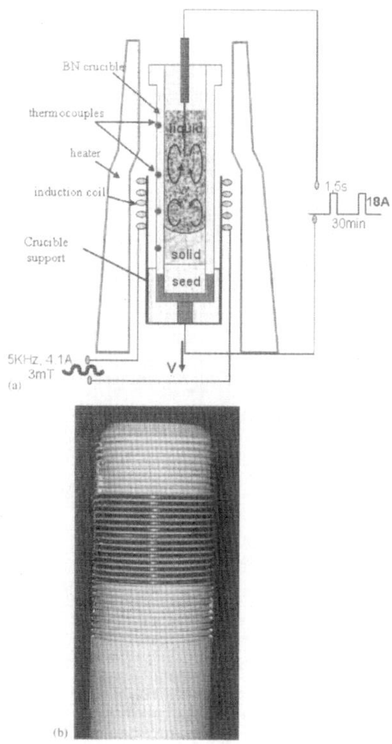

Fig. 19.18 (a) Sketch of the experimental setup, (b) Picture of the electromagnetic coil. (reprinted from Ref. 116 with permission, copyright 2006 Elsevier).

plied a thermochemical analysis for the vertical Bridgman method to grow homogeneous $In_{1-x}Ga_xSb$ crystals.

$In_{1-x}Ga_xSb$ bulk crystals have also been grown by the solution growth method. Kumagawa et al.[117-121] have used a GaSb/InSb/GaSb sandwich structure arrangement in the gradient freezing method. Hikida et al.[121] have developed a way to measure the growth rate by the thermal pulse method.

Sukegawa et al.[106] have developed a YO-YO solute feeding method to grow thick $In_{1-x}Ga_xSb$ bulk crystals by liquid phase epitaxial growth as shown in Fig. 19.19.

19.3.4 $InP_{1-x}As_x$

Thompson and Wagner[122] have grown $InP_{1-x}As_x$ ($0 < x < 0.10$) single crystals throughout the entire composition range by the Gremmelmaier method[123] (closed or sealed CZ method, Sec. 2.2.3(1) in Vol. 1) as shown in Fig. 19.20 and the conventional LEC method. Distribution coefficients for the main dopants such as S, Se, Te, Sn and Zn have been determined.

Watanabe et al.[124] have grown $InP_{1-x}As_x$ ($0.1 < x < 0.3$) single crystals by the LEC method. InP and $InP_{1-x}As_x$ seed crystals have been used depending on the desired composition. It was found that a growth rate less than 100 μm/h is necessary to prevent polygonization. The structural defects in grown crystals have been evaluated using synchrotron X-ray diffraction topography.[125]

Bonner[28] has grown $InP_{1-x}As_x$ ($0 < x < 0.10$) single crystals using InP seed crystals and $InP_{1-x}As_x$ seed crystals by the LEC method. Typical crystals were 20-30 mm in diameter and 40-80 mm in length.

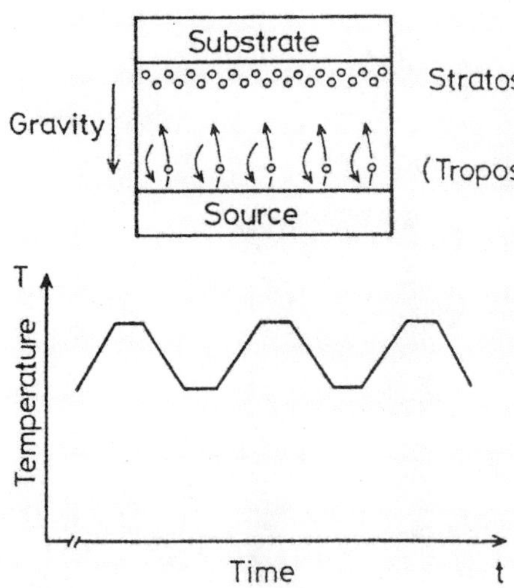

Fig. 19.19 Growth principle and temperature profile of the YO-YO solute feeding method (from Ref. 106 with permission).

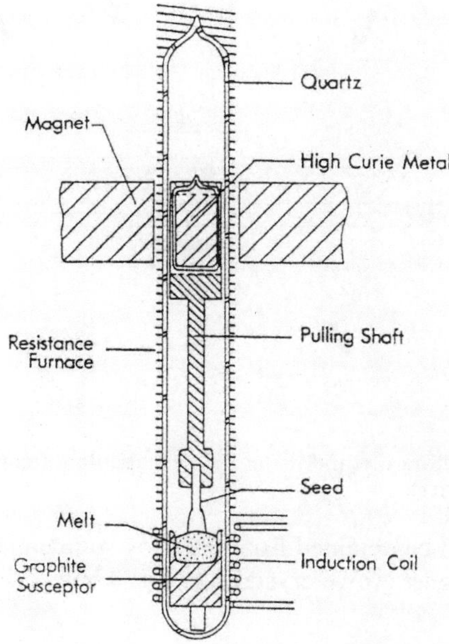

Fig. 19.20 Diagram of apparatus used for pulling crystals of $InP_{1-x}As_x$ by the sealed Czochralski technique (reprinted from Ref. 123 with permission, copyright 1956 Verlag der Zeitschrift für Naturforschung).

19.3.5 $GaAs_{1-x}P_x$

By the LEC method, $GaAs_{1-x}P_x$ bulk crystals have been grown from a Ga-rich solution using (111) GaAs seed crystals.[126] The composition x varied from 0.3 to 0.1 over an increasing range of lengths.

Hibiya et al.[127,128] have grown $GaAs_{1-x}P_x$ single crystals of the diameter 50 mm and length of 10 mm by the LEC method from the mixed melt of GaAs and GaP.[128]

$GaAs_{1-x}P_x$ crystals of length 45 mm have been grown at 1050°C by the THM method using a BN crucible. Compositionally uniform crystals (x = 0.6) could be obtained since the composition could be kept constant in the travelling zone.[129]

Draper[49] has grown $GaAs_{1-x}P_x$ crystals by the zone leveling method using B_2O_3 as the liquid encapsulant as seen in Fig. 19.21. The zone leveling method has the facility of controlling the composition.

Pizzarello[130] and Rubenstein[42] have grown $GaAs_{1-x}P_x$ crystals from GaP and GaAs powders by the vapor transport method using iodine as the transport agent. Relatively homogeneous $GaAs_{1-x}P_x$ could be obtained because there is no supercooling effect as in the case of melt growth.

Takeda et al.[131] have also applied iodine vapor transport using the growth apparatus as shown in Fig. 2.15 of Vol.1. $GaAs_{1-x}P_x$ single crystals (x=0.4 and 0.6) of uniform

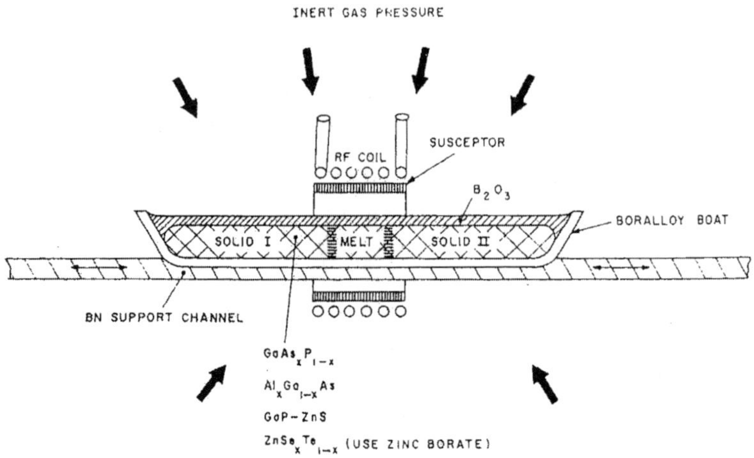

Fig. 19.21 Zone leveling method using B_2O_3 encapsulant (reprinted from Ref. 49 with permission, copyright 1980 Elsevier).

composition could be obtained from the alloy source material at a growth temperature of 850°C. The largest grown crystal size was $3 \times 4 \times 5$ mm^3.

19.3.6 $AlGa_{1-x}As_x$

Daniele et al.[132-134] have applied the liquid phase electroepitaxy (LPEE) method to the growth of $AlGa_{1-x}As_x$ crystals. Electric currents of 10-30 A/cm^2 were passed through the solid-liquid interface. It was found that the growth rate increased proportional to the current as shown in Fig. 19.22. $AlGa_{1-x}As_x$ ($0.03 < x < 0.38$) thick epitaxial layers of thickness up to 1.4 mm could be grown on (100) GaAs substrates.

19.3.7 $InSb_{1-x}Bi_x$

$InSb_{1-x}Bi_x$ single crystals up to composition $x = 0.026$ have been grown by the CZ method.[135] In order to improve the growth conditions, a flat seed which has a similar diameter to the grown crystal was used. By this technique, higher x-value compositions could be obtained from the region close to the seed crystal.

$InSb_{1-x}Bi_x$ ($x = 0.008$) bulk crystals of thickness 15 mm and diameter of 14 mm have been grown on InSb seed crystals by continuous feeding of InSb in the rotational Bridgman method[136] using the same furnace as shown in Fig. 19.12.

19.3.8 Quaternary III-V Mixed Crystals

The growth of quaternary III-V mixed crystals from melt and solution is reviewed by Bachmann et al.[13]

Backmann et al.[137] have grown $Ga_xIn_{1-x}P_yAs_{1-y}$ crystals by the gradient freezing method for the composition range $0 \leq x \leq 0.07$. $GaP_xAs_ySb_{1-x-y}$ crystals have been grown from a mixture of GaP, GaAs and GaSb by the Bridgman method but only a small crystal could be obtained.[138]

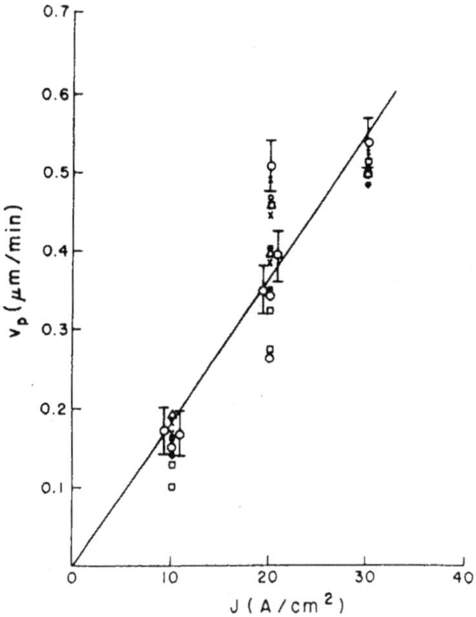

Fig. 19.22 Plot of the average growth rate versus current density taken from several areas of the grown layers (reprinted from Ref. 132 with permission, copyright 1975 The American Institute of Physics).

Quaternary $Al_xGa_yIn_{1-x-y}Sb$ single crystals have been grown from an alloyed mixture of the binary materials by directional solidification.[139] $In_xGa_{1-x}Sb_yBi_{1-y}$[140] and $In_xGa_{1-x}Sb_yAs_{1-y}$[136] bulk crystals have been grown on InSb substrates by the rotational Bridgman method.

19.4 APPLICATIONS

Optoelectronic devices such as semiconductor lasers and light emitting diodes are mainly made using direct transition type ternary and quaternary III-V mixed crystals such as III group (Al, Ga, In) and V group (N, P, As, Sb). These ternary and quaternary crystals are grown epitaxially on binary III-V crystal substrates and other hetero substrates. There is however lattice-mismatching between these epitaxial layers and substrates which give rise to large leakage currents due to poor interface regions. Ternary and quaternary substrates are therefore required as lattice-matching substrates, which can provide a continuum of lattice constants ranging from GaP to InSb, with a wavelength range of range of 0.3-12 μm. Some examples of the application of ternary and quaternary III-V mixed crystal substrates follow.

Since $In_xGa_{1-x}P$ is a promising material for green light emitting diodes (LEDs) and lasers in applications such as high-density recording and measurement, plastic fiber communications and mobile short distance communications, bulk crystal growth of $In_xGa_{1-x}P$ has been achieved as explained in Sec. 19.3.1.

The theoretical gain of strained quantum wells on an $In_{1-x}Ga_xAs$ ternary substrate was calculated and an optical gain greater than 750 cm^{-1} is expected.[141] Shouji et al.[1] have fabricated $In_{0.25}Ga_{0.75}P/In_xGa_{1-x}As_yP_{1-y}$ strained single quantum well lasers using $In_{0.05}Ga_{0.95}As$ ternary substrates.[2] Otsubo et al.[142] have fabricated infrared laser diodes on $In_{0.22}Ga_{0.78}As$ substrates and have shown laser oscillations of 1.22 μm with improved temperature characteristics.

Substrate crystals were grown by the LEC method with supplying GaAs source material. The lasers lased at 1.03 μm with a low threshold current density of 222 A/cm^2 and showed an excellent characteristic temperature of 221 K. Menna et al.[2] have grown 1.75 μm strained InGaAs multi quantum well lasers on an $InAs_{0.08}P_{0.92}$ ternary substrate. This laser showed that a pulsed power output of 60 mW at wavelength of λ =1.75 μm could be achieved. Arai et al.[143] have fabricated 1.31 μm ridge waveguide lasers using InGaAs substrate grown by the TLZ method and demonstrated a 10 Gbs data transmission.

$In_xGa_{1-x}Sb$ is an important material for thermo-photovoltaic devices, photodetectors for optical communication beyond 2 μm with non-silica fibers[144, 145] and sensors in the infrared region with the wavelength from 1.7 to 6.8 μm. $In_xGa_{1-x}Sb$, $In_xGa_{1-x}As_ySb_{1-y}$ and $InAs_xSb_yP_{1-x-y}$ are used as bulk substrates for thermo-photovoltaic devices.[146] P-n junction thermo-photovoltaic cells have been fabricated by diffusing Zn to Te-doped $Ga_{1-x}In_xSb$ substrates. Peak external quantum efficiencies greater than 60 % are reported.[147]

InP_xAs_{1-x} has the bandgap covering the near IR spectral region of the wavelengths for the neodium-doped glass lasers and yttrium aluminum garnet (YAG) lasers. It is also promising as a highly efficient photocathode.[148] The Gunn oscillators has also been shown as an application candidate of this material.[149]

$GaAs_xP_{1-x}$ bulk crystals are required for $In_xGa_{1-x}As_yP_{1-y}$ laser diodes (LDs) for the 630-700 nm wavelength region, since they avoid the time-consuming graded epitaxy process for lattice matching, and for epitaxy-free red $GaAs_xP_{1-x}$ LEDs.[127] Using $GaAs_xP_{1-x}$ substrates, the highest direct energy-gap double hetero structure devices with GaInP/AlInP become possible.[131]

$InSb_xBi_{1-x}$ is an interesting material since InBi has semimetallic properties and the bandgap of InSb can be reduced by introducing Bi into Sb sites. $InSb_xBi_{1-x}$ is therefore useful for detection devices in the middle infrared field (7.4-16 μm).[150] On the other hand, fluoride glasses are potential mateiral for ultra low-loss fibers, with the intrinsic minimum loss estimated to be about 10^{-3} dB/km in the 2-3 μm wavelength region.[151, 152] Since the wavelength of $In_xGa_{1-x}As_ySb_{1-y}$ quaternary alloys can be controlled between 0.88 and 6.8 μm, these quaternary bulk crystals are also valuable.[136]

REFERENCES

1. H. Shoji, T. Uchida, T. Kusunoki. M. Matsuda, H. Kurakake, S. Yamaski. K. Nakajima and H. Ishikawa, *IEEE Photon. Technol. Lett.* **6** (1994) 1170.
2. R.J. Menna, R.U. Martinelli, D. Garbuzov, R. Paff, J.S. Vermaak, G.H. Olsen and W.A. Bonner, *Electron. Lett.* **31** (1995) 188.
3. V.T. Bublik, M.B. Karman, V.F. Kleptsin and V.N. Leikin, *phys. stat. sol. (a)* **32** (1975) 631.
4. V.T. Bublik and V.N. Leikin, *phys. stat. sol. (a)* **46** (1978) 365.
5. M.B. Panish, *J. Chem. Thermodynam.* **2** (1970) 319.
6. G.B. Stringfellow, *J. Electrochem. Soc.* **117** (1970) 1301.
7. L.M. Foster and J.E. Scardefield, *J. Electrochem. Soc.* **117** (1970) 534.
8. A. Lewis, *J. Crystal Growth* **53** (1981) 530.

9. J.C. Woolley and B.A. Smith, *Proc. Phys. Soc. (London)* **72** (1958) 214.
10. M.S. Abrahams, R. Braunstein and F.D. Rosi, *J. Phys. Chem. Solids* **10** (1959) 204.
11. I.V. Bodnar, E.E. Matyas, *Zh. Neorg. Khim.* **22** (1977) 796.
12. W. Köster and W. Ulrich, *Z. Metallk.* **49** (1958) 365.
13. K.J. Bachmann, F.A. Thiel and H. Schreiber, Jr., *Prog. Crystal Growth and Characterization, Vol. 2* (Pergamon Press, Oxford, 1979) p. 171.
14. V.B. Ufimtzev, A.S. Timoshin and G.V. Kostin, *Izv. Akad. Nauk SSSR, Neorg. Mater.* **7** (1971) 2088.
15. J.C. Woolley and B.A. Smith, *Proc. Phys. Soc. (London)* **72** (1958) 214.
16. J.C. Woolley and D.G. Less, *J. Less Common Metals* **1** (1959) 192.
17. K. Nakajima, K. Osamura, K. Yasuda and Y. Murakami, *J. Crystal Growth* **41** (1977) 87.
18. E.K. Müller and J.L. Richards, *J. Appl. Phys.* **35** (1964) 1233.
19. S.A. Abagyan, S.M. Gorodetskii, T.B. Zhukova, A.I. Zaslavskii, A.V. Lishina and V.K. Subashiev, *Sov. Phys. Solid State* **7** (1965) 153.
20. M.E. Staumanis, J.P. Krummme and M. Rubenstein, *J. Electrochem. Soc.* **114** (1967) 640.
21. A. Onton, M.R. Lorenz and W. Reuter, *J. Appl. Phys.* **42** (1971) 3420.
22. E.F. Hockings, I.Kudman, T.E. Seidel, C.M. Schmalz and E.F.Steigmeier, *J. Appl. Phys.* **37** (1966) 879.
23. G. Gieseke and H. Pfister, *Acta. Cryst.* **11** (1958) 369.
24. D. Auvergne, J. Camassel, H. Mathiew and A. Joullie, *J. Phys. Chem. Solids* **35** (1974) 133.
25. N.A. Goryunova and N.N. Fedorova, *Zhur. Tekh. Fiz.* **25** (1955) 1339.
26. *Landolt-Börnstein Vol. 17, Semiconductors*, ed. O. Madelung (Springer-Verlag, 1982) p. 333.
27. J.C. Woolley, in *Compound Semiconductors, Vol.1, Preparation of III-V Compounds*, eds. R.K. Willardson and H.L. Goering (Reinhold, Publ. Co., 1962) p. 3.
28. W.A. Bonner, *Proc. 9th Inter. Conf. InP Related Mater.* (Hyaniss, 1997) p. 205.
29. T. Onda, R. Ito and N. Ogasawara, *Jpn. J. Appl. Phys.* **25** (1986) 82.
30. T. Onda and R. Ito, *J. Crystal Growth* **78** (1986) 479.
31. T. Nishinaga and K. Hiramatsu, *Oyo Buturi (Monthly Publ. Jpn. Soc. Appl. Phys.) (in japanese)* **50** (1981) 1065.
32. G.B. Stringfellow, *J. Crystal Growth* **27** (1974) 21.
33. G.B. Stringfellow. *J. Electron. Mater.* **11** (1982) 903.
34. S. Yamazaki, M. Kishi and T. Katoda, *phys. stat. sol. (b)* **113** (1982) 421.
35. K.A. Jones, W. Porod and D.K. Ferry, *J. Phys. Chem. Solids* **44** (1983) 107.
36. K. Onabe, *Jpn. J. Appl. Phys.* **22** (1983) 663.
37. W. Nakwaski, *J. Appl. Phys.* **64** (1988) 159.
38. T. Sukegawa and A. Tanaka, *Rep. Res. Inst. Electron. Shizuoka Univ.* **19** (1984) 55.
39. P.S. Dutta, *J. Crystal Growth* **275** (2005) 106.
40. G.R. Antell and D. Effer, *J. Electrochem. Soc.* **106** (1959) 509.
41. H.T. Minden, *J. Electrochem. Soc.* **112** (1965) 300.
42. M. Rubenstein, *J. Electrochem. Soc.* **112** (1965) 426.
43. W.F. Leverton, *J. Appl. Phys.* **29** (1958) 1241.
44. S.V. Airapetyabts and G.I. Shmelev, *Fiz. Tver. Tela* **2** (1960) 747.
45. I. Shindo, *Semiconductor World* **January** (1985) 1.
46. T. Sukegawa, M. Yamashita, H. Ohshima and A. Tanaka, *Rep. Res. Inst. Electron. Shizuoka Univ.* **20** (1985) 193.
47. I. Shindo, N. Ii, K. Kitamura and S. Kimura, *J. Crystal Growth* **46** (1979) 307.
48 W.G. Pfann, in *Zone Melting Ch. 10* (John Wiley & Sons, New York, 1966).
49. A.G. Fisher, in *Crystal Growth,* ed. by B.P. Pamplin, (Pergamon Press, Oxford, 1975) p. 521.
50. E.M. Swiggard, R.L. Henry and H. Lessoff, *Inst. Phys. Conf. Ser.* **65** (1983) 1.
51. W. G. Pfann. *Trans. AIME* **194** (1952) 747.
52. K. Kinoshita, H. Kato, M. Iwai, T. Tsuru, Y. Muramatsu and S. Yoda, *J. Crystal Gowth* **225** (2001) 59.
53. G.A. Wolff, P.H. Keck and J.D. Border, *Phys. Rev.* **94** (1954) 753, H. Sonomura, T. Nanmori and T. Miyauchi, *J. Appl. Phys.* **46** (0975) 2693.
54. S.E. Blum and R.J. Chicotka, *J. Electrochem. Soc.* **117** (1970) 94C.
55. S.E. Blum and R.J. Chicotka, *Electrochemical Society Annual Meeting* Abstract No.73 (Los Angels 1970) p. 202.
56. G. Bischopink and K.W. Benz, *J. Crystal Growth* **97** (1989) 245.
57. T. Kato, A. Shimizu and T. Ishida, *Jpn. J. Appl. Phys. 13* (1974) 1481.
58. K. Gillesen and A.J. Marshall, *J. Crystal Growth* **38** (1977) 82.
59. Y. Okuno, K. Suto and J. Nishizawa, *Jpn. J. Appl. Phys.* **10** (1971) 388.
60. J. Nishizawa and S. Yoshida, *J. Crystal Growth* **76** (1986) 1.
61. D. Mateika, *J. Crystal Growth* **9** (1971) 249.
62. J.J. Daniele and A. Lewis, *J. Electron. Mater.* **12** (1983) 1015.
63. T. Bryskeiwicz, C.F. Boucher, Jr., J. Lagowski and H.C. Gatos, *J. Crystal Growth* **82** (1987) 279.

64. T. Ozwa, Y. Ohnishi, T. Koyama, Y. Hayakawa and M. Kumagawa, *Tech. Rep. IEICE* **ED97-37, CPM97-25** (1997-05) p. 73.
65. P.S. Dutta and T.R. Miller, *J. Electron. Mater.* **29** (2000) 956.
66. R.D. Burnham, N. Holonyak, Jr., D.L. Keune, D.R. Scifres and P.D. Dapkus, *Appl. Phys. Lett.* **17** (1970) 430.
67. H. Itoh, K. Hara, A. Tanaka and T. Sukegawa, *Appl. Phys. Lett.* **19** (1971) 348.
68. D.J. Lawrence and L.F. Eastman, *J. Crystal Growth* **30** (1975) 267.
69. J.J. Daniel, D.A. Cammack and P.M. Asbeck, *J. Appl. Phys.* **48** (1977) 914.
70. J.W. Wagner, *J. Electrochem. Soc.* **117** (1970) 1193.
71. R. Gremmelmeier, *Z. Natureforsch.* **11a** (1956) 511.
72. W.A. Bonner, B.J. Skromme, E. Berry, H.L. Gilchrist and R.E. Nahory, *Inst. Phys. Conf. Ser.* **96** (1989) 337.
73. W.A. Bonner, R.E. Nahory, H.L. Gilchrist and E. Berry, *Proc. 6th Conf. Semi-Insul. III-V Mater.* ed. by A. Milnes and C.J. Miner (Adam Hilger, 1990) p. 199.
74. W.A. Bonner, B. Lent, D.J. Freschi and W. Hoke, *Proc. SPIE* **2228** (1994) 33.
75. B. Lent, W.A. Bonner and W. Hoke, *"LEC Growth and Characterization of Substrate-Quality GaInAs"*, presented at the AACG/west, 13th Conf. Crystal Growth, (Fallen Leaf Lake, CA, 1994).
76. K. Sugiura, Y. T.akeda, Y. Kuroda and A. Sasaki, *Record 7th Symp. Alloy Phys. Electron.* (1987) p. 181.
77. K. Nakajima, T. Kusunoki, and C. Takenaka, *J. Crystal Growth.* **113** (1991) 485.
78. K. Nakajima and T. Kusunoki, *J. Crystal Growth* **169** (1996) 217.
79. T. Kusunoki, C. Takenaka and K. Nakajima, *J. Crystal Growth* **112** (1991) 33.
80. T. Suzuki, T. Kusunoki, T. Katoh abnd K. Nakajima, *Proc.7th Int. Conf. InP Related Mater.* (Sapporo, 1995) p. 45.
81. K. Nakajima, T. Kusunoki and K. Otsubo, *J. Crystal Growth* **173** (1997) 42.
82. Y. Nishijima, K. Nakajima, K. Otsubo and H. Ishikawa, *J. Crystal Growth* **197** (1999) 769.
83. Y. Nishijima, K. Nakajima, K. Otsubo and H. Ishikawa, *J. Crystal Growth* **208** (2000) 171.
84. Y. Nishijima and K. Nakajima, *J. Assoc. Crystal Growth* **32** (2005) 36.
85. T. Ozawa, H. Ono, Y. Ohnishi, T. Koyama, Y. Hayakawa and M. Kumagawa, *Tech. Rep. IEICE* **ED96-16, CPM96-1**(1996-05) p. 1.
86. W.A. Tiller, *J. Appl. Phys.* **34** (1963) 2757.
87. K. Kinoshita, Y. Ogata, N. Koshikawa, S. Adachi, S. Matsumoto, M. Iwai, T. Tsuruta, Y. Muramatsu, H. Nakamura, T. Maekawa and S. Yoda, *J. Assoc. Crystal Growth (in japanese)* **29** (2002) 349.
88. K. Kinoshita, Y. Hanaue, H. Nakamura, S. Yoda, M. Iwai, T. Tsuru and Y. Muramatsu, *J. Crystal Growth* **237-239** (2002) 1859.
89. H. Nakamura, Y. Hanaue, H. Kato, K. Kinoshita and S. Yoda, *J. Crystal Growth* **258** (2003) 49.
90. K. Kinoshita, Y. Ogata, N. Koshikawa, S. Adachi, M. Iwai, T. Tsuru, Y. Muramatsu and S. Yoda, *J. Assoc. Crystal Growth* **30** (2003) 52.
91. K. Kinoshita, S. Adachi, Y. Ogata, T. Tsuru, H. Miyata, Y. Muramatsu and S. Yoda, *J. Assoc. Crystal Growth (in japanese)* **31** (2004) 109.
92. K. Kinoshita, Y. Ogata, S. Adachi, T. Ishiukawa, T. Masaki, M. Takayanagi, S. Yoda, M. Arai and Y. Kondo, *J. Assoc. Crystal Growth (in japanese)* **32** (2005) 115, K. Kinoshita, Y. Ogata, S. Adachi, S. Yoda, T. Tsuru, H. Miyake and Y. Muramatsu, *Abb. New York Acad. Sci.* **1077** (2005) 161.
93. K. Hashio, M. Tatsumi, H. Kato and K. Kinoshita, *J. Crystal Growth* **210** (2000) 471.
94. S. Kodama, Y. Furumura, K. Kinoshita, H. Kato and S. Yoda, *J. Crystal Growth* **208** (2000) 165.
95. S. Fujiwara, T. Araki, M. Tatsumi, M. Irikura, K. Matsumoto, S. Murai and T. Shirakawa, *J. Crystal Growth* **167** (1996) 357.
96. K. Nakajima, S. Yamazaki and I. Umebu, *Jpn. J. Appl. Phys.* **23** (1984) L26.
97. K. Nakajima and S. Yamazaki, *J. Electrochem. Soc. 132* (1985) 904.
98. K. Nakajima and S. Yamazaki, *J. Crystal Growth* 74 (1986) 39.
99. K. Nakajima, *J. Appl. Phys.* **61** (1987) 4626.
100. T. Sukegawa, S. Watabe, H. Momoi, M. Kimura and A. Tanaka, *Tech. Rept. 8th Meeting Monbusho Special Project Res., Alloy Semicon. Phys. Electron.* (1988) p. 7.
101. T. Bryskiewicz, M. Bugajski, B. Bryskiewicz, J. Lagowski and H.C. Gatos, *Inst. Phys. Conf. Ser.* **91** (1988) 259.
102. T. Tsuruta, N. Nishida, Y. Hayakawa and M. Kumagawa, *Tech. Rep. IEICE* **ED90-20, CPM90-14** (1990) 35.
103. T. Ashley, J.A. Beswick, B. Cockayne and C.T. Elliott, *Inst. Phys. Conf. Ser.* **144** (1995) 209.
104. S.C. Tsar and S. Kou, *J. Crystal Growth* **307** (2007) 268.
105. K.J. Bachmann, F.A. Thiel, H. Schreiber, Jr. and J.J. Rubin, *J. Electron. Mater.* **9** (1980) 445.
106. T. Sukegawa, M. Kimura, H. Katsuno, S. Kegeyama and A. Tanaka, *Shizuoka Univ. Res. Rep.* 21 (1986) 119.

107. A. Tanaka, M. Ishino, A. Okugaki and T. Sukegawa, *Proc. 7th Rec. Alloy Semicon. Phys. Electron. Symp.* (1987) 189.
108. A. Watanabe, A. Tanaka and T. Sukegawa, *Shizuoka Univ. Res. Rep.* **27** (1992) 101.
109. A. Watanabe, A. Tanaka and T. Sukegawa, *Jpn. J. Appl. Phys.* **32** (1993) L793.
110. A. Watanabe, A. Tanaka and T. Sukegawa, *J. Crystal Growth* **128** (1993) 462.
111. A. Watanabe, H. Katsuno, A. Tanaka and T. Sukegawa, *Tech, Rep. IEICE*, **ED93-120, CPM93-91** (1993-11) 1.
112. A. Tanaka, A. Watanabe, M. Kikuma and T. Sukegawa, *J. Crystal Growth* **135** (1994) 269.
113. A. Tanaka, T. Yoneyama, M. Kimura and T. Sukegawa, *J. Crystal Growth* **186** (1998) 305.
114. A. Tanaka, J. Shintani, M. Kimura and T. Sukegawa, *J. Crystal Growth* **209** (2000) 625.
115. P.S. Dutta and A.G. Ostrogorsky, *J. Crystal Growth* **194** (1998) 1.
116. A. Mitric, T. Duffar, C. Diaz-Guerra, V. Corregidor, L.C. Alves, C. Garnier and G. Vian, *J. Crystal Growth* **287** (2006) 224
117. M. Murakami, T. Kimura, K. Arafune, T. Ozawa, Y. Okano, Y. Hayakawa and M. Kumagawa, *J. Assoc. Crystal Growth (in japanese)* **30** (2003) 53.
118. T. Ozawa, M. Murakami, M. Suzuki, K. Arafune, Y. Okano, Y. Hayakawa and M. Kumagawa, *J. Assoc. Crystal Growth (in japanese)* **30** (2003) 54.
119. N. Murakami, K. Arafune, T. Koyama, M. Kumagawa and Y. Hayakawa, *J. Assoc. Crystal Growth C* **31** (2004) 108.
120. N. Murakami, K. Arafune, T. Koyama, M. Kumagawa and Y. Hayakawa, *J. Crystal Growth* **275** (2005) 433.
121. T. Hikida, N. Murakami, T. Koyama, Y. Momose, M. Kumagawa and Y. Hayakawa, *J. Assoc. Crystal Growth (in japanese)* **33** (2006) 178.
122. A.C. Thompson and J.W. Wagner, *J. Phys. Chem. Solids* **32** (1971) 2613.
123. R. Gremmelmaier, *Z. Naturf.* **11A** (1956) 511.
124. H. Watanabe, H. Watanabe, M. Yoshida and J. Matsui, *Ext. Abst. 15th Conf. Solid State Devices Mater.* (Tokyo, 1983) p. 169.
125. I. Minato, H. Hashizume, H. Watanabe and J. Matsui, *Jpn. J. Appl. Phys.* **25** (1986) 1485.
126. F. Cerrina, D. Margadonna and P. Perfetti, *J. Crytal Growth* **18** (1973) 202.
127. T. Hibiya, H. Wtanabe, H. Matsumoto and N. Iwata, *J. Electrochem. Soc.* **134** (1987) 981.
128. T. Hibiya, H. Wtanabe, H. Matsumoto, H. Ono and N. Iwata, *J. Electrochem. Soc.* **23** (1988) 425.
129. G.A. Wolff, H.E. LaBelle, Jr. and B.N. Das, *Trans. AIME* **242** (1968) 436.
130. F.A. Pizzarello, *J. Electrochem. Soc.* **111** (1962) 226.
131. Y. Takeda, S. Fugita and A. Sasaki, *Inst. Phys. Conf. Ser.* **74** (1984) 23.
132. J.J. Daniele, *Appl. Phys. Lett.* **27** (1975) 373.
133. J.J. Daniele, *Inst. Phys. Conf. Ser.* **24** (1975) 155.
134. J.J. Daniele and A.J. Hebling, *J. Appl. Phys.* **52** (1981) 4325.
135. B. Joukoff and A.M. Jean-Louis, *J. Crystal Growth* **12** (1972) 169.
136. T. Ozawa, Y. Hayakawa and M. Kumagawa, *J. Crystal Growth* **109** (1991) 212.
137. K.J. Bachmann, F.A. Thiel and S. Ferris, *J. Crystal Growth* **43** (1978) 752.
138. E. Lendvay, *J. Crystal Growth* **60** (1982) 447.
139. K. Zbitnew and J.C. Woolley, *J. Appl. Phys.* **52** (1981) 6611.
140. Y. Hayakawa, T. Matsumoto, M. Ando, S. Adachi, K. Takahashi and M. Kumagawa, *Shizuoka Univ. Res. Rep.* **28** (1993) 7.
141. H. Ishikawa, *Appl. Phys. Lett.* **63** (1993) 712.
142. K. Otsubo, H. Shoji, T. Knsunoki, T. Sizuki, T. Uchida, Y. Nishijina and H. Ishikawa, *Electron. Lett.* **33** (1997) 1795.
143. M. Arai, W. Kobayashi, T. Fujisawa, M. Yuda, T. Tadokoro, K. Kinoshita, S. Yoda and Y. Kondo, *Appl. Phys. Express* **2** (2009) Art. No. 022101.
144. P.S. Dutta, H.L. Bhat and V. Kumar, *J. Appl. Phys.* **81** (997) 5821.
145. K. Onabe, *Jpn. J. Appl. Phys.* **22** (1983) 287.
146. O.V. Sublima, A.W. Bett, M.G. Mauk, B.Ya. Ber and P.D. Dutta, *in Thermophotovoltaic Generation of Electricity; 5th Conference*, eds. T.J. Coutts, G. Guazzoni and J. Luther (2003, AIP) p. 402.
147. P.S. Dutta, J.M. Borrego, H. Ehsani, G. Rajagopalan, I.B. Bhat, R.J. Gutmann, G. Nicholas and P.F. Baldasaro, in *Thermophotovoltaic Generation of Electricity; 5th Conference*, eds. T.J. Coutts, G. Guazzoni and J. Luther (2003, AIP) p. 392.
148. H. Sonnenberg, *Appl. Phys. Lett.* **16** (1970) 245.
149. C. Hilsum and H.D. Rees, *Electron. Lett.* **6** (1970) 277.
150. A.M. Jean-Louis, B. Ayrault and J. Vargas, *phys. stat. sol.* **34** (1969) 341.
151. T. Kanamori and S. Takahashi, Japan. *J. Appl. Phys.* **24** (1985) L758.
152. S. Shibata, M. Moriguchi, K. Jinguji, S. Mitachi, T. Kanamori and T. Manabe, *Electron. Lett.* **17** (1981) 775.

20. Nitride and Other III-V Compounds

20.1 INTRODUCTION

Other than the III-V materials previously described in Vol.1 such as GaP, GaAs, GaSb, InP, InAs and InSb, there are further interesting III-V materials such as nitrides (BN, AlN, GaN, InN) and other compounds (BP, BAs, BSb, AlP, AlAs, AlSb). Recently, because of the interest in nitride devices such as light emitting diodes (LEDs), laser diodes (LDs), detectors and electronic devices [1-4], there has been an increase in activity in the growth of GaN and AlN single crystals. It is however difficult to grow these materials, mainly because of physical properties such as very high melting temperatures and extremely high decomposition vapor pressures. These materials however have many interesting characteristics which can not be matched by already industrialized materials. Among various application possibilities, much attention is focussed on the crystal growth of GaN since highly luminous blue LEDs have been realized by GaN based materials. These LEDs have up to now been fabricated using sapphire or SiC substrates but this is because GaN and AlN substrates are not commercially available. These nitride materials are also promising for high temperature electronic devices which are expected to have applications for automobiles, communications, power systems and devices in harsh atmospheres. Various applications for LEDs, LDs, electronic devices and others are reviewed in Refs. 1-7.

20.2 PYSICAL PROPERTIES

Physical properties of these nitride materials are reviewed in Refs. 1-15 and some of them are summarized in Table 20.1.

The wurzite polytype of AlN, GaN and InN form a continuous alloy system whose direct bandgap ranges from 6.2 eV (AlN), 3.4 eV (GaN) to 1.9 eV (InN). The bandgap of InN has recently been found to be as low as 0.7 eV (Sec.20.4.3(4)).[16] These nitrides are chemically stable at high temperatures and have wide bandgaps so that they are promising for device operation in high temperature and caustic environments. In Fig. 20.1, the bandgap as a function of lattice constant is summarized from various data for AlN-GaN-InN.[3] By changing the composition of nitride alloys, $Al_{1-x}Ga_{1-y}In_{1-x-y}N$, it is possible to cover the wide range of the bandgap, so that it is possible to realize photonic devices such as LEDs, LDs and photodetectors covering a wide range of wavelengths from ultraviolet to infrared. Therefore, nitride materials can cover most photonic devices.

Since nitrides, AlN, GaN and InN, have such high melting points as 3500 K, 2800 K and 2150 K respectively[17] and the decomposition pressure, especially for GaN and InN is very high, it is difficult to grow these crystals from their melts.[3, 18, 19] In Fig. 20.2, the vapor pressure and the melting point of AlN, GaN and InN are shown as a function of

Table 20.1 Physical properties of nitrides

	BN	AlN	GaN	InN
Crystal Structure	zincblende	wurzite	wurzite	wurzite
Lattice Constant (Å)	3.6160 Å	a = 3.112	a = 3.189	a = 3.548
		c = 4.892	c = 5.185	c = 5.760
Density (g/cm³)	3.45	3.255	6.085	6.81
Melting Point (°C)	2973	3000	>2000	1100
Linear Expansion c⊥	1.9	4.15	5.59	3.40-5.70
Coefficient (×10⁻⁶/deg) c ∥		5.27	3.17	2.70-3.70
Thermal Conductivity (W/cm•K)		7.4	2	1.3 0.8
Dielectric Constant		8.5	8.9-9.5	15.3
Refractive index	2.1	2.15	2.33-2.67	2.85-3.05
Bandgap at Room Temperature (eV)	6.4	6.2	3.39	0.7
Optical Transition Type	indirect	direct	direct	direct
Electron Mobility at R.T. (cm²/V•sec)	200	300	900	4400
Hole Mobility at R.T. (cm²/V•sec)	14	150	39	

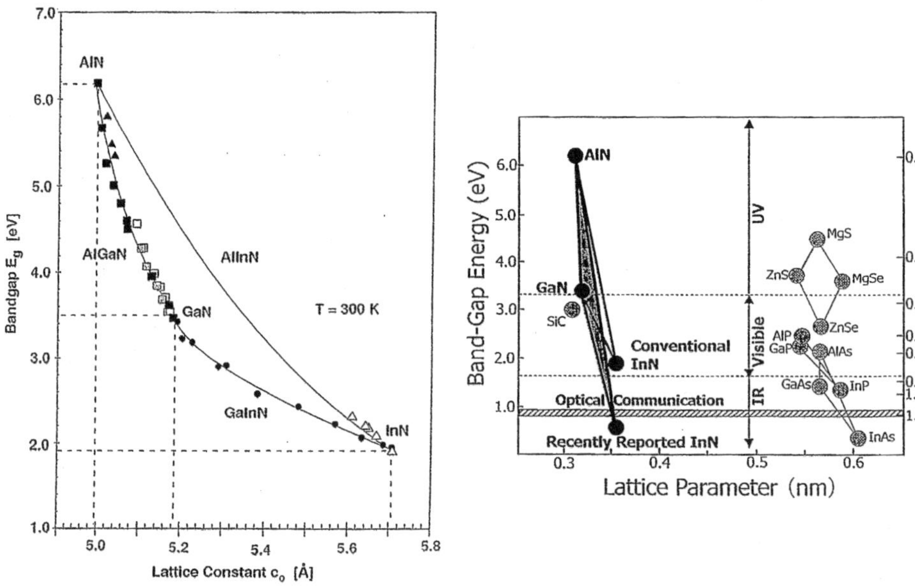

(a) (b)

Fig. 20.1 (a) Experimental results of bandgaps of hexagonal III-nitrides versus lattice constant C_0 at room temperature (reprinted from Ref. 3 with permission, copyright 1998 Institute of Physics). (b) The band gap of InN has recently been found to be 0.7 eV (courtesy of Profs. Y. Nanishi and T. Arai).

temperature, which are obtained from calculations and experiments.[3] Recently, the phase diagrams for GaN and InN under high pressure have been determined. It was found that the congruent melt growth of GaN can be achieved at above 2220°C under pressure higher than 6 GPa[20,21], and that of InN at above 1900°C under pressure higher than 10 GPa.[22,23]

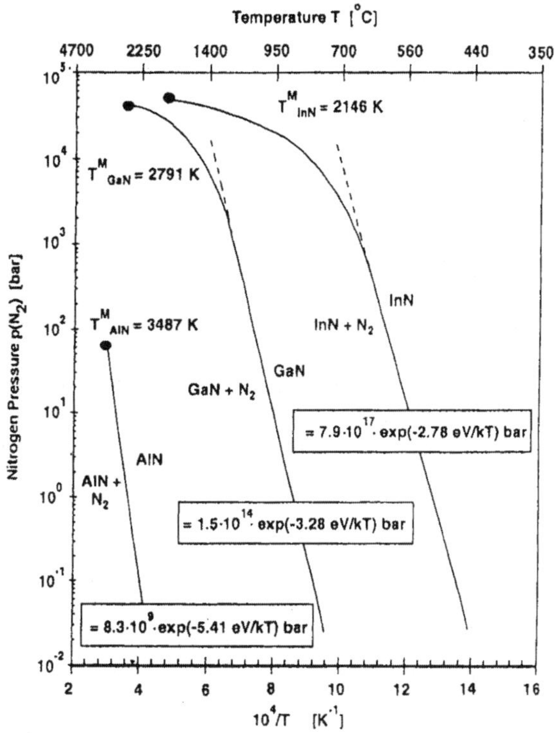

Fig. 20.2 Equilibrium N_2 pressure and melting points of III-nitrides (reprinted from Ref. 3 with permission, copyright 1998 Institute of Physics). References for the data in this figure are in Ref. 3.

20.3 CRYSTAL GROWTH

20.3.1 Boron Nitride

BN is a potential wide-bandgap semiconductor for electrical and optical applications.[24, 25] At low temperatures less than 1300°C, the cubic phase (c-BN) is stable while at higher temperatures, the hexagonal phase (h-BN) is stable. There are however only a few reports on the growth of BN single crystals. The growth of BN crystals was only performed by the high-pressure high temperature solution growth method. Alkali or alkali-earth metal boron nitrides have been used as the solvent for BN.[26, 27] Mishima et al.[28, 29] have first grown c-BN crystals, using $LiCaBN_2$ as the solvent at up to 1800°C under 5.5 GPa. Kagamida et al.[30] have grown 2.6 mm diameter crystals, using Li_3BN_2 as the solvent at 1700-1950°C under 5.5-6 GPa. Taniguchi et al.[31-33] have grown 1-3 mm single crystals, using Li_3BN_2, $Ba_3B_2N_4$, $Sr_3B_2N_4$ and a mixture of them as the solvent, at high temperatures (1500-1700°C) under high pressure (5.5 GPa) with a growth time of 20-80 h, by spontaneous nucleation. The effect of rare earth metal doping on the electrical properties was also studied.[34] The growth of c-BN on various substrates such as chemical vapor deposition (CVD) grown diamond, B_4C and surface treated refractory metals has been performed.[35]

Various metal solvents such as Ni, Ni-Cr-Mo, Ni-Cr-Mo-Al, Hasteloy C276 alloys have been also used for growing c-BN single crystals.[36-43] The addition of Cr, Mo and Al was found to increase the solubility of nitrogen and the yield of c-BN against h-BN. These Ni-based alloys also have the advantage of not being affected by oxygen and water as is the case with alkali-based solvents.

BN crystals have been hydrothermally synthesized from an aqueous solution of H_3BP_3 and NaN_3 by the reaction-coupling effect at temperatures of 300-550°C by adding white phosphorus and NH_4Cl in order to increase the growth yield and to improve the crystal quality.[44] Feigelson et al.[45] have grown h-BN from a nitride-salt solution by the temperature gradient growth method, at temperatures 760-900°C under 0.2 MPa nitrogen pressure.

Ishii and Sato[46] have obtained 2 mm size h-BN crystals of thickness 20 μm from a Si flux at 1850°C under N_2 atmosphere using a h-BN crucible. Yano et al.[47] have grown small h-BN crystals (about 2 μm) by the Na flux method at a temperature of 800°C and a nitrogen pressure of 100 atm. Dwilinski et al.[48] have applied the ammonothermal method to grow BN crystals at relatively low temperature and pressure conditions (below 550 and 5 kbar) and obtained small needle like crystals.

20.3.2 Aluminum Nitride

Because of its excellent electrical, optical and thermal properties, AlN is now an attractive candidate for many applications in electronic and optical devices.[3] In fact, AlN has a smaller lattice mismatching (~ 2.4% in a-axis[17]) with III-nitrides than sapphire and 6H-SiC, and has an almost identical thermal expansion coefficient to that of GaN. It is therefore expected that lower defect densities of GaN can be obtained when AlN is used as substrate for its epitaxial growth. In addition, its high thermal conductivity (3.4 W/Kcm) is suitable for high power device applications.[3] AlN is also promising for high power microwave devices and ultraviolet (UV) detectors due to its high resistivity (10^7-10^{13} Ω•cm) and wide bandgap energy (6.2 eV).

Since AlN melts congruently at 2750°C with a dissociation pressure of 9 atm.[49,50], it is therefore difficult to grow AlN crystals from the melt. The crystal growth of AlN is summarized in Refs. 51-53 and Mymrin et al.[54] have recently discussed engineering the bandgap for various devices using AlN substrates.

AlN is now being reconsidered as a good substrate material for various devices because it has good thermal conductivity combined with electrical insulating properties. AlN is also promising as substrate for AlGaN epitaxial layers which are promising for ultraviolet LEDs and lasers. The research activity in the crystal growth of AlN is now increasing, mainly by the physical vapor transport (PVT) method. Crystal growth research from earlier[55-66] to recent ones are summarized in Table 20.2.

(1) Melt growth

Class[65] attempted the melt growth of AlN crystals for the first time and found that the melting point is about 2800°C. Vinogradove et al.[49,50] found it as 2750°C and found that the nitrogen pressure for controlling the excessive aluminum vaporization is about 100 bar and the dissociation pressure is about 9 bar.

Table 20.2 Crystal growth of AlN

Method	Contents	Authors	Year	Ref.
PVT	Sublimation in N_2 using a electrically heated tungsten furnace	Ficher et al.	1915	55
PVT	Small crystal growth using a graphite crucible	Kohn et al.	1956	56
AD	Preparation by arc-discharge between aluminum electrodes in N_2	Long et al.	1959	57
PVT	Growth of 7 mm^3 crystals using a graphite crucible	Taylor et al.	1960	58
PVT	Growth of rod-like crystals smaller than 1 mm^3	Matsumura et al.	1960	59
Reaction	Growth by the reaction between Al_2O_3, C and N_2	Kleber et al.	1961	60
PVT	Growth of 3 mm^3 crystals using a graphite crucible	Witzke	1961	61
VM	0.4 mm^3 crystals by the vaporization method	Pastrnak	1964	62
PVT	Growth using AlN crucibles	Kawabe et al.	1967	63
PVT	0.1 mm^3 crystals using graphite or TiN crucibles	Cox et al.	1967	64
MG	Melting point temperature measurement	Class	1968	65
PVT	0.06 mm^3 crystals using a graphite crucible	Berezhkova et al.	1972	66
Flux	24 mm^3 crystals using Ca_3N_2 as a flux	Dugger	1974	67
PVT	0.06 mm^3 crystals using a graphite crucible	Ishii et al.	1975	79
CSVT	0.4 mm^3 crystals using a graphite crucible	Rutz	1976, 1978	145, 146
PVT	A few mm size crystals and purification	Slack et al.	1976, 1977	52, 80
ATG	1 mm thick crystals with HNH_2 mineralizer	Peters	1990	70
MG	Melting point temperature measurement	Vinogradov et al.	1991, 1992	49, 50
PVT	Estimation of maximum growth rate	Dryburgh	1992	84
PVT	Seeded-sublimation on 6H-SiC substrates	Balkas et al.	1997	81, 82
PVT	7 mm^3 needle-shaped crystal, FWHM = 39 arcsec	Tanaka et al.	1997	83
PVT	Theoretical consideration of gaseous species	Yu et al.	1997	85
ATG	Growth needle like crystals with $LiNH_2$ mineralizer	Dwilinski et al.	1998, 1999	48, 68
ATG	Growth of 0.4 mm crystals with KNH_2, NaH_2	Kolis et al.	1998	69
HVPE	Thick AlN on 6H-SiC and Si substrates	Abrecht et al.	1999	174
PVT	Theoretical analysis considering thermal convection and Stefan flow	Liu et al.	2000	87
PVT	Comparison of freely-nucleated and seeded crystals	Edger et al.	2000	93
PVT	Mechanism of sublimation growth	Segal et al.	2000	86
Flux	Growth of AlN on Al wires at 800°C under N_2 100 atm	Yano et al.	2000	47
HVPE	Freestanding AlN of thickness 1 mm eparated from Si	Nikolaev et al.	2000	175
HP-SG	Growth of 1 cm × 1 mm plates at 2000 K under 1 GPa	Bockwoski et al.	2001	71
PVT	11 mm in diameter, 1 mm in length, FWHM < 10 arcsec, EPD = 0-500 cm^{-2}	Rojo et al.	2001	97-99
MOCVD	Preparation of AlN/sapphire templates up to 100 mm diameter	Shibata et al.	2001-2005	149-163
VM	50 mm^3 plate-like crystals grown at 2100°C	Schlesser et al.	2002	143
PVT	Seeded-sublimation, centimeter-size single crystal with FWHM = 25 arcsec	Schlesser et al.	2002	103
PVT	Seeded-sublimation on 6H-SiC substrates	Edger et al.	2002	94
PVT	Analysis of temperature distribution, thermal stress and growth rate	Wu et al.	2003-2007	88-91
VM	Vaporization of Al with AlN	Sitar et al.	2003	144
PVT	Growth of crystals 12 mm in diameter	Schowalter et al.	2003	100
CSVT	12.5 mm diameter, 8 mm length at 0.5 mm/h	Mokhov et al.	2003	113
PVT	Cystals 25 mm in diameter and 10 mm length	Bickermann et al.	2003	118

Table 20.2 Crystal growth of AlN (continued)

Method	Contents	Authors	Year	Ref.
PVT	Platelet of 14 × 7 × 2 mm^3, growth habit examination	Epelbaum et al.	2003	120
PVT	Growth of 44 mm diameter crystals on AlN seeds	Melnik et al.	2003	131
HVTE	Growth using $AlCl_{3-x}$-NH_3 (x = 1-6) adducts	Bliss et al.	2003-2005	196, 197
Microwave	Platelets and needles of several mm by direct heating of source materials with microwaves	Zhuang et al.	2004	200
Flux	Growth from Cu-Ti alloy solution	Kamei et al.	2005-2007	72-74
HVPE	Thick AlN layers on AlN/sapphire templates	Liu et al.	2005	190
PVT	Thick AlN layers on SiC seed crystals	Dalmau et al.	2005	104
PVT	Crucibles for sublimation growth	Schlesser et al.	2005	105
PVT	25 mm diameter, 10-12 mm length from Ta crucibles	Mokhov et al.	2005	114
PVT	Seeded regrowth on AlN seeds, 0.1-0.3 mm/h	Noveski et al.	2005	92
CSVT	Growth on bare SiC and AlN buffered SiC seeds	Yakimova et al.	2005	126
HVPE	50 mm freestanding AlN substrates	Kovalenkov et al.	2005	176
PVT	Growth of freestanding AlN crystals of length 25 mm	Epelbaum et al.	2005-2007	121-124
HVPE	Freestanding AlN with the thickness of 112 μm, void formation for separating AlN	Kumagai et al.	2005-2009	184-188
HVPE	Thick AlN on AlN/sapphire templates	Miyake et al.	2005-2009	189-193
PVT	Growth expansion from 5 to 15 mm in diameter	Herro et al.	2006	106-108
PVT	Freestanding crystals with self-separation from SiC seeds	Wang et al.	2006	132
PVT	Effect of oxygen on the growth morphology	Kazan et al.	2006	133
NM	Conversion of Al_2O_3 substrate to AlN	Furuyama et al.	2006	147
HVPE	Thick AlN on sapphire at 40-60 μm/h	Kamber et al.	2006	182
HVPE	Thick AlN on 50 mm sapphire combined wih migration enhaced MOCVD (MEMOCVD)	Jain et al.	2006	183
HVPE	Thick AlN on up to 100 mm diameter SiC and and preparation of freestanding substrate	Soukhoveev et al.	2006-2009	178-180
PVT	Growth 25 mm diameter and 4 mm long crystals	Mizuhara et al.	2006-2007	128-130
PVT	Seeded-sublimation on 4H- and 6H-SiC substrates	Lu et al.	2007	95
PVT	In situ preparation of a TaC crucible and purification of AlN source	Helava et al.	2007	115
PVT	Origin of oxygen contamination	Edger et al.	2008	96
PVT	Growth of 50 mm diameter crystals with EPD about 10^4 cm^{-2}, FWHM < 28-32 arcsec	Scujman et al.	2008	101, 102
PVT	Growth of 50 mm diameter and 10 mm long crystals	Markov et al.	2008	116, 117
PVT	Growth of 15 mm in diameter and 15 mm in length	Siter et al.	2008-2010	110-112
PVT	Freestanding crystals by mirorods on AlN seeds	Yazdi et al.	2008	127
PVT	Growth of 35 mm diameter and 25 mm long crystal	Hartmann et al.	2008	134
SHVPE	Growth using solid $AlCl_3$ source	Eriguchi et al.	2008	195
HTCVD	Reaction of Al source with Cl_2 and growth of AlN	Claudel et al.	2008-2009	198, 199
PVT	Growth of a-plane crystals by spontaneous nucleation	Letts et al.	2009	135
NM	Nitridation of (111) Al by ammonia	Honda et al.	2009	201
Flux	Growth of $Al_{1-x}Ga_xN$ (0.22<x<0.61) from Ga solution at 1800°C under 10 kbar	Belousov et al.	2009, 2010	76, 77
PVT	Seeded growth and dislocation distribution	Nagai et al.	2010	136
Flux	Growth from Li-Al-N solution at 1050°C	Kanagawa et al.	2010	75

Table 20.2 Crystal growth of AlN (continued)

Method	Contents	Authors	Year	Ref.
PVT	Growth of AlN-SiC single crystals	Filip	2010	137
PT	Growth of AlN by pyrolytic transportation	Hironaka et al.	2010	148
PVT	Seeded growth on AlN seed crystal	Filip	2011	125
HVPE	Thick AlN on off-axis 6H-SiC	Volkova et al.	2011	181

PVT: Physical Vapor Transport, HVPE: Horizontal Vapor Phase Epitaxiy, SHVPE: Solid source HVPE, HVTE: Halide Vapor Transport Epitaxy, HTCVD: High Tempeature Chemical Vapor Deposition, CSVT: Closed Space Vapor Transport, PT: Pyrolytic Transportation, VM: Vaporization Method, NM: Nitridation Method, ATG: Ammonothermal Growth, MG:Melt Growth

(2) Solution growth

Dugger[67] employed molten Ca_3N_2 as a flux for AlN solution growth. Ca_3N_2 is stable at high temperatures, with low viscosity and with low melting point (1195°C). Ca_3N_2 can be dissolved easily by dilute HCl and AlN crystals can be removed. The solubility of AlN in Ca_3N_2 is as shown in Fig. 20.3. By this flux method, it is reported that small bipyramidal (1.1 mm × 0.3 mm) single crystals could be grown.

The ammonothermal method which has been extensively studied for growing GaN (Sec. 20.3.3(2)) has also been applied for growing AlN.[48,68,69] Peters[70] has grown AlN single crystals of thickness 1 mm by the ammonothermal method in ammonia using KNH_2 as the mineralizer. The growth was performed at temperatures of 500-600°C under ammonia pressure of 2 kbar.

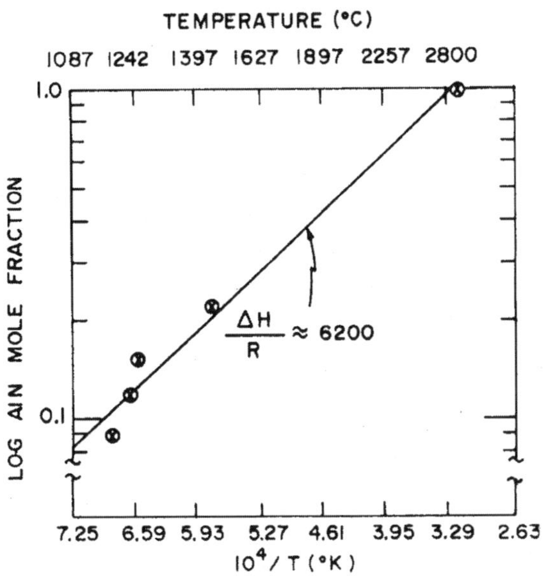

Fig. 20.3 Partial solubility data of the AlN-Ca_3N_2 system under atmospheric pressure (reprinted from Ref. 67 with permission, copyright 1974 Elsevier).

Boskowski et al.[71] have grown AlN crystals from an Al solution using a high pressure furnace at temperatures up to 2000 K under the high nitrogen pressure of 1 GPa. Bulk AlN single crystals measuring 1 × 10 mm could be obtained.

Recently, Kamei et al.[72-74] have found that AlN can be grown from an alloy solution such as Cu-Al-Ti. Kanagawa and Kakimoto[75] have grown AlN crystals from Li-Al-N solution. Belousov et al.[76, 77] have grown $Al_xGa_{1-x}N$ of high Al content ($0.5 < x < 0.9$) from a Ga solution under high pressure up to 10 kbar. Yano et al.[47] have applied the Na flux method to grow AlN crystals.

Taniguchi et al.[34] have applied a high pressure solution growth method, the method mainly applied for BN crystal growth (Sec. 20.31), using Li_3AlN_2 and $Ba_3B_2N_4$ solvents to grow AlN crystals and studied the effect of rare earth metal doping on the optical properties.

(3) Vapor phase growth

- Physical Vapor Transport (PVT) method

As the crystal growth method for AlN, the PVT method (also called as the sublimation method) is the most common as reviewed[78] and studied extensively[52, 79-136] as shown in Table 20.2. Small AlN single crystals could be grown by several researchers in the past.[79-83] Seeded bulk growth of single crystalline AlN (001) platelets was achieved on 6H-SiC (0001) substrates in the temperature range of 1950-2250°C under 500 Torr nitrogen atmosphere using an assembly as shown in Fig. 20.4.

The vapor and dissociation pressures which are essential data for PVT growth have been studied as a function of temperature as shown in Fig. 20.5 [52] and the growth rate was estimated based on the growth kinetics.[84] However, for long period, the mechanism of sublimation growth of AlN was not clarified. This is because of the underestimated role of N_2 adsorption/desorption kinetics on the AlN growth surface.[85] The drawback of the former growth mechanism was improved by Segal et al.[86] and Liu and Edger[87] have studied the two-dimensional model for sublimation crystal growth and estimated the growth rate more precisely. Wu et al.[88-90] made a growth simulation and analyzed the thermal stress in grown crystals. Wang et al.[91] have made a simulation to find the conditions to increase the growth rate.

Growth of AlN is performed by the furnace arrangement as shown in Fig. 20.6[92], an arrangement like SiC crystal growth (Ch. 23) and that like the closed sandwich arrangement. The growth is performed at much higher temperatures (1800-2300°C) than the case of SiC so that the furnace material is a key issue from the viewpoint of durability and impurity contamination. The nitrogen pressure is controlled to 200-700 mbar. Both spontaneous nucleation and seeded growth on SiC substrates are studied.

The group at Kansas University[93, 94] have compared the growth by self-nucleation and by seeded nucleation on bare 6H-SiC and on AlN-SiC layer on 6H-SiC. It was found that self-nucleated AlN single crystals showed higher crystal quality than seeded grown ones. The quality of seeded growth was however improved by using AlN-SiC layered seeds. They have also examined the growth mode of SiC seeded growth by changing the polytype and the off angle of the seed crystal and found that basal plane

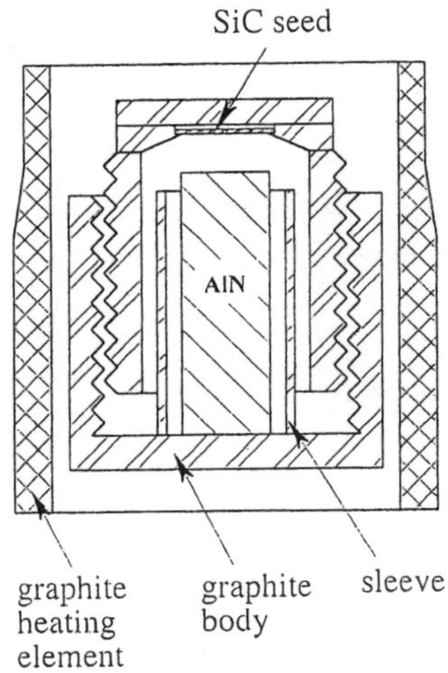

Fig. 20.4 Schematic of the graphite growth assembly employed in the sublimation growth of AlN (reprinted from Ref. 81 with permission, copyright 1997 Elsevier).

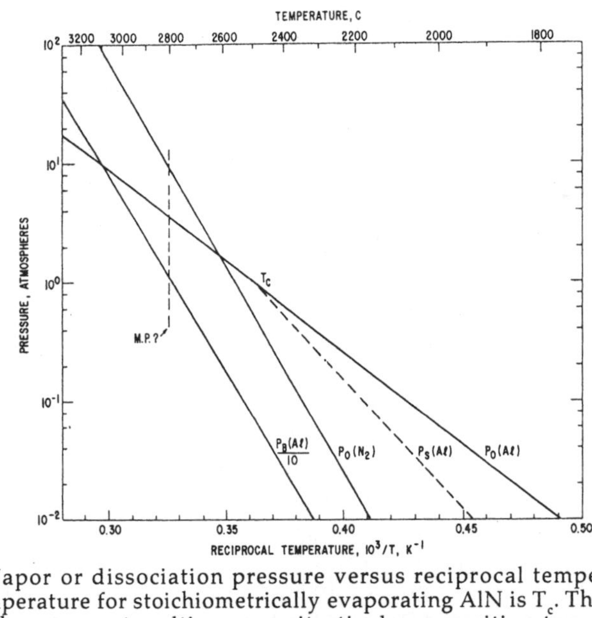

Fig. 20.5 Vapor or dissociation pressure versus reciprocal temperature for AlN. The critical temperature for stoichiometrically evaporating AlN is T_c. The dashed vertical line indicates the apparent melting or peritectic decomposition temperature of AlN. The several labelled curves are: $P_0(N_2)$ = pressure of N_2 over the two-phase system AlN(s) + Al(l), P_0(Al) = pressures of Al(g) over Al(l), P_s(Al) = pressure of Al(g) over AlN when the gas is stoichiometric, Pb_B(Al) = pressure of Al(g) over AlN when $P(N_2)$ is held at 1 atm. The curve shown is for 1/10 of P_B(Al) (reprinted from Ref. 52 with permission, copyright 1976 Elsevier).

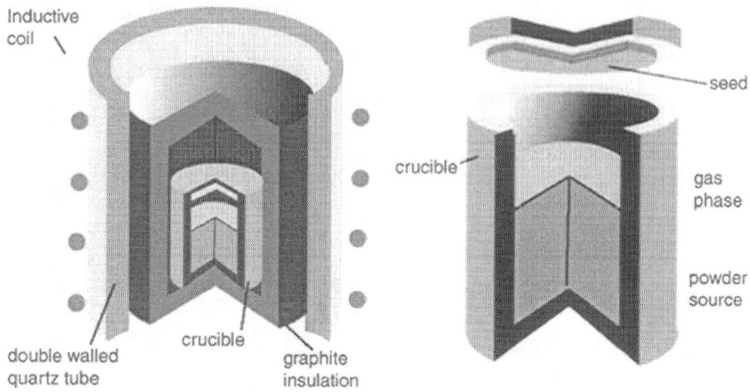

Fig. 20.6 RF heated reactor (left) and sandwich-sublimation configuration (reprinted from Ref. 92 with permission, copyright 2005 Elsevier).

dislocations are generated only on the 8° off-axis seeds.[95] Edgar et al.[96] have shown that source AlN powders include oxides and hydroxides at the powder surface which are the origin of oxygen contamination in AlN single crystals.

The group at Crystal IC[97-102] have succeeded in growing an AlN single crystal boule 15 mm in diameter and several mm in length using the self-seeded PVT method (Fig. 20.7). Single crystal grains more than 80 mm^2 with dislocation density less than 10^3 cm^{-2}, the lowest 500 cm^{-2} could be obtained with the growth rate between 0.5 mm/h and 1 mm/h. The best full width at half maximum (FWHM) of the X-ray diffraction rocking curve (XRC) was reported to be lower than 10 arcsec.[99] Using AlN seeds cut from thus obtained AlN single crystals, they finally succeeded in growing 50 mm diameter AlN crystals with the FWHM-XRC of 28-32 arcsec and of dislocation density 10^3 cm^{-3}.[102]

Schlesser et al.[103] at North Calorina State University grew AlN single crystals by sublimation of AlN powders at 2100-2300°C in an open crucible under 400 Torr nitrogen atmosphere in the arrangement as shown in Fig. 20.8. Single crystalline AlN c-platelets were used as seed crystals. Seeded growth in the c-direction occurred with the growth rate exceeding 500 μm/h. Dalmau et al.[104] have studied the growth of the initial stage of the on-axis and off-axis SiC seed growth. Various crucible materials such as refractory metals, nitrides, carbides and borides have been compared and they concluded that sintered TaC is appropriate without Ta contamination and low O and C contamination of the level of 50 ppm.[105] Noveski et al.[92] have grown AlN crystals using AlN seeds, by the multiple step regrowth process at a growth rate of 0.1-0.3 mm/h. Seeded growth on N- and Al-polar AlN seeds has been developed and the crystal diameter was expanded from 5 mm seed to 15 mm[106-108] with a height of 4-12 mm and with an expansion angle of 45°. Cai et al.[109] have made a simulation for optimizing the thermal environment for obtaining a large and thick AlN single crystal. The evolution of the temperature distribution has been calculated and the phenomenon of cessation of

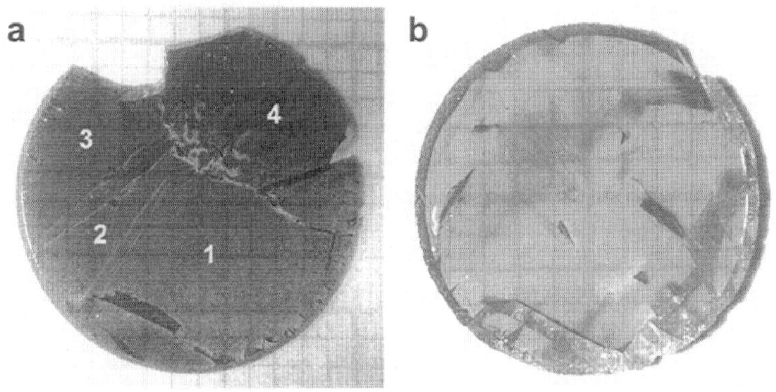

Fig. 20.7 Large AlN substrates cut from AlN boules (grids are in mm). The substrate on the left (a) exhibits three main single crystal grains (1,2 and 3) and a polycrystal region (4). (b) does not present polycrystal regions but displays severe cracking area close to the periphery (reprinted from Ref. 98 with permission, copyright 2001 Cambridge University Press).

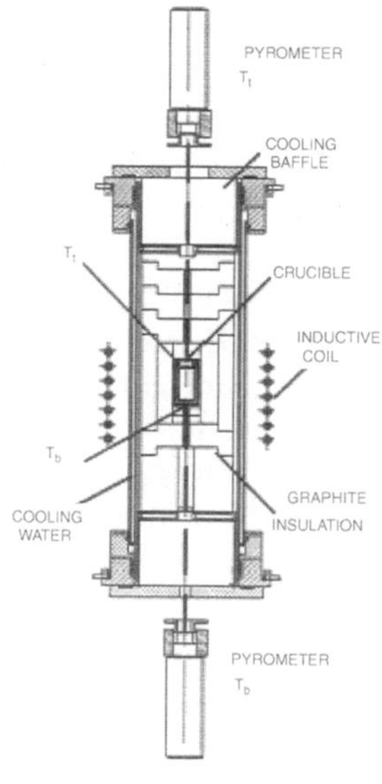

Fig. 20.8 Schematic of the RF heated AlN reactor (reprinted from Ref. 108 with permission, copyright 2006 Elsevier).

growth was discussed. Large diameter (15-20 mm) and thick (1-2 mm) AlN crystals were grown on 3.5° off-axis 6H-SiC (0001) seeds and the dislocation density was found to be reduced from 10^8 to 10^6 cm^{-2}.[110] Using AlN seeds, single crystals 15 mm in diameter and 15 mm long could be obtained and the FWHM-XRC decreased from 78 arcsec to 13 arcsec as the growth proceeded.[111] Lu et al.[112] have grown m-orientation and c-orientation AlN single crystals using m- and c-oriented AlN seed crystals cut from freestanding AlN single crystals. The growth rate was 150-170 μm/h with expansion angles of 22-27°, and the dislocation density was 10^2-10^5 cm^{-2}.

The group at N-Crystals[113] have obtained AlN crystals 12.5 mm in diameter and up to 8.5 mm in length at a growth rate of 0.5 mm/h using 4H- and 6H-SiC seed crystals. They used a Ta crucible and achieved growth 45 mm in diameter and 4 mm long by nucleation on the crucible lid through polycrystalline boules. 4H- and 6H-SiC seeds and AlN layers grown on SiC seeds were also used for the seeded growth and AlN crystals 12.5 mm in diameter and up to 10-12 mm in length were obtained.[114] They have developed a technology[115] combined with an in situ crucible preparation of TaC from Ta metal by carbonization, the purification of AlN source powders by recrystallization and the usage of SiC seed substrates. The use of AlN seeds which are prepared by separating AlN layers from SiC seed substrates has led to further improvement and the growth of 50 mm diameter, 10 mm long AlN crystals with a single crystal core of 40 mm diameter has become possible[116, 117] (Fig. 20.9).

Bickermann et al.[118] at Erlangen University have grown low oxygen content AlN single crystals by the temperature reversal method[119] where the seed was partially resublimed after the growth temperature was reached. Epelbaum et al.[120-122] studied the growth conditions for self-nucleation and the subsequent growth of bulk AlN crystals and obtained well-faceted AlN single crystals (Fig. 20.10). With increasing growth temperature, the natural habit of AlN crystals became asymmetric planes. The best formed platelet crystals up to $14 \times 7 \times 2$ mm^3 could be obtained. The most successful seeded growth was achieved at growth temperatures exceeding 2200°C and along Al-terminated (0001), (-1012) and nonpolar (-1010) faces. They also studied three approaches, (i) the growth of freestanding crystals up to 15 mm in length, (2) the grain selection in sequential growth of bulk 50 mm in diameter and up to 15 mm in height and (3) the seeded growth on 25 mm diameter (0001) 4H-SiC substrates. The effect of oxygen contamination of the source material was also discussed. The influence of Al_2O_3 at the initial growth stage of AlN on SiC substrates was studied and it was found that the addition of Al_2O_3 enhances the smooth surface of AlN in the 2D layered growth mode. Epelbaum et al.[123] have grown freestanding single crystals up to 25 mm in length using a WC crucible and analyzed the growth habits. They also compared the growth of AlN with that of SiC and the difficulty of AlN is concerned with the poor compatibility with high temperature crucible materials.[124] Filip et al.[125] have grown AlN single crystals using AlN seeds of very different orientations and the effect of growth orientations on the crystal properties has been studied.

Yakimova et al.[126] at Lingoping University have studied the various growth parameters on the growth mode on SiC seeds both bare and with a metal organic chemical vapor deposition (MOCVD) AlN buffer layer using a high purity graphite furnace ma-

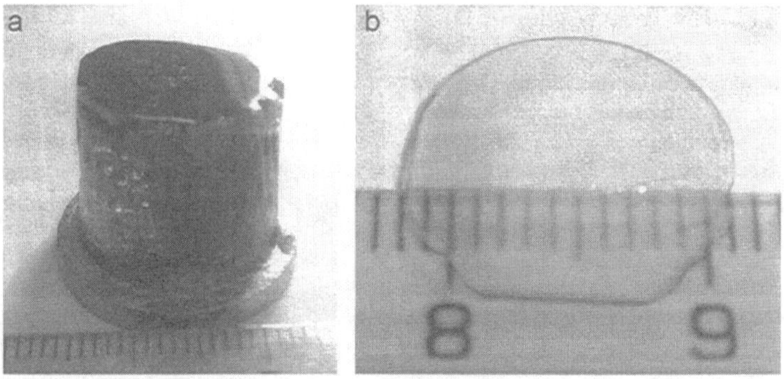

Fig. 20.9 (a) A 15 mm diameter and 15 mm long AlN single crystal grown with the technology developed and (b) AlN substrate produced from a similar crystal (reprinted from Ref. 116 with permission, copyright 2008 Elsevier).

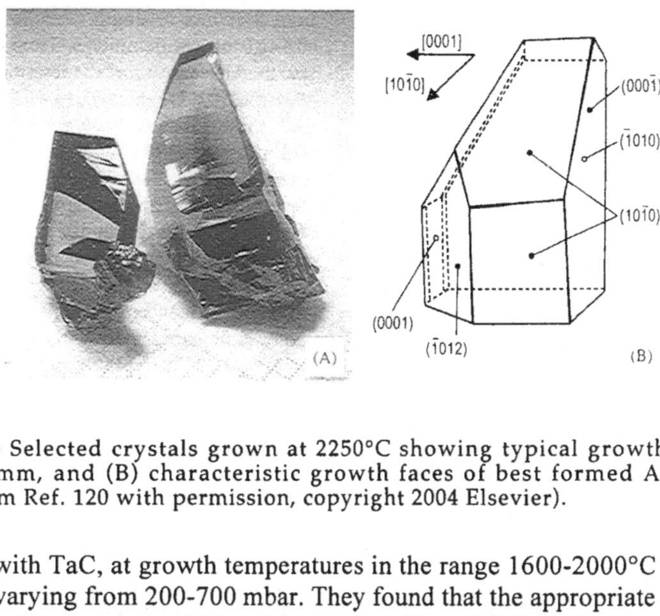

Fig. 20.10 (A) Selected crystals grown at 2250°C showing typical growth habit, grid spacing is 1 mm, and (B) characteristic growth faces of best formed AlN platelets (reprinted from Ref. 120 with permission, copyright 2004 Elsevier).

terial coated with TaC, at growth temperatures in the range 1600-2000°C under nitrogen pressure varying from 200-700 mbar. They found that the appropriate growth condition is in a narrow range and impurities such as Ta and C from the coating material greatly affect the growth kinetics. Yazdi et al.[127] have fabricated freestanding AlN wafers of diameter 10 mm and about 120 μm thickness, using 4H-SiC seed substrates and by growing initial AlN microrods which facilitate the separation of AlN layers.

AlN single crystals were grown on 10-25 mm diameter SiC substrates from 3 μm to 4 mm thickness[128-130] (Fig. 20.11) by Sumitomo Electric Industry and the dislocation density was evaluated as a function of the thickness (Fig. 20.12). It was found that the

Fig. 20.11 Photograph of a 4 mm thick AlN single crystal grown on SiC (reprinted from Ref. 130 with permission, copyright 2007 Elsevier).

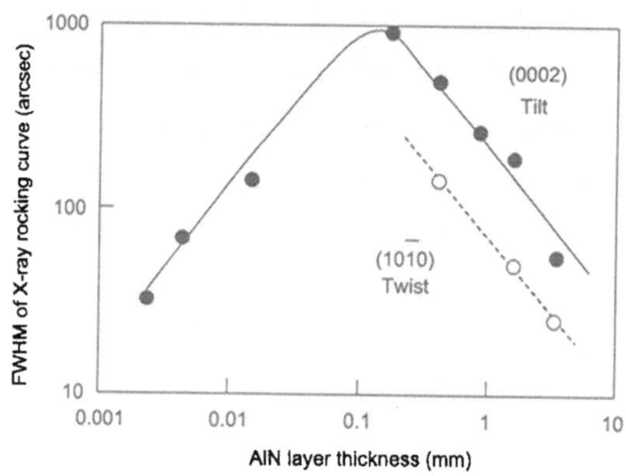

Fig. 20.12 Dependence of AlN crystallinity on growing film thickness (reprinted from Ref. 130 with permission, copyright 2007 Elsevier).

dislocation density and the FWHM-XRC decreased as a function of the thickness with a maximum at 0.1 mm.

Melnik et al.[131] have grown 44 mm diameter AlN single crystals using AlN seeds and showed resistivity exceeding 10^8 Ωcm. Wang et al.[132] have grown AlN single crystals on (0001) C-face SiC seeds and obtained 4 × 5 mm^2 freestanding AlN wafers which separated spontaneously from the seed because of the mismatch of the thermal expansion coefficients.

Kazan et al.[133] have examined the effect of oxygen on the growth surface morphology and Hartmann et al.[134] have grown 35 mm diameter and 25 mm long AlN crystals on AlN seeds. They examined the growth morphology and the behavior of the incorporation of oxygen and discussed the role of oxygen in increasing the size of AlN single crystals. Letts et al.[135] have grown AlN single crystals by spontaneous nucleation and obtained a-plane platelets having FWHM-XRC of 17-60 arcsec. They also found that the addition of In has the effect of improving the crystal size. Nagai et al.[136] have grown AlN single crystals on SiC seed substrates using a TaC crucible (Fig. 20.13) and found

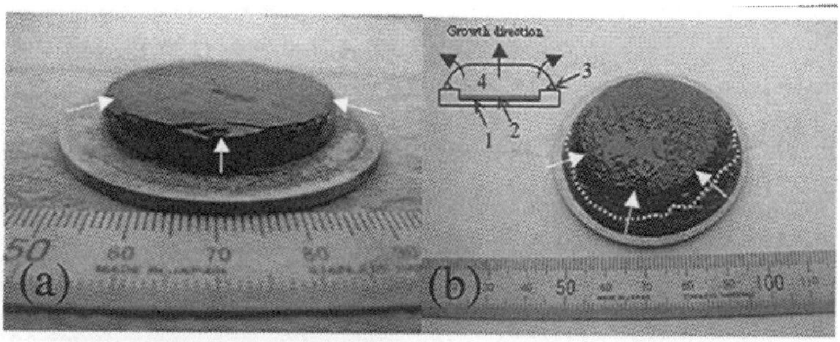

Fig. 20.13 Photographs of two AlN bulk single crystals grown measuring (a) 35 mm in diameter and 6 mm in thickness and (b) 43 mm in diameter and 10 mm in thickness and inset shows the schematic cross-section: 1. TaC cap, 2. SiC substrate, 3. AlN polycyrstalline and 4. AlN singe crystal. The white arrows indicate (10-1N) faces. For the grown crystal of (b), the dotted line shows the boundary between a single crystal (inner part) and a polycrystal (outer part) - a slight AlN polycrystal deposited on lateral side of the AlN single crystal near the TaC cap (reprinted from Ref. 136 with permission, copyright 2010 Elsevier).

that the dislocation density reaches a minimum of 7×10^4 cm^{-2} as a function of growth thickness.

Filip et al.[137] have grown AlN-SiC mixed crystals on 6H-SiC seed substrates, at temperatures 1900-2050°C with different off-axis seeds inclined up to 42° from the basal plane towards the (01-10) plane and using different source materials. Depending on the growth conditions, crystals with different Si/C content could be obtained.

The epitaxial growth of high quality AlN and AlGaN on PVT-grown AlN substrates has been performed successively[138-142] even without buffer layers. The crystal quality was evaluated by XRC, photoluminescence (PL) and atomic force microscopy (AFM) and it was found that the quality is much better than that by heteroepitaxial growth. It was also found that pseudomorphic layers with a thickness far exceeding the critical thickness could be obtained.[142]

- *Other vapor phase growth methods*

Crystal growth by the vaporization of aluminium in nitrogen atmosphere was first reported by Pastrnak et al.[62] Schlesser et al.[143] have studied this vaporization method and found that c-oriented AlN plates up to 50 mm^2 could be grown by spontaneous nucleation within 2 h. Sitar et al.[144] have used metallic Al as a source material and have grown c-plates at a growth rate exceeding 10 mm/h, a rate much faster than when AlN is used as the source material. Rutz et al.[145,146] pioneered the closed space vapor transport (CSVT) method and grew 1×1 cm epitaxial AlN single crystals with a thickness of several μm on sapphire substrates and measured the optical properties.

Fukuyama et al. [147] have converted a-Al_2O_3 substrates to AlN films by the nitridation of the substrate surface based on the thermodynamical reaction under N_2-CO gas flow. Hironaka et al.[148] have grown AlN single crystals using a reaction based on the pyrolytic transportation method. α-Al_2O_3 powders were used as the source material instead of AlN powders and AlN single crystals were grown on the AlN sintered seeds according to the reaction based on the phase stability diagram of the Al_2O_3-AlN-C-N-CO system. Needle like single crystals with a length of 48 mm at a high growth rate of 1.6 mm/h could be grown.

(4) Metal Organic Chemical Vapor Deposition (MOCVD)

Shibata et al.[149-154] proposed an AlN template which is a thin AlN layer grown by MOCVD on a sapphire substrate, used as a AlN substrate for GaN epitaxial growth. When AlN templates were used as substrates for GaN epitaxial growth, it was found that the dislocation density of GaN was lowered one order of magnitude compared with GaN grown directly on sapphire substrates. This reduction was found to be because only edge dislocations thread through AlN layers and screw dislocations are eliminated at the interface between AlN and GaN layers. These AlN templates have been applied for blue LEDs[155, 156] and ultraviolet (UV)-LEDs[157] and it is found that the light emission efficiency is increased due to the reduction of dislocations. AlN templates up to 100 mm in diameter have been applied to high electron mobility transistor (HEMT) structure epitaxial growth[158-163] and it was found that electron mobility can be increased due to the reduction of dislocation densities in the HEMT structures.

Imura et al.[164] and Osaka et al.[165] have grown thick AlN layers on sapphire substrates by MOCVD at a growth rate of 6 μm/h[164], varying the V/III ratio with the thickness in order to reduce the dislocations.[165]

(5) Hydride Vapor Phase Epitaxy (HVPE)

Since the HVPE method yields a high growth rate in GaN epitaxial growth (Sec. 20 3.3(3)), this method is also promising for the preparation of AlN substrates. In this method, $AlCl_3$ gas, prepared by the reaction of metal Al in the furnace with HCl gas, reacts with NH_3, and AlN layers are grown on the seed substrate according to the following reaction.

$$AlCl_3 + NH_3 = AlN + 3HCl \qquad (20.1)$$

HVPE has been applied to grow AlN films on sapphire[166, 167], quartz glass[168] and SiC[169] in the past and the appropriate growth conditions were defined by thermodynamical theoretical study.[170, 171] The growth of AlN layers on sapphire substrates and AlN templates was performed based on these theoretical considerations.[172, 173]

Alebrecht et. al.[174] have grown thick AlN on 6H-SiC and Si substrates and analyzed the reduction of dislocation densities. Freestanding AlN substrates have been obtained by several researchers.[175-181] Nikolaev et al.[175] have obtained AlN substrates with a maximum thickness of 1 mm by removing the Si seed substrates. Kovalenkov et al.[176] have obtained 50 mm diameter AlN substrates by removing SiC seed substrates. Volkova et al.[181] have grown 20 nm thick AlN layers on off-axis 6H-SiC substrates as the templates for AlGaN/GaN HEMT structures. Kamber et al.[182] have grown AlN layers on sapphire substrates with a growth rate of 40-60 μm/h and studied the effect of the

V/III ratio on the crystal quality. Jain et al.[183] have grown AlN films with a thickness ranging 1-25 μm with growth rate as high as 200 μm/h.

Kumagai et al.[184] have grown AlN films on Si substrates and were able to obtain freestanding AlN substrates of thickness of 112 μm by removing the Si substrate by chemical etching. Kumagai et al.[185-188] have grown crack-free AlN template layers on sapphire substrates at 1450°C and were able to separate AlN films from substrates by the formation of voids.

The growth of thick AlN films by the low pressure HVPE (LP-HVPE) method on AlN/sapphire templates prepared by MOCVD (Sec. 20.3.2(4) has been reported by the group at Mie university.[189-193] The growth has been performed at temperatures of 1100-1200°C[189,190] and 1400-1500°C[191,192] and produced thick AlN films with the (0002) and (10-12) FWHM-XRC of 351 and 781 arcsec respectively at a growth rate of 14.3 μm/h. The growth on a trench-patterned AlN template was also performed in order to improve the crystal quality.[193]

The disadvantage of the HVPE method is in the reactivity of HCl gas with the quartz tube and that of Al with oxygen and the difficulty of the realization of high purity crystal growth. In order to counter this disadvantage of HVPE, solid $AlCl_3$ powders were used as the Al source and no HCl gas was used as the carrier gas.[194] Eriguchi et al.[195] have used a solid $AlCl_3$ source using N_2 + H_2 gas carrier and were able to grow AlN layers at higher temperatures of 1500°C. $AlCl_3$ however still had the reactivity with the quartz tube. In order to reduce this difficulty further, the halide vapor transport epitaxy (HVTE) has been studied[196,197] where adducts such as $AlCl_{3-x}{\cdot}NH_3$ (x =1-6) which were presynthesized were used as the Al source. In this method, no HCl gas is used and the reactivity of $AlCl_3$ was reduced by using adduct materials. AlN layers of thickness up to 50 μm could be grown at a growth rate up to 60 μm/h.

To increase the growth rate, the high temperature CVD (HT-CVD) method[198,199] has been studied. In this method, the Al source metal is reacted with Cl_2 gas to produce $AlCl_x$ gas and this is reacted with NH_3 to grow AlN at higher temperatures as shown in Fig. 20.14.

(6) Other growth methods

Zhuang et al.[200] have grown AlN single crystal platelets up to 2 × 3 mm^2 and needles 1 mm in diameter and 1 mm in length by heating the source materials directly in a microwave atmosphere as in the arrangement shown in Fig. 20.15. The growth was performed at 2080°C at a constant pressure of 910 Torr. The growth rate was 300 μm/h in the c-direction. Honda et al.[201] have made AlN layers by ammonia nitridation of (111) Al substrates. In this method, the removal of surface oxides on Al substrates prior to nitridation was important in achieving the nitridation.

20.3.3 Gallium Nitride

Various methods for growing GaN have been tried over many years[202-207] and they have been reviewed by Rabenau.[51] The melt growth of GaN requires a very high vapor pressure to prevent its decomposition so that it is impossible to grow GaN crystals by the melt growth method. This is the reason why various solution growth and vapor phase

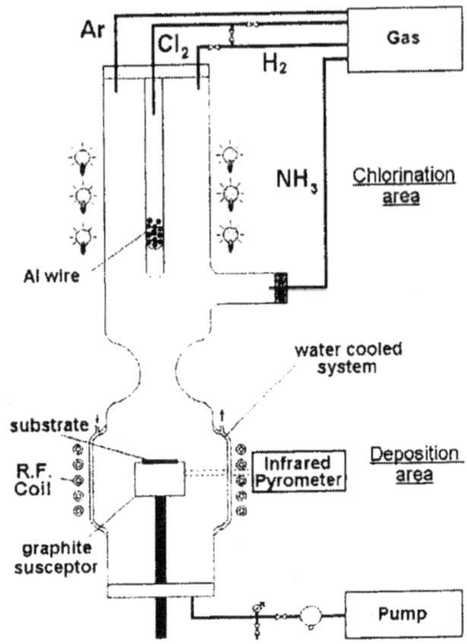

Fig. 20.14 High temperature CVD quartz reactor (reprinted from Ref. 198 with permission, copyright 2009 Elsevier).

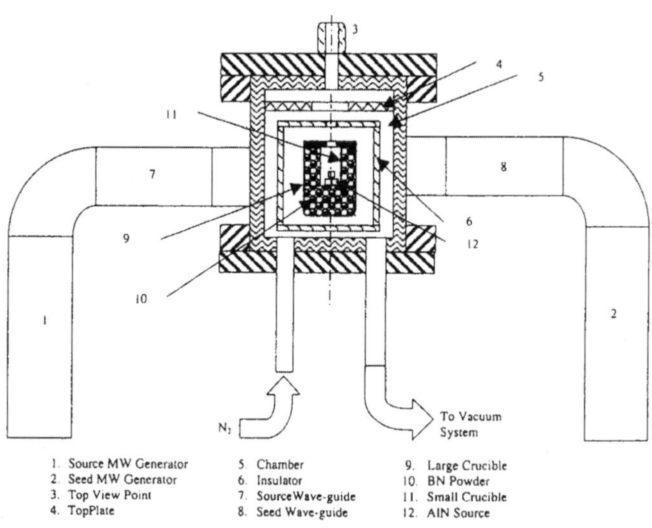

Fig. 20.15 Cross section view of growth reactor (reprinted from Ref. 200 with permission, copyright 2004 Elsevier).

growth methods are being developed as explained below. Since wurtzite GaN has a strong electric field in the [0001] direction due to polarization, there is a considerable interest in using semipolar and nonpolar directions to grow nitrides.[208,209] Therefore, there are also efforts not only towards obtaining polar substrates but also semipolar and nonpolar substrates. In Table 20.3, crystal growth by solution growth and in Table 20.4, that by vapor phase growth are summarized. Crystal growth of GaN is reviewed in Refs. 210-213.

(1) Melt growth

Utusmi et al.[20-22] have grown GaN from a congruent melt under 7.7 GPa by gradual freezing from 2250°C using a multi-anvil high pressure apparatus for the in-situ X-ray diffraction measurement by synchrotron radiation at the SPring-8 facility. They have also grown $Al_{0.1}Ga_{0.9}N$ mixed crystals, under 6.5 GPa by gradual freezing from 2500°C.

(2) Solution growth

- High Pressure Solution Growth (HP-SG)

Since GaN has a high decomposition pressure, it is necessary to impose a high pressure up to about 10 GPa in order to obtain a reasonable nitrogen solubility in the Ga melt. To achieve this high pressure, a specially designed high pressure furnace is required.

The group at the High Pressure Institute in Poland first started to grow GaN single crystals by using an ultrahigh pressure furnace as seen in Fig. 20.16, under pressures ranging 10-30 GPa at 1500 K. Solution growth was performed by decreasing the temperature under a temperature gradient. They showed that GaN single crystals can be

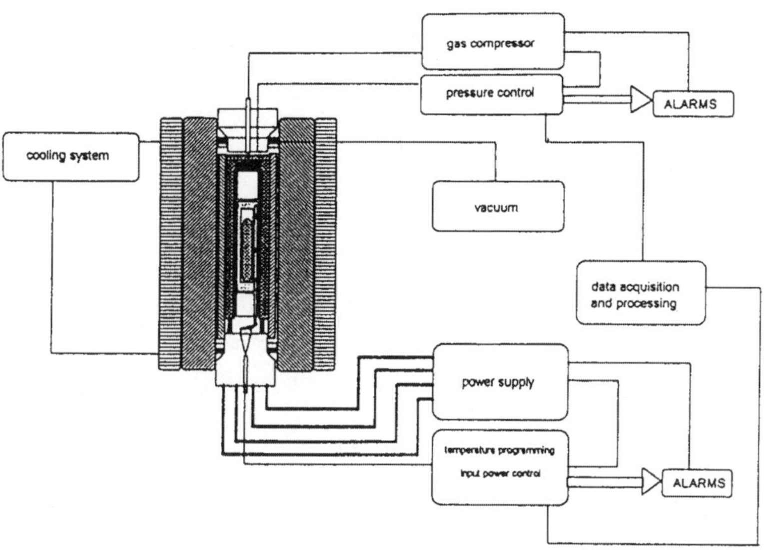

Fig. 20.16 Experimental system for crystal growth of GaN at N_2 pressure up to 15 kbar (reprinted from Ref. 219 with permission, copyright 1997 Elsevier).

Table 20.3 Crystal growth of GaN by solution growth

Method	Contents	Authors	Year	Ref.
LPE	Growth of GaN on sapphire substrates	Logan et al.	1972	349
HP-SG	First GaN crystal growth under ultrahigh pressure of N_2.	Karpinski et al.	1984	17
HP-SG	Growth in a high pressure furnace, thermodynamical consideration	Porowski et al.	1989-1996	214-217
HP-SG	Growth of GaN measuring 10 mm and as substrates for MOCVD and MBE	Grzegory et al.	1996	218-219
ATG	Growth of small GaN crystals below 550°C and 5 kbar	Dwilinski et al.	1995-2011	48,68, 307-309
SG	Small GaN crystals grown from Ga based melt under low pressure	Ivantsov et al.	1997	350
Flux	Spontaneous and seeded growth of GaN crystals from Na flux with NaN_3	Yamane et al.	1997-2000	247-252
ATG	Growth by KNH_2 and KI mineralizers	Kolis et al.	1998	69, 304 315, 316
HP-SG	Effect of Mg, Ca and Zn doping	Grzegory et al.	1998	220
Flux	GaN growth from Na flux with NaN_3	Yano et al.	1998-1999	47, 266, 267
HP-SG	Growth mechanism under high pressure	Boskowski et al.	1999	221, 222
HP-SG	Annealing under high pressure nitrogen	Porowski et al.	1999	227
ATG	Growth from NH_2NH_3Cl mineralizer	Collade et al.	1999	343
PC-SG	Growth of GaN crystals up to 10 mm using a multi-crucible system under constant temperature	Inoue et al.	1999-2002	238-246
ATG	Growth using Li mineralizer	Lan et al.	2000	344
Flux	Growth of 0.8-1 mm GaN crystals from Na flux under N_2 gas	Aoki et al.	2000-2002	253-258
SG	Growth of about 20 mm diameter 15 mm long ingots from Ga-based solution at 800-1000°C under 2 atm.	Sukhoveyev et al.	2000	351
LPE	Growth of GaN on various substrates	Klemenz et al.	2000	352
ATG	Growth at 400°C and 2.4 kbar	Ketchum et al.	2001	315
Flux	Growth of 5 × 3 mm^2 crystals from Na	Aoki et al.	2001	254
Flux	Morphologies of grown crystals from Na	Aoki et al.	2001	255
HP-SG	Growth using GaN, SiC, sapphire GaN/sapphire seed crystals	Bockowski et al.	2002-2008	228-233
HP-SG	Theoretical consideration of growth mechanism	Grzegory et al.	2002	223
Flux	Growth of 3 × 1.5 × 1 mm^3 crystals by seeding	Aoki et al.	2002	259
Flux	Growth of h-GaN and c-GaN by Na or K flux	Sekiguchi et al.	2002	264
Flux	Growth under N_2 gas using Li-Na, Li-Ca flux and LPE growth on GaN/sapphire substrates	Kawamura et al.	2002-2004	268-276
Flux	Growth from Li-Ga-N flux	Song et al.	2003-2004	287, 288
MG	Growth of GaN and $Al_xGa_{1-x}N$ crystals in a multi -anvil high pressure furnace	Utsumi et al.	2003-2004	20, 21
ASG	Growth using NH_4X (X = Cl, Br, I) mineralizers	Yoshikawa et al.	2004-2007	321-323
ATG	Growth in a high pressure furnace	d'Evelyn et al.	2004-2007	345, 346
Flux	Growth from Na-Ga melts with Li_3N	Aoki et al.	2004	260
Flux	Growth from a flux composed of Li_3N and fluoride salts	Feigelson et al.	2005	290
HP-SG	Growth of $Al_xGa_{1-x}N$ crystals in a cubic anvil cell	Geiser et al.	2005	236
HP-SG	Growth from molten GaN under 6 GPa	Utsumi et al.	2005	20-22
Flux	Growth under Na vapor	Yamada et al.	2005-2006	261, 262
Flux	Growth from a Li-Ga flux	Chen et al.	2005	289
Flux	Growth from LiF-BaF_2-L_3N solvent by a multi-anvil high pressure apparatus	Feigelson et al.	2005	290

Table 20.3 Crystal Growth of GaN by Solution Growth (continued)

Method	Contents	Authors	Year	Ref.
Flux	LPE growth from Li-Na, Li-Ca flux by a shaking furnace	Kawamura et al.	2005-2006	277-280
Flux	Growth from Ca_3N_2 flux at 800°C	Jian et al.	2006	291
Flux	Growth from alkali metal flux	Shin et al.	2006	292
ATG	Growth using $NaNH_2$ mineralizer	Hashimoto et al.	2005-2006	331-333
ATG	Growth on HVPE-grown seed crystals	Hashimoto et al.	2005-2010	334-339
ATG	Growth on HVPE-grown seed crystals	Callahan et al.	2006-2011	317-320
Flux	Growth from molten halides wih $GaCl_3$ and Li_3N as precursors	Wadrip	2007	353
HP-SG	Growth mechanism of platelets and needle like crystals	Bockowski et al.	2007	225, 226
Flux	a-plane crystal from Sr-Na flux and m-plane crystal from Na flux	Iwahashi et al.	2007	281, 282
ATG	Growth up to 50 mm diameter on HVPE grown seeds	Ehrentraut et al.	2007-2008	324-326
ATG	Spontaneous seeded growth	Hashimoto	2007-2008	340, 341
Flux	Effect of carbon addition and growth of 50 mm diameter wafer of EPD less than 10^4 cm^{-2}	Kawamura et al.	2008	283, 284, 286
LPE	Growth by the reaction of Ga melt with activated N	Ogawa et al.	2008	354
ATG	Growth of 25-37.5 mm diameter crystal	Dwilinski et al.	2008-	303, 310, 311
HP-SG	Growth of $Al_xGa_{1-x}N$ crystals from Ga solution	Belousov et al.	2010	237
HP-SG	Use of Ca_3N_2 as a flux	Bockowski et al.	2010	234
Flux	Growth from Ga and Li-Ga flux at near atmospheric pressure	Feigleson et al.	2010	293
ATG	Preparation of nonpolar and semipolar substrates	Dwilinski et al.	2010	313, 314
ATG	Solubility of NH_4X (X = Cl, Br, I) mineralizers and numerical simulation of the growth process	Kagamitani et al.	2010-2011	327-330
HP-SG	Effect of In addition	Grzegory et al.	2011	235

SG: Solution Growth, HP-SG: High Pressure SG, PC-SG: Pressure Controlled SG, ATG: Ammonthermal Growth, MG: Melt Growth, LPE: Liquid Phase Epitaxy

grown measuring from 4-6 mm^2 up to about 10×10 cm^2 as seen in Fig. 20.17. The details of this HP-SG method are presented in Refs. 214-226. They found that high quality GaN single crystals almost free from dislocations (density of 10-100 cm^{-2})[223,225,226] with FWHM-XRC as low as 20-40 arcsec could be grown.[220,225,226] The growth mechanism in the solution and the surface of the melt, including the solution kinetics of nitrogen have been discussed.[222,224] The effect of the addition of Mg, Ca and Zn on the electrical properties and the structural quality has been studied.[220] Porowski et al.[227] studied the effect of annealing and found that it can improve the crystal quality, remove the radiation-induced damage to implanted layers and activate Mg and Zn acceptors.

Various seed crystals, such as HP-SG grown substrates, SiC, sapphire, GaN/sapphire MOCVD templates have been also used in the HP-SG growth method[228-233] for the directional crystallization of GaN single crystals. Bockowski et al.[234] have grown GaN single crystals using Ca_3N_2 as a flux. Grzegory et al.[235] have studied the effect of In on the growth of GaN single crystals.

Geiser et al.[236] have grown $Al_xGa_{1-x}N$ single crystals using a cubic anvil cell with a solid pressure medium. The growth was performed at 1750°C under 30 kbar. Single

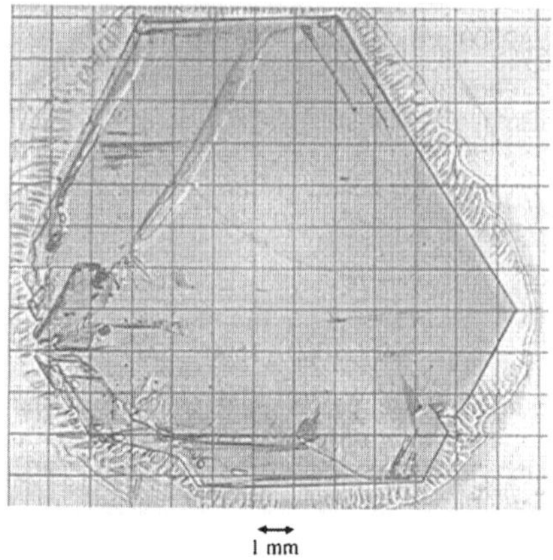

Fig. 20.17 Plate-like GaN crystals grown by high nitrogen pressure solution growth method (reprinted from Ref. 222 with permission, copyright 1999 John Wiley & Sons).

crystals up to 0.6 × 0.4 × 0.2 mm^3 in size with the aluminum content as high as x = 0.27 could be grown. Belousov et al.[237] have then grown $Al_xGa_{1-x}N$ single crystals with 0.5 ≤ x ≤ 0.9 from a Ga solution at 1800°C under 10 kbar nitrogen pressure. Single crystals measuring 0.8 × 0.8 × 0.8 mm^3 could be grown at a growth rate of 0.1 mm/h.

- *Pressure-Controlled Solution Growth (PC-SG) method*

Inoue et al.[238-246] pioneered the growth of GaN single crystals by using an ultrahigh pressure furnace equipped with a piston mechanism (Fig. 20.18) in such a way that crystal growth can be performed at a constant temperature while increasing the pressure as shown in Fig. 20.19. This method is quite different from the conventional solu-

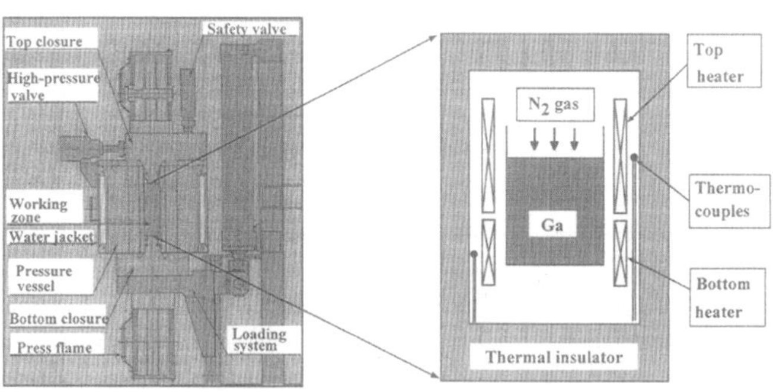

Fig. 20.18 Schematic cross-sectional view of the ultra high-pressure furnace and schematic diagram of the inside of the crystal growth furnace (reprinted from Ref. 239 with permission, copyright 2000 The Japan Society of Applied Physics).

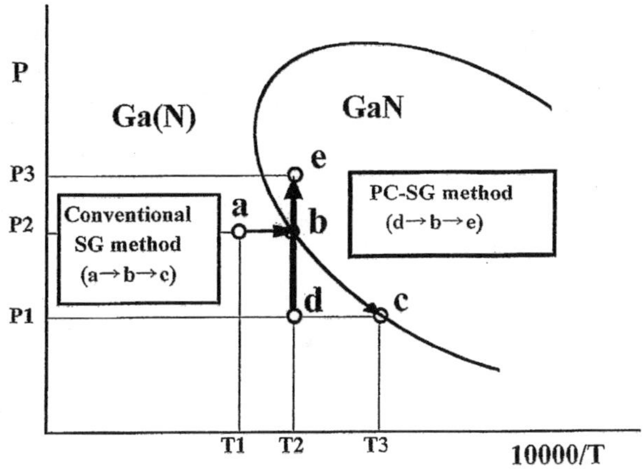

Fig. 20.19 Schematic Ga-N pressure-temperature phase diagram. In the conventional SG method, the temperature is lowered according to a → b → c. In the PC-SG method, pressure is increased as d → b → e and held at a constant temperature (T2) during crystal growth (reprinted from Ref. 239 with permission, copyright 2000 The Japan Society of Applied Physics).

tion growth in which the temperature is decreased in order to grow crystals under supersaturation. In this method, instead of decreasing the temperature to grow crystals under supersaturation as a→b→c, the pressure is increased as d→b→e in order to achieve the supersaturation to grow crystals. This method is called the pressure controlled solution growth (PC-SG) method. The PC-SG method can be applied only for the case when one of the constituents of the grown compound is in the vapor phase at room temperature. GaN is such a compound to which the PC-SG method could be first applied among various compound semiconductor materials.

The advantage of this method is as follows.

(i) The solubility is held constant because the temperature can be held constant, while in the conventional solution growth it reduces as the temperature decreases.

(ii) The solute composition can be held constant since the nitrogen pressure can be held constant by the piston mechanism, while in conventional solution growth it decreases because the solute cannot be fed during crystal growth.

(iii) The growth rate can be held constant because the temperature, solubility and the solute composition can be held constant as mentioned above.

(iv) Growth without a temperature gradient can relieve the thermal stress during crystal growth.

In fact, the growth rate reached a maximum at a temperature just beneath the equilibrium vapor pressure as shown in Fig. 20.20. The advantage of this method is to make possible the growth of many crystals in the arrangement as shown in Fig. 20.21 since crystal growth can occur without a temperature gradient. By using this method, Inoue et al.[244] have grown many GaN single crystals measuring about 3.35 cm^2 in one growth run, one of which is shown in Fig. 20.22.

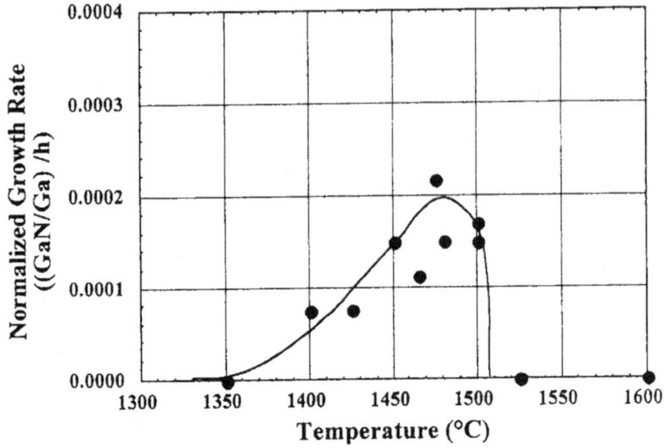

Fig. 20.20 Normalized growth rate as a function of temperature (reprinted from Ref. 239 with permission, copyright 2000 The Japan Society of Applied Physics).

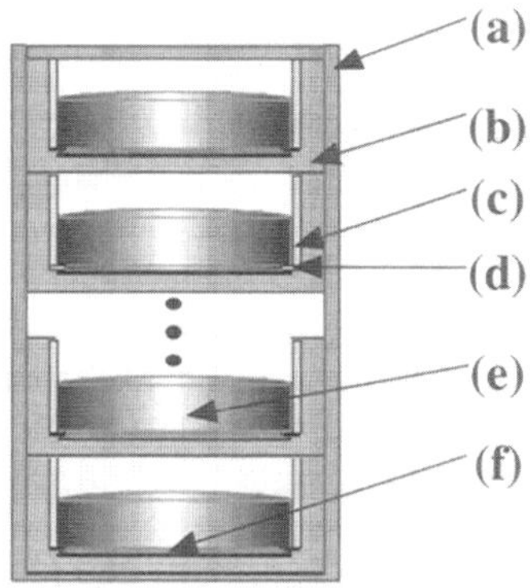

Fig. 20.21 Schematic cross-sectional view of the multi-crucible system: (a) graphite susceptor, (b) graphite crucible, (c) pBN tube, (d) pBN mask , (e) Ga and (f) hetero-seed (sapphire substrate) (reprinted from Ref. 244 with permission, copyright 2001 John Wiley & Sons).

- *Flux method*

In order to grow GaN single crystals at low temperatures and pressures, the Yamane group at Tohoku university were the first to find that GaN crystals can be grown from a Ga-Na flux[247] and then have developed this flux method.[248-264] In the beginning, nitrogen gas was supplied by the thermal decomposition of NaN_3 and nitrogen was dissolved to the Ga-Na flux in a BN crucible and small size crystals were grown at temperatures of 570-840°C for 6-300 h.[247-252] The arrangement of the reactor is shown in Fig. 20.23(a). The morphology of grown crystals was affected by the Na to Ga ratio. In this case, the nitrogen concentration in the flux decreases during growth and the growth

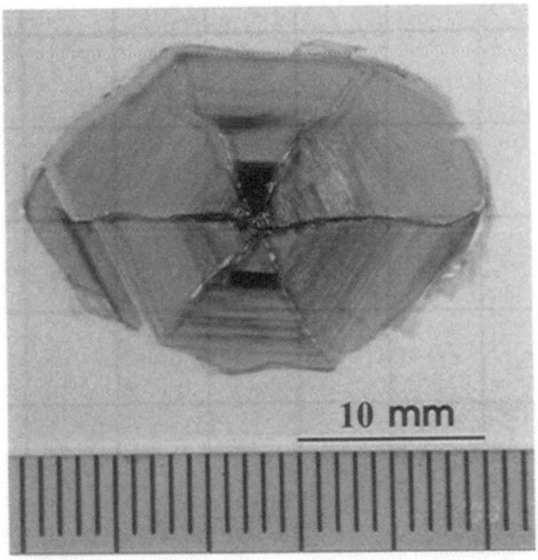

Fig. 20.22 GaN single crystal grown by the PC-SG method (reprinted from Ref. 244 with permission, copyright 2001 John Wiley & Sons).

rate slows. To deal with this problem, they started to use nitrogen under pressure (1-5 MPa) to dissolve nitrogen in the flux as shown in Fig. 20.23(b).[253-258] Na was also purified by distillation.[257] By using this method, GaN single crystals measuring up to 0.1 mm thick 10 mm platelets with the FWHM of XRC 45-55 arcsec could be grown at temperatures of 750-775°C under 5 MPa.[257] The grown crystals were also used as the seed and single crystals with a maximum area of 3.0 × 1.5 mm^2 and thickness 1.0 mm were obtained.[259] In Fig. 20.24, grown crystal morphologies are summarized as a func-

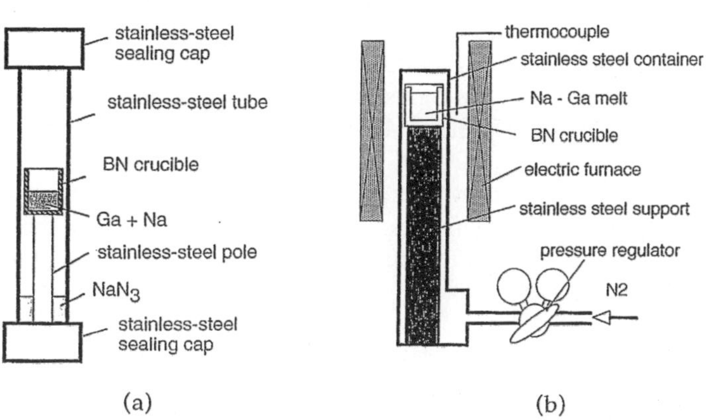

Fig. 20.23 Schematic view of the reaction furnace. (a) Nitrogen is supplied by the decomposition of NaN_3 (reprinted from Ref. 250 with permission, copyright 1999 The Ceramic Society of Japan). (b) Nitrogen is supplied as high pressure gas (reprinted from Ref. 256 with permission, copyright 2001 The Japanese Association for Crystal Growth).

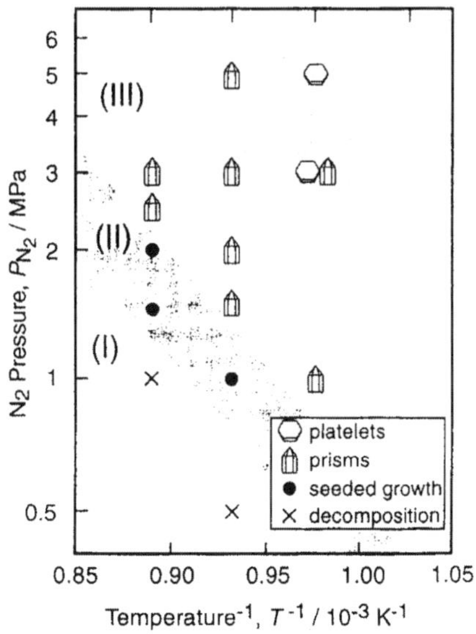

Fig. 20.24 Nucleation regions and morphology of GaN grown by the Na flux method. GaN seed crystals decomposed in region (I). GaN nucleated only on seed crystals in region (II) and GaN nucleated on both seed crystals and crucible wall in region (III) (reprinted from Ref. 259 with permission, copyright 2002 Elsevier).

tion of the N_2 pressure and the reciprocal of temperature. The addition of Li_3N to the flux was found to have the effect of obtaining colorless single crystals.[260] The modification of this method where Ga is melted under Na vapor, was also studied.[261,262] The yield of various crystal morphologies was studied.[263] h- and c-GaN single crystals could be grown from a Na or K flux.[264] Pr-doped GaN crystals were prepared from a Na flux.[265]

The group at Osaka University[266-286] also developed this flux method using an apparatus as shown in Fig. 20.25. They first traced the growth by the Ga-NaN_3 flux method [47,266,267] with spontaneous seeding[47,266] and with various seed crystals[47,267] and obtained small sized GaN single crystals of about 0.5 mm. N_2 gas pressure control was introduced[268] instead of using GaN_3 in order to control the supersaturation and to increase the crystal size up to 3 mm by reducing the nucleation sites. The addition of Ca[269,273,278] or Li[270,273,278] to the Na flux was studied to increase the solubility of nitrogen (Fig. 20.26) and the yield of grown crystals (Fig. 20.27). In order to grow large diameter GaN substrates, the liquid phase epitaxy (LPE) method has been applied using MOCVD-GaN/sapphire templates.[271,273-284,286] The influence of supersaturation [276], thermal convection[274,275] and the III-V ratio on the color of grown crystals[277,278], on the crystal growth and the growth

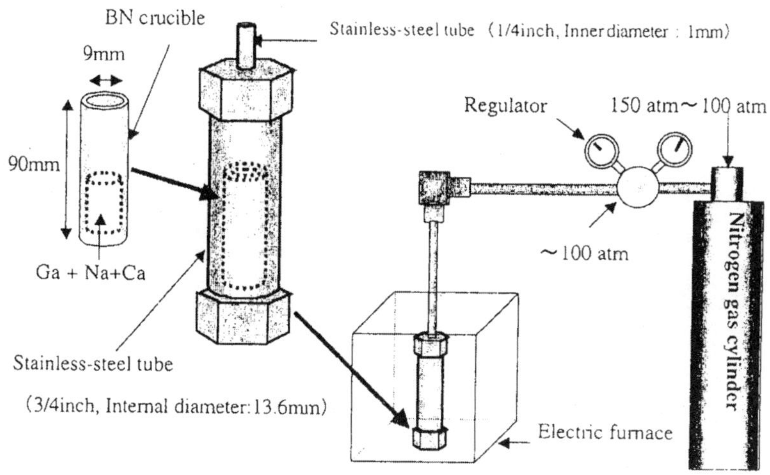

Fig. 20.25 Schematic illustration of experimental setup (reprinted from Ref. 269 with permission, copyright 2002 The Japan Society of Applied Physics).

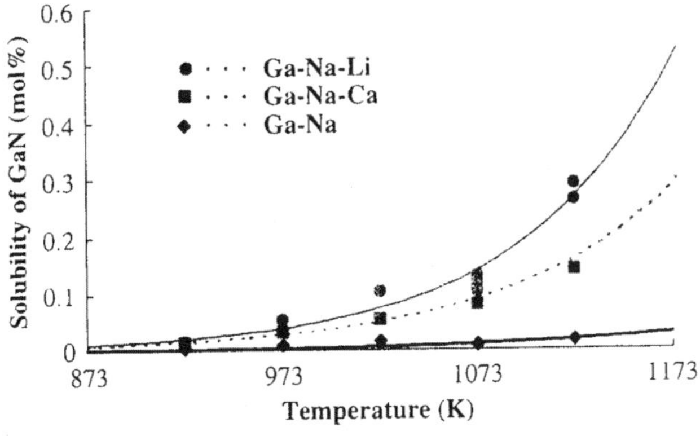

Fig. 20.26 Solubility curves of solid GaN in Ga-Na, Ga-Na-Li and Ga-Na-Ca (reprinted from Ref. 278 with permission, copyright 2005 Elsevier).

mechanism[272,273,277] has been studied. It was also found that pressurization of the N_2 gas has a similar effect as in the PC-SG method in that crystal growth proceeds without the growth temperature dropping. Supersaturation could be achieved by pressurizing the N_2 gas and GaN could be grown without lowering the temperature.[273,275,277,278]

They have also applied a shaking furnace system to increase the growth rate.[277,278,280] It was found that during LPE growth, the dislocation density was drastically reduced because of facet formation[279] and that the addition of carbon has the effect of suppressing polycrystal formation.[283] The growth rate was also found to be strongly dependent

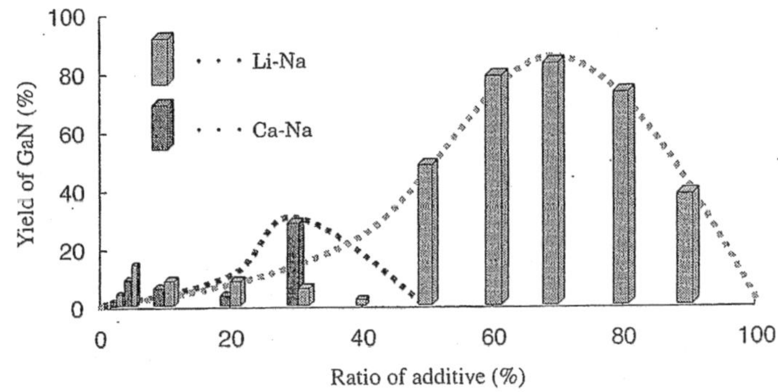

Fig. 20.27 Relationship between flux composition and yield of GaN grown in various compositional fluxes. Yield is indicated as the ratio of GaN crystals grown in flux against metallic Ga of the starting material (reprinted from Ref. 270 with permission, copyright 2003 The Japan Society of Applied Physics).

on the temperature difference which is related to the thermal convection[284] as predicted by simulation based on radiative, convective and conductive heat and mass transfer including nitrogen transfer.[285] They finally were able to grow 50 mm diameter GaN crystals of thickness 2 mm and dislocation density of the order of 10^4 cm^{-2} (down to a minimum of 10^2 cm^{-2} over a certain area) (Fig. 20.28).[280,283,284,286] The mechanism of this drastic dislocation reduction is as shown in Fig. 20.29, where the stable growth of large facets of (10-11) promotes the elimination of dislocations taking over from the MOCVD-grown GaN templates. a-plane GaN from Sr-Na flux[281] and m-plane GaN[282] from Na flux have been grown measuring up to 10 × 10 × 0.5 mm^3 by controlling the flux molar ratio to optimize the growth preference of the desired face.

Song et al.[287, 288] have grown hexagonal GaN platelet crystals measuring 1-4 mm by the Li-based flux method. Chen[289] have grown GaN crystals from a Li-Ga-N solution. Feigelson et al.[290] have grown GaN crystals at 800°C under 25 bar nitrogen pressure in a solvent of LiF-BaF_2-Li_3N using a multi-anvil split sphere high pressure apparatus as shown in Fig. 20.30. Jian et al.[291] have used Ca_3N_2 as flux, with which they have grown GaN crystals at 900°C under a nitrogen pressure of about 0.2 MPa. Shin et al.[292] have grown GaN single crystals from an alkali metal flux and Feigelson et al.[293] have grown GaN crystals from a Li based solution under near atmospheric pressure.

- Ammonothermal method

This method is similar to the hydrothermal method for growing quartz crystals in supercritical water.[294,295] It is a solvo-thermal method using ammonia as a fluid instead of water as in the hydrothermal method.[296] In the case of GaN, the crystal growth is performed in a supercritical ammonia fluid using basic, neutral or acidic mineralizers. Mineralizers are used to increase the solubility of Ga in the ammonia fluid. This

Fig. 20.28 50 mm GaN single crystal with a thickness of 2 mm which was grown in a large chamber. The order of the dislocation density over three-fourth of the surface area was less than 10^2 cm^{-2} and the order of the dislocation density of the concentrated region was 10^4 cm^{-2} (reprinted from Ref. 286 with permission, copyright 2009 Elsevier).

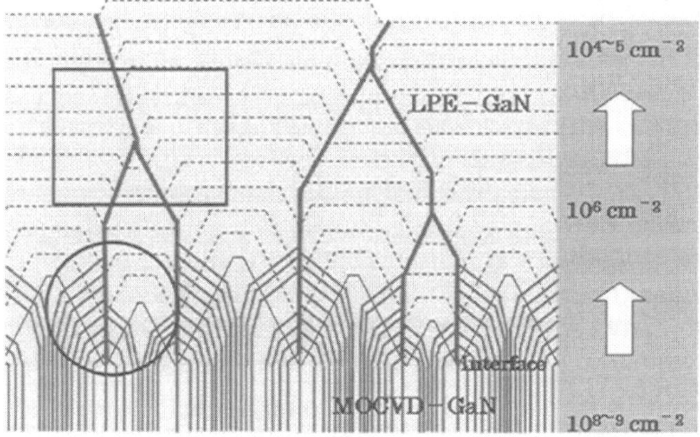

Fig. 20.29 Dislocation reduction model of LPE growth in Na flux method. Concentration and bending of bundled dislocations cause a drastic decrease in the dislocation density (reprinted from Ref. 286 with permission, copyright 2009 Elsevier).

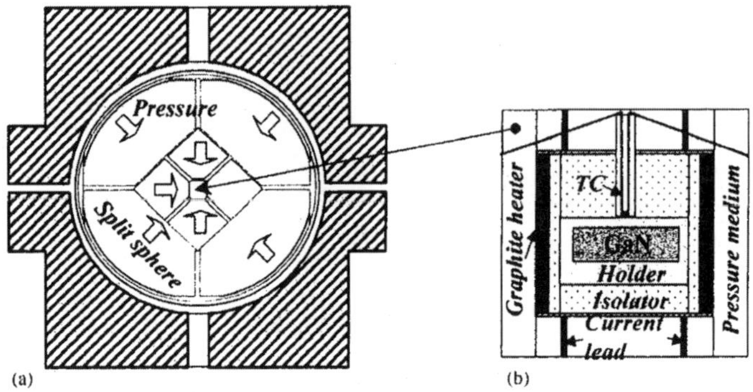

Fig. 20.30 Schematic of the (a) multi-anvil split sphere high-pressure apparatus and (b) high pressure cell (reprinted from Ref. 290 with permission, copyright 2005 Elsevier).

method is known as the ammonothermal method. The chemical basis of this method can be seen in Refs. 297-301 and the results of GaN crystal growth have been reviewed.[210-212,302] The advantage of this method is that many crystals can be grown in one growth cycle and that crystals can be grown with a zero temperature gradient which ensures the growth of low-strain materials.

It is known that basic mineralizers show retrograde solubility (Fig. 20. 31)[303,304] while acidic mineralizers have normal solubility (Fig. 20.32).[305] Therefore, in the former case crystals are grown in the higher temperature zone, while in the latter case, crystals are grown in the lower temperature zone as shown in Fig. 20.33 [304,306], granted that the solubility can be raised in the latter case by increasing the temperature, while in the former case the solubility is lowered with increasing temperature. It is known that the pressure for growth is higher in the former case than in the latter case. Since the growth rate increases with increasing solubility and growth temperature, the latter case allows the option of increasing the growth rate.

The ammonothermal method was first applied to the growth of GaN by Dwilinski et al.[48,68,303,307-309], using GaN powders as the source material with $LiNH_2$ and/or KNH_2 as basic mineralizers. Growth was performed at 550°C and 0.2-0.5 GPa and small GaN crystals (25 μm long) were obtained with a higher mineralizer ratio $LiNH_2$:NH_3 (1:10). In the presence of rare earth (RE) metals at a ratio of 1:10 to Ga, grown crystals showed well resolved bound exciton emissions by PL and electron concentrations as low as 5×10^{15} cm^{-3}.[48,68,309]

They have recently made great progress in growing GaN single crystals of diameter from 25 mm to up to 37.5 mm[303,310,311] as shown in Fig. 20.34. The grown crystals had a dislocation density as low as 5×10^3 cm^{-3}, FWHM of (0002) XRC as low as 17 arcsec and radius of curvature higher than 1000 m. Recently, they reported the growth of GaN single crystals measuring 50 mm[312] and showed a highly resolved PL spectrum with

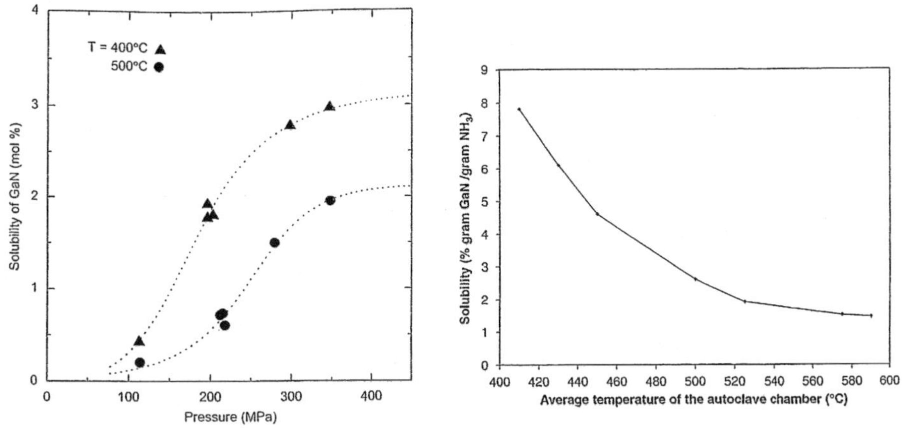

(a) (b)

Fig. 20.31 Solubility of basic ammonothermal methods. (a) solubility of GaN in the NH_3 + KNH_2 solution (molar ratio KNH_2:NH_3 = 0.07 (reprinted from Ref. 303 with permission, copyright 2008 Elsevier), (b) retrograde solubility of GaN measured in KNH_2-NH_3 solutions. Potassium amide concentration is about 3.5 ± 0.5 M, the temperature difference 10°C/cm, pressure 1.2-2.4 kbar (reprinted from Ref. 304 with permission, copyright 2006 Elsevier).

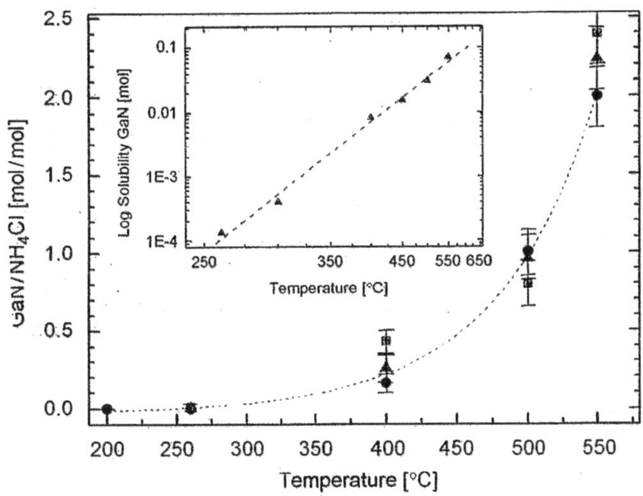

Fig. 20.32 Solubility of GaN is shown for three different molar rations of GaN and NH_4Cl: 0.0127 (■), 0.03 (▲) and 0.127 (●). The exponential fit is made to the latter curves. The inset shows the Arrhenius plot drawn for the molar ration of 0.032 (reprinted from Ref. 305 with permission, copyright 2008 Elsevier).

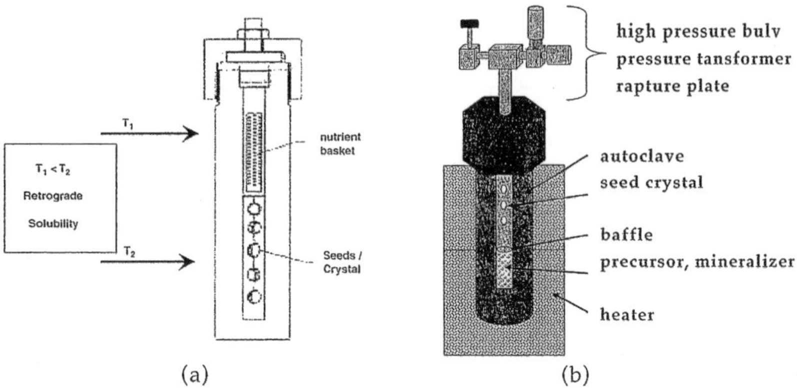

Fig. 20.33 Schematic configuration of ammonothermal methods. (a) Schematic of the basic ammonothermal method. The seed crystals are located in the lower part of the furnace where the temperature is higher than the upper part where GaN source material is located (reprinted from Ref. 304 with permission, copyright 2006 Elsevier). (b) Schematic of the acidic ammonothermal method. The seed crystals are located in the upper part of the furnace where the temperature is lower than the lower part where GaN source material is located (reprinted from Ref. 306 with permission, copyright 2011 The Japanese Association for Crystal Growth).

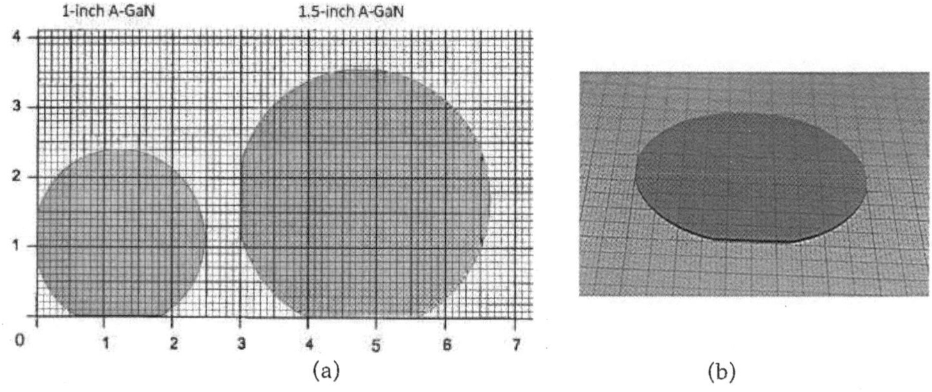

Fig. 20.34 GaN bulk crystals grown by the ammonothermal methods. (a) Photograph of 25 mm and 37.5 mm diameter A-GaN wafer manufactured by the basic ammonothermal method (reprinted from Ref. 311 with permission, copyright 2009 Elsevier). (b) Photograph of a 50 mm size GaN crystal (reprinted from Ref. 312 with permission, copyright 2011 John Wiley & Sons).

free-exciton peaks for an epitaxially grown layer on a semi-insulating (SI) GaN substrate.[312] They also fabricated m- and a- plane substrates.[313,314] It is also noteworthy that the conductivity type can be controlled from n-type (10^{18}-10^{19} cm^{-3}), SI (up to 10^{11} Ωcm) to p-type (up to 10^{18} cm^{-3}). The type of the energy band bending due to conductivity was evaluated by contactless electroreflectance spectroscopy (CER).[313]

Kolis et al.[69,304,315,316] applied this method using a mixture of KNH_2 and KI as the mineralizer and GaN powders as the feedstock. Crystal growth was performed at 400°C and 2.4 kbar and produced single crystals measuring 0.5 × 0.2 × 0.1 mm. The reaction is supposed to proceed as follows.

$$\begin{aligned} &GaN + NH^{-2} + 3NH_3 \rightarrow \text{soluble Ga-imide complex} \\ &\rightarrow \text{GaN crystals} + NH^{-2} + 3NH_3 \end{aligned} \quad (20.2)$$

The growth was also performed on HVPE-grown GaN seed crystals[304,317-320] and GaN single crystals up to 10 × 10 × 1 mm³ were obtained at a growth rate up to 50 μm/day. Wang et al.[318] have analyzed the formation of inversion domains in GaN crystals grown on seeds. The grown crystals have a crystal quality with FWHM of (002) XRC of 100 arcsec.[319] Adekore et al.[320] have grown GaN crystals doped with Er in order to getter oxygen impurities by Er. The oxygen concentrations were mitigated down to 1-7 × 10^{19} cm^{-3} and a strong edge emission and a blue luminescence peak could be observed by PL.

Fukuda's group at Tohoku university[305,306,321-330] have developed the ammonothermal method using acidic mineralizers (ammonium halogenides, NH_4X (X = Cl, Br, I)). It was found that the solubility of Ga is regular, increasing as a function of temperature as shown in Fig. 20.35. The mineralizer NH_4Cl is found to be the best one from the viewpoint of lattice distortion. Crystals have been grown in an autoclave at relatively low temperature (<550°C) and under low pressure (<150 MPa). The growth rate could be increased from zero to >50 μm/day by increasing the content of mineralizer from zero

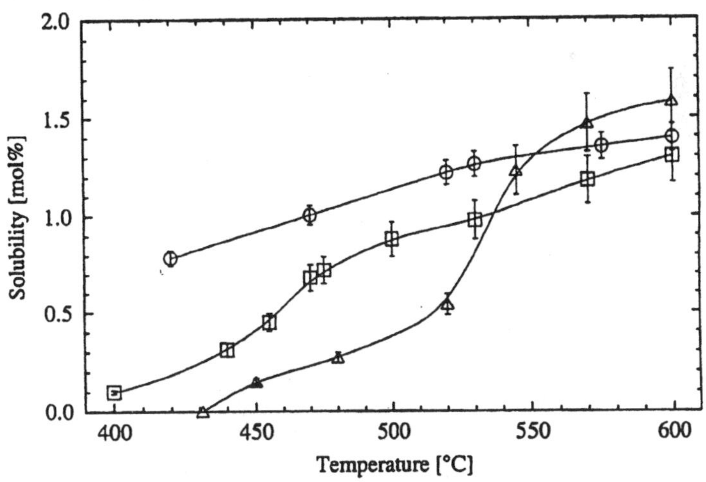

Fig. 20.35 Temperature dependence of the solubility of GaN in supercritical ammonia with the following mineralizers (100 MPa and 3.1±0.1 % amount-of-substance fraction): circle; NH_4Cl, square; NH_4Br and triangle; NH_4I (reprinted from Ref. 329 with permission, copyright 2011 Elsevier).

to >1 mol%. Large diameter crystals up to 50 mm could be grown as shown in Fig. 20.36 using HVPE grown GaN seed crystal with a thickness of 200 μm (Fig. 20.37)

Hashimoto et al.[331-333] have compared basic, neutral and acidic mineralizers, and concluded that the basic mineralizer, sodium amide ($NaNH_2$), is the best one to grow hexagonal phase GaN only, i.e. without the cubic phase. They showed the growth of GaN on HVPE grown freestanding GaN seed crystals.[334-339] The growth was performed at 500-600°C under about 100-300 MPa on 3 × 4 cm^2 seed crystals at a growth rate of 10-400 μm/day depending on the growth conditions. The threading dislocation density

Fig. 20.36 GaN bulk crystals grown by the ammonothermal method. Photograph of 50 mm diameter GaN wafer grown on a HVPE seed crystal (the thickness of GaN is 0.5 mm) (reprinted from Ref. 326 with permission, copyright 2008 Elsevier).

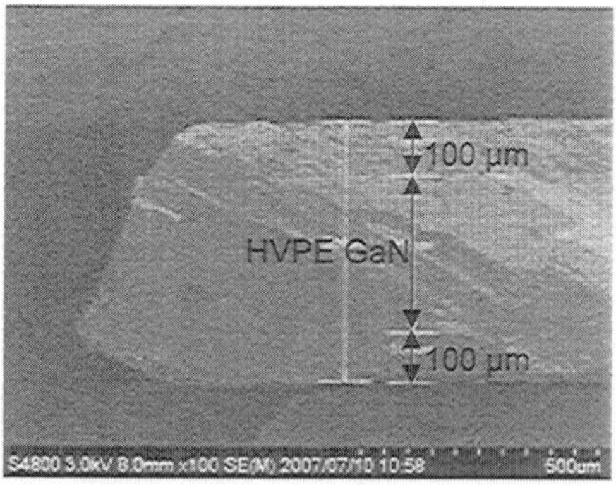

Fig. 20.37 Cross-sectional SEM image showing the Ga- and N-polar face and strong faceting in the direction toward a plane. Growth conditions were T = 425/525°C and p = 145 MPa (reprinted from Ref. 326 with permission, copyright 2008 Elsevier).

was 1×10^6 cm^{-2} for the Ga face and 1×10^7 cm^{-2} for the N face. No dislocation generation was observed on the Ga face although a few defects were generated on the N face. The FWHM of (0002) XRC was about 53-62 arcsec. By spontaneous nucleation, polyhedron-shaped crystals measuring approximately 5 mm could be obtained[340, 341] growing along the c direction at more than 50 μm/day. The narrowest FWHM of XRC for sliced wafers was 220 arcsec. The homoepitaxial growth of GaN on GaN substrates was performed and it was shown that the homoepitaxial layers have fewer impurities such as oxygen, carbon and hydrogen than those in GaN layers on sapphire substrates. This is attributed to the low dislocation density of the GaN substrates. Because of this high purity, lower carrier concentrations of around 1×10^{16} cm^{-3} preferable for high performance vertical Schottky barrier diodes (SBDs) could be obtained.[342]

Collado et al.[343] have reported the growth of GaN, using NH_2NH_3Cl as the mineralizer and Ga as the precursor. Growth was performed at temperatures of 400-800°C under 100-250 MPa. Lan et al.[344] have used lithium as a mineralizer and have grown GaN crystals in the temperature range 350-500°C.

D'Evelyn et al.[345, 346] have grown bulk GaN, using a capsule in which GaN seeds obtained by cutting HVPE GaN grown on sapphire substrates, polycrystalline GaN raw materials, a mineralizer and ammonia were sealed. This capsule is then placed in a high pressure apparatus where pressure can be increased by the piston system. Crystal growth has been performed at temperatures of 600-1000°C under 5-20 kbar. It was demonstrated that crystals measuring up to 15×18 mm^2 could be grown on HVPE-GaN seeds, with dislocation densities of 10^6-10^8 cm^{-2}. They however also showed that spontaneously grown crystals and crystals grown laterally with respect to the C-oriented seed had dislocation density less than 500 cm^{-2}, typically below 100 cm^{-2}. Even though these crystals are small ones, it shows the promise of the ammonothermal method.

Simulation studies of the ammonothermal method have been reported.[347,348] Chen et al.[347] have studied the effect of the charge particle size on the flow pattern and the temperature distribution in the autoclave. Pendurti et al.[348] have analyzed the transport-determined growth rate by simulating crystallization and dissolution kinetics in order to optimize the growth equipment and growth conditions.

Using both basic and acidic mineralizer methods, crystals larger than 50 mm have been grown and the ammonothermal method is thought to have every likelihood of being able to supply GaN bulk substrates at low cost because of its scalability.

- *Other solution growth methods*

Logan et al.[349] used liquid phase epitaxy (LPE) to grow GaN on sapphire substrates in molten Ga under NH_3/H_2 atmosphere at temperatures of 850-1050°C. They examined the effect of NH_3 partial pressure, growth melt position and the effect of adding Bi on the growth rate and the morphology of grown layers.

Sukhoveyev et al.[350, 351] have reported the growth of ~ 20 mm diameter 15 mm long GaN ingots (Fig. 20.38) from a Ga-based solution in the temperature range of 800-1000°C under less than 2 atm. Growth was performed on a graphite seed at ~ 2 mm/h growth rate.

Klemenz et al.[352] studied GaN epitaxial growth in molten Ga under NH_3 gas atmo-

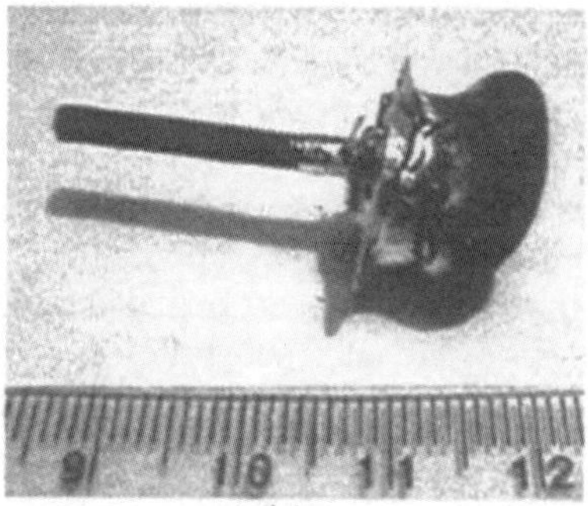

(a)

(b)

Fig. 20.38 As-grown ~ 20 mm diameter 15 mm in length GaN ingot (a). The same ingot after etching in $HF+HNO_3$ mixture and sawing in ~ 2 mm thick wafers (b) (reprinted from Ref. 351 with permission, copyright 2000 Materials Research Society).

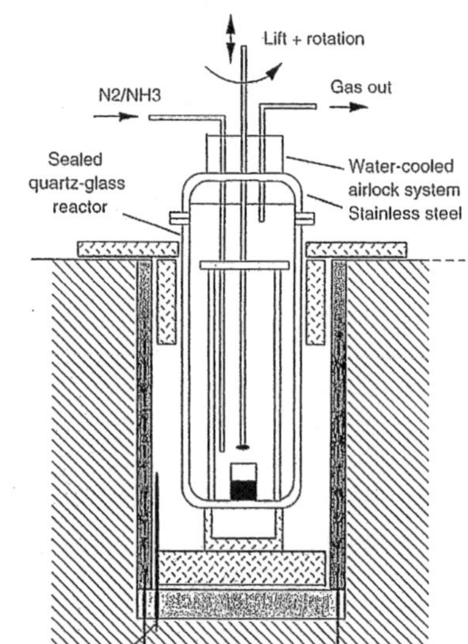

Fig. 20.39 Schematic view of the LPE-reactor system. The introduction of the growth solution and of the substrates proceeds from the top of the reactor through an airlock system. Forced convection is applied for higher growth rates (reprinted from Ref. 352 with permission, copyright 2000 Elsevier).

sphere using equipment as seen in Fig. 20.39. They have grown GaN crystals by LPE on $LiGaO_2$ (LGO) and $LiAlO_2$ (LAO) substrates and on HVPE GaN on sapphire substrates at 900°C.

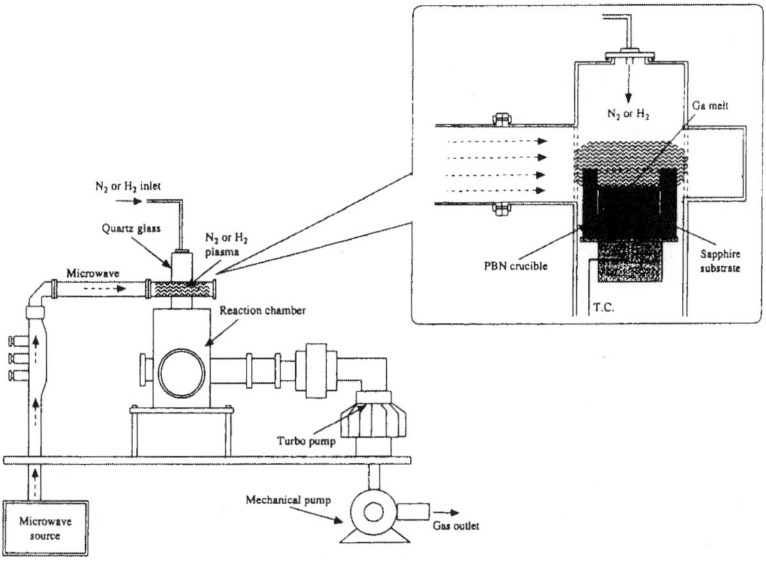

Fig. 20.40 Schematic drawing of the experimental setup for the melt growth of GaN under nitrogen plasma (reprinted from Ref. 354 with permission, copyright 2008 Elsevier).

Waldrip[353] has invented a novel method in which $GaCl_3$ and Li_3N precursors are dissolved in molten halide salts such as LiCl and LiCl+KCl, and GaN crystals were grown in an electrochemical apparatus. The high pressure Chochralski method was also combined with this method to grow single crystals.

Ozawa et al.[354] have grown GaN crystals by reacting the Ga metal with atomic nitrogen in a microwave plasma at a low temperature of 610-700°C and under a pressure of 200-400 Pa using an apparatus as shown in Fig. 20.40.

(3) Vapor phase growth

Various vapor phase growth methods such as physical vapor transport (PVT), direct reaction and others have been studied as shown in Table 20.4.

- Physical Vapor Transport (PVT) method

The PVT method is also known as the sublimation method. The growth of GaN needle crystals by sublimation of GaN powders in ammonia gas was first performed by Zetterstrom.[355] Balkas et al.[82,356] have grown up to 3 mm long GaN single crystals by the sublimation of GaN under ammonia gas flow. Kurai et al.[357-359] studied the growth of GaN on various substrates by the sublimation technique under NH_3/H_2 atmosphere. A typical growth furnace is shown in Fig. 20.41. Thick GaN measuring 10-30 μm was

Table 20.4 Crystal growth of GaN vapor phase growth

Method	Contents	Authors	Year	Ref.
PVT	Growth of GaN crystals of 1 × 5 mm^2	Zettersrom	1970	355
VPR	Growth by the reaction of Ga with NH_3	Ejder	1974	364
VPR	Growth by the reaction of Ga or Ga_2O_3 with NH_3	Ogino et al.	1979	365
HVPE	Growth of GaN on NGO substrates	Okazaki et al.	1994-1997	398, 399
SSM	Growth on sapphire and SiC by the reaction of Ga with NH_3	Fischer et al.	1996	376
PVT	Sublimation of GaN under NH_3 atmosphere	Kurai et al.	1996-1997	357-359
HVPE	Freestanding GaN separated from SiC by RIE	Melnick et al.	1998	526
HVPE	Freestanding GaN by laser lift-off separation	Wong et al.	1998	493
HVPE	Growth on SLGO substrates	Simamura et al.	1998-2002	470-472
SSM	Growth of GaN on SiC and sapphire substrates	Vodakov et al.	1980 1998	360, 361
HVPE	Freestanding GaN up to 45 mm using NGO substrates	Wakahara et al.	1998-2000	240, 516-521
HVPE	Freestanding GaN by polishing of sapphire substrates	Lee et al.	1998 1999	494, 495
HVPE	Freestanding GaN by laser lift-off from sapphire	Kelly et al.	1999	496
HVPE	Freestanding GaN using SiC seed substrates	Nikitina et al.	1999	527
HVPE	Freestanding GaN using LGO substrates	Kryliouk et al.	1999	535-538
PVT	Lateral overgrowth by PVT on MOCVD-grown GaN	Faredd et al.	1999	362
VPR	Growth by the reaction of Ga with NH_3	Kamler et al.	2000	366
PVT, VPR	Growth of 3 mm long crystal on 6H-SiC substrates	Balkas et al.	2000	35
VPR	Reaction of Ga with microwave activated nitrogen	Tatsumi et al.	2000	384, 385
VPR	Growth of $Ga_{1-x}Mn_xN$	Szyszko	2001	374
HVPE	Freestanding GaN by laser lift-off from sapphire	Freitas et al.	2001	497
HVPE	Freestanding GaN using GaAs substrates	Motoki et al.	2001-2002	522, 523
HVPE	Growth of GaN using GaN seed substrates	Xu et al.	2002	405
VPR	Growth by the reaction of Ga with NH_3	Shin et al.	2002	367
VPR	Growth of GaN on MOCVD GaN substrates	Nishio et al.	2002	368
HVPE	Freestanding GaN by laser lift-off from sapphire	Paskova et al.	2002	499
AL	Reaction of Ga/Ga_2O under Ni or B_2O_3 catalysts	Bae et al.	2002	371
HVPE	Freestanding GaN using a TiN interlayer with void assisted separation	Oshima et al.	2002-2007	501-503
HVPE	Growth of thick GaN on (111)A GaAs and preparation of freestanding Fe-doped SI GaN	Kumagai et al.	2002 2006	524, 525
AL	Mg incorporation during the reaction	Cheung et al.	2003	372
VPR	Growth by the reaction of Ga_2O_3 with NH_3	Shimada et al.	2004	378
AL	Comparison of Ga, Ga/Ga_2O_3, Ga_2O_3 precursors	Schwenzer et al.	2004	373
HVPE	Freestanding GaN using LAO substrates	Reed et al.	2005	539
HVPE	Freestanding GaN by laser lift-off from sapphire	Monemar et al.	2005	505
VPR	Growth of $Ga_{1-x}Mn_xN$ crystals by reacting Ga and Mn sources with ammonia gas	Szysko et al.	2005	375
HVPE	Growth using HP-SG GaN seed substrates	Lucznik et al.	2005	532
VPR	Preparation of GaH_x gas by the reaction of Ga with H_2 and growth by the reaction of GaH_x with NH_3	Imade et al.	2005 2006	382, 383
VPR	Growth by the reaction of Ga from GaN powders	Wu et al.	2006	369
VPR	Growth by the reaction of Ga_2O with NH_3	Konkapaka et al.	2006	379
VPR	Reduction of Ga_2O_3 to Ga_2O by C for the source gas	Miura et al.	2006	380
VPR	Reduction of Ga_2O_3 to Ga_2O by C for the source gas	Raghothamachar	2006	381
PVT	Growth using various GaN powder sources	Kallinger et al.	2007	363
VPR	Spontaneous growth by the reaction of Ga vapor with NH_3 gas	Rost et al.	2007	370
SSM	Growth on MOCVD-grown GaN/sapphire seeds	Kaminski et al.	2007	377
HVPE	Freestanding GaN by self-separation from substrates	Willimams et al.	2007	508
HVPE	Freestanding GaN using a NH_4Cl interlayer	Lee et al.	2007	509

Table 20.4 Crystal Growth of GaN by Vapor Phase Growth (continued)

Method	Contents	Authors	Year	Ref.
HVPE	Freestanding GaN using a ZnO buffer	Lee et al.	2007	510
HVPE	50 mm diameter freestanding GaN by selfseparation based on ELO with a tungsten based mask	Henning et al.	2007	511
HVPE	Freestanding GaN by selfseparation from patterned substrate	Lai et al.	2007	512
HVPE	Freestanding GaN using LAO substrates	Richer et al.	2007	540
VPR	Reaction of Ga vapor with plasma-activated nitrogen	Siche et al.	2008, 2011	386, 387
MBE	Growth of cubic GaN by plasma-assisted MBE	Novikov et al.	2008-2011	541-546
HVPE	Growth of semipolar GaN on patterned Si	Sawaki et al.	2009	528-530
HVPE	Freestanding m-plane GaN using a AlC buffer	Sasaki et al.	2009	513
HVPE	Reduction of bowing by regrowth process	Chen et al.	2009	514
HVPE	Bulk GaN of diameter 52 mm and thickness 5.8 mm	Fujito et al.	2009	515

PVT: Physical Vapor Transport, VPR: Vapor Phase Reaction, HVPE: Hydride Vapor Phase Epitaxy, SSM: Sublimation Sandwitch Method, AL: Ammonolysis, MBE: Molecular Beam Epitaxy

grown in 0.5-1 h.

The sublimation sandwich method (SSM)[360,361] has been applied to grow thick GaN on SiC or sapphire substrates and 15 × 15 mm^2 GaN layers of thickness 300 μm at growth rates up to 0.3 mm/h could be grown and freestanding GaN substrates were obtained by separating the seed substrates. Fareed et al.[362] have performed the lateral epitaxial overgrowth of GaN by the PVT method. Kallinger et al.[363] have carried out PVT growth using different powder sources and studied the behavior of GaN powder by thermogravimetry and predicted the appropriate growth conditions.

- ***Reaction of Ga with NH_3***

Vapor phase growth by reacting Ga with NH_3 has been tried by several researchers.

Ejder[364] have tried to grow GaN crystals by the reaction of Ga with NH_3 gas. Aoki et al.[365] also tried the growth of GaN by the reaction of molten Ga with NH_3 gas in an open

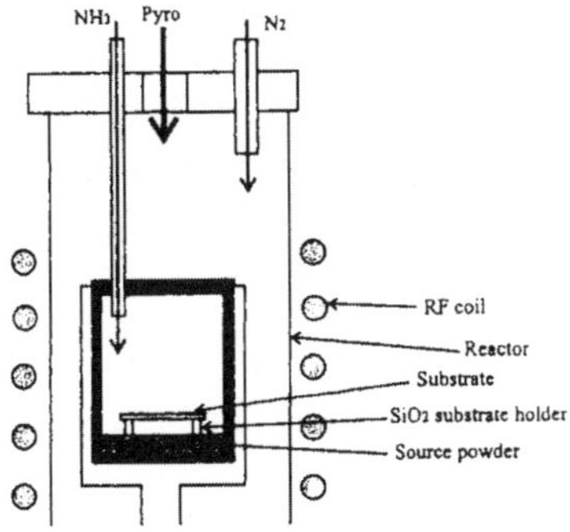

Fig. 20.41 The system used for sublimation growth (reprinted from Ref. 359 with permission, copyright 1997 The Japan Society of Applied Physics).

tube furnace. Balkas et al.[356] have grown up to 3 mm long GaN single crystals by the evaporation of Ga under ammonia gas flow. Kalmer et al.[366] prepared GaN powders by the reaction of Ga with NH_3, and obtained GaN single crystals of 3 × 2 × 0.2 mm by the sublimation of GaN powders in a horizontal furnace at temperatures of 1200-1250°C. Shin et al.[367] studied the growth of GaN by the reaction of Ga vapor with ammonia gas using a growth assembly as shown in Fig. 20.42. GaN platelets were obtained at 1130°C with ammonia gas flow of 25-75 sscm under 430 Torr.

Nishio et al.[368] have grown GaN thick layers by the reaction of Ga with NH_3 at 1050-1200°C on sapphire substrates covered with MOCVD grown GaN buffer layers. Wu et al.[369] have used GaN powders as the Ga source. The Ga vapor was transported to the seed substrate by the decomposition of GaN powders at 1100°C and the Ga vapor was reacted with NH_3 gas. Growth rate of 500 μm/h could be achieved. Rost[370] has grown GaN by the reaction of Ga with NH_3 at 1000-1300°C under NH_3 gas pressure varying between 200-800 hPa. Under the appropriate conditions, growth rates higher than 10 μm/h could be achieved.

By an ammonolysis process using Ga-containing powders such as Ga_2O_3, Ga_2O_3 + elemental Ga and elemental Ga, GaN nanostructures have been prepared[371-373] and their structural properties evaluated. A mixture of Ga and Mn was heat treated in an NH_3 stream at temperatures of 1200-1250°C. GaN single crystals containing 2 % Mn by wt. measuring 2.7 × 1.5 × 0.5 mm were obtained.[374, 375]

The sublimation sandwich method (SSM) was applied to the growth of GaN by the reaction of Ga with ammonia.[376, 377] Growth rates as high as 250 μm/h were achieved. Instead of evaporating the Ga vapor, Ga_2O_3 oxide powders with carbon were heated to vaporize Ga_2O gas as the Ga source and GaN crystals were grown by the reaction of

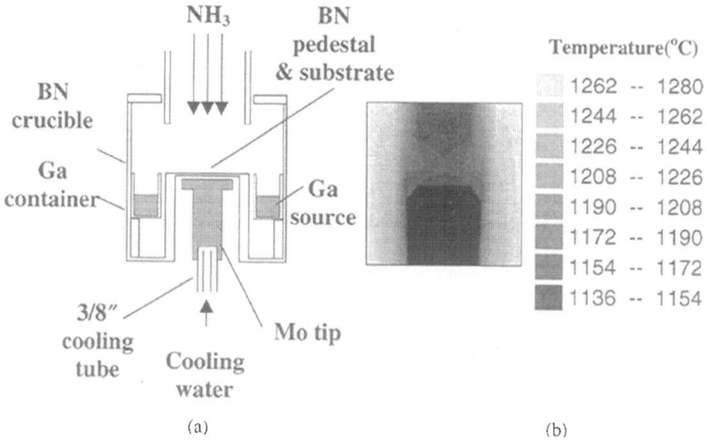

Fig. 20.42 (a) Schematic diagram of the BN growth cell and interfacing 3/8" cooling tube. (b) A typical temperature profile in the growth cell at the BN substrate temperature of 1130°C (reprinted from Ref. 367 with permission, copyright 2002 Elsevier).

Ga_2O gas with ammonia.[378-381] In order to improve the reaction with NH_3, gallium hydride (GaH_x) was supplied instead of Ga.[382, 383] The GaH_x was produced by bubbling hydrogen gas through the Ga source as shown in Fig. 20.43.

- *Reaction of Ga with activated nitrogen*

Tatsumi et al.[384, 385] have experimented with open-tube growth in which the gallium source and the sapphire substrate were heated to about 1000°C and nitrogen gas flowed under 200 Pa pressure. It was found that GaN could not be synthesized because the dissociation pressure of GaN is higher than 200 Pa at this temperature, but GaN films could be grown where nitrogen gas is activated by a microwave of 1 kW.

Siche et al.[386,387] have developed a plasma source based on short microwave pulses and a method to grow GaN crystals by reacting the Ga vapor evaporated from the melt with reactive nitrogen generated from the decomposition of ammonia. Spontaneously nucleated millimeter-sized crystallites were grown at a rate of 50 μm/h.

(4) Hydride Vapor Phase Epitaxy (HVPE)

Since bulk GaN crystal growth from the melt and from solution both proved difficult, the HVPE method was developed to obtain thick GaN substrates. The HVPE method itself had been developed for preparing GaAs and other III-V materials and then it was developed as an epitaxial growth method for GaN.[388-397] A typical HVPE method is as shown in Fig. 20.44. High purity gallium melt is set in the furnace and hydrogen gas and hydrogen chloride gas are passed through. GaN is grown on the seed substrate by the following reaction.

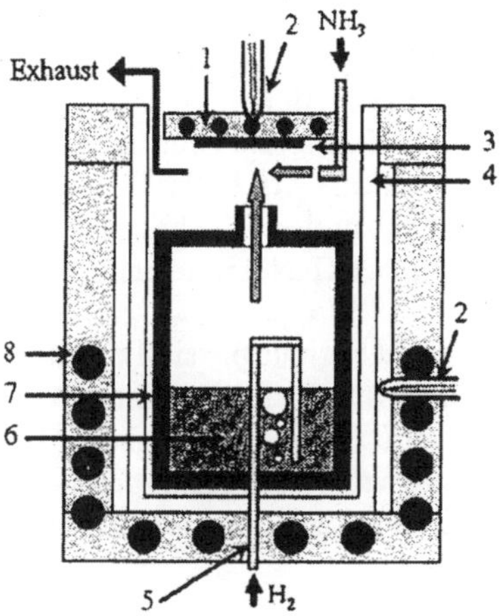

Fig. 20.43 Schematic drawing of the growth apparatus used in this study: (1) substrate heater, (2) thermocouple, (3) substrate, (4) quartz tube, (5) injection pipe, (6) Ga melt, (7) BN vessel, (8) source heater (reprinted from Ref. 383 with permission, copyright 2005 Elsevier).

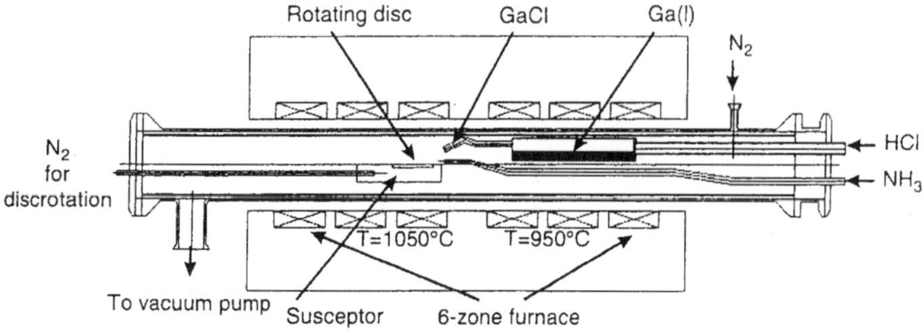

Fig. 20.44 Schematic drawing of a typical Hydride Vapor Phase Epitaxy (HVPE) reactor (reprinted from Ref. 395 with permission, copyright 2006 Elsevier).

$$GaCl_3 + NH_3 = GaN + 3HCl \quad (20.3)$$

The advantage of HVPE is in the fact that the growth rate is faster than other epitaxial growth methods such as MOCVD and molecular beam epitaxy (MBE), reaching about 100 μm/h at reasonable temperatures. This high growth rate makes it possible to prepare freestanding GaN substrates if GaN is first grown on appropriate foreign substrates.

The first trial to prepare freestanding GaN substrates by HVPE was performed by Okazaki et al.[398,399] using $NdGaO_3$ (NGO) as the seed substrate. Because of the large difference of thermal expansion coefficient, the separation of GaN layers from the seed crystal occurred spontaneously during the cooling process.

Due to the facility of crystal growth by HVPE, much research has been reported since then using various seed substrates. The key point is to use a seed substrate whose lattice coefficient matches that of GaN and to remove the grown GaN layer from the seed crystal.

Seed substrates have included various possibilities as reviewed[1-3,5,400-404] and summarized in Table 20.5.[405-474] These substrates have also been used for thin GaN epitaxial growth by MOCVD and MBE. Among them, the following studies have been performed for thick GaN film growth by HVPE in order to obtain GaN substrates with a seed substrate and then separating the seed substrate to prepare freestanding GaN substrates.

- *Sapphire*

It is known that dislocations can be reduced by the epitaxial lateral overgrowth (ELO or ELOG) method.[475-478] This method is based on the lateral overgrowth of GaN on patterned sapphire substrates. Various modified ELO methods such as the facet initi-

Table 20.5 Various Substrates for GaN Epitaxial Growth

Substrate	Crystal structure	Δλ/λ	Δα/α	Thermal stability	Cleava -bility	Ref.
Al_2O_3	corundum	-13.8	1.9	○	×	1
GaN	wurtzite	0	0	○	○	400, 405
AlN	wurtzite	2.35	-0.06	○	○	78
Si	diamond	20.1	-2.0	○	○	407-409
GaAs	zinc blende	25.3	0.4	△	○	410, 411
GaP	zinc blende	20.7	-0.9	△	○	412, 413
6H-SiC	ZnS 6H	-3.4	-1.4	○	○	414-419
4H-SiC	ZnS 4H	3.53	0.03	○	○	404
3C-SiC	ZnS 3C	3.55	0.09	○	○	404
MgO	rock salt	-6.5	4.9	△		400, 420, 421
MnO	rock salt	-1.4		×		400
CoO	rock salt	-5.4		×		400
NiO	rock salt	-7.6		×		400
ZnO	zinc blende	2.0	-2.7	△		389, 422-426
Ga_2O_3	monoclinic			○		427-430
ZrB_2	AlB_2	0.5				431-438
$MoSe_2$	layered					439, 440
$NbSe_2$	layered					439-441
MoS_2	layered	0.94				440, 442-444
WS_2	layered					445
$MgAl_2O_4$	spinel	-10.3	1.9	○		420, 446-452
$NdGaO_3$(NGO)	perovskite	-1.2	1.9	△	×	398, 399, 453
$LiAlO_2$(LAO)	b-$NaFeO_2$	1.7	1.7	△		454-459
$LiGaO_2$(LGO)	b-$NaFeO_2$	-0.1	1.9	△		460-466
$LiNbO_3$	illumenite	-6.7	9.9	×		400
$LiTaO_3$	illumenite	-6.8	10.6	△		400
$YAlO_3$	$GdFeO_3$			○		420
$SrLaGaO_4$	spinel $YbFe_2O_4$					467
$ScAlMgO_4$	spinel $YbFe_2O_4$	1.8	0.6			452, 468, 469
$(La,Sr)(Al,Ta)O_3$	perovskite	0.0	0.0			470-473
$Ca_8La_2(PO_4)_6O_2$	apertite	0.0	0.0			474

ated epitaxial overgrowth (FIELO)/facet controlled epitaxial lateral overgrowth (FACELO)[479], PENDEO (from latin, hang or suspended) epitaxy[480], antisurfactant[481], application of SiN[482], double buffer[483], air-bridge lateral epitaxial growth (ABLEG)[484], and buried regrowth in stripes[485] have been studied including MOCVD growth.[486-491] These ELO methods have been applied to the HVPE method for the reduction of dislocation densities. Nagahama et al.[492] have actually performed ELO growth by MOCVD on a freestanding GaN substrate prepared by HVPE by the ELO process.

HVPE growth has also been performed on sapphire substrates.[493-515] In order to separate the grown GaN from the seed substrate, the laser lift-off method is applied.[493,496,497,499,505,510] Mechanical polishing of the substrate is also possible.[494,495,512] The self-separation of grown GaN could be achieved using various interlayers such as TiN with void formation around it[501-503], NH_4Cl[509], WSiN[511], AlC[513], patterned substrates[512] and by spontaneous separation.[508] 50 mm freestanding GaN substrates are prepared using TiN interlayers and are industrially produced for LD fabrication. Vertical HVPE has been developed as shown in Fig. 20.45 and freestanding bulk GaN crystals of 52 mm diameter and 5.8 mm thickness could be successfully grown (Fig. 20. 46).[515]

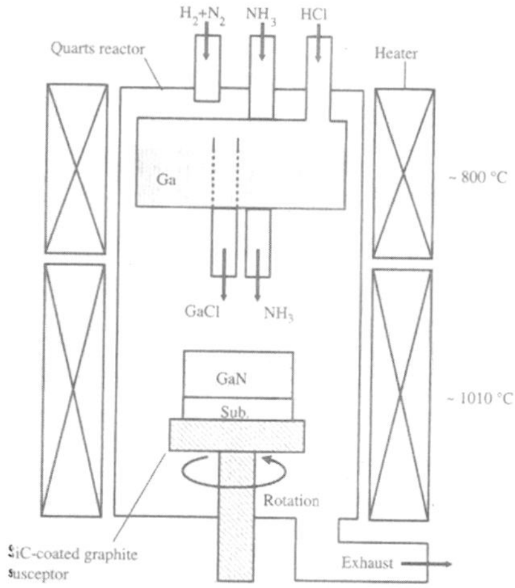

Fig. 20.45 Schematic drawing of the vertical HVPE system (reprinted from Ref. 515 with permission, copyright 2009 Elsevier).

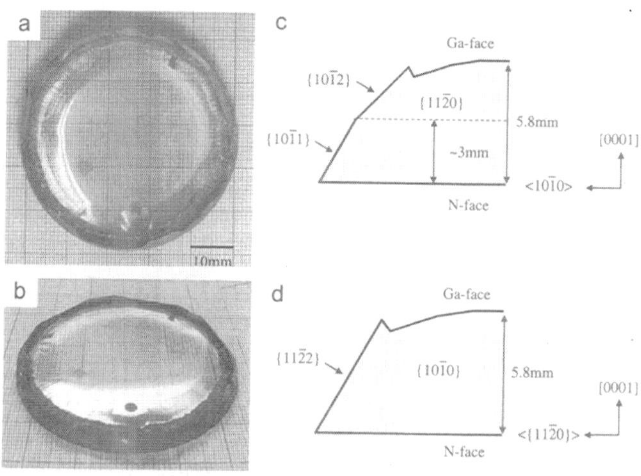

Fig. 20.46 (a) Freestanding bulk GaN crystal grown by HVPE at 1010°C for 55 h with a thickness of 5.8 mm and diameter 52 mm. (b) Bird-eye-view photograph of the bulk GaN crystal. Cross-sectional schematic images of the bulk GaN crystal for (c) {11-20} and (d) {10-10} (reprinted from Ref. 515 with permission, copyright 2009 Elsevier).

- *$NdGaO_3$ (NGO)*

Okazaki et al.[398, 399] were the first to find that NGO substrates have good lattice matching with GaN and showed that thick GaN layers could be grown on NGO substrates. Because of the difference in the thermal expansion coefficients, thick GaN layers could be spontaneously removed from NGO substrates, so that freestanding GaN substrates could be obtained. The study using this system has been continued since then[240,516-521] and GaN substrates up to 45 mm size have been feasible.

- *GaAs*

Motoki et al.[522,523] have used GaAs as the seed substrate. Since GaAs can be separated off by an etching procedure, separation is easily performed. In order to reduce the dislocation density which is rather high due to the difference in lattice constants, the surface of GaAs is etched to prepare etch pits. GaN layers then grow preferentially on the surface without pits, and dislocations are concentrated around the pits, so that low dislocation areas can be obtained.

Kumagai et al.[524,525] have also grown thick GaN on (111)A GaAs substrates and were able to prepare Fe-doped semi-insulating (SI) freestanding GaN substrates.

- *SiC*

Melnick et al.[526,527] have grown thick GaN layers on 30 mm diameter 6H-SiC substrates and removed the substrate by reactive ion etching (RIE) to obtain 200 μm thick freestanding GaN substrates. The FWHM-XRC was less than 140 arcsec. The dislocation distribution in the GaN substrates was studied by transmission electron microscopy (TEM).

- *Si*

Sawaki et al.[528-531] have developed a growth technology for semipolar thick GaN layers on Si substrates. On patterned (001) and (11-3) Si substrates with (111) facets, semipolar (1-101) and (11-22) GaN layers were first laterally overgrown by MOCVD [528,530,531] and by HVPE[529] and thick GaN layers of thickness up to 230 μm could be grown by HVPE at a low temperature of 870°C. It was also shown that the two-step growth reduced the dislocation density to the order of 10^7 cm^{-2}.[530]

- *GaN*

The best substrate for HVPE growth is GaN because homoepitaxial growth without lattice mismatching is possible. Xu et al.[405] have grown thick GaN layers exceeding 3 mm on GaN seed substrates which were prepared also by HVPE. It was found that dislocation densities could be lowered to the level of 10^4 cm^{-2}. Lucznik et al.[532] have grown 2 mm thick GaN layers on undoped and Mg-doped GaN seed substrates prepared by the HP-SG method, whose dislocation density was less than 100 cm^{-2}. The dislocation density of the homoepitaxially grown GaN increased as a critical thickness was exceeded and decreased as a function of the thickness. Grzegory et al.[533] have studied the crystal quality of HVPE grown crystals of a few mm thickness using HP-SG grown platelet and needle crystals as the seeds. Kalmer et al.[534] have developed a photochemical etching procedure for preparing patterned HVPE GaN substrates and have regrown thick GaN layers on them. The dislocation density was as low as 3 x 10^5 cm^{-2} in the laterally grown region.

- *Oxides*

Detchprohm et al.[389] were the first to grow thick GaN on ZnO substrates. Kryliouk et al.[535-538] demonstrated the growth of thick GaN layers on $LiGaO_2$ (LGO) substrates. They developed a system in which HVPE and MOVPE growth can be performed in the same reactor (H-MOVPE). GaN layers were grown on LGO substrates and then thick GaN layers were grown on them. Thin GaN MOVPE layers were then grown on the HVPE layers in order to obtain a smooth surface. LGO substrates could be self-separated easily by nitridation of the substrate and the cooling process. Reed et al.[539] applied the same method using $LiAlO_2$ (LAO) substrates and have obtained 35-40 μm thick GaN substrates. Richer et al.[540] have obtained freestanding 200 μm substrates by HVPE growth on LAO seeds by spontaneous separation.

(5) Others

Recently, molecular beam epitaxy (MBE) has been applied to the growth of cubic bulk GaN and $Al_{1-x}Ga_xN$ single crystals free from polarization effect.[541,542] By radio ferquency (RF)-MBE, freestanding cubic (zinc-blende) GaN substrates of thickness 100 μm could be obtained.[543-545] The first zinc-blend LEDs have been fabricated using these substrates.[546]

A novel growth method was reported in which nanocrystal-assembled (NAB) bulk GaN crystals were grown by the reaction of Ga metal in liquid ammonia with NH_4Cl at 350-500°C.[547] These workers could obtain small platelike GaN single crystals measuring 0.32-0.42 mm^2 and having average thickness of 0.2-0.4 mm. The conversion of a Ga_2O_3 substrate surface to GaN by nitridation by the reaction with NH_3 gas has been performed.[548,549] The synthesis of GaN by the reaction of Li_3B with Ga_2O_3 and Ga under NH_3 atmosphere has been performed at a relatively low temperatures of about 700°C.[550,551]

20.3.4 Indium Nitride (InN)

There are fewer reports on the bulk crystal growth of InN. Since InN has a small bandgap of 1.89 (recent data is 0.7) eV and is promising for high speed electronic devices, there is a demand for the preparation of InN. InN has however a higher decomposition pressure than that of GaN and AlN[552] and bulk crystals are much more difficult to grow. Utusmi et al.[21] have determined the pressure-temperature phase diagram for InN using a multi-anvil high pressure apparatus. They found that congruent melt growth can be realized under pressures above 10 GPa and at temperatures above 1900°C. Under these conditions, only the rock salt crystal structure is stable because of the pressure induced phase transformation. It is thus difficult to grow bulk InN crystals because of the high decomposition pressure. Only a few reports have been published of the HP-SG method[215,219,221] and only small crystals (5-50 μm) were obtained.

Waldrip[353] has invented a novel method in which $InCl_3$ and L_3N precursors are dissolved in molten halide salts such as LiCl and LiCl+KCl, and GaN crystals are grown electrochemically. The high pressure Chochralski method was also combined with this method to grow single crystals. Murali et al.[553] have synthesized InN by nitridation of In_2O_3 under NH_4 atmosphere and studied the reaction kinetics. Schwenzer et al.[554] have prepared InN nanoparticles by the ammonolysis of indium oxide by reaction with ammonia.

20.3.5 Other III-V Compounds

Other III-V materials such as BP, BAs, BSb, AlP, AlAs and AlSb are also interesting but there are hardly any reports on their bulk crystal growth.

In the case of BP, since it has a extremely high melting point exceeding 3000°C and decomposes even at 1550°C under one atmosphere, the customary crystal growth methods can not be applied. The literature on BP is therefore very scanty. Several old studies of the preparation of BP are summarized by Williams.[555] The growth of BP from a molten solution of dinickel phosphide[556] and of copper[557] and from the vapor phase[555-564] has been reported. BP has the bandgap of 1.50-1.56 eV and the future possibility of BP for the application in luminescent devices has been discussed.[564]

Boron arsenide such as $B_{12}As_2$ which is an excellent candidate for neutron detectors, thermoelectric converters and radioisotope batteries, has been grown by precipitation from a nickel based metal solution from 1150°C with a cooling rate of 3.5 h.[565]

The growth of AlSb which is promising for use in radiation detectors, has been performed by various methods, such as Bridgman [566-570], traveling heater method (THM) [566], Czochralski and liquid encapsulated Czochralski (LEC).[566,571-575] The main problem was the crystal sticking to the material. This was recently solved by using LiCl-KCl eutectic as the encapsulant.[576]

20.4 CHARACTERIZATION

20.4.1 Purity

(1) BN

c-BN crystals grown from a Li_3BN_2 solvent by high pressure solution growth showed an amber color and it is known that this is because of carbon and oxygen impurities, respectively 10^{19} cm^{-3} and 10^{20} cm^{-3}, concentrations similar to those in the h-BN source. When a $Ba_3B_2N_4$ solvent was used, colorless crystals could be grown and the concentrations of carbon and oxygen could be reduced to around 10^{17}-10^{18} cm^{-3}.[33]

(2) AlN

To grow AlN crystals by the PVT method, it is important to improve the purity since the growth temperature is relatively high at 2000-2300°C. Purity is impaired by contamination from the crucible material. Purity from the viewpoint of crucible materials[105] has been discussed. It is also known that AlN has a large concentration of C, O and H impurities.[102] Carbon is deposited as Al_4C_3[52] and is incorporated in AlN as a solid solution of Al_2OC. Carbon is considered to be transported from the furnace material as gaseous CN, C_2N_2, CO and AlCN species.[577] By selecting appropriate furnace materials, the impurity levels of carbon are improved.[80] The presintering of AlN powder sources is also effective in reducing carbon impurities.[578,579] The use of TaC crucibles fabricated using a binderless process achieved the reduction of C contamination to the level of 50 ppmw.[105] Carbon concentration as low as 8×10^{18} cm^{-3} was reported.[111]

Oxygen is also a major impurity in AlN crystals due to the affinity of aluminium with oxygen and because at low temperatures, oxygen assisted transport of Al species such as Al_3O_3N takes places.[120] Oxygen is incorporated as a substitutional impurity on the nitrogen site[580] and the concentration is in the region of 10^{19}-10^{20} cm^{-3}.[577] Presintering of AlN powders has been successful in reducing the oxygen concentrations[578,579] and oxygen levels of about 10^{19} cm^{-3} have been reported.[92,111] Oxygen has the effect of contaminating the seed with small AlO_xN_x particles which can be avoided by the temperature reversal method.[118] The present oxygen level was reported to be less than 10^{18} cm^{-3} by secondary ion mass spectroscopy (SIMS) measurement.[102] Edgar et al.[96] have examined the origin of oxygen in AlN crystal growth as oxides and hydroxides in AlN source powers and found that they are reduced by preheat treatment, at low temperatures (950-1000°C) by the removal of these oxygen-related materials and at high temperatures by reducing the surface area of AlN powders by sintering. The addition of hydrogen to the N_2 atmosphere was also effective in reducing the oxygen contamination.[581] The role of oxygen in the growth mode has also been studied.[122] The formation of oxy-nitride layers detrimental to the growth of AlN can be avoided by presintering of the source powders.[134] Slack et al.[582] have studied the effect of oxygen on various physical properties such as luminescence, optical absorption, thermal conductivity and crystal perfection.

The behavior of other impurities such as Be, C, Mg, Mn, S and Si has been studied and their reduction by the preparation of pure AlN powders from AlF_3 has been discussed.[52] Presintering of AlN power sources also had the effect of reducing various impurities.[92] Markov et al.[116] have applied the preprocessing technique, that is, annealing AlN powders in N_2 atmosphere and sublimation-recrystallization of source powders. The impurity level could be reduced to a level below 100 ppm for all impurities including oxygen.

(3) GaN

Dwilinski et al.[48,68,309] have shown that carrier concentrations as low as 5×10^{15} cm^{-3} could be obtained for ammonothermally grown GaN because of the high degree of purity. It was pointed out that various impurities such as Fe and Ni come from the mineralizers[302] and their treatment is important. Amides are sensitive to oxygen and moisture while azides are preferable for low oxygen contamination.[302] Callahan et al.[317] have analyzed the incorporation of the various impurities. Oxygen was in the level of 10^{20} cm^{-3} and C, K and Si were in the level of low 10^{19} cm^{-3}. The incorporation of Fe, Ni, K and Li was dependent on the growth face. Adekore et al.[320] have shown that the oxygen content can be mitigated to the level of 1-7 $\times$ 10^{19} cm^{-3}. D'Evelyn et al.[346] have shown that the total impurity level could be reduced to 10^{18} cm^{-3} by improving the purity of raw materials. The effect of oxygen on various physical properties such as luminescence, optical absorption, thermal conductivity and crystal perfection has also been studied for GaN.[582]

20.4.2 Defects

(1) Dislocations

The dislocation density can be revealed not only by the transmission electron microscopy (TEM)[583] and atomic force microscopy (AFM)[584] but also by selective etching. There are several dislocation etchants for nitrides, a hot (160°C) H_3PO_4 solution [585], molten KOH-NaOH eutectic (E etchant)[586, 587] and molten KOH-NaOH eutectic doped with 10 wt.% MgO (E + M etchant)[588] firstly confirmed for GaN. There is a photochemical etching method for GaN[589-591] but this cannot be applied for AlN whose bandgap is as high as 6.28 eV. Cathodoluminescence (CL) mapping was also found to be effective in evaluating dislocation density.[592] Various characterization methods were compared for GaN.[593] Zhang et al.[594] have confirmed that E and E + M etchants can be applied for AlN. The etchant for AlN is reviewed by Zhang and Edger.[595]

- BN

c-BN single crystals have been grown by the high pressure solution growth method (Sec. 20.3.1) and the dislocation density was evaluated by molten KOH at 450°C.[596] Amber color crystals grown from a Li_3BN_2 solvent showed a density of 5×10^6 cm^{-2} while that of colorless crystals grown from a $Ba_3B_2N_4$ solvent showed a density of 4×10^4 cm^{-2}.[596]

- AlN

It was found that the dislocation density was at the level of 10^7 cm^{-3} for AlN crystals grown on Si-face 6H-SiC (0001) substrates while it was in the order of 10^3 cm^{-2} for self-seeded AlN single crystals.[594]

Bondokov et al.[597] have found that the dislocation density on the prismatic planes is about 6×10^4 - 5×10^5 cm^{-2}, two orders of magnitude higher than that on the basal planes, where it was 1×10^3 -3×10^4 cm^{-2}. This observation implies that the prismatic glide system plays an important role in the elastic-plastic transitions at elevated temperatures. They showed the difference of etch pit patterns for screw and edge dislocations.

It is known that etching of the freestanding AlN crystals exhibits natural habits with well-developed faces.[120, 121] Bickermann et al.[598,599] have analyzed the etch pits on the basal plane revealed by the E etchant for freestanding AlN single crystals grown by the PVT method and classified the etch pits for screw, edge and mixed dislocations. The lowest dislocation density reported was less than 500 cm^{-2} for a crystal size of about 10 $\times$ 10 mm^2.[99] Shujman et al.[600] have evaluated the dislocation density of 50 mm diameter AlN wafers to be of the order of 3×10^4 cm^{-2}.

- GaN

Elsner et al.[601] have calculated the electrical properties and line energies of threading screw and edge dislocations. The electrical properties of dislocations were also studied.[602, 603]

Baranowski et al.[604] showed that etch pits due to dislocations can be revealed by treating in liquid gallium at a temperature slightly above the equilibrium temperature for Ga-GaN-N_2. The dislocation level of bulk GaN grown by the solution method was in the region of 10^5 cm^{-2} and was lower than that for epitaxial layers on sapphire substrates. In many cases, the dislocation densities are measured by hot (160 °C) H_3PO_4 solution[585,605], molten KOH-NaOH eutectic (E etchant)[586, 587] and molten KOH-NaOH eutectic doped with 10 wt.% MgO (E + M etchant)[588] and H_3PO_3 + H_2SO_4.[606] There is however an argument that hexagonal etch pits are only concerned with hollow core screw dislocations[605,607,608] so that methods other than selective etching must be employed to evaluate total dislocation densities.

The dislocations in the heterostructure of GaN on sapphire substrates were in the region of 10^8-10^9 cm^{-2}. The density could be reduced by MOCVD and HVPE methods using ELO and other methods as explained in Sec. 20.3.3(4) to the order of 10^4-10^6 cm^{-2}.[609] The double ELO method achieved a reduction in the dislocation density and a prolongation of the LD lifetime.[492] Contrasting with these high dislocation density levels, bulk GaN grown by the flux method[286] and by the ammonothermal method[310] showed densities as low as 10^3 cm^{-3}. Gregory et al.[610] have shown that the dislocation density could be reduced even down to a few tens cm^{-2} for HP-SG grown GaN by Mg doping.

The effect of dislocations on device performance is important especially in the case of high brightness LEDs, UV-LEDs and LDs. It is known that the dislocation density in GaN/sapphire at 10^9 cm^{-2} does not affect the luminosity of blue LEDs[611], a surprising phenomenon which is quite different from other color LEDs where the dislocation density greatly affects the luminosity when it exceeds about 5×10^4 cm^{-2}.[612] This phenomonon has been explained by the confinement effect of carriers in island like structures of InGaN layers[611,613], by the nature of the bonding[612] and by low dislocation mobility.[614] This is however different for the case of high power LEDs and LDs where the carrier concentration is much higher and carriers overflow the confined area. In fact, low dislocation density GaN prepared by HVPE methods is used industrially for blue-violet LDs for digital versatile disks (DVDs).[492, 615] For short wavelength LEDs and LDs, the role of dislocations is important since the carrier lifetime is much longer than that for long wavelength devices.[616] Kozodoy et al.[617] have shown that the reverse leak current becomes three orders of magnitude higher where there are threading dislocations. Threading dislocations are also known to be electrically active donor centers[618], highly active non-radiative recombination centers[619], over mid-gap trap states assisting charge tunneling[620] and seeds for V-defects.[621] The effect of dislocations on electronic devices, especially on the breakdown voltage and leak current is discussed.[622, 623] From these points of view, it is essential to obtain GaN substrates with low dislocation densities.

(2) Structural defects

- BN

Even amber color c-BN single crystals grown by the high pressure solution growth method showed the FWHM-XRC of 80 arcsec. Colorless c-BN crystals showed the smallest FWHM of the Raman spectrum of 3.0 cm^{-1}.[596]

- *AlN*

The structural defects in AlN crystals have been evaluated mainly by synchrotron white beam X-ray topography (SWBXT) (an example is shown in Fig. 20.7)[624], by high resolution X-ray diffraction (HRXRD) and by TEM. The freely nucleated AlN platelets showed low dislocation densities[93] but PVT-grown AlN bulk crystals showed grains and cracks,

The FWHM-XRC was also used as an index of the crystal perfection. The FWHM is reported to be normally in the range of 10-83 arcsec. AlN crystals grown by the PVT method in the early stage already had a value of about 10 arcsec even for the 10 mm diameter size.[99] The FWHM values for 50 mm diameter AlN wafers were reported as 28-32 arcsec as shown in Fig. 20.47.

The structure of single crystals of diameter 10 mm grown in the directions of the m-plane and c-plane was evaluated by SWBXT[112] as shown in Fig. 20.48. For the m-plane crystal there was no orientation contrast due to low angle grain boundaries or mosaic blocks and dislocations were found to lie in (0002) basal planes and have edge components. The c-plane crystal had dislocations with <11-20>(0001) slip system and most

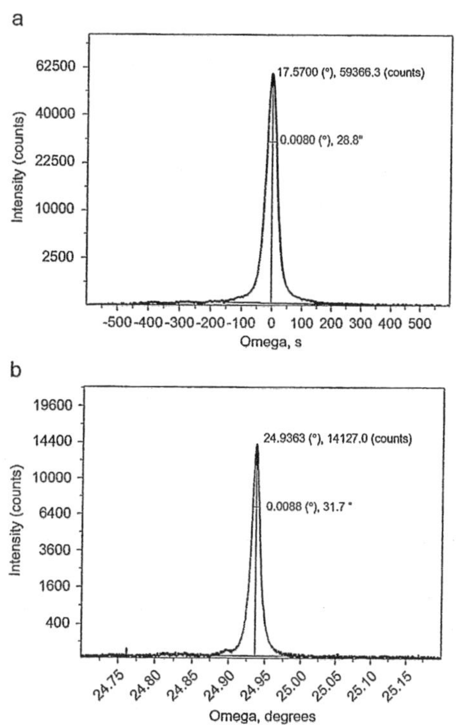

Fig. 20.47 AlN wafer rocking curve: (a) symmetric (0002) curve and (b) asymmetric (10.2) curve for 40 mm diameter AlN (reprinted from Ref. 102 with permission, copyright 2008 Elsevier).

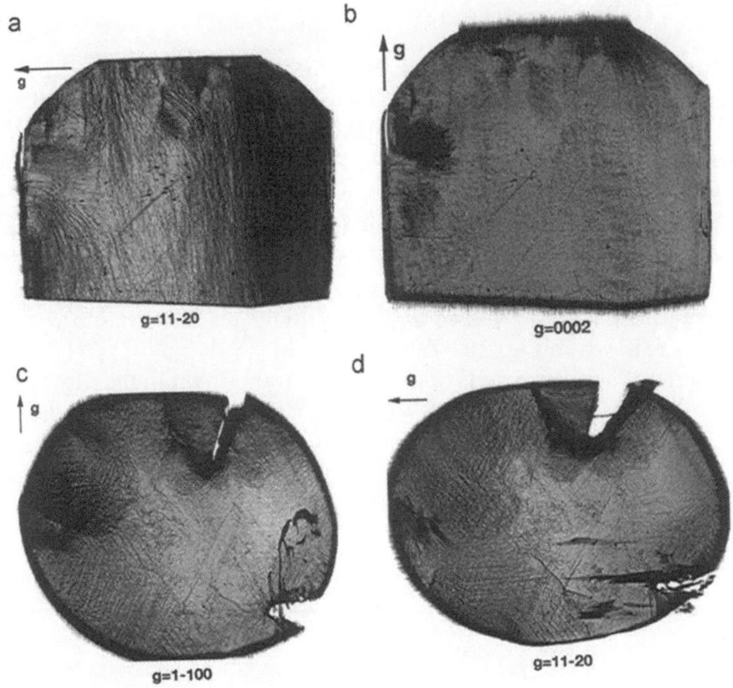

Fig. 20.48 X-ray topographs of (a) an m-plane crystal, g=11-20, (b) an m-plane crystal, g = 0002, (c) a c-plane crystal, g = 1-100, (d) a c-plane crystal, g = 11-20 (reprinted from Ref. 112 with permission, copyright 2009 Elsevier).

of them were screw type. The dislocation density was 10^2-10^3 cm^{-2} in the center region.[113]

- GaN

Electronic structure of stacking faults, inversion domain and stacking mismatch boundaries have been studied theoretically.[625, 626]

By the HVPE method, low FWHM-XRC of a few tens of arc seconds can be obtained for GaN in the case of symmetric reflections while they are several hundreds arc seconds for the unsymmetrical reflections. This is because with the HVPE method, cell structures are formed because nuclei are generated at various surface points.[627]

On the other hand, in the case of bulk crystal growth methods, the FWHM-XRC is of the order of a few tens of arc seconds even for the unsymmetric reflections. This is because the generation of nuclei occurs selectively in the case of bulk crystal growth methods.

Structural defects in HP-SG grown GaN single crystals have been evaluated by X-ray topography[628] and the formation mechanism of low angle grain boundaries on the N-plane and the Ga-plane has been discussed. Kawamura et al.[279] have evaluated the propagation of dislocations from a seed GaN MOCVD grown on sapphire to bulk GaN

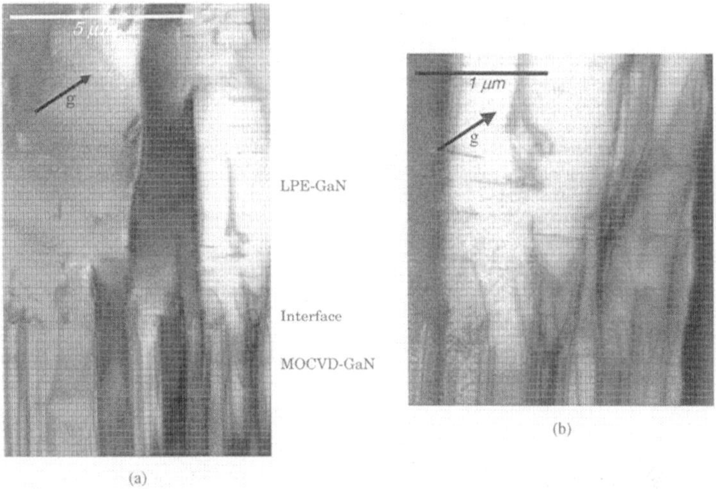

Fig. 20.49 (a) TEM [1-100] photograph taken around the interface between MOCVD-GaN and LPE-GaN, and (b) enlarged view of TEM [1-100] photograph around the interface between MOCVD-GaN and LPE-GaN (reprinted from Ref. 279 with permission, copyright 2006 The Japan Society of Applied Physics).

grown by LPE from a flux. It was observed that most dislocations were eliminated during crystal growth as shown in Fig. 20.49. This is the basis of the dislocation reduction model shown in Fig. 20.29.

Structural defects in ammonothermal GaN and PVT-grown GaN[629, 630] have been studied. Basal plane stacking faults have been observed by TEM and SWBXT in basic-ammonothermally grown GaN crystals.[630] The FWHM-XRC was reported to be about 16 arcsec as shown in Fig. 20.50 for ammonothermally grown 25 mm diameter GaN single crystals, with dislocation density of the order of 10^3 cm^{-2}.[310] Hashimoto et al. have studied the growth interface between ammonothermally grown GaN and a GaN seed crystal by TEM as shown in Fig. 20.51.[331] It was found that at the Ga face, the dislocations are bent due to faceting growth and it was concluded that the Ga face is preferable for dislocation reduction.

(3) Point defects

- BN

Native point defects in Be-doped c-BN crystals have been studied by high frequency electron spin resonance (ESR) measurement and the identification of several intrinsic defects has been discussed.[631]

- AlN

The nitrogen vacancy V_N was theoretically predicted as a shallow donor with an energy level of about E_C-0.3 V[632] and was experimentally observed by PL at E_C-0.26 V.[633]

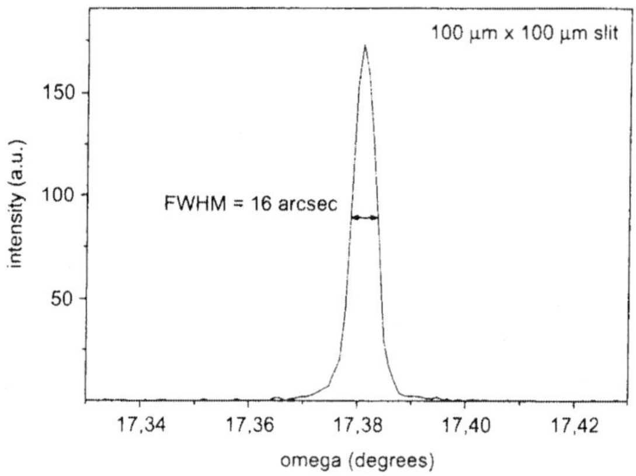

Fig. 20.50 (0001) X-ray rocking curve of AMMONO-GaN (reprinted from Ref. 310 with permission, copyright 2009 Elsevier).

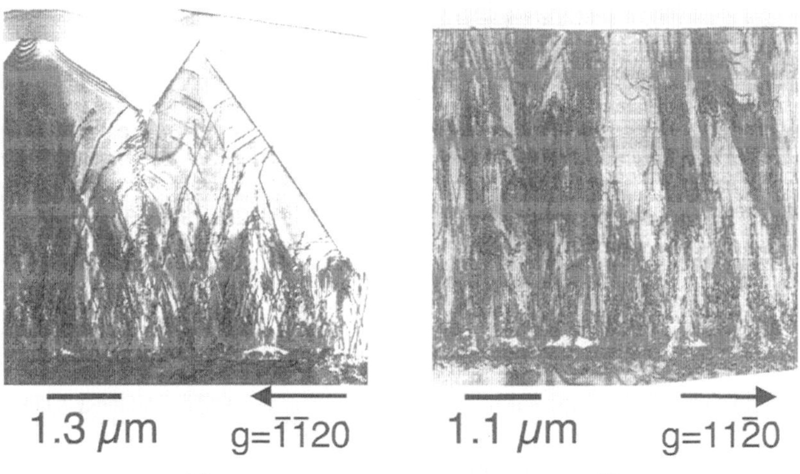

Fig. 20.51 Cross-sectional TEM images of the films on Ga face (a) and N face (b). Both are bright field images with g inclined in the figure (reprinted from Ref. 331 with permission, copyright 2005 Cambridge University Press).

Since the energy level and the formation energy are large, it is concluded that V_N does not act as an n-type donor.

The Al vacancy, V_{Al} was studied by positron annihilation spectroscopy (PAS) and was found to exist in PVT-grown bulk AlN crystals at a concentration of 10^{17} cm^{-3}.[634] Negatively charged defects were also observed and were attributed to oxygen related defects based on theoretical calculations.[635]

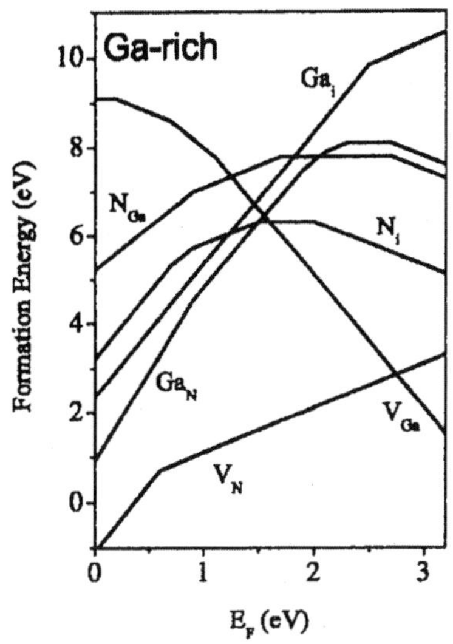

Fig. 20.52 Formation energies as a function of Fermi level for native point defects in GaN. Ga-rich conditions are assumed. The zero of Fermi level corresponds to the top of the valence band. Only segments corresponding to the lowest energy charge states are shown (reprinted from Ref. 643 with permission, copyright 2004 Elsevier).

- *GaN*

The native point defects in GaN have been theoretically studied and their formation energies and the ionization energies have been determined.[636-647] The formation energies of various point defects under Ga-rich conditions are as shown in Fig. 20.52[643] and it emerges that self-interstitial and antisite defects are unlikely to occur at least in n-type GaN since the formation energies are very high.

(i) Vacancies

Gallium vacancy V_{Ga} dominates for p-type GaN and nitrogen vacancy dominates for n-type GaN. V_{Ga} is a native defect and acts as an acceptor and three ionization energies are reported.[643,644,648] The migration barrier for V_{Ga}^{3-} is relatively low[643] so that V_{Ga} is mobile and likely to form complex defects.

Gorczyca et al.[648] have calculated the effect of hydrostatic pressure on the formation energy and energy level of V_{Ga}. Saarinenm et al.[649,650] have used PAS in order to clarify the relationship between the concentration of V_{Ga}, the stoichiometry and Mg-doping. Uedono et al.[651] have also shown that PAS is effective to study vacancy-like defects. Tumisto et al.[652] have studied vacancies in ammonothermally grown bulk GaN by PAS and found a high concentration of V_{Ga} in n-type crystals and no vacancies in p-type GaN crystals.

The ionization energy of the nitrogen vacancy is close to or inside the conduction band[636,637,640] so that n-type conductivity was attributed to V_N.[636] Further investigation showed that a detectable concentration of V_N can be achieved only under Ga-rich conditions[637,640] and V_N acts as a donor. V_N is predicted to appear at E_C-0.3 eV and its effect on the optical properties was studied.[653]

(ii) Divacancies

The formation energy of divacancy $V_{Ga}V_N$ is relatively high and unlikely to exist in high concentrations. This defect will act as a deep level, a double acceptor in n-type and a double donor in p-type.[644] Chen et al.[654] have studied vacancy-type defects in Er doped GaN by PAS and the formation of complex defects such as $V_{Ga}V_N$ was suggested.

(iii) Interstitials

Formation energies of interstitials are rather high and have low probability to exist. Gallium interstitials, Ga_i have a large formation energy and are stable only for the octahedral site.[643] Ga_i acts as a donor with the resonance (+/0) state and a deep level with the resonance (3+/+) state.[640, 643]

Nitrogen interstitials (Ni) forming N-N bonds[637,640,643,646] have high formation energy especially in Ga-rich conditions. Ni will act as an acceptor in n-type GaN.[643] Chow et al.[655] have observed Ga_i by electron paramagnetic resonance (EPR) and found that at room temperature this defect disappears.

(iv) Antisite defects

Gallium antisite defects, Ga_N, and nitrogen antisite defects, N_{Ga}, have been reported to form deep levels.[640, 643] Nitrogen antisite defects introduce three[643] or four[656] deep levels and act as double donors in p-type GaN and acceptors in n-type GaN. Neutral N_{Ga} exhibits metastable behavior as in the case of EL2 in GaAs.[657,658]

(v) Complex defects

Shallow donor - gallium vacancy complexes such as V_{Ga}-O_N, V_{Ga}-Si_N and V_{Ga}-C_N are reported.[639,644,659] As shallow acceptor - nitrogen vacancy complexes, Mg_{Ga}-V_N[651,660-662], Be_{Ga}-V_N[663,664], Mg_i-V_N have been discussed.[661,665] Hydrogen related complex defects such as $(V_{Ga}H^n)^{n-3}$ have been reported.[666] In p-type GaN, Mg-H[666], Be-H[667], $(V_NH^n)^{n-3}$ complexes are suggested.[668,669] Other complex defects, such as $Mg_{Ga}O_N$, $Be_{Ga}O_N$[667,670], $Be_{Ga}O_NBe_{Ga}$[667] are also discussed.

(4) Deep levels

- AlN

Polyakov et al.[671] have observed a trap at E_C-0.26 eV at a concentration of 6.7×10^7 cm^{-3} and a trap at E_C-0.65 eV at a concentration of 5×10^{15} cm^{-3} by resistivity and C-V measurements. They also observed a trap at E_C-3.3 eV by microcathodo-luminescence (MCL) and attributed it to an Al vacancy-oxygen acceptor complex[672], whose intensity is enhanced by low angle dislocation boundaries. Si doped AlN has been found to have a DX-like center with a large persistent photoconductivity.[673]

Sotamov et al.[674] have studied deep level defects in AlN single crystals by EPR and found two types of shallow donors and one deep level. One shallow donor was presumed O_N and another was C_{Al} or Si_{Al}. The deep donor level was found to be V_N with an energy level of 0.75 eV.

- *GaN*

There are various deep levels reported so far not only in bulk materials but also in epitaxial materials and irradiated materials. The deep levels in GaN have been studied extensively [675-709] by various methods such as deep level transient spectroscopy (DLTS) [676-692], isothermal current transient spectroscopy (ICTS) [676], photocapacitance (PHCAP) [686], optical DLTS (ODLTS) [677], photoinduced current transient spectroscopy (PICTS) [693], deep level optical spectroscopy (DLOS) [691,694], thermoelectric effect spectroscopy (TEES) [695], thermally stimulated luminescence (TSL)[696], photocapacitance spectroscopy (PCS) [690,697,698], persistent photoconductivity (PPC) measurement [685,700,701,702], electron spin resonance (ESR) [703], photoluminescence (PL) [685,704-707], Hall measurement [684,685,699], C-V measurement [708] and resistivity measurement.[709]

Theoretical predictions of deep levels have also been performed for p-type GaN [710], carbon impurity[711] and isovalent P and As impurities.[712]

These deep levels are summarized in Table 20.6, and are also reviewed in Refs. 713-715.

(i) n-type GaN

There are many deep levels in intentionally and unintentionally doped n-type GaN. The main deep levels can be classified as follows [713]:

- 0.1-0.2 eV levels

These levels are considered to be V_N or V_N related defects.[637,640]

- 0.3-0.4 eV levels

These levels are also considered to be related to V_N.

- 0.5-0.6 eV levels

These levels are often called E2 levels according to Hacke et al.[676] Independent of the growth method, E2 is observed at a high concentration of $1\text{-}9 \times 10^{14}$ cm^{-3}. This level is considered to be a complex of V_{Ga} with residual impurities such as C, Si and O.[688] This level is known to accompany a strong lattice relaxation.[716]

- 0.7-0.9 eV levels

These levels are considered as complex defects related to nitrogen or gallium interstitials because these levels are increased by irradiation.[688,717] They are also considered as complex defects related to impurities because they correlate with residual impurities and are detected only in undoped GaN.[692,718]

The concentrations of these deep levels observed in n-GaN is about 10^{14}-10^{15} cm^{-3}, much lower than those observed in GaAs and AlGaAs.[713]

(ii) p-type

There are a few reports of deep levels in p-type GaN[677,719-721] such as E_V+0.2-0.4 eV[677]

Table 20.6 Deep Levels in GaN

Material	Growth method	Deep level energy (eV)	Author	Year	Ref.
n-type	VPE	E_C-0.264, -0.580, -0.665	Hacke et al.	1994	676
	MOVPE	Ev+1.0, +1.2, -1.8, 3.1	Yi et al.	1996	697
	MBE	E_C-0.234, -0.578, -0.657, -0.961 -0.240	Wang et al.	1998	678
	MBE	E_C-0.44	Kim et al.	1998	679
	MOCVD	E_C-~2.2	Shalish et al.	1999	705
	MOCVD	E_C-1.42, -2.9	Kang et al.	1999	701
	MBE	E_C-2.9	Yuldashev et al.	2000	702
	MOCVD	E_V+2.8	Shalish et al.	2001	708
	MOCVD	E_C-(0.19-0.23), -(0.50-0.60)	Cho et al.	2001	681
	MOCVD	E_C-0.569, -1.013	Lee et al.	2002	682
	HVPE	E_C-0.61	Wu et al.	2003	683
	MOCVD	E_C-(0.10+0.11), -0.18, -(0.24-0.27) -(0.37-0.40), -(0.59-0.63)	Soh et al.	2004	684
	MOCVD	E_C-0.1, Ec-(0.2-0.24), E_V+0.87	Soh et al.	2004	685
	MOCVD, HVPE	E_C-0.17, -0.52, -0.53, -0.94	Py et al.	2005	686
	HVPE	E_C-0.60, -1.0	Look et al.	2005	688
	MOCVD	E_C-0.23, -0.58	Tokuda et al.	2005	689
	MOCVD	E_C-(2.1-2.2)	Wang et al.	2006	707
	MBE	E_C-2.44, -3.28	Armstrong et al.	2007	694
	MOCVD	E_C-(0.21-0.28), -(0.55-0.58) -(1.2-1.3), -1.6, -2.2, -3.1	Nie et al.	2008	670
	MBE	E_C-0.25, -0.62, -1.28, -3.28	Arehart et al.	2008	691
	MBE	E_C-0.21, -0.31, -0.84, -1.13	Mitsunaga et al.	2008	692
	MOCVD	E_C-(0.21-0.22), -(0.43-0.51), -(0.93-1.02)	Ito et al.	2008	699
	MOCVD	E_C-1.4, -1.70, -2.08, -2.64, -2.90	Nakano et al.	2011	698
p-type	MOCVD	E_V+0.21, +0.39, +0.41	Götz et al.	1996	677
	MOCVD	E_V+1.1, 1.40, +2.04	Qui et al.	1997	700
	MOCVD	E_V+0.41, +0.49, +0.59	Zhu et al.	2000	680
	MBE	E_V+3.05, +3.22, +3.26, E_C-2.97	Armstrong	2007	694
HR or SI	MBE	E_C-0.09, E_V+0.167	Desnica et al.	2002	695
	HVPE	E_C-0.51	Vaudo et al.	2003	709
	MBE	E_C-0.11, -2.05, -3.0, -3.28, E_V+0.9	Amstrong et al.	2005	687
	MOCVD	E_C-0.12, -0.62	Gai et al.	2009	696
	MOCVD	E_C-0.6, -0.7, 0.25	Polyakov et al.	2008	693

and E_V+0.4 eV.[719] Since the Mg acceptor level is about E_V+0.2 eV, these data are considered to be questionable because of the poor responsivity of carriers in the depletion layer.[720] By analysis of photocurrent, a level at E_V+1.85 eV is reported.[721]

(iii) Semi-insulating (SI)

SI-GaN is obtained not only by MOCVD but also by MBE. The deep level responsible for this semi-insulating behavior is not yet clearly understood. Amrstrong et al.[694] have proposed that the mechanism is the compensation of Si shallow donors by C deep acceptors. It is known that SI GaN can be obtained when the concentration of Si shallow donors is reduced to a level below 10^{16} cm^{-3} for MOCVD grown GaN/Si epitaxial layers.[699] It is also reported that when carbon is intentionally doped in MOCVD grown GaN/Si, semi-insulating GaN can be obtained.[694]

Carbon was believed to be a shallow acceptor[633,634,639,642,659,687,722-725] and the semi-insulating behavior was explained as the self-compensation between a C_N acceptor and a C_{Ga} donor.[723,725] Carbon was also considered as a deep level from interstitial carbon[659] and a deep acceptor as C_N.[726] Recently, C_N has been proposed as a deep acceptor, not a shallow acceptor as conventionally believed and is calculated to have an energy level of E_V+0.9 eV.[711]

(iv) Fe and Cr doping

As in the case of Cr doped GaAs (Sec. 8.7.2(1)) and Fe doped InP (Sec. 10.4.3(2)), the doping of transition metals whose energy level is sufficiently deep to pin the Fermi level to the middle region of the bandgap can realize semi-insulating materials. The doping of Fe in GaN has therefore been studied. It was found that GaN:Fe has an energy level of E_C-0.59 eV and resistivity as high as 10^9 Ωcm at room temperature.[704,709]

20.4.3 Electrical Properties

For various device fabrications, pn control is indispensable. This control is now can be achieved for several nitride materials. The electrical properties of nitrides are reviewed in several Refs. 5 and 727-729

(1) BN

Wentrof[25] was the first to report the growth of p type and n type crystals. p-type was achieved by Be doping[25,730] and n-type by Si or S doping. Amber color c-BN crystals grown from Li_2BN with high carbon and oxygen concentrations showed n-type conductivity[731] while colorless c-BN crystals had high resistivity and pn conductivity could not be determined.[32] Energy levels of various impurities have been determined theoretically and experimentally.[730-732] S-doped crystal had an activation energy of 0.32 eV and that of unintentionally doped crystal was 0.47 eV, probably due to an oxygen donor.[731] Oba et al.[732] have calculated the ionization energies of donors (Si, S, O) and acceptors (Be, Mg) for c-BN and h-BN. Park and Chadi[733] have studied the atomic and electronic structures of various shallow and acceptor impurities. Mishima et al.[734] were the first to fabricate pn junctions, using Be as the p-type dopant and Si as the n-type dopant. The activation energy for each dopant was found to be 0.23 eV and 0.24 eV respectively.

(2) AlN

Since the bandgap of AlN is 6.2 eV, it behaves as insulating material with resistivity 10^{11}-10^{13} Ωcm[64] and very low carrier concentration and mobility, $<10^{16}$ cm^{-3} and <2

cm^2/Vs respectively, though the theoretical mobility is much higher.[729] It is however reported that AlN becomes conductive (n-type, ~ 10^3 Ωcm) when crystals are grown by the closed space vapor transport (CST) method.[145] It is also reported that AlN becomes n-type conductive with Si doping.[735] With C doping, a hole concentration of 3×10^{18} cm^{-3} has been achieved.[736] Park and Chadi[733] have studied the atomic and electronic structures of various shallow and acceptor impurities.

(3) GaN

The electron mobility is shown in Fig. 20.53 as a function of carrier concentration. The

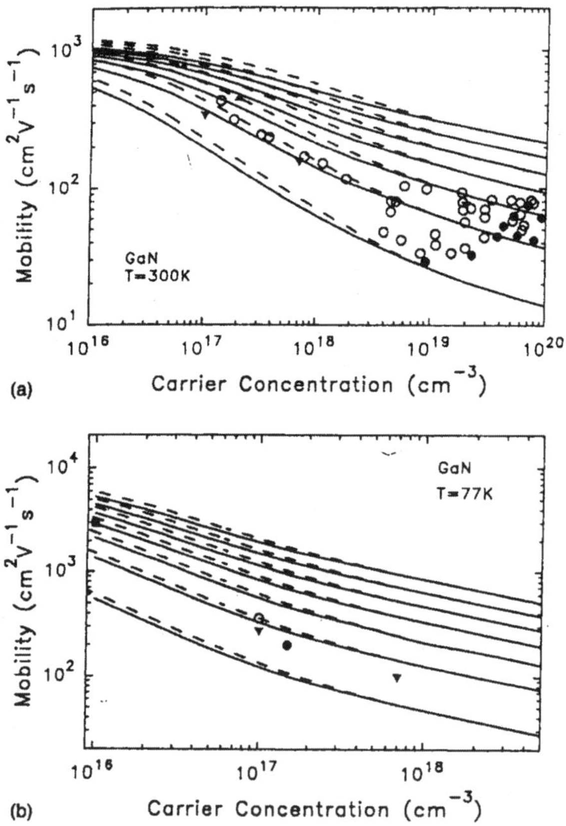

Fig. 20.53 The electron drift (solid curves) and Hall (dashed curves) mobility of GaN calculated as a function of carrier concentration with compensation ratio of 0.00, 0.15, 0.30, 0.45, 0.60, 0.75 and 0.90. (a) 300 K and (b) 77 K (reprinted from Ref. 729 with permission, copyright 1994 The American Institute of Physics). The references of the experimental data indicated by circles and triangles are found in Ref. 729. The horizontal axis represents actual electron concentrations for drift mobilities, Hall concentrations for Hall mobility.

highest electron mobility is as high as several thousands of cm^2/Vs at room temperature.[729] For bulk materials, an electron mobility of 441 cm^2/Vs at a carrier concentration of 1.0×10^{18} cm^{-3} [737] was reported for HVPE freestanding substrates. Using MOCVD epitaxially-grown AlGaN/GaN on Si substrates, an electron mobility of 3000 cm^2/Vs which is the highest value ever reported was obtained.[738] The effect of dislocations on carrier mobility was discussed.[602,603]

Drift mobility is known to show the highest value as high as 2.8×10^7 cm/s in an electric field of 1.5×10^5 V/cm[739-742] as shown in Fig. 20.54. This value itself is a higher drift mobility than that of SiC at this electric field though SiC exhibits higher values at electric fields above 2.4×10^5 V/cm. GaN is therefore a good material for electronic power devices under medium electric fields of $1\text{-}2 \times 10^5$ V/cm.

The free carrier concentration of undoped GaN bulk crystals grown by the solution method was in the range of $10^{18}\text{-}10^{19}$ cm^{-3} with n-type conductivity. The dominant shallow donor for nominally undoped GaN is considered to be nitrogen vacancies [743-745] with activation energy ranging 17-49 meV.[746-748] Residual impurities such as Si, Ge and O are also considered to be shallow donors.[403,749-751] The atomic and electronic structures of various shallow and acceptor impurities were also studied.[733] Various types of n- and p-doping as explained above were also reviewed more precisely in Ref. 13.

Donor dopants, Si and Ge are mainly used for epitaxial layers.[751] The ionization energy is reported as 12-17 meV by Hall measurement and 22 meV by photoluminescence (PL) measurement for Si.[750] For Si doping, carrier concentration as high as 2×10^{19} cm^{-3} could be obtained. These shallow donors were also studied by PL.[639,645,659,752,753]

The acceptor dopant Mg has mainly been studied in epitaxial layers but it is known to need low energy electron beam irradiation or thermal treatment after high quality

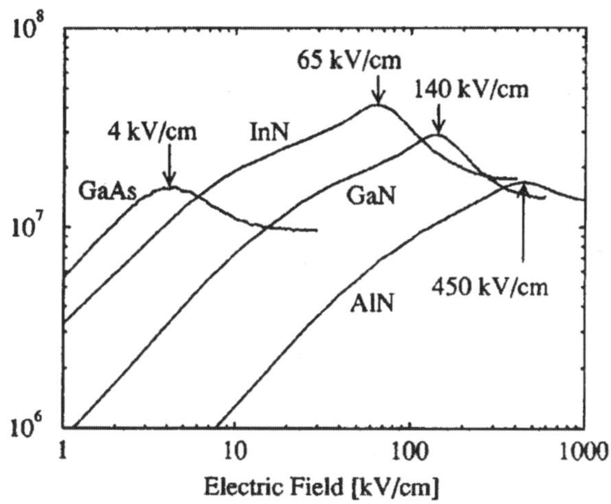

Fig. 20.54 The velocity-field characteristics associated with wurtzite GaN, InN , AlN and zincblende GaAs. In all cases, the temperature is set to 300 K and the doping concentration to 10^{17} cm^{-3} (reprinted from Ref. 728 with permission, copyright 1999 The American Institute of Physics).

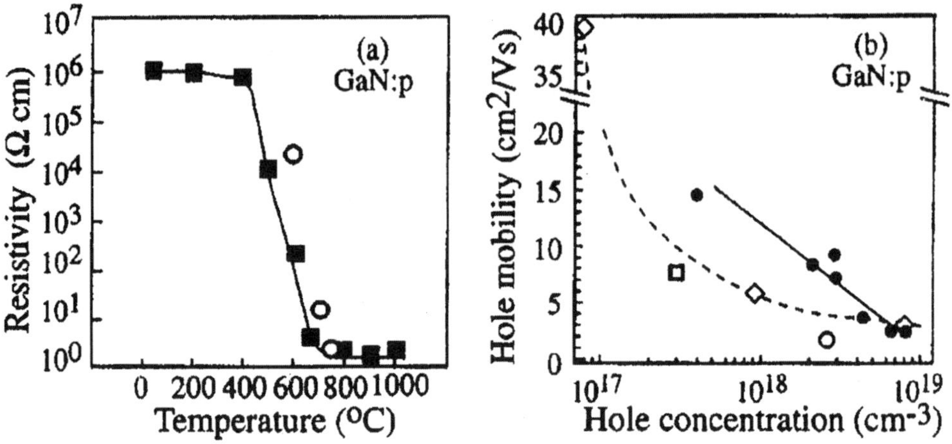

(a) (b)

Fig. 20.55 (a) Effect of thermal annealing on the resistivity of GaN:Mg as a function of annealing temperature, (b) Hall mobility in Mg-doped GaN as a function of hole concentration (reprinted from Ref. 5 with permission, copyright 2000 The American Institute of Physics). The references for the gathered data are found in Ref. 5.

crystal growth.[754-762] In Fig. 20.55 the resistivity and hole mobility of Mg-doped GaN are shown.

Other acceptors such as various impurities including Zn, Cd, Ca, Be and C have been studied[92,96,763] and similar binding energies of about 225 meV were reported.[3] The ionization energy of Mg is reported as 150-182 meV by Hall measurement[760,764-766] and as 209-224 meV by PL measurement.[750,767] The difference is discussed by Hughes and Jain.[768]

The highest hole mobility of 39 cm^2/Vs was reported[762] as shown in Fig. 20.55. It is also reported that codoping of Be and oxygen increased the mobility up to 150 cm^2/Vs.[769] This is attributed to the Be-O pairs which do not greatly scatter hole carriers. Mg-O codoping is also reported to have a similar effect.[770] PL of shallow acceptors has been widely studied.[402,641,645,659,667,670,733,753,771-778]

For a long time, carbon and carbon related defects were believed to be shallow acceptors[639,659,687,722-726,753,779-783] and it was believed that self-compensation between the C_N acceptor and the C_{Ga} donor was the mechanism of the semi-insulating behavior.[723,725,784] This argument is now being reconsidered based on the theoretical calculation that carbon forms a deep acceptor[711] which may have a decisive role in the semi-insulating (SI) behavior.

Mg doping in HP-SG grown GaN has the effect of increasing resistivity.[220,222] Semi-insulating GaN has been reported for HVPE grown GaN[709,785] doped with Fe, which has resistivity as high as 2×10^9 Ω.cm as the extrapolated value to room temperature and activation energy of 0.51 eV. Murakami et al.[786] also obtained SI GaN by MOHVPE doped with Fe. They showed that the resistivity increases as a function of the Fe/Ga ratio. Evelyn et al.[346] have shown that SI GaN can be obtained by Fe or Co doping.

From recent experimental results as described in Sec. 20.4.2(4) about deep levels and from the calculation that carbon acts as a deep level with energy level $E_V + 0.9$ eV [711], nominally undoped SI-GaN can be achieved as predicted from the Schockley diagram (Vol.1, Sec. 1.6.1(3)) when the condition that $N_{aa} > N_d > N_a$ is satisfied. Here, N_{aa} is the concentration of a deep acceptor, N_d, that of a shallow donor and N_a, that of a shallow acceptor. This means that it is essential that $N_d > N_a$ so that a certain amount of a shallow donor is indispensable for the realization of SI-GaN.

(4) InN

The bandgap of InN was at first reported as 1.7 eV but recent study has found that it is 0.7-0.8 eV[787-794] because of the crystal quality improvement of InN films grown by MBE. Carrier concentration of 4×10^{17} cm^{-3} and mobility of 2100 cm^2/Vs have been reported. The electron drift velocity is estimated as 4×10^7 cm/s in an electric field of 64 kV/cm as shown in Fig. 20.51. Because of this narrow bandgap and high mobility, InN is now becoming a candidate for electronic devices.

20.4.4 Optical Properties

Optical properties of nitride materials are reviewed in Refs. 1-5.

(1) BN

Colorless c-BN crystals grown by the high pressure solution growth method showed an edge emission at about 200 nm in CL which was ascribed to a free exciton emission and the bandgap was estimated as 6.17 eV with indirect transition.[33,596,795] h-BN crystals grown together with c-BN crystals from a $Ba_3B_2N_4$ solvent showed a strong far UV emission at 215 nm, a peak 1000 times stronger than that of c-BN crystals due to direct transition. This emission was apt to be influenced by impurities of carbon and oxygen.[795,796] The bandgap of h-BN was estimated to be 5.95 eV with direct transition.[797] When the impurity concentration is over 10^{18} cm^{-3}, a 325 nm emission is observed instead of the 215 nm emission.[33] Time-resolved PL showed that the 215 nm emission has a very short decay time of about 70 ps while the 227 nm emission has a relatively slow decay time of a few ns and is assigned as a bound exciton luminescence.[798] By PL, the origin of the strong emission of 215 nm was studied and was ascribed to self-trapped exciton states due to the strong exciton-phonon interactions.[799] Eu implanted c-BN[800] and Eu and Tb doped c-BN crystals[801] showed visible emissions due to intra 4f electron transitions.

(2) AlN

The bandgap of AlN was determined as 6.2 eV by optical absorption [802], though other data gave lower values due to oxygen contamination.[1] The temperature dependence of the bandgap has been measured by Silveira et al.[803] A broad PL peak in the range 2-3 eV with a peak at 2.8 eV has been observed[52,64] and was attributed to oxygen impurities.[33] Regarding impurities, Mg[804-809] and rare earth metals[806,808] have been observed by luminescence. The infrared reflectivity has also been measured.[810] Raman scattering analysis has been performed for AlN[811] to evaluate stress related point defects. Raman scattering is thus effective for evaluating crystal perfection.[52,82,93] The effect of oxygen on the photoluminescence spectra has been studied as a function of oxygen concentration and a defect-cluster model was presented to explain the oxygen-related defects.[580] Ultraviolet electroluminescence emitting in a broad band extending from 215 nm to the blue end was observed by Rutz.[145] The temperature dependence of the bandgap has been determined by measuring the optical properties of a PVT-grown AlN single crystal.[812]

(3) GaN

Raman scattering analysis has been performed for GaN [813,814] to evaluate the phase transition under high pressure [813], strains in crystals[814] and the stress related point defects.

The PL of GaN has been widely studied and is reviewed in Ref. 1 and 10. Especially, Götz[750] and Reshchikov and Morkoc [815] have reviewed it very precisely. The edge emission of PL has been extensively studied [816-822] and the main assignments are as shown in Table 20.7. At low temperatures, exciton peaks bound to impurities are dominating and free exciton peaks can be observed if the purity of grown crystals is improved. From

Table 20.7 Assignments of PL Peaks in GaN

Peak (eV)	Assignments
Exciton peaks	
3.474-3.475	free excitons
3.480-3.481	free excitons
3.49	free excitons
3.385	free exciton LO replica
3.293	free exciton 2-LO replica
Impurity related peaks	
3.44-3.47	donor bound exciton
3.377-3.378	LO replica
3.286	2-LO replica
3.400	TO replica
3.455	acceptor bound exciton
3.454-3.455	Cd acceptor bound exciton
3.364	LO replica
3.355	2-LO replica
3.26-3.27	donor-acceptor pair
3.17-3.18	LO replica
3.08-3.09	2-LO replica
2.99-3.00	3-LO replica
2.5-2.8	GaN:Zn (ZnGa)
1.8-2.2	GaN:Zn (ZnN)
2.95-3.2	GaN:Mg
2.7-2.85	GaN:Cd
-2.2	GaN:Be
2.43-2.9	GaN:Hg
2.15	GaN:C
2.23	GaN:li
2.85	GaN:P
2.58	GaN:As

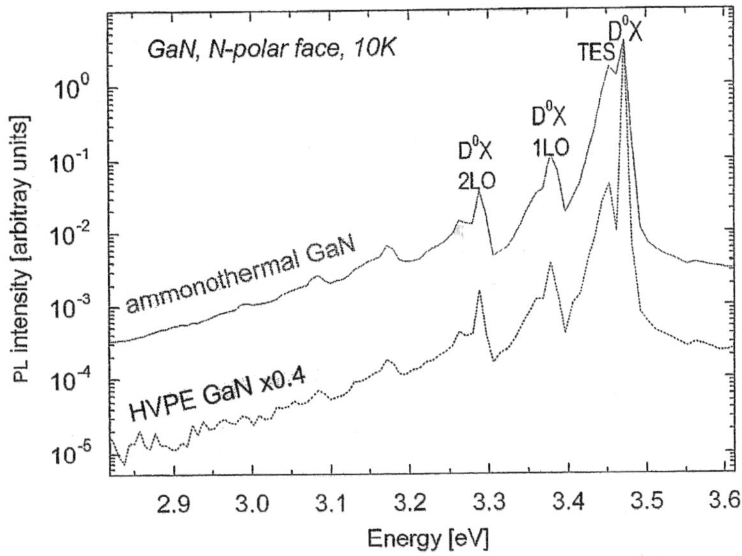

Fig. 20.56 Photoluminescence from an ammonothermal and HVPE GaN sample at 10 K. Note that the intensity from the HVPE sample is reduced by a factor of 0.4 (reprinted from Ref. 326 with permission, copyright 2008 Elsevier).

the edge emission spectra, the purity level can therefore be predicted. In Fig. 20.56, a typical PL spectrum taken from an ammonothermally grown bulk GaN crystal is shown. Fujii et al.[823] have also observed a well resolved edge emission at 3.357 eV from an ammonothermally grown bulk GaN.

In the energy range 2.95-3.27 eV, donor-acceptor pair (DAP) peaks can be observed.[749,750] The broad band PL peak observed at around 2.2-2.3 eV [749, 750], called the yellow luminescence (YL), often known as the yellow band, has been observed in GaN independent of the growth method and it is considered to be due to deep levels. The YL has been extensively studied in the past[824, 825] and is ascribed to point defects such as Ga and nitrogen vacancies.[749] Ogino and Aoki[826] first proposed that the origin of this band is the complex defect, V_{Ga}-C_N. This complex however is difficult to be formed because V_{Ga} and C_N are acceptors. Neugebauer and van de Walle[639] have explained the origin as V_{Ga} or complex defects of V_{Ga} with donor impurities such as O (V_{Ga}-O_N) as supported by experiments.[706,827,828] It is actually known that the yellow band increases with increasing Si and Ge doping concentrations.[751] The existence of this band has an adverse effect on carrier lifetime and on the intensity of the edge emission.

Li et al.[829] have shown the intensity of YL is related to extended defects such as dislocations. Kang et al.[830] have used CL to observe the intensity distributions of YL and energy dispersive X-ray spectroscopy (EDS) to show that YL is related to C- or O-related defects. Kucheev et al.[831] have also reported that the origin of YL may be C- or O-related complex defects. Chang et al.[832] have shown the metastability character of YL. Mg doping in HP-SG grown crystals had the effect of reducing the yellow band.[220, 222]

It is reported that semi-insulating material which may not include V_{Ga} also showed YL.[828,833] Correlations between C and YL have been reported.[687,723,724,833,834] Recently, it was proposed that C_N is not a shallow acceptor as normally believed but a deep acceptor with energy level $E_V + 0.9$ eV[711] and the origin of the YL was proposed to be this deep level together with the possibility of complex defects with C_N.[711]

Armitage et al.[835] have studied the origin of the blue band at 2.86 eV in carbon-doped GaN as a function of C-doping level, temperature and excitation energy. The results were consistent with a C_{Ga}-C_N deep donor-deep acceptor recombination mechanism proposed by Seager et al.[723]

Various red luminescence (RL) phenomena were also observed at 1.8-2.0 eV for Mg-doped GaN[815,836,837], O-contaminated GaN[838], ammonothermally grown GaN[839] and MBE grown GaN.[837,838] Fujii et al.[839] have observed PL from an ammonothermally grown bulk GaN. This red band is different from those previously observed and was attributed to oxygen contamination during growth.

Bauer et al.[840] have observed the infrared luminescence of Fe at 1.30 eV and Cr at 1.19 eV, omnipresent trace impurities in GaN epitaxial layers. They suggested that these impurities act as minority-carrier lifetime killers.

The PL of bulk GaN crystals has been evaluated for the HP-SG growth[219, 220], top-seeded solution growth[841], Na or Na-Ca flux growth[842] and ammonothermally grown crystals.[326] The ammonothermally grown crystals showed a highly refined PL spectrum with emissions from neutral donors (D^0X) and their phonon replicas as shown in Fig. 20.56. h- and c-GaN single crystals grown by a Na or K flux have been evaluated by CL.[264]

20.5 APPLICATIONS

20.5.1 Substrates for Devices

(1) BN

Since the h-BN single crystal has an atomically smooth surface, relatively free of dangling bonds and charge traps, and the lattice constant is similar to graphenes, h-BN is an appealing substrate for graphene devices. The high-quality graphene characteristics on h-BN substrates have recently been reported.[843,844] Graphene on h-BN showed mobility and carrier homogeneity of one order of magnitude better, electron-hole charge fluctuation two orders of magnitude better, than that on conventional SiO_2 substrates.

(2) AlN

AlN templates (Sec. 20.4.2(4)) based on sapphire substrates can be used as substrates for visible LEDs[155-157] and ultraviolet LEDs.[845] Since GaN layers grown on AlN templates have fewer dislocation densities because screw dislocations are eliminated at the interface between GaN and AlN, higher light emission efficiency can be obtained. Nisida et al.[845] have fabricated GaN-free UV-LEDs on PVT-grown AlN substrates and found that high power LEDs exceeding 300 mA could be realized due to good thermal conductivity of the AlN substrate. Ren et al.[846] have fabricated UV-LEDs based on $Al_{1-x}Ga_xN$ on AlN substrates and have shown improved efficiency and thermal characteristics. Electronic devices such as AlGaN channel HEMTs have been fabricated on PVT-grown AlN substrates[847,848] and it was found that the sheet resistance of the channel could be reduced due to low threading dislocations.

(3) GaN

For the application of GaN based devices to LEDs, LDs, detectors and electronic devices as explained below, presently sapphire, SiC and Si substrates are mainly used for the epitaxial growth of GaN based materials. There is however a strong demand to develop bulk GaN substrates[849] because of the necessity of low dislocation density substrates and nonpolar and semipolar substrates. For this reason, various growth methods have been extensively studied as described hitherto.

In fact, bulk GaN substrates are now applied for epitaxial growth[850] and for various devices such as LEDs [851-853], LDs [854-860] and electronic devices.[861-866] Among them, blue LDs using HVPE grown substrates are commercialized and are actually used for blue-ray DVD systems. Other devices are still under investigation and further cost reduction of GaN bulk substrates is required for their commercialization.

20.5.2 *Light Emitting Diodes (LEDs)*

(1) Red LEDs

Red LEDs based on nitride and other materials are not yet realized on an industrial scale. However, red LEDs based on GaN have been fabricated by Shottkey diodes using GaN quantum well structures doped with RE metals [867,868] and by pn junctions using GaN quantum dot structures grown by MBE[869] have been reported. High intense emission of red light by external excitation has been reported from GaN quantum dot structures doped with Er on AlN/sapphire template.[870] Red light emission from RE metal doped MOCVD quantum well structures has also been reported.[871] These results shows the future possibility of red LEDs combined with green and blue LEDs all based on nitride materials. However, all of these results have been obtained on sapphire substrates or AlN/sapphire templates. For further improvement of emission efficiency, GaN or AlN substrates of low dislocation density are desired.

(2) Green LEDs

GaN based green LEDs were realized just after the realization of blue LEDs, by increasing the content of In in $In_xGa_{1-x}N$ active layers from 0.05 to 0.5.[872,873] When the In content is increased, the growth temperature has to be reduced in order to avoid the decomposition of InN, so that the crystal quality is impaired.[874] The higher In content also increases the higher compressive strains[875] and degrades the reliability of fabricated LEDs.[876] Recent progress has been made by using nonpolar and semipolar substrates[852] in order to increase the emission efficiency by preventing the quantum-confined Stark effect (QCSE) and the separation of the electron and hole wave functions [208,209] due to polarization. Green LEDs have been fabricated using nonpolar and semipolar GaN substrates such as m-plane, a-plane and other orientation substrates mainly prepared by the HVPE method[877-882] and the emission efficiency was reported to be improved. Various other possibilities such as patterned sapphire substrates [883], Si substrates [884], temperature ramping technology [885], growth of p-GaN[886] and of p-InGaN[887] were also studied. There is a demand for pure green LEDs for application to full-color displays based on red-green-blue (RGB) components and they are also promising for optical plastic fiber (OPF) communications.[888] Nonpolar and semipolar GaN substrates are therefore sought after for the realization of green LEDs. These substrates actually are being developed by the ammonothermal method.[889]

(3) Blue and violet LEDs

Blue LEDs were first developed using ZnSe and SiC substrates as described in Ch. 17 in Vol. 1 and Ch. 23 in this volume. Blue LEDs based on GaN were first reported as metal-insulating-semiconductor (MIS) structure GaN based LEDs[890] but their emission efficiency was too weak for industrialization. The crystal quality of GaN on sapphire substrates was improved by MBE growth [891], HVPE growth[892] and then greatly by MOCVD.[754, 893] p-type GaN was first realized by electron beam irradiation[756] followed by thermal annealing[894] and pn junction GaN based LEDs have been realized using InGaN as the emission layer grown on sapphire substrates.[895-899] These developments are reviewed in Refs. 900 and 901. These blue LEDs showed a significant emission efficiency and this was attributed to the carrier confinement into In islands formed due to the segregation of In in InGaN as mentioned in Sec. 20.4.2(1). Since the In island density is about 10^{20} cm^{-2} which is much higher than the dislocation density 10^9 cm^{-2} and the carrier diffusion length is very short with this In content, it seems apparent that dislocations do not affect the emission efficiency. It is however known that when the carrier injection rate is very high and carriers overflow the band barrier of In islands, overflowed carriers recombine at dislocations, and then the dislocation density affects the emission efficiency. It is therefore necessary to use low dislocation density substrates in order to increase the emission efficiency of high power LEDs and LDs. For this purpose, low dislocation density bulk GaN or AlN substrates are sought after. Egawa et al.[155] have shown that by using AlN/sapphire templates, higher currents can flow and higher luminosity can be obtained.

(4) White LEDs

White LEDs were first realized by combining blue LED chips with yttrium aluminum garnet (YAG) phosphors (blue/YAG LEDs) as seen in Fig. 20.57.[902,903] Since then, various types of white LEDs have been developed, using blue LEDs on sapphire[904] or SiC substrates [905], various phosphors [906-911] and packaging structures [912], and the emission efficiency and chromaticity have greatly improved.[902,913]

Instead of using blue LED chips, when UV-LED chips are used with RGB phosphors (UV white LEDs) (Fig. 6.10 in Vol.1), it gives much more variety of chromaticity and natural colors. This concept of UV white LEDs was first proposed by the 21th century program[914,915] and was widely studied, and some realization has been reported.[916,917]

The application of white LEDs is expanding to very wide fields, liquid crystal

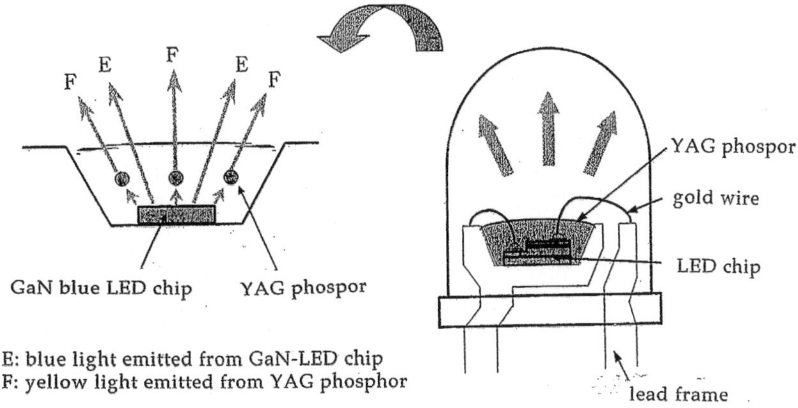

YAG; Yttriunm Aluminium Garnet $(Y_aGd_{1-a})_3(Al_b, Ga_{1-b})_4O_{12}:Ce^{3+}$

Fig. 20.57 The structure of a InGaN/YAG white LED. White light is emitted by mixing the yellow light of YAG with blue light of the InGaN/GaN LED (reprinted from Ref. 903 with permission, copyright 1996 The Japan Society of Applied Physics).

display (LCD) backlighting for personal computors, cellular phones and personal digital assistants (PDAs), car lighting, medical applications. The final goal of the application of white LEDs is for general lightning, which will contribute to a tremendous reduction in global energy consumption. The recent progress of white LEDs is reviewed in the literature[904,918-923] and the advantage of bulk GaN substrates is discussed.[924]

Other efforts to realize white LEDs involve quantum wells doped with rare earth (RE) atoms.[867,868] By this structure white light emission with RGB emission was reported but it was due to Schottky diodes and the applied voltage was too high. By using quantum dots or quantum wells grown by MBE, it was reported that white light emission can be achieved. RGB emission was realized by controlling the quantum dot size[869] or the well size.[925] The disadvantage of this structure is its low chromaticity. In order to overcome this drawback, RE doping of quantum dots has been investigated.[926]

(5) Violet and ultraviolet (UV) LEDs

The applications of UV-LEDs are very wide, one of which is the light source for white LEDs using three color phosphors as explained above. Short wavelength LEDs of the order of 300 nm have applications in the medical field, chemical industry, rapid decomposition of environmentally harmful materials, biosensors and chemical sensors. Using LEDs, high-efficiency and compact light sources can be fabricated which can replace large devices such as excimer gas lasers, second harmonic generating (SHG) lasers and excimer lamps. Therefore, the exploitation of UV-LEDs is now under extensive investigation. The development of UV-LEDs is reviewed in Refs. 927 and 928.

Near UV-LEDs (λ = 382-409 nm) have been extensively developed since they are indispensable for manufacturing white LEDs using three color phosphors. Several research reports have been produced.[929-933] Cao et al.[853] have shown that the internal quantum efficiency of near-UV-LEDs can be increased by using freestanding GaN substrates.

In order to realize short wavelength LEDs below 360 nm using nitride materials, it is necessary to use AlGaN materials instead of InGaN materials. When the Al content is increased in order to shorten the emission wavelength, the emission efficiency is greatly reduced as seen in Fig. 20.58.[934] This is firstly because of the large lattice mismatching and secondly because of the lack of carrier confinement due to In segregation. AlGaN-based LEDs have been developed by MOCVD[935-939] and HVPE.[940] In order to increase the emission efficiency, Hirayama et al.[941-943] have developed UV-LEDs (l = 284-360 nm) based on quaternary AlInGaN layers on sapphire substrates in order to achieve lattice matching and carrier confinement due to In segregation.[613] Even shorter wavelength AlGaN based UV-LEDs (λ = 230-280 nm) have also been developed.[944-947]

Sumiya et al.[948], have shown that AlN/sapphire templates are effective for deep-UV LEDs (λ = 265-277 nm). Kyono et al.[854] have applied HVPE-grown GaN substrates to UV-LEDs (λ = 352 nm). Tyagi et al.[949] have fabricated violet LEDs on semipolar (10-1-1) substrates cut from HVPE-grown bulk GaN. Nishida et al.[950] have fabricated ten-mW UV-LEDs on HVPE-grown substrates.

Further short wavelength (less than 260 nm) LEDs can be achieved by AlN emission layers. These LEDs have recently been developed by Kasu's group [951-953], achieving p-doping and n-doping. These LEDs have been fabricated using a-plane SiC substrates but AlN substrates promise to improve the emission efficiency.

As an application of BN crystals, Mishima et al.[29,730,734,954] have demonstrated UV-LEDs using c-BN with pn junctions fabricated in a high pressure furnace. The n-type dopant was Si and p-type dopant was Be and UV emission at 215 nm was achieved.

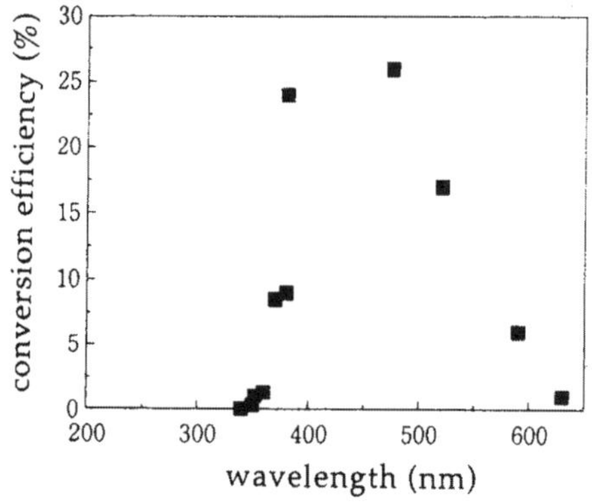

Fig. 20. 58 Emission peak wavelength versus conversion efficiency for nitride LEDs under the standard operation (20 mA DC) (reprinted from Ref. 934 with permission, copyright 2002 The Japan Society of Applied Physics).

Kubota et al.[39,955] have shown that h-BN also emits a deep ultraviolet emission.

Watanabe et al.[956] have fabricated a far-ultraviolet plane-emission compact device using h-BN as the fluorescent screen, equipped with a field-emission array as an excitation source. The h-BN material was prepared with the equipment described in Ref. 799. An output power of 0.2 mW at 225 nm has been stably achieved.

20.5.2 Lasers

(1) Blue and violet LDs

Blue-violet continuous-wave (CW) LDs (406 nm) operating at room temperature were first achieved by Nakamura et al.[957-959] based on ELO structure for reducing the dislocation density. Subsequently much research has been carried out in order to realize high power, various lasing wavelengths and high reliability LDs as reviewed in Refs. 960-963 and they are commercialized as the light source[615] for optical data storage players using HVPE grown GaN substrates.[522] These lasers are also finding further applications in laser printers, laser beam lithography, and LDs in the field of medical and biological use.

In the case of LDs, since many carriers are injected, the carriers overflow the In islands in the InGaN layers and the dislocations act as recombination centers. In order to achieve long lifetime, it is necessary to reduce the dislocation densities, and much more so for shorter wavelength LDs.

For these reasons, low dislocation density GaN substrates are eagerly desired and the development of bulk GaN materials which may have lower dislocation densities than ELO materials is desired. In fact, the fabrication of LDs on bulk GaN substrates has been studied.[860,964,965]

(2) Ultraviolet (UV) LDs

Shorter wavelength LDs less than 390 nm have various new applications such as higher storage-density next-generation DVDs, higher photographic resolution and an excitation light source for phosphors.

The first UV-LDs at 376 nm were reported by Akasaki et al.[966] In order to realize short wave length oscillation, $Al_xIn_yGa_{1-x-y}N$ layers are used as the active layers.[492,967-971] Because of the carrier overflow, dislocations have a more marked effect than with blue LDs. These violet LDs (366-369 nm) were therefore fabricated using ELO GaN substrates. The dislocation density could be reduced by ELO growth on freestanding ELO layers separated from sapphire substrates. UV-LDs based on GaN/AlGaN multiquantum wells (MQWs) at 350-361 nm have been developed using hetero-epiaxial overgrowth technology.[972,973]

In order to achieve shorter wavelength in the deep UV (DUV) region (205-210 nm), Kawanishi et al.[974-977] have developed the alternate source-feeding epitaxy (ASFE) method and have achieved the shortest wavelength LD at 241 nm using SiC substrates.

(3) Green LDs

Green LDs are desired for compact projectors where three color lasers are necessary.

Since green LDs were not realized because of low lasing intensity due to confined quantum Stark shift and separation of electron and hole wave functions, SHG devices have been used for green LDs.

Recently, green lasing has been reported even for c-plane sapphire substrates[978,979] and the device performance has been improved using semipolar GaN substrates prepared from HVPE grown GaN bulk crystals.[855-858]

20.5.3 Detectors

GaN and AlGaN based photodetectors are attracting much attention as solar blind UV detectors since they have various applications as biological and chemical sensors, flame sensors for fire alarms or missile threat warning, optical communications and UV emitter calibrations. Since wide bandgap materials such as GaN and AlGaN are chemically and thermally stable, they are suitable for harsh environments. Detector structures including photoconductors, Schottky photodiodes, metal-semiconductor-metal (MSM) photodiodes, pn junction photodiodes, pin photodiodes, Avalanche photodiodes, field effect and bipolar phototransistors have been investigated as summarized in Refs. 980-983.

The responsivity of various detector structures has been compared.[983] In photovoltaic detectors, the visible rejection was limited to about 10^4 by the high density of defects and dislocations. It is seen that Schottky diodes fabricated on ELO GaN showed high responsivity.[983] It is known that the detectivity as high as 10^{11} $mW^{-1}Hz^{1/2}$, comparable with other commercial detectors, is markedly reduced when the Al content is increased. In order to achieve photodetectors with shorter wavelengths, there is a need to grow AlGaN material on high quality AlN/sapphire substrates, so that the development of bulk AlN materials is also desired in this field.

Recently, InAlN based photodetectors have been developed[984-986] because of good lattice matching to GaN with the composition x = 0.17-0.20. AlN templates as described in Sec. 20.3.2(4) were found to be effective for preparing UV shottkey photodiodes.[987]

AlGaN/GaN Schottky diodes with Pd Shottky electrodes have been fabricated as hydrogen sensors.[988,989] These sensors are much in demand because hydrogen energy is being developed as a renewable and clean energy source.

20.5.4 Electronic devices

Nitride semiconductors such as AlN, GaN and InN have great potential for high-frequency and high-power electronic devices because of their wide bandgap, large breakdown field one order of magnitude larger, and high saturated drift velocity, two times higher than Si and other III-V materials (Sec. 20.4.2). Especially AlGaN/GaN heterostructure FETs, first demonstrated by Khan [990], have a large two-dimensional electron gas density of 1×10^{13} cm^{-2} induced by spontaneous polarization and piezoelectric polarization at the heterointerface formed by a large band offset of about 1 eV. Because of the large dielectric breakdown electric field, it is possible to apply a high voltage of up to several hundred V between the source and drain electrodes and because of the high carrier concentration, high power devices can be realized. Since the mobility and the saturation drift velocity are also large because of the suppression of impurity scattering due to AlGaN/GaN heterostructures, high frequency devices are also possible. Due to the strong chemical bonds in its crystal structure, GaN based devices are receiving attention as hard electronic materials which can be applied in high temperature and high radiation environments. As seen in Table 6.3 in Vol.1, these priorities are seen compared with other materials as important factors in the various figures of merit.

High frequency AlGaN/GaN heterostructure HEMTs on sapphire substrates for applications including cell phones, satellites and TV broadcasting have been extensively studied as reviewed by Mishra[991] and high-frequency devices even up to f_T=121 GHz are fabricated.[992] Various types of electronic devices such as HEMTs, hetero bipolar transistors (HBTs), bipolar junction transistors (BJTs), Schottky and pin diodes and metal-oxide-semiconductor field effect transistors (MOSFETs) have been reported.[993, 994]

In order to realize high power devices based on nitrides, it is necessary to use high thermal conductivity substrates such as Si, SiC, GaN and AlN. Electronic devices using SiC substrates [995-998], Si[999-1003] have already been explored extensively and AlGaN/GaN HEMTs on SiC substrates are already commercialized as high frequency power devices for wireless base stations especially for mobile worldwide interpretability for microwave access (WiMAX). AlGaN/GaN HEMTs on Si substrates are now being explored aiming at commercialization as power devices for home electrics, personal computer power supplies and power devices for automobiles on a world wide scale. AlGaN/GaN HEMTs on 100 mm diameter Si substrates have been already achieved[999-1003] and high power devices have been fabricated (Fig. 20.59).[1004] Various HEMT device structures have been developed aimed at commercialization.[1005] Further development for 150 mm[1006] and 200 mm[1007] GaN/Si substrates is starting world wide.

The lack of large diameter low cost GaN and AlN substrates is hampering the development of electronic devices on these substrates. Some device fabrications on bulk substrates, such as HEMTs on HP-SG GaN [861], Fe-doped HVPE GaN [862], Schottky bar-

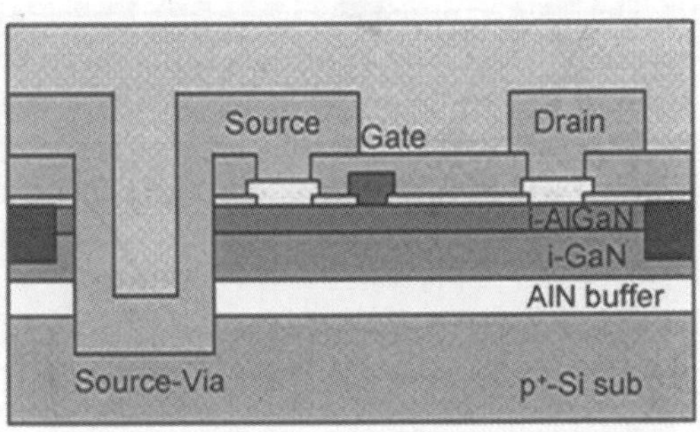

Fig. 20.59 A cross sectional view of the AlGaN/GaN HEMT with sourc-via grounding (SVG) structure (reprinted from Ref. 1004 with permission, copyright 2004 IEEE).

rier diodes (SBDs) [863], trench gate FETs[864-866] and HEMTs on PVT grown AlN[847] are reported. The production of GaN and AlN substrates of large diameter up to at least 200 mm and of low cost comparable with GaAs and Si substrates is much needed for electronic devices because of the good thermal conductivity and the possibility of vertical structure devices for high power applications.

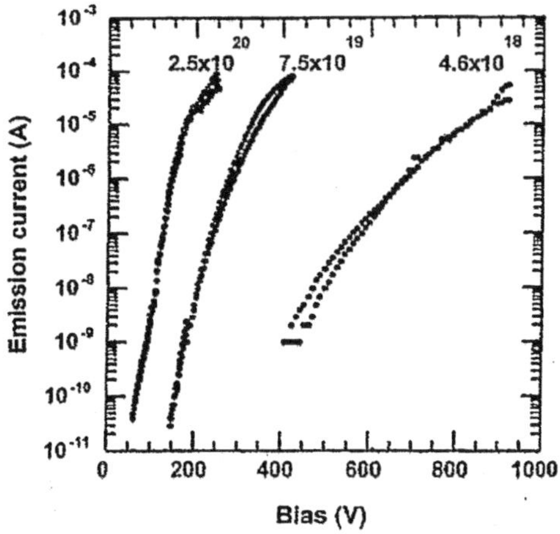

Fig. 20.60 Field emission current as a function of the applied bias voltage of AlN with different Si dopant densities (N_{Si}, cm^{-3})(reprinted from Ref. 735 with permission, copyright 2000 Elsevier).

20.5.5 Field emitters

Benjamin et al.[1008] first found that the electron affinity of $Al_xGa_{1-x}N$ ($x > 0.75$) was negative. After this finding, $Al_xGa_{1-x}N$ has received attention for field emitters. The threshold field of AlN is known to be the lowest among various field emitting materials such as Si and diamond. The field emitting characteristics have been studied[735,1009] and found to be largely dependent on the Si impurity concentration as seen in Fig. 20.60 [735], and at the Si concentration of 1×10^{21} cm^{-3}, the highest current density of 60 mA/cm^2 was reported. This value was two times higher than that of the reported highest value for diamond. A display using AlN field emitters to excite phosphors deposited on the cathode has been tested and it was found that 1200 cd/m^2 could be obtained.[1009]

REFERENCES

1. S. Strite and H. Morkoç, *J. Vac. Sci. Technol.* **B10** (1992) 1237.
2. H. Morkoc, S. Strite, G.B. Gao, M.E. Lin, B. Sverdlov and M. Bruns, *J. Appl. Phys.* **76** (1994) 1363.
3. O. Ambacher, *J. Phys.* **D31** (1998) 2653.
4. B. Monemar, *J. Mater. Sci.* **10** (1999) 227.
5. S.C. Jain, M. Willander, J. Narayan and R. Van Overstraten, *J. Appl. Phys.* **87** (2000) 965.
6. *III-Group Nitride Semiconductors (in japanse)*, ed. I. Akasaki (Baifukan, 1999).
7. H. Morkoc, *Handbook of Nitride Semiconductors and Devices (Vol. 1-3)* (Wiley-Vch, 2008, 2009).
8. J.A. Van Vechten, *Phys. Rev. B* 7 (1973) 1479.
9. *Landolt-Börnstein Vol.17* (Springer, New York, 1982) p. 160.

10. *Landolt-Börnstein Vol.17* (Springer, New York, 1982) p. 178.
11. *Landolt-Börnstein Vol.2A2* (Springer, Berlin, 2002) p. 118.
12. *Landolt-Börnstein Vol.17* (Springer, New York, 1982) p. 278.
13. S.J. Pearton, J.C. Zolper, R.J. Shul and F. Ren, *J. Appl. Phys.* **86** (1999) 1.
14. *Properties of Advanced Semiconductor Materials*, ed. M.E. Levinshtein, S.L. Ramyantsev and M.S. Shur (Wiley, New York, 2001) p.31.
15. R. Mohammad and S. Katirciolu, *J. Alloys Compound.* **478** (2000) 531.
16. T. Matsuoka, H. Okamoto, M. Nakao, H. Harima and E. Kurimoto, *Appl. Phys. Lett.* **81** (2002) 1246.
17. J. Karpinski, J. Jun and S. Porwoski, *J. Crystal Growth* **66** (1984), 1.
18. S. Krukowski, *Diamond Relat. Mater.* **6** (1997) 1515.
19. *Properties of Group III Nitrides*, ed. J.H. Edgar (EMIS, London, 1994)
20. W. Utsumi, H. Saitoh, H. Kaneko, T. Watanuki, K. Aoki and O. Shimomura, *Nature Mater.* **2** (2003) 735.
21. W. Utsumi, H. Saitoh, H. Kaneko, K. Kiriyama and K. Aoki, *J. Jpn. Assoc. Crystal Growth (in japanese)* **46** (2004) 297.
22. W. Utsumi, H. Saitoh, T. Taniguchi and K. Aoki, *Ouyou Butsuri (Appl. Phys.)(in japanese)* **74** (2005) 593.
23. H. Saitoh, W. Utsumi, H. Kaneko and K. Aoki, *J. Crystal Growth* **300** (2007) 26.
24. R.H. Wentorf, Jr., *J. Chem. Phys.* **26** (1957) 956.
25. R.H. Wentorf, Jr., *J. Chem. Phys.* **36** (1962) 1990.
26. R.C. Devries, J.F. Fleisher, *Mater. Res. Bull.* **12** (1969) 433.
27. R.C. DeVries, J.F. Fleiscer, *J. Crystal Growth* **13/14** (1972) 88.
28. O. Mishima, S. Yamaoka and O. Fukunaga, *J. Appl. Phys.* **61** (1987) 2822.
29. O. Mishima, *J. Jpn. Assoc. Crystal Growth (in japanese)* 17(1990) 183.
30. M. Kagamida, H. Kanda, M. Akaishi, A. Nukai, T. Ohsawa, S. Yamaoka, *J. Crystal Growth* **94** (1989) 261.
31. T. Taniguchi, *Jpn. J. Assoc. Crystal Growth (in japanese)* **25** (1998) 141.
32. T. Taniguchi and S. Yamaoka, *J. Crystal Growth* **222** (2001) 549.
33. T. Taniguchi and K. Watanabe, *J. Crystal Growth* **303** (2007) 525.
34. T. Taniguchi, K. Watanabe, A. Nakayama, F. Ob and I. Tanaka, *Materia (in japanese)* **48** (2009) 311.
35. B. Lux, W. Kalss, R. Haubner and T. Taniguchi, *Diamond Relat. Mater.* **8** (1999) 415.
36. H. Saito, M. Ushio and S. Nagano, *Yogyo Kyokai-shi (in japanese)* **74** (1970) 7.
37. C. Kowanda and T. Taniguchi, *Jpn. J. Appl. Phys.* **48** (2003) 1073.
38. T. Taniguchi, *New. Diam. Front. Carbon Technol.* **14** (2004) 289.
39. Y. Kubota, K. Watanabe and T. Taniguchi, *Jpn. J. Appl. Phys.* **46** (2007) 311.
40. Y. Kubota and T. Taniguchi, *Jpn. J. Appl. Phys.* **47** (2008) 8375.
41. Y. Kubota, K. Watanabe, O. Tsuda and T. Taniguchi, *Chem. Mater.* **20** (2008) 1661.
42. Y. Kubota and T. Taniguchi, *Jpn. J. Appl. Phys.* **48** (2009) 071004.
43. T. Taniguchi, *Jpn. J. Assoc. Crystal Growth (in japanese)* **38** (2011) 19.
44. S. Dong, M. Yu, X. Hao, D. Cui, Q. Wang, K. Li and M. Jiang, *J. Crystal Growth* **254** (2003) 229.
45. B.N. Feigelson, R.M. Frazeir and M. Twigg, *J. Crystal Growth* **305** (2007) 399.
46. T. Ishii and T. Sato, *J. Crystal Growth* **61** (1983) 689.
47. Y. Yano, M. Okamoto, Y.K. Yap, M. Yoshimura, Y. Mori and T. Sasaki, *Diamond Relat. Mater.* **9** (2000) 512.
48. R. Dwilinski, R. Doradzinski, J. Garczynski, L. Sierzputowski, M. Palczewska, A. Wysmolek and M. Kaminska, *MRS Internet J. Nitride Semicon. Res.* **3** (1999) Article 25.
49. V.L. Vinogradov, A.V. Kostanoviskii and A.V. Kirillin, *High Temp. High Press.* **23** (1991), 685.
50. V.L. Vinogradov, A.V. Kostanoviskii and A.V. Kirillin, *Tepl. Vyso. Temper.* **30** (1992) 731.
51. A. Rabenau, in *Compound Semiconductors, Vol.1. Preparation of III-V Compounds*, eds. by R.K. Willardson and H.L. Goering (Reinhold Publ, Corp., New York, 1962) p.174.
52. G.A. Slack and T.F. McNelly, *J. Crystal Growth* **34** (1976), 263.
53. O.V. Avdeev, T.Yu. Chemekova, H. Helva, M.G. Ramm, Yu.N. Makarov, E.N. Mokhov, S.S. Nagalyuk, A.S. Segal and A.I. Zhmakin, in *Materials, Preparation and Properties*, ed. R. Fornari (Elsevier, 2011) p. 282.
54. V.F. Mymrin, K.A. Bulashevich, N.I. Podolskaya and S.Yu. Karpov, *J. Crystal Growth* **281** (2005) 115.
55. F. Fichter and G. Oesterheld, *Zeit. f. Electrochem.* **21** (1915) 50.
56. J.A. Kohn, P.G. Cotter and R.A. Potter, *Amer. Mineral.* **41** (1956) 355.
57. G. Long and L.M. Foster, *J. Am. Ceram. Soc.* **42** (1959) 53.
58. K.M. Taylor and C. Leine, *J. Electrochem. Soc.* **107** (1960) 308.
59. T. Matsumura and Y. Tanabe, *J. Phys. Soc. Japan* **15** (1960) 203.
60. W. Kleber and H.D. Witzke, *Z. Krist.* **116** (1961) 126.
61. H.D. Witzke, *phys. stat. sol.* **2** (1962) 1109.

62. J. Pastrnak and L. Roscovkova, *phys. stat. sol.* 7 (1964), 331.
63. K. Kawabe, R.H. Tregold and Y. Inuishi, *Electr. Engineer. Jpn. (in japanese)* **87** (1967) 62.
64. G.A. Cox, d.O. Cummins, K. Kawabe and R.H. Tregold, *J. Phys. Chem. Solids* **28** (1967) 543.
65. W. Class, NASA-CR-1171(1968), *Chem. Abstr.* 69 (1968) 99062f.
66. G.V. Berezhkova, I.N. Tsvetkova, N.D. Zakarov, V.N. Rozhanskii and V.I. Koryukin, *Sov. Phys. Cryst.* **16** (1972) 848.
67. C.O. Dugger, *Mater. Res. Bull.* **9** (1974) 331.
68. R. Dwilinski, R. Doradzinski, J. Garczynski, L. Sierputowski, J.M. Baranowski and M. Kaminska, *Diamond. Relat. Mater.* 7 (1998) 1348.
69. J.W. Kolis, S.W. Wilcinski and R.A. Laudise, *Mat. Res. Soc. Sym. 495* (1998) 367.
70. D. Peters, *J. Crystal Growth* **104** (1990) 411.
71. M. Bockowski, M. Wroblewski, B. Lucznik and I. Grzegory, *Mater. Sci. Semicon. Process.* **4** (2001) 543.
72. K. Kamei, S. Inoue and Y. Ito, *J. Jpn. Assoc. Crystal Growth (in japanese)* 32 (2005) p. 20.
73. K. Kamei, Y. Shirai, T. Tanaka, N. Okada, A. Yauchi and H. Amano, *phys. stat. sol.* **4** (2007) 2211.
74. T. Tanaka, N. Yashiro, Y. Shirai, K. Kamei and A. Yauchi, *phys. stat. sol. (c)* **4** (2007) 2227.
75. Y. Kanagawa and K. Kakimoto, *J. Crystal Growth* **312** (2010) 2569.
76. A. Belousov, S. Katrych, J. Jun, J. Zhang, D. Günther, R. Sobolewski and J. Karpinski, *J. Crystal Growth* **311** (2009) 3971.
77. A. Belousov, S. Katrych, K. Hametner, D. Günther, J. Karpinsk and B. Batlogg, *J. Crystal Growth* **312** (2010) 2585.
78. R. Dalmau and Z. Sitar, *Encycl. Mater.; Sci. Technol.* (2005) 1.
79. T. Ishii, T. Sato and M. Iwata, *Mineralogical J. (in japnese)* **8** (1975) 1.
80. G.A. Slack and T.F. McNelly, *J. Crystal Growth* **42** (1977), 560.
81. C.M. Balkas, Z. Sitar, T. Zheleva, L. Bergman, R. Nemanich and R.F. Davis, *J. Crystal Growth* **179** (1997) 363.
82. C.M. Balkas, Z. Sitar, T. Zheleva, L. Bergman, I.K. Schmagin, J.F. Muth, R. Kolnas, R. Nemanich and R.F. Davis, *Mat. Res. Soc. Symp. Proc.* **449** (1997) 41.
83. M. Tanaka, S. Nakahata, K. Sogabe, H. Nakata, M. Tobioka, *Jpn. J. Appl. Phys.* **36** (1997) L1062.
84. P.M Dryburgh, *J. Crystal Growth* **125** (1992) 65.
85. S. Yu. Karpov, Yu.N. Makarov, M.S. Ramm and R.°A. Talalaev, *MRS Internet J. Nitride Semicond. Res.* **2** (1997) Article 45.
86. A.S. Segal, S. Yu. Karpov, Yu.N. Makarov, E.N. Mokhov, A.D. Roenkov, M.G. Ramm and Yu.A. Vodakov, *J. Crystal Growth* **211** (2000) 68.
87. L. Liu and J.H. Edgar, *J. Crystal Growth* **220** (2000) 243.
88. B. Wu, R. Ma, H. Zhang, M. Dudley, R. Schless and Z. Sitar, *J. Crystal Growth* **253** (2003) 326.
89. B. Wu, R. Ma, H. Zhang and V. Prasad, *J. Crystal Growth* **266** (2004) 303.
90. B. Wu and H. Zhang, *Int. J. Heat Mass Transfer* **47** (2004) 2839.
91. X. Wang, D. Cai and H. Zhang, *Int. J. Heat Mass Trans.* **50** (2007) 1221.
92. V. Noveski, R. Schlesser, B. Raghothamachar, M. Dudley, S. Mahajan, S. Beaudoin and Z. Sitar, *J. Crystal Growth* **279** (2005) 13.
93. J.H. Edgar, L.H. Robins, S.E. Coatney, L. Liu, J. Chaudhuri, K. Ignatiev and Z. Rek, *Mater. Sci. Forum* **338-342** (2000) 1599.
94. J.H. Edgar, L. Liu, B. Liu, D. Zhuang, J. Chaudhuri, M. Kuball and S. Rajasingam, *J. Crystal Growth* **246** (2002) 187.
95. P. Lu, J.H. Edgar, R.G. Lee and J. Chaudhuri, *J. Crystal Growth* **300** (2007) 336.
96. J.H. Edgar, L. Du, L. Nyakiti, J. Chaudhuri, *J. Crystal Growth* **310** (2008) 4002.
97. J.C. Rojo, G.A. Slack, K. Morgan, B. Raghothamachar, M. Dudley and L.J. Schowalter, *J. Crystal Growth* **231** (2001) 317.
98. J.C. Rojo, L.J. Schowlter and K. Morgan, *Mat. Res. Soc. Symp. Proc.* **680E** (2001) E.2.1.1.
99. J.C. Rojo, L.J. Schowalter, G. Slack, K. Morgan, J. Barani, S. Schujman, S. Biswas, B. Raghothamachar, M. Dudley, M. Shur, R. Gaska, N.M. Johnson and M. Kneissl, *Mat. Res. Soc. Symp. Proc.* **722** (2002) 5.
100. L.J. Showalter, G.A. Slack, J.B. Whitlock, K. Morgan, S.B. Shujman, B. Raghothamachar, M. Dudley and K.R. Evans, *phys. stat. sol. (c)* **0** (2003) 1997.
101. S.B. Schujman, L.J. Schowalter, R.T. Bondokov, K.E. Morgan, W. Liu, J.A. Smart and T. Bettles, *J. Crystal Growth* **310** (2008) 887.
102. R.T. Bondokov, S.G. Mueller, K.E. Morgan, G.A. Slack, S.Schujman, M.C. Wood, J.A. Smart and L. Schowalter, *J. Crystal Growth* **310** (2008) 4020.
103. R. Schlesser, R. Dalmau and Z. Sitar, *J. Crystal Growth* **241** (2002) 416.
104. R. Dalmau, R. Schlesser, B.J. Rodriguez, R.J. Nemanich and Z. Sitar, *J. Crystal Growth* **281** (2005) 68.

105. R. Schlesser, R. Dalmau, D. Zhuang, R. Collazo and Z. Sitar, *J. Crystal Growth* **281** (2005) 75.
106. Z.G. Herro, D. Zhuang, R. Schlesser, R. Collazo and Z. Sitar, *J. Crystal Growth* **286** (2006) 205.
107. Z.G. Herro, D. Zhuang, R. Schlesser, R. Collazo and Z. Sitar, *Mat. Res. Soc. Symp. Proc.* **892** (2006) 499.
108. D. Zhuang, Z.G. Herro, R. Schlesser and Z. Sitar, *J. Crystal Growth* **287** (2006) 372.
109. D. Cai, L.L. Zheng, H. Zhang, d. Zhang, Z.G. Herro, R. Schlesser and Z. Sitar, *J. Crystal Growth* **306** (2007) 39.
110. P. Lu, J.H. Edgar, C. Cao, K. Hohn, R. Dalmau, R. Schlesser and Z. Sitar, *J. Crystal Growth* **310** (2008) 2464.
111. Z.G. Herro, D. Zhuang, R. Shclesser and Z. Siar, *J. Crystal Growth* 312 (2010) 2519.
112. P. Lu, R. Collazo, R.F. Dalmau, G. Durkaya, N. Diets, B. Raghotahmachar, M. Dudley and Z. Sitar, *J. Crystal Growth* **312** (2009) 58.
113. E.N. Mokhov, S.A. Smirnov, Y.N. Makarov and H. Helava, *Proc. 5th Int. Conf. Nitride Semicon.* (Nara, Japan, 2003) p. 58.
114. E.N Mokhov, O.V. Avdeev, I.S. Barash, T.Yu. Chemekova, A.D. Roenkov, A.S. Segal, A.A. Wolfson, Yu.N. Makarov, M.G. Ramm and H. Helave, *J. Crystal Growth* **281** (2005) 93.
115. H. Helava, S.J. Davis, G.D. Huminic, M.G. Ramm, O.V. Avdeev, I.S. Barash, T. Yu. Chemekova, E.N. Mokhov, S.S. nagalyk, A.D. Roenkov, A.S. Segal, Yu.A. Vodakov and Yu.N. Makarov, *phys. stat. sol. (c)* **4** (2007) 2281.
116. Yu.N. Makarov, O.V. Avdeev, I.S. Barash, D.S. Bazarevskiy, *J. Crystal Growth* **310** (2008) 881.
117. T.Yu Chemekova, O.V. Avdeev, I.S. Barash, E.N. Mokhov, S.S. Nagalyuk, A.D. Roenkov, A.S. Segal, Yu.N. Makarov, M.G. Ramm, S. Davis, G. Huminic and H. Helava, *phys. stat. sol. (c)* **5** (2008) 1612.
118. M. Bickermann, B.M. Epelbaum and A. Winnacker, *Proc. 5th Int. Conf. Nitride Semicond.* (Nara, Japan, 2003) p. 59.
119. M. Schieber, I. Beinglass, G. Dishon and A. Holzer, in *Crystal Growth and Materials*, ed. E. Kaldis and H. Scheel (North Holand, Amsterdam, 1977) p. 280.
120. B.M. Epelbaum, C. Seitz, A. Magerl, M. Bickermann and A. Winnacker, *J. Crystal Growth* **265** (2004) 577.
121. B.M. Epelbaum, M. Bickermann and A. Winnacker, *J. Crystal Growth* **275** (2005) e479.
122. P. Heimann, C. Pfeiffer, M. Bickermann, B.M. Epelbaum, S. Nagata and A. Winnecker, *phys. stat. sol. (c)* **4** (2007) 2223.
123. B.M. Epelbaum, S. Nagata, M. Bickermann, P. Heimann and A. Winnacker, *phys. stat. sol. (b)*, 244(2007) 1780.
124. B.M. Epelbaum, M. Bickermann, S. Nagata, P. Heimann, O. Filip and A. Winnacker, *J. Crystal Growth* **305** (2007) 317.
125. O. Filip, B.M. Epelbaum, M. Bickermann, P. Heimann and A. Winncaker, *J. Crystal Growth* **318** (2011) 427.
126. R. Yakimova, A. Kakanokova-Georgieva, G.R. Yazdi, G.K. Georguiev and M. Syvajarvi, *J. Crystal Growth* **281** (2005) 81.
127. G.R. Yazdi, R. Vasiliauskas, M. Syväjärvi and R. Yakimova, *J. Crystal Growth* **310** (2008) 935.
128. N. Mizuhara, M. Miyanaga, S. Fujiwara, H. Nakahaa and T. Kawase, *J. Jpn. Assoc. Crystal Growth (in japanese)* **33** (2006) 190.
129. N. Mizuhara, M. Miyanaga, S. Fujiwara, H. Nakahaa and T. Kawase, *phys. stat. sol. (c)* **4** (2007) 2244.
130. M. Miyanaga, N. Mizuhara, S. Fujiwara, M. Shimizu, H. Nakahata and T. Kawase, *J. Crystal Growth* **300** (2007) 45.
131. Y. Melnik, V. Soukhoveev, V. Ivantsov, V. Sizov, A. Pechnikov, K. Tsvekov, O. Kovalenko, V. Dmitriev, A. Nikolaev, N. Kuznetsov, E. Silveira and J. Freitas, Jr., *phys. stat. sol. (a)* **200** (2003) 22.
132. S. Wang, B. Raghothamachar, M. Dudley and A.G. Timmerman, *Mat. Res. Soc. Symp. Proc.* **892** (2006) 775.
133. M. Kazan, R. Nadar, E. Moussaed and P. Masri, *J. Crystal Growth* **290** (2006) 44.
134. C. Hartmann, J. Wollweber, C. Seits, M. Albrecht and R. Fornari, *J. Crystal Growth* **310** (2008) 930.
135. E.R. Letts, J.S. Speck and S. Nakamura, *J. Crystal Growth* **311** (2009) 1060.
136. I. Nagai, T. Kato, T. Miura, H. Kamata, K. Naoe, K. Sanada and H. Okumura, *J. Crystal Growth* **312** (2010) 2699.
137. O. Filip, M. Bickermann, B.M. Epelbaum, P. Heimann and A. Winnacker, *J. Crystal Growth* **312** (2010) 2522.
138. L.J. Schowalter, Y. Shusterman, R. Wang, L. Bhat, G. Arunmozhi, G.A. Slack, *MRS Internet J. Nitride Semicond. Res.* **4S1** (1999) G3.76.
139. L.J. Schowalter, Y. Shusterman, R. Wang, L. Bhat, G. Arunmozhi, G.A. Slack, *Appl. Phys. Lett.* **76** (2000) 985.

140. L.J. Schowalter, J.C. Rojo, G.A. Slack, Y. Shusterman, R. Wang, I. Bhat and G. Arumozhi, *J. Crystal Growth* **211** (2000) 78.
141. Q. Fareed, R. Jain, R. Gaska and M.S. Shur, *Proc. 5th Int. Conf. Nitride Semicon.* (Nara, Japan, 2003) p. 89.
142. J.R. Grandusky, J.A. Smart, M.C. Mendrick, L.J. Schowalter, K.X. Chen and E.F. Shubert, *J. Crystal Growth* **311** (2009) 2864.
143. R. Schlesser and Z. Sitar, *J. Crystal Growth* **234** (2002) 349.
144. Z. Sitar, R. Dalmau, B. Raghothamachar, M. Dudley and R. Schlesser, *Proc. 5th Int. conf. Nitride Semicon.* (Nara, Japan, 2003) p. 57.
145. R.F. Rutz, *Appl. Phys. Lett.* **28** (1976) 379.
146. P.B. Perry and R.F. Rutz, *Appl. Phys. Lett.* **33** (1978) 319.
147. H. Fukuyama, S. Kusunoki, J.A. Hakomor and K. Hiraga, *J. Appl. Phys.* **100** (2006) 024905.
148. K. Hironaka, T. Nagashima, S. Ikeda, M. Azuma, K. Takada and H. Fukuyama, *J. Crystal Growth* **312** (2010) 2527.
149. T. Shibata, Y. Kia, H. Miyake, K. Hiramatsu, Y. Hori, K. Asai, T. Nagai, S. Sumiya, M. Tanaka and O. Oda, *Inst. Phys. Conf. Ser.* **170** (2002) 795.
150. M. Sakai, H. Ishikawa, T. Egawa, T. Jimbo, M. Umeno, T. Shibat, K. Asai, S. Suiyam Y. Kuraoka, M. Tanaka and O. Oda, *Inst. Phys. Conf. Ser.* **170** (2002) 783.
151. M. Tanaka, T. Shibata, M. Sakai, O. Oda, H. Miyake, K. Hiramatsu, H. Ishikawa, T. Egawa and T. Jimbo, *Tech. Rep. IECIE* **102** (2002) 9.
152. M. Sakai, H. Ishikawa, T. Egawa, T. Jimbo, M. Umeno, T. Shibata, K. Asai, S. Sumiya, Y. Kuraoka, M. Tanaka and O. Oda, *J. Crystal Growth* **244** (2002) 6.
153. T. Shibata, K. Asai, S. Sumiya, M. Mouri, O. Oda, H. Katsukawa, H. Miyake and K. Hiramatsu, *phys. stat. sol. (a)* **195** (2003) 1.
154. M. Sakai, K. Asano, S. Arulkumaran, H. Ishikawa, T. Egawa, T. Jimbo, T. Shibata, M. Tanaka and O. Oda, *IEICE Trans. Electon.* **E86-C** (2003) 2071.
155. T. Egawa, H. Ohmura, H. Ishikawa and T. Jimbo, *Appl. Phys. Lett.* **81** (2002) 292.
156. B. Zhang, T. Egawa, Y. Liu, H. Ishikawa and T. Egawa, phys. stat. sol. (c) **0** (2003) 2244.
157. T. Nishida, N. Kobayashi and T. Ban, *Appl. Phys. Lett.* **82** (2003) 1.
158. M. Sakai, K. Asano, S. Arulkumaran, H. Ishikawa, T. Egawa, T. Jimbo, T. Shibata, M. Tanaka and O. Oda, *Tech. Rep. IEICE* **ED2002-92** (2002) 77.
159. M. Miyoshi, H. Ishikawa, T. Egawa, K. Asai, M. Mouri, T. Shibata, M. Tanaka and O. Oda, *Appl. Phys. Lett.* **85** (2004) 1710.
160. M. Miyoshi, M. Sakai, H. Ishikawa, T. Egawa, T. Jimbo, M. Tanaka and O. Oda, *J. Crystal Growth* **272** (2004) 293.
161. M. Miyoshi, M. Sakai, S. Arulkumaran, H. Ishikawa, T. Egawa, M. Tanaka and O. Oda, *Jpn. J. Appl. Phys.* **43** (2005) 7939.
162. M. Miyoshi, A. Imanishi, T. Egawa, H. Ishikawa, K. Asai, T. Shibata, M. Tanaka and O. Oda, *Jpn. J. Appl. Phys.* **44** (2005) 6490.
163. M. Miyoshi, T. Egawa, H. Ishikawa, K. Asai, T. Shibata, M. Tanaka and O. Oda, *J. Appl. Phys.* **98** (2005) 063713.
164. M. Imura, K. Nakano, T. Kitano, N. Fujimoto, N. Okada, K. Balakrishnan, M. Iwaya, S. Kamiyama, H. Amano, I. Akasaki, K. Shimono, T. Noro, T. Takagi and A. Bandoh, *phys. stat. sol. (a)* **203** (2006) 1626.
165. N. Okasa, N. Kato, S. Sato, T. Sumii, T. Nagai, N. Fujimoto, M. Imura, K. Balakrishnan, M. Iwaya, S. Kamiyama, H. Amano, I. Akasaki, H. Maruyama, T. Takagi, T. Noro and A. Bandoh, *J. Crystal Growth* **298** (2007) 349.
166. W.M. Yim, E.J. Stoko, P.J. Zanzucchi, J.I. Pankove, E. Ettenberg and S.L. Gibert, *J. Appl. Phys.* **44** (1973) 292.
167. M.P. Callaghan, E. Patterson, B.P. Richards and C.A. Wallagce, *J. Crystal Growth* **22** (1979) 85.
168. H. Komiyama and O. Osawa, *Jpn. J. Appl. Phys.* **24** (1985) L795.
169. Y. Melnik, D. Tsvetkov, A. Pechnikov, I. Nikitina, N. Kuznetsov and V. Dmitriev, *phys. stat. sol. (a)* **188** (2001) 463.
170. Y. Kumagai, T. Yamane. T. Miyaji, H. Murakami, Y. Kanagawa and A. Koukitu, *phys. stat. sol. (c)* **0** (2003) 2498.
171. Y. Kumagai, T. Yamane and A. Koukitu, *J. Crystal Growth* **281** (2005) 62.
172. T. Nagashima, M. Harada, H. Yanagi, H. Fukuyama, Y. Kumagai, A. Koukitu and K. Takeda, *J. Crystal Growth* **305** (2007) 355.
173. T. Nagashima, M. Harada, H. Yanagi, Y. Kumagai, A. Koukitu and K. Takeda, *J. Crystal Growth* **300** (2007) 42.
174. M. Albrecht, I.P. Nikitina, A.E. Nikolaev, Yu.A. Melnik, V.A. Dmitriev and H.P. Strunk, *phys. stat. sol. (a)* **176** (1999) 453.

175. A. Nikolaev, I. Nikitina, A. Zubrilov, M. Mynbaeva, Y, Melnik and V. Dmitriev, *Mat. Res. Soc. Symp.* **595** (2000) W6.5.1.
176. O. Kovalenkov, V. Soukhoveev, V. Ivantsov, A. Usikov and V. Dmitriev, *J. Crystal Growth* **281** (2005) 87.
177. H. Mauk, B. Amastatt, d. Turover, E. Bellet-Amalric, B. Duadin, v. Ivantsov, V. Dmitriev and V. Maslennikov, *phys. stat. sol. (c)* **3** (2006) 1448.
178. V. Soukhoveev, O. Kovalenkov, V. Ivantsov, A. Syrkinm A. Usikov, V. Matlennikov and V. Dmitriev, *phys. stat. sol. (c)* **3** (2006) 1653.
179. A. Usikov, V. Souhoveev, L. Shapovalova, A. Syrkin, O. Kovalenkov, A. Volkova, V. Sizov, V. Ivantsov and V. Dmitrev, *phys. stat. sol. (c)* **5** (2008) 1825.
180. V. Soukhoveev, A. Volkova, V. Ivantsov, O. Kobaenkov, A. Syrkin and A. Usikov, *phys. stat. sol. (c)* **6** (2009) 5333.
181. A. Volkova, V. Ivantsov and L. Leung, *J. Crystal Growth* **314** (2011) 113.
182. D.K. Kamber, Y. Wu, B.A. Haskell, S. Newman, S.P. DenBaars, J.S. Speck and S. Nakamura, *J. Crystal Growth* **297** (2006) 321.
183. R.B. Jain, Y. Gao, J. Zhang, R.S.Q. Fareed, R. Gaska, J. Li, A. Arjunan, E. Kukstis, J. Yang and M.A. Khan, *phys. stat. sol. (c)* **3** (2006) 1491.
184. Y. Kumagai, T. Nagashima, H. Murakami, K. Takeda and A. Koukitsu, *phys. stat. sol. (c)* **5** (2008) 1512.
185. T. Yamane, H. Murakami, Y. Kanagawa, Y. Kumagai and A. Koukitsu, *phys. stat. sol. (c)* **2** (2005) 2062.
186. Y. Kumagai, J. Tajima, M. Ishizuki, T. Nagashima, H. Murakami, K. Takada and A. Koukitsu, *Appl. Phys. Express* **1** (2008) 045003.
187. J. Tajima, H. Murakami, Y. Kumagai, K. Takada and A. Koukitu, *J. Crystal Growth* **311** (2009) 2837.
188. J. Tajima, Y. Kuboa, M. Ishizuki, T. Nagashima, R. Togashi, H. Mirakami, Y. Kumagai, K. Takeda and A. Koukitsu, *phs. stat. sol.* **6** (2009) 5447.
189. K. Tsujisawa, T. Tanabe, A. Abe, Yu-H. Liu, H. Miyake and K. Hiramatsu, *IEICE Tech. Rep.* **105** (2005) 1.
190. Yu-H. Liu, T. Tanabe, H. Miyake, K. Hiramatsu, T. Shibata, M. Tanaka and Y. Masa, *Jpn. J. Appl. Phys.* **44** (2005) L505.
191. Y.-H. Liu, T. Tanabe, H. Miyake, K. Hiramatsu, T. Shibata and M. Tanaka, *phys. stat. sol. (c)* **3** (2006) 1479.
192. K. Tsujisawa, S. Kishino, Y.H. Liu, H. Miyake, K. Hiramatsu, T. Shibata and M. Tanaka, *phys. stat. sol. (c)* **4** (2007) 2252.
193. Y. Katagiri, S. Kishino, K. Okuura, H. Miyake and K. Hiramatsu, *J. Crystal Growth* **311** (2009) 2831.
194. K. Nakamura, T. Ostvold and H. Oye, *Phys. Chem.* **90** (1986) 141.
195. K. Eriguchi, T. Hiratsuka, H. Murakami, Y. Kumagai and A. Koukitsu, *J. Crystal Growth* **310** (2008) 4016.
196. D.F. Bliss, V.L. Tassev, D. Weyburne and J.S. Bailey, *J. Crystal Growth* **250** (2003) 1.
197. D. Cai, L.L. Zheng, H. Zhang, V.L. Tassev and D.F. Bliss, *J. Crystal Growth* **276** (2005) 182.
198. A. Claudel, E. Blanquet, D. Chaussend, M. Audier, D. Pique and M. Pons, *J. Crystal Growth* **311** (2008) 3371.
199. A. Claudel, E. Blanquet, D. Chaussend, M. Audier, D. Pique and M. Pons, *Mater. Sci. Forum* **600-603** (2009) 1269.
200. D. Zhuang, J.H. Edgar, B. Liu, H.E. Huey, H.X. Jiang, J.Y. Lin, M. Kuball, F. Mogal, J. Chaudhuri and Z. Rek, *J. Crystal Growth* **262** (2004) 168.
201. T. Honda, H. Yamamoto, M. Sawadaishi, S. Taguchi and K. Sasaya, *J. Crystal Growth* **311** (2009) 2844.
202. W.C. Johnson, J.B.Parsons and M.C. Crew, *J. Phys. Chem.* **36** (1932) 2651.
203. R. Juza and H. Hahn, *Z. anorg. u. allgem. Chem.* **239** (1938) 282.
204. R. Juza and H. Hahn, *Z. anorg. u. allgem. Chem.* **244** (1938) 133.
205. R.J. Sime and J.L. Margrave, *J. Phys. Chem.* **60** (1956) 810.
206. H.G. Grimmeiss, R. Groth and J. Maak, *Z. Naturforsch.* **14a** (1959) 264.
207. T. Renner, *Z. anorg. u. allgem. Chem.* **298** (1959) 22.
208. P. Waltereit, O. Barndt, A. Tramert, H.T. Grahn, J. Menninger, M. Ramsteiner, M. Reiche and K.H. Ploog, *Nature* **406** (2000) 865.
209. T. Takeuchi, S. Sota, M. Katsuragawa, M. Komori, H. Takeuchi, H. Amano and I. Akasaki, *Jpn. J. Appl. Phys.* **36** (1997) L382.
210. A. Denis, G. Goglio and G. Demazeau, *Mater. Sci. Engineer. R* **50** (2006) 167.
211. B. Feigelson and T. Paskova, in *Comprehensive Semiconductor Science and Technology*, eds. P. Bhattacharya, R. Fornari and H. Kamimura (Elsevier, 2011) p. 232.
212. R. Fornari, in *Comprehensive Semiconductor Science and Technology*,, eds. P. Bhattacharya, R. Fornari and H. Kamimura (Elsevier, 2011) p. 1.
213. S. Sarayama, *Ricoh Quat. Rev.* **31** (2009) 49.

214. S. Porowski, I. Grzegory and J. Jun, in *High Pressure Chemical Synthesis*, eds. by J. Jurczak and B. Baranowski, (Elsevier, Amsterdam, 1989) p. 21.
215. I. Grzegory, J. Jun, M. Bockowski, S. Kurkowski, M. Wroblewski, B. Lucznik and S. Porowski, *J. Phys. Chem. Solds* **56** (1995) 639.
216. S. Porowski, *J. Crystal Growth* **166** (1996) 583.
217. S. Krukowski, *MRS Internet J., Nitride Semicon. Res.* **V.1** (1996) Artcile 27.
218. I. Grzegory, M. Bockowski, B. Lucznik, S. Kurkowski, M. Wroblewski and S. Porowski, *MRS Internet J., Nitride Semicon. Res.* **V.1** (1996) Artcile 20.
219. S. Porwoski and I. Grzegory, *J. Crystal Growth* **178** (1997) 174.
220. I. Grzegory, M. Bockowski, B. Lucznik, M. Wroblewski, S. Kurkowski, J. Weyer, G. Nowak, T. Suski, M. Leszcynski, E. Litwin-Staszewska and S. Porowski, *Mat. Res. Soc. Symp. Proc.* **482** (1998) 15.
221. M. Bockowski, *Physica B* **265** (1999) 1.
222. S. Krukowski, *Cryst. Res. Technol.* **34** (1999) 785.
223. I. Grzegory, M. Bockowski, B. Lucznik, S. Kurkowski, Z. Romanowski, M. Wroblewski and S. Porowski, *J. Crystal Growth* **246** (2002) 177.
224. I. Grzegory, M. Bockowski, S. Porowski, in *Bulk Crystal Growth of Electonic, Optical and Opto-electronic Materials*, (Wiley, 2005) p. 173.
225. M. Bockowski, I. Grzegory, B. Lucznik, G. Kamler, S. Kurkowski, M. Wroblewski, P. Kwiatkowski, K. Jasik and S. Porowski, *phys. stat. sol. (c)* **4** (2007) 2236.
226. M. Bockowski, I. Grzegory, G. Kamler, B. Lucznik, S. Krukowski, M. Wroblewski, P. Kwiatkowski, K. Jasik and S. Porowski, *J. Crystal Growth* **305** (2007) 414.
227. S. Porowski, J. Jin, S. Krukowski, I. Grzegory, M. Leszczynski, T. Suski, H. Teisseyre, C.T. Foxon and D. Korakakis, *Physica B* **265** (1999) 295.
228. M. Bockowski, I. Grzegory, S. Kurkowski, B. Lucznik, Z. Romanowski, M. Wroblewski, J. Borysiuk, J. Weyher, P. Hageman and S. Porowski, *J. Crystal Growth* **246** (2002) 194.
229. M. Bo'ckowski, I. Grzegory, S. Kurkowski, B. Lucznik, M. Wroblewski, G. Kamler, J. Borysiuk, P. Kwiatkowski, K. Jasik and S. Porowski, *J. Crystal Growth* **270** (2004) 409.
230. M. Bo'ckowski, I. Grzegory, S. Kurkowski, B. Lucznik, M. Wroblewski, G. Kamler, J. Borysiuk, P. Kwiatkowski, K. Jasik and S. Porowski, *J. Crystal Growth* **274** (2005) 55.
231. M. Bockowski, I. Grzegory, J. Borysiuk, G. Kamler, B. Lucznik, M. Wroblewski, P. Kwiatkowski, K. Jasik, S. Kurkowski and S. Porowski, *J. Crystal Growth* **281** (2005) 11.
232. M. Bockowski, P. Strak, P. Kempisty, I. Grzegory, S. Kurkowski, B. Lucznik and S. Porowski, *J. Crystal Growth* **307** (2007) 259.
233. M. Bockowski, P. Strak, I. Grzegory, B. Lucznik and S. Porowski, *J. Crystal Growth* **310** (2008) 3924.
234. M. Bockowski, I. Grzegory, A. Kchahapuridze, S. Gierlotka and S. Porowski, *J. Crystal Growth* **312** (2010) 2574.
235. I. Grzegory, N. Bockowski, P. Strak, S. Krukowski and S. Porowski, *J. Crystal Growth* **312** (2010) 2593.
236. P. Geiser, J. Jin, S.M. Kazkov, P. Wägll, J. karpinski and B. Batlogg, *Appl. Phys. Lett.* **86** (2005) 081908.
237. A. Belousov, S. Kätrych, J.Jun, J. Zhang, D. Gunther, R. Sobolweski and J. Karpinski, *J. Crystal Growth* **311** (2009) 3971.
238. T. Inoue, Y. Seki, O. Oda, S. Kurai, Y. Yamada and T. Taguchi, *Ext. Abst. of the 1999 Int. Conf. Solid State Dev. & Mater.* (Tokyo, 1999) p. 60.
239. T. Inoue, Y. Seki, O. Oda, S. Kurai, Y. Yamada and T. Taguchi, *Jpn. J. Appl. Phys.* **39** (2000) 2394.
240. O. Oda, T. Inoue, Y. Seki, A. Wakahara, A. Yoshida, S. Kurai, Y. Yamada and T. Taguchi, *IEICE Trans. Electron.* **E83-C** (2000) 639.
241. O. Oda, T. Inoue, Y. Seki, K. Kainosho, S. Yaegashi, A. Wakahara, A. Yoshida, S. Kurai, Y. Yamada and T. Taguchi, *phys. stat. sol. (a)* **180** (2000) 51.
242. T. Inoue, Y. Seki, O. Oda, S. Kurai, Y. Yamada and T. Taguchi, *J. Jpn. Assoc. Crystal Growth (in japanese)* **27** (2000) 54.
243. T. Inoue, Y. Seki, O. Oda, S. Kurai, Y. Yamada and T. Taguchi, *Jpn. J. Appl. Phys.* **229** (2001) 35.
244. T. Inoue, Y. Seki, O. Oda, S. Kurai, Y. Yamada and T. Taguchi, *phys. stat. sol. (b)* **223** (2001) 15.
245. T. Inoue and T. Taguchi, *Oyo Buturi (Monthly Publ. Jpn. Soc. Appl. Phys.) (in japanese)* **70** (2001) 567.
246. K. Kainosho, T. Inoue, A. Yokohata, S. Kurai, Y, Yamada and T. Taguchi, *J. Jpn. Assoc. Crystal Growth (in japanese)* **29** (2002) 439.
247. H. Yamane, M. Shimada, S.J. Clarke and F.J. DiSalvo, *Chem. Mater.* **9** (1997) 413.
248. H. Yamane, M. Shimada, T. Sekiguchi and F.J. DiSalvo, *Oyo Buturi (Monthly Publ. Jpn. Soc. Appl. Phys.) (in japanese)* **186** (1998) **8**.
249. H. Yamane and M. Shimada, *J. Jpn. J. Assoc. Cryst. Growth (in japanese)* **25** (1998) 152.
250. H. Yamane, D. Kinno, M. Shimada and F.J. DiSalvo, *J. Ceram. Soc. Jpn.* **107** (1999) 925.

251. H. Yamane, M. Shimada and F.J. DiSalvo, *Mater. Lett.* **42** (2000) 66.
252. H. Yamane, D. Kinno, M. Shimada, T. Sekiguchi and F.J. DiSalvo, *J. Mater. Sci.* **35** (2000) 801.
253. M. Aoki, H. Yamane, M. Shimada, T. Sekiguchi, T. Hanada, T. Yao, S. Sarayama and F.J. DiSalvo, *J. Crystal Growth* **218** (2000) 7.
254. M. Aoki, H. Yamane, M. Shimada, S. Sarayama and F.J. DiSalvo, *Crystal Growth & Des.* **1** (2001) 119.
255. M. Aoki, H. Yamane, M. Shimada, S. Sarayama and F.J. DiSalvo, *J. Ceram. Soc. Jpn.* **109** (2001) 858.
256. H. Yamane, M. Aoki and S. Shimada, *J. Jpn. Assoc. Crystal Growth (in japanese)* **28** (2001) 316.
257. M. Aoki, H. Yamane, M. Shimada, S. Sarayama and F.J. DiSalvo, *J. Crystal Growth* **242** (2002) 70.
258. H. Yamane, M. Aoki and S. Sarayama, *Oyo Buturi (Monthly Publ. Jpn. Soc. Appl. Phys.) (in japanese)* **71** (2002) 548.
259. M. Aoki, H. Yamane, M. Shimada, S. Sarayama and F.J. diSalvo, *Mater. Lett.* **56** (2002) 660.
260. M. Aoki, H. Yamane, M. Shimada, S. Sarayama, H. Iwata and F.J. DiSalvo, *J. Crystal Growth* **266** (2004) 461.
261. T. Yamada, H. Yamane, H. Iwata and S. Sarayama, *J. Crystal Growth* **281** (2005) 242.
262. T. Yamada, H. Yamane, H. Iwata and S. Sarayama, *J. Crystal Growth* **286** (2006) 494.
263. H. Yamane, M. Aoki, t. Yamada, M. Shimada, H. Goto, T. Goto, H. Makino, T. Yao, S. Sarayama, H. Iwata and F.J. DiSalvo, *Jpn. J. Appl. Phys.* **44** (2005) 3157.
264. T. Sekiguchi, H. Yamane, M. Aoki, T. Araki and M. Shimada, *Sci. Technol. Adv. Mater.* **3** (2002) 91.
265. E. Brown, H. Hömmerich, T, yamada, H. Yamane and J.M. Zavada, *J. Alloys Comp.* **488** (2009) 628.
266. Y. Mori, M. Yano, M. Okamoto, T. Sasaki and H. Yamane, *Bull. Solid State Phys. Appl. (in japanese)* **4** (1998) 198.
267. M. Yano, M. Okamoto, Y.K. Yap, M. Yoshimura, Y. Mori and T. Sasaki, *Jpn. J. Appl. Phys.* **38** (1999) L1121.
268. M. Onda, T. Iwahashi, M. Okamoto, Y.K. Yap, M. Yoshimura, M. Mori and T. Sasaki, *J. Alloys Comp.* **237-239** (2002) 2112.
269. K. Kawamura, M. Morishita, T. Iwahashi, M. Yoshimura, Y. Mori and S. Sasaki, *Jpn. J. Appl. Phys. Part2* **41** (2002) L1440.
270. M. Morishita, F. Kawamura, T. Iwahashi, M. Yoshimura, Y. Mori and T. Sasaki, *Jpn. J. Appl. Phys.* **42** (2003) L565.
271. F. Kawamura, T. Iwahashi, K. Omae, M. Morishita, M. Yoshimura, Y. Mori and T. Sasaki, *Jpn. J. Appl. Phys.* **42** (2003) L4.
272. M. Morishita, F. Kawamura, M. Yoshimura, Y. Mori and T. Sasaki, *J. Jpn. Assoc. Crystal Growth (in japanese)* **30** (2003) 100.
273. F. Kawamura, T. Iwahashi, M. Morishita, K. Omae, M. Yoshimura, Y. Mori and T. Sasaki, *J. Jpn. Assoc. Crystal Growth (in japanese)* **30** (2003) 96.
274. F. Kawamura, Y. Yamada, H. Umeda, M. Morishita, M. Kawahara, M. Yoshimura, Y. Mori and T. Sasaki, *J. Jpn. Assoc. Crystal Growth (in japanese)* **31** (2004) 216.
275. F. Kawamura, Y. Yamada, H. Umeda, M. Morishita, M. Kawahara, M. Yoshimura, Y. Mori and T. Sasaki, *J. Jpn. Assoc. Crystal Growth (in japanese)* **31** (2004) 215.
276. M. Morishita, F. Kawamura, M. Kawahara, M. Yoshimura, Y. Mori and T. Sasaki, *J. Crystal Growth* **270** (2004) 402.
277. F. Kawamura, M. Kawahara, M. Morishita, H. Umeda, Y. Yamada, R. Shimojo, M. Yoshimura, Y. Mori and T. Sasaki, *J. Jpn. Assoc. Crystal Growth (in japanese)* **32** (2005) 3.
278. M. Morishita, F. Kawamura, M. Kawahara, M. Yoshimura, Y. Mori and T. Sasaki, *J. Crystal Growth* **284** (2005) 91.
279. F. Kawamura, H. Umeda, M. Kawahara, M. Yoshimura, Y. Mori and T. Sasaki, H. Okada, K. Arakawa and H. Mori, *Jpn. J. Appl. Phys.* **45** (2006) 2528.
280. F. Kawamura, H. Umeda, M. Morishita, M. Kawahara, M. Yoshimura, Y. Mori, T. Sasaki and Y. Kitaoka, *Jpn. J. Appl. Phys.* **45** (2006) L1136.
281. T. Iwahashi, Y. Kitaoka, M. Kawahara, F. Kawahara, M. Yoshimura, Y. Mori, T. Sasaki, R. Armitage and H. Hirayama, *Jpn. J. Appl. Phys.* **45** (2006) L103.
282. T. Iwahashi, Y. Kitaoka, F. Kawahara, M. Yoshimura, Y. Mori, T. Sasaki, R. Armitage and H. Hirayama, *Jpn. J. Appl. Phys.* **45** (2007) L227.
283. F. Kawamura, M. Morishita, M. Tanpo, M. Imade, M. Yoshimura, Y. Kitaoka, Y. Mori and T. Sasaki, *J. Crystal Growth* **310** (2008) 3946.
284. T. Sasaki, Y. Mori, F. Kawamura, M. Yoshimura and Y. Kitaoka, *J. Crystal Growth* **310** (2008) 1288.
285, D. Kashiwagi, R. Gejo, Y. Kanagawa, L. Liu, F. Kawamura, Y. Mori, T. Sasaki and K. Kakimoto, *J. Crystal Growth* **310** (2008) 1790.
286. F. Kawamura, M. Tanpo, N. Miyoshi, M. Imade, M. Yoshimura, Y. Mori, Y. Kitaoka and T. Sasaki, *J. Crystal Growth* **311** (2009) 3019.
287. Y.T. Song, W.J. Wang, W.X. Yuan, X. Wu and X. Chen, *J. Crystal Growth* **247** (2003) 275.
288. Y.T. Song, X.L. Chen, W.J. Wang, W.X. Yuan, Y.G. Cao and X. Wu, *J. Crystal Growth* **260** (2004) 327.

289. X.L. Chen, Sci. Technol. *Adv. Mater.* **6** (2005) 766.
290. B.N. Feigelson and R.L. Henry, *J. Crystal Growth* **281** (2005) 5.
291. J.K. Jian, G. Wang, C. Wang, W.X. Yan and X.L. Chen, *J. Crystal Growth* **291** (2006) 73.
292. T.I. Shin, H.J. Lee, J.-H. Lee, S.-W. Kim, S.J. Suh and D.H. Yoon, *J. Crystal Growth* **292** (2006) 216.
293. B.N. Feigelson, J.K. Hite, N.Y. Garces, J.A. Freitas, Jr., J.G. Tischler and P.B. Klein, *J. Crystal Growth* **312** (2010) 2551.
294. K. Byrappa, in *Handbook of Crystal Growth, Bulk Crystal Growth,* ed. D.T.J. Hurle (North-Hol land, Amsterdam, 1994) p. 467.
295. F. Iwasaki and H. Iwasaki, *J. Crystal Growth* **237-239** (2002) 820.
296. D. Nicholls, in *Inorganic Chemistry in Liquid Ammonia; Topics in Inorganic and General Chem istry, Monograph 17*, ed. R.J.H. Clark (Elsevier, Amsterdam, 1979).
297. H. Jacobs and D. Schmidt, *Curr. Top. Mater. Sci.* **8** (1982) 381.
298. J.F. Jenik and R.L. Wells, *Chem. Mater.* **8** (1996) 8.
299. R. Guarino and J. Rouxel, *Bull. Soc. Chim. Fr.* **2** (1969) 2284.
300. M. Roos, J. Wittrock, G. Meyer, S. Fritz and J. Strahle, *Z. Anorg. Allg. Chem.* **625** (2000) 1179.
301. A.P. Purdy and R.J. Jouet and C. Geroge, *Cryst. Growth Des.* **2** (2002) 141.
302. B. Wang and M. Callahan, *Cryst. Growth Des.* **6** (2006) 1227.
303. R. Dwilinski, R. Doradzinski, J. Garczynski, L.P. Sierzputowski, A. Puchalski, Y. Kanbara, K. Yagi, H. Minakuchi and H. Hayashi, *J. Crystal Growth* **310** (2008) 3911.
304. B. Wang, M.J. Callahan, K.D. Rakes, L.O. Bouthillette, S.-Q. Wang, D.F. Bliss and J.W. Kolis, *J. Crystal Growth* **287** (2006) 376.
305. D. Ehrentraut, Y. Kagamitani, C. Yokoyama and T. Fukuda, *J. Crystal Growth* **310** (2008) 891.
306. Y. Kagamitani, T. Ishiguro, C. Yokoyama and T. Fukuda, *Jpn. J. Assoc. Crytal Growth (in japanese)* **38** (2011) 27.
307. R. Dwilinski, A. Wysmolek, J. Baranowski and M. Kaminska, *Acta Phys. Pol.* **A88** (1995) 833.
308. R. Dwilinski, J.M. Baranowski, M. Kaminska, R. Doradzinski, J. Garczynski, L. Sierzputowski, *Acta Physica Polonica* **A90** (1996), 763.
309. R. Dwilinski, R. Dradzinski, J. Garczynski, L. Sierzputowski, M. Palczewska, A. Wysmolek and M. Kaminska, *Mater. Sci. Engineer. B* **50** (1997) 46.
310. R. Dwilinski, R. Doradzinski, J. Garczynski, L.P. Sierzputowski, A. Puchalski, Y. Kanbara, K. Yagi, H. Minakuchi and H. Hayashi, *J. Crystal Growth* **311** (2009) 3015.
311. R. Dwilinski, R. Doradzinski, J. Garczynski, L.P. Sierzputowski, M. Zajac and M. Rudzinski, *J. Crystal Growth* **311** (2009) 3058.
312. R. Dwilinski, R. Doradzinski, J. Garczyski, L. Sierzputwoski, R. Kucharski, M. Zajac, M. Rudzinski, R. Kudrawiec, W. Srupinski and J. Mizjewicz, *phys. stat. sol. (a)* **0** (2011) 1489.
313. R. Dwilinski, r. Doradzinsku, J. Garczyski, L. Sierzputowski, R. Kucharski, M. Zajac, M. Rudzinsku, R. Kudrawiec, J. Serafinczuk and W. Strupinski, *J. Crystal Growth* **312** (2010) 2499.
314. M. Rudzinski, R. Kudrawiec, L. Janicki, J. Serafinczuk, R. Kucharski, M. Zajac, J. Misiewicz, R. Dorandzinski, R. Dwilinski and W. Strupinsku, *J. Crystal Growth* **328** (2011) 5.
315. D.R. Ketchum and J.W. Kolis, *J. Crystal Growth* **222** (2001) 431.
316. M.J. Callahan, B. Wang, L.O. Bouthillette, S.-Q. Wang, J.W. Kolis and D.F. Bliss, *Mat. Res. Soc. Symp. Proc.* **798** (2004) 263.
317. M.J. Callahan, B.-G. Wang, K. Rakes, D. Bliss, L. Bouthellette, M. Suscavage and S.Q. Wang, *J. Mater. Sci.* **41** (2006) 1399.
318. B. Wang, M. Suscavage, D.F. Bliss and J. Jimenez, *J. Crystal Growth* **312**(2010) 2507.
319. B. Wang, D. Bliss, M. Suscavage, S. Swider, R. Lancto, C. Lynch, D. Weyburne, T. Li and F.A. Ponce, *J. Crystal Growth* **318** (2011) 1030.
320. B.T. Adekore, M.J. Callahan, L. Bouthillette, R. Dalmau and Z. Sitar, *J. Crystal Growth* **308** (2007) 71.
321. A. Yoshikawa, E. Ohshima, T. Fukuda, H. Tsuji and K. Oshima, *J. Crystal Growth* **260** (2004) 67.
322. Y. Kagamitani, D. Ehrentraut, A. Yoshikawa, N. Hoshino, T. Fukuda, S. Kawabata and K. Inaba, *Jpn. J. Appl. Phys.* **45** (2006) 4018.
323. D. Ehrentraut, N. Hoshino, Y. Kagamitani, A. Yoshikawa, T. Fukuda, H. Itoh and S. Kawabata, *J. Mater. Chem.* **17** (2007) 886.
324. D. Ehrentraut, *J. Jpn. Assoc. Crystal Growth (in japanese)* **34** (2007) 3.
325. T. Fukuda and D. Ehrentraut, *J. Crystal Growth* **305** (2007) 304.
326. D. Ehrentraut, Y. Kagamitani, T. Fukuda, F. Orito, S. Kawabata, K. Katano and S. Terada, *J. Crystal Growth* **310** (2008) 3902.
327. Y. Kagamitani, T. Kuribayashi, K. Hazu, T. Onuma, D. Tomida, R. Simura, S.F. Chichibu, K. Sugiyama, C. Yokohama, T. Ishiguro and T. Fukuda, *J. Crystal Growth* **312** (2010) 3384.
328. Y. Masuda, A. Suzuki, Y. Miyakawa, Y. kagamitani, T. Ishiguro, C. Yokoyama and T. Tsukada, *Int. J. Heat and Mass Trans.* **53** (2010) 940.

329. D. Tomida, T. Kuribayashi, K. Suzuki, Y. Kagamitani, T. Ishiguro, T. Fukuda and C. Yokoyama, *J. Crystal Growth* **325** (2011) 52.
330. D. Tomida and C. Yokoyama, *J. Jpn. Assoc. Crystal Growth (in japanese)* **38** (2011) 31.
331. T. Hashimoto, K. Fujito, F. Wu, B.A. Haskell, P.T. Fini, J.S. Speck and S. Nakamura, *Mat. Res. Soc. Symp. Proc.* **831** (2005) 81.
332. T. Hashimoto, K. Fujito, B.A. Haskell, P. Fini, J.S. Speck and S. Nakamura, *J. Crystal Growth* **275** (2005) e525.
333. T. Hashimoto, K. Fujito, R. Sharma, E.R. Letts, P.T. Fini, J.S. Speck and S. Nakamura, *J. Crystal Growth* **291** (2006) 100.
334. T. Hashimoto, K. Fujito, F. Wu, B.A. Haskell, P.T. Fini, J.S. Speck and S. Nakamura, *Jpn. J. Appl. Phys.* **44** (2005) L797.
335. T. Hashimoto, K. Fujito, M. Saito, J.S. Speck and S. Nakamura, *Jpn. J. Appl. Phys.* **44** (2005) L1570.
336. T. Hashimoto, F. Wu, J.S. Speck and S. Nakamura, *Nat. Mater.* **6** (2007) 568.
337. T. Hashimoto, F. Wu, J.S. Speck and S. Nakamura, *Jpn. J. Appl. Phys.* **46** (2007) L525.
338. T. Hashimoto, F. Wu, M. Saito, K. Fujito, J.S. Speck and S. Nakamura, *J. Crystal Growth* **310** (2008) 876.
339. T. Hashimoto, E. Letts, M. Ikari and Y. Nojima, *J. Crystal Growth* **312** (2010) 2503.
340. T. Hashimoto, F. Wu, J.S. Speck and S. Nakamura, *Jpn. J. Appl. Phys.* **46** (2007) L889.
341. T. Hashimoto, F. Wu, J.S. Speck and S. Nakamura, *J. Crystal Growth* **310** (2008) 3907.
342. S. Hashimoto, Y. Yoshizumi, T. Tanaka and M. Kiyama, *J. Crystal Growth* **298** (2007) 871.
343. C. Collado, G. Demazeau, B. Berdeu, A. Largeteau, J.C. Carcia, J. Guyaux and J. Massies, C.R. Acad. Sci. Pris. t.2 Serie II(1999) 483.344. Y.C. Lan, X.L. Chen, Y.P. Xu, Y.G. Cao and F. Huang, *Mater. Res. Bull.* **35** (2000) 2325.
345. M.P. D'Evelyn, N.J. Narang, D.-S. Park, H.C. Hong, M. Barbar, S.A. Tysoe, J. Leman, J. Balch, V.L. Lou, S.F. Leboeuf, Y. Gao, J.A. Teetsov, P.J. Codella, P.R. Tavernier, D.R. Clarke and R.J. Molmar, *Mat. Res. Soc. Symp. Proc.* **798** (2004) 275.
346. M.P. D'Evelyn, H.C. Hong, D.-S. Park, H. Lu, E. Kaminsky, R.R. Melkote, P. Perlin, M. Lesczynski, S. Porowski and R.J. Molmar, *J. Crystal Growth* **300** (2007) 11.
347. Q.-S. Chen, V. Prasad and W.R. Hu, *J. Crystal Growth* **258** (2003) 181.
348. S. Pendurti, Q.-S. Chen and V. Prasad, *J. Crystal Growth* **296** (2006) 150.
349. R.A. Logan and C.D. Thurmond, *J. Electrochem. Soc.* **119** (1972) 1727.
350. V.A. Ivantsov, V.A. Sukhoveev and V.A. Dmitriev, *Mat. Res. Soc. Symp. Proc.* **468** (1997) 143.
351. V. A. Sukhoveyev, V.A. Ivantsov, I.P. Nikitina, A.I. Babanin, A.Y. Polyakov, A.V. Govorkov, N.B. Smirov, M.G. Mil'vidskii and V. Dmitriev, *MRS Internet J. Nitride Semicon. Res.* **5SI** (2000) W6.6.
352. C. Klemenz and H.J. Scheel, *J. Crystal Growth* **211** (2000) 62.
353. K.E. Waldrip, *Sandia Report* **SAND2007-5210** (2007).
354. T. Ozawa, M. Dohi, T. Matsuura and Y. Hayakawa, *J. Crystal Growth* **310** (2008) 1785.
355. R.B. Zettersrom, *J. Mater. Sci.* **5** (1970) 1102.
356. C.M. Balkas, Z. Sitar, L. Bergman, I.K. Shmagin, J.F. Muth, R. Kolbas, R.J. Nemanich and R.F Davis, *J. Crystal Growth* **208** (2000) 100.
357. S. Kurai, T. Abe, Y. Naoi and S. Sakai, *Jpn. J. Appl. Phys. Part1* **35** (1996) 1637.
358. S. Kurai, Y. Naoi, T. Abe, S. Homi and S. Sakai, *Jpn. J. Appl. Phys.* **36** (1996) L77.
359. S. Kurai, K. Nishino and S. Sakai, *Jpn. J. Appl. Phys.* **36** (1997) L184.
360. Yu.A. Vodakov, M.I. Karklina, E.N. Mokhov and A.D. Roenkov, *Inorganic Mater.* **17** (1980) 537.
361. Yu. A. Vodakov, E.N. Mokhov, M.G. Ramm, M.S. Ramm, A.D. Roenkov, A.G. Ostroumov, A.A. Wolfson, S.Yu. Karpov, Yu.N. Makarov and H. Jurgensen, *Mat. Res. Soc. Symp. Proc.* **482** (1998) 27.
362. R.S.Q. Faredd, S. Tottori, T. Inaoka, K. Nishino and S. Sakai, *J. Crystal Growth* **207** (1999) 174.
363. B. Kallinger, E. Messner, D. Seng, G. Sun, S. Hussy, J. Friedrich and G. Müller, *phys. stat. sol. (c)* **4** (2007) 2264.
364. E. Ejder, *J. Crystal Growth* **22** (1974) 44.
365. T. Ogino and M. Aoki, *Oyo Buturi (Monthly Publ. Jpn. Soc. Appl. Phys.) (in japanese)* **48** (1979) 269.
366. G. Kamler, J. Zachara, S. Podsiadl, O. Adamowicz and W. Gebick, *J. Crystal Growth* **212** (2000) 39.
367. H. Shin, D.B. Thomson, R. Schlesser, R.F. Davis and Z. Sitar, *J. Crystal Growth* **241** (2002) 404.
368. K. Nishio, D. Kikuta and S. Sakai, *J. Crystal Growth* **237-239** (2002) 922.
369. H. Wu, J. Spinelli, P. Konkapaka and M.G. Spencer, *Mat. Res. Soc. Symp. Proc.* **892** (2006) 757.
370. H.-J. Rost, D. Siche, R. Müller, D. Gogova, T. Schulz, M. Albrecht and R. Fornari, *phys. stat. sol.* **4** (2007) 2219.
371. S.Y. Bae, H.W. Seo, J. Park, H. Yang, S.A. Song, *Chem. Phys. Lett.* **365** (2002) 525.
372. G. Cheng, A. Kolmakov, Y. Zhang, M. Moskovits, R. Munden, M.A. Reed G. Wang, D. Moses and J. Zhang, *Appl. Phys. Lett.* **83** (2003) 1578.

373. B. Schwenzer, J. Hu, R. Seshadri, S. Keller, S.P. DenBaars and U.K. Mishra, *Chem. Mater.* **16** (2004) 5088.
374. T. Szyszko, g. Kalmer, B. Strojek, G. Weisbrod, S. Podsiadlo, L. Adamowicz, W. Gebicki, J. Szczytko, A. Twardowski and K. Sikorski, *J. Crystal Growth* **233** (2001) 631.
375. T. Szyzsko, M. Kaminski, S. Podsialdlo, K. Wozniak, L. Dobrzycki and W. Gebicki, *J. Crystal Growth* **276** (2005) 419.
376. S. Fisher, C. Wentzel, W.L. Hansen, E.D. Bourret-Courchesne, B.K. Meyer and E.E. Haller, *Appl. Phys. Lett.* **69** (1996) 2716.
377. M. Kaminski, S. Podsiadlo, K. Wozniak, L. Dobrzycki, R. Jakiela, A. Barcz, M. Psoda and J. Mizera, *J. Crystal Growth* **303** (2007) 395.
378. S. Shimada and R. Taniguchi, *J. Crystal Growth* **263** (2004) 1.
379. P. Konkapaka, B. Raghothamachar, M. Dudley, Y. Makarov and M.G. Spencer, *J. Crystal Growth* **289** (2006) 140.
380. A. Miura, S. Kamoshida, T. Sekiguchi, *Jpn. J. Assoc. Crystal Growth (in japanese)* **33** (2006) 13.
381. B. Rahothamachar, P. Konkapaka, H. Wu, M. Dudley and M. Spencer, *Mat. Res. Soc. Symp. Proc.* **892** (2006) 781.
382. M. Imade, Y. Kitano, N. Yamada, M. Kawahara, F. Kawamura, M. Yoshimura, Y. Kitaoka, Y. Mori and T. Sasaki, *Jpn. J. Assoc. Crystal Growth (in japanese)* **33** (2006) 186.
383. M. Imade, M. Kawahara, F. kawamura, M. Yoshimura, Y. Mori and T. Sasaki, *Mater. Lett.* **59** (2005) 4026.
384. M. Tatsumi, R. Hirota and T. Kusano, *Jpn. J. Assoc. Crystal Growth (in japanese)* **27** (2000) 15.
385. Hirota and M. Ttsumi, *Abstr. First Asian Conf. Cryst. Growth Cryst. Tech. CGCT-1* (Sendai, 2000) p. 453.
386. D. Siche, H.-J. Rost, K. Böttcher, D. Gogova, R. Fornari, *J. Crystal Growth* **310** (2008) 916.
387. D. Siche, D. Gogova, S. Lehmann, T. Fizia, R. Fohrnari, M. Andrashc, A. Pipa and J. Ehlbeck, *J. Crystal Growth* **318** (2011) 406.
388. A.V. Kuznetsov, E.V. Rakova, S.S. Lee, P.J. Chong, *J. Crystal Growth* **167** (1996) 458.
389. T. Detchprohm, K. Hiramatsu, H. Amano and I. Akasaki, *Appl. Phys. Lett.* **61** (1992) 2688.
390. H. Lee and J.S. Harris, Jr., *J. Crystal Growth* **169** (1996) 689.
391. E. Aujol, J. Napierala, A. Trassoudaine, E. Gil-Lafon and R. Cadoret, *J. Crystal Growth* **222** (2001) 548.
392. E. Aujol, A. Trassoudaine, L. Siozade, A. Pimpinelli and R. Cadoret, *J. Crystal Growth* **230** (2001) 372.
393. R. Paskova, E. Valcheva, J. Birch, S. Tungasmita, P.O.A. Persson, P.P. Paskov, S. Evtimova, M. Abrashev and B. Monekar, *J. Crystal Growth* **230** (2001) 381.
394. A.S. Segal, A.V. Kondartyev, S. Yu, Karpov, D. Martin, V. Wagner and M. Ilgems, *J. Crystal Growth* **270** (2004) 384.
395. P. Kempisty, B. Lucznik, B. Pastuszka, I. Grzegory, M. Bockowski, S. Krukowsi and S. Porowsi, *J. Crystal Growth* **296** (2006) 31.
396. A. Koukitu, F. Satoh, T. Yamane, H. Murakami and Y. Kumagai, *J. Crystal Growth* **305** (2007) 335.
397. J.F. Wang, X.J. Hu, Y.M. Zhang, Y. Xu, H.B. Wang, B.S. Zhang, K. Xu and H. Yang, *J. Crystal Growth* **311** (2009) 3033.
398. H. Okazaki, *41th Spring Meeting, Jpn. Soc. Appl. Phys. and Related Socs.* (1994) 19p-MG-4.
399. H. Okazaki, A. Arkawa, T. Asahi, O. Oda and K. Aiki, *Proc. Topical workshop III-V Nitrides* (Nagoya, 1995) P. 219. *Solid State Electon.* **41** (1997) 263.
400. A. Kuramata, K. Horiono, K. Domen and T. Tanahashi, *Oyo Buturi (Monthly Publ. Jpn. Soc. Appl. Phys.) (in japanese)* **65** (1996) 936.
401. T. Shibata, *Res. Rep. on Blue/Ultraviolet Emission Devices (in japanese)* (JRCM, 1990) p. 63.
402. C.R. Abernathy, in *GaN and Related Materials*, ed. S.J. Pearton (Gordon and Breach, New York, 1997) p. 11.
403. G. Popovici, H. Morkoc and G.-B. S.N. Mohammad, in *Group III Nitride Semiconductor Com pounds*, ed. B. Gil (Clarendon, Oxford, 1998) p. 19.
404. H. Morkoc, *Handbook of Nitride Semiconductors and Devices (Vol. 1)* (Wiley-Vch, 2008) p. 323.
405. X. Xu, R.P. Vaudo, C. Loria, A. Salant, G.R. Brandes and J. Chaudhuri, *J. Crystal Growth* **246** (2002) 223.
406. F.A. Ponce, *MRS Bull.* **22** (1997) 51.
407. T. Takeuchi, H. Amano, K. Hiramatsu, N. Sawaki and I. Akasaki, *J. Crystal Growth* **115** (1991) 634.
408. A. Watanabe, T. Takeuchi, K. Hirosawa, H. Amano, K. Hiramatsu and I. Akasaki, *J. Crystal Growth* **128** (1993) 391.
409. K. Hirosawa, K. Hiramatsu, N, Sawaki and I. Akasaki, *Jpn. J. Appl. Phys.* **32** (1993) L1039.
410. S. Fujieda, M. Mizuta and Y. Matsumoto, *Jpn. J. Appl. Phys.* **27** (1988) L296.
411. A. Nakadaira and H. Tanaka, *Appl. Phys. Lett.* **70** (1997) 2720.

412. P.J. Born and D.S. Robertson, *J. Mater. Sci.* **15** (1980) 3003.
413. Y. Morimoto, K. Uchiho and S. Ushio, *J. Electrochem. Soc.* **120** (1973) 1783.
414. M.E. Lin, S. Strite, A. Agarwai, A. Salvador, G.L. Zhou, N. Teraguchi, A. Rockett and H. Morkoc, *Appl. Phys. Lett.* **62** (1993) 702.
415. M.E. Lin, B. Sverdlov, G.L. Zhou and H. Morkoc, *Appl. Phys. Lett.* **62** (1993) 3479.
416. T.W. Weeks, Jr., M.D. Bremser, K.S. Ailey, E. Carlson, W.G. Perry and R.F. Davis, *Appl. Phys. Lett.* **67** (1995) 401.
417. F.A. Ponce, B.S. Krusor, J.S. Major, Jr., W.E. Plano and D.F. Welch, *Appl. Phys. Lett.* **67** (1995) 410.
418. A.S. Zubrilov, V.I. Nikolaev, D.V. Tsvetkov, V.A. Dmitriev, K.G. Irvine, J.A. Edmond and C.H. Carter, Jr., *Appl. Phys. Lett.* **67** (1995) 533.
419. N. Onojima, J. Suda and H. Matsunami, *Jpn. J. Appl. Phys., Part 2* **42** (2003) L445.
420. T. Yonezawa, *Proc. Melt Growth Meeting (Tohoku University)* **14** (1998) 105.
421. R.C. Powell, G.A. Tomasch, Y.-W. kim, J.A. Thronton and J.E. Greene, *Mat. Res. Soc. Symp. Proc.* **162** (1990) 525.
422. T. Matsuoka, N. Yoshimoto, T. Sasaki and A. Katsui, *J. Electron. Mater.* **21** (1992) 157.
423. E.S. Hellman, D. Buchanan, D. Wiesmann and I. Brener, *MRS. Int. J. Nitride Semicond. Res.* **1** (1996) 16.
424. H. Fujioka, A. Kobayashi and J. Ohta, *IEICE Technical Rep.* **ED2005-51** (2005) 91.
425. Z. Sitar, M.J. Paisley, B. Yan and R.F. Davis, *Mat. Res. Soc. Symp. Proc.* **162** (1990) 263.
426. R.F. Davis, M.J. Paisley, Z. Sitar, D.J. Kester, K.S. Ailey, K. Linthicum, L.B. Lowland, S. Tanaka and R.S. Kern, *J. Crystal Growth* **178** (1997) 87.
427. Y. Tomm, P. Reiche, D. Klimm and T. Fukuda, *J. Crystal Growth* **220** (2000) 510.
428. E.G. Villora, K. Shimamura, K. Aoki and N. Ichinose, *Jpn. J. Assoc. Crystal Growth (in japanese)* **31** (2004) 105.
429. E.G. Villora, K. Shimamura, Y. Yoshikawa, K. Aoki and N. Ichinose, *J. Crystal Growth* **270** (2004) 420.
430. N. Suzuki, S. Ohira, M. Tanaka, T. Sugawara, N. Nakajima and T. Shisjido, *phys. stat. sol. (c)* **4** (2007) 2310.
431. S. Otani and Y. Ishizawa, *J. Crystal Growth* **165** (1996) 319.
432. H. Kinoshita, S. Otani, S. Kamiyama, H. Amano, I. Akasaki, J. Suda and H. Matsunami, *Jpn. J. Appl. Phys.* **40** (2001) L1280.
433. J. Suda and H. Matsunami, *J. Crystal Growth* **237-239** (2002) 1114.
434. R. Liu, A. Bell, F.A. Ponce, S. kamiyama, H. Amano and J. Akasaki, *Appl. Phys. Lett.* **81** (2002) 3182.
435. K. Shiraishi, J. Iwata and A. Oshiyama, *Jpn. J. Assoc. Crystal Growth (in japanese)* **30** (2003) 68.
436. J. Suda and H. Matsunami, *Oyo Buturi (Monthly Publ. Jpn. Soc. Appl. Phys.) (in japanese)* **72** (2003) 318.
437. Y. Kawaguchi, J. Ohta, A. Kobayashi and H. Fujioka, *Appl. Phys. Lett.* **87** (2005) 221907.
438. R. Armitage, K. Nishizono, J. Suda and T. Kimoto, *J. Crystal Growth* **284** (2005) 369.
439. K. Ueno, K. Sakai, T. Shimada and A. Koma, *J. Vac. Sci. Technol. A* **8** (1980) 68.
440. T. Ueno, H. Yamamoto, K. Saiki and A. Koma, *Appl. Surface Sci.* **113/114** (1997) 33.
441. A. Koma, K. Sunouchi and T. Miyajima, *J. Vac. Sci. Technol. B* **3** (1985) 724.
442. A. Yamada, K.P. Ho, T. Maruyama and K. Akimoto, *Appl. Phys. A* **69** (1999) 89.
443. A. Yamada, K.P. Ho, T. Akaogi, T. Maruyama and K. Akimoto, *J. Crystal Growth* **201/202** (1999) 332.
444. A. Koma and K. Yoshimura, *Surf. Sci.* **174** (1986) 556.
445. J.-W. Chung and F.S. Ohuchi, *Mat. Res. Soc. Symp. Proc.* **449** (1997) 379.
446. A. Kuramata, K. Horino, K. Domen, K. Shinohara and T. Tanahashi, *Appl. Phys. Lett.* **67** (1995) 2521.
447. S. Nakamura, M. Senoh, S. Nagahama, N. Iwasa, T. Yamada, T. Matsushita, H. Kiyoku and Y. Sugimoto, *Appl. Phys. Lett.* **68** (1996) 2105.
448. T. Geroge, E. Jacobsohn, W.T. Pike, P. Chang-Chien, M.A. Kahn, J.W. Yang and S. Mahajan, *Appl. Phys. Lett.* **68** (1996) 337.
449. C.J. Sun, J.W. Yang, Q. Chen, M.A. Khan, T. George, P. Chang-Chien and S. Mahajan, *Appl. Phys. Lett.* **68** (1996) 1129.
450. E.S. Hellman, C.D. Brandle, L.F. Schneemeyer, D. Wiesmann, I. Brener, T. Siegrist, G.W. Berkstesser, D.N.E. Buchanan and E.H. Hartford, Jr., *Mat. Res. Soc. Symp., Proc.* **395** (1996) 51.
451. C.J. Sun, J.W. Yang, B. W. Lim, Q. Chen, M.Z. Anwar and M.A. Khan, *Appl. Phys. Lett.* **70** (1997) 1444.
452. E.S. Hellman, C.D. Brandle, L.F. Schneemeyer, D. Wies-mann, I. Brener, T. Siegrist, G.W. Berkstresser, D.N.E. Buchanan and E.H. Hartford, *MRS Internet J. Nitride Semicon. Res.* **1** (1996) 1.
453. V.V. Mamutin, A.A. Toropov, N.F. Kartenko., S.V. Ivanov, A. Wagner and B. Monemar, *Mater. Sci. Engineer. B* **59** (1999) 56.
454. S. Fischer, A. Gisbertz, B.K. Meyer, M. Topf. S. Koynov, I. Dirnstorfer, D. Volm, R. Uecker, P. Reiche, S. Ganschow and Lilienal-Weber, *Symp. Proc.* (Rigi, Switzerland, 1996) EGN-1.

455. E.S. Hellman, Z. Lilental-Weber and D.N.E. Buchmann, *MRS Internet J. Nidtride Semicon. Res.* **2** (1998) 30.
456. K. Xu, J Xu, P. Deng, R. Qiu and Z. Fang, *phys. stat. sol.(a)* **176** (1999) 589.
457. E. Eiper, A. Hofmann, J.W. Gerlach, B. Rayshenbach and J. Keckes, *J. Crystal Growth* **284** (2005) 561.
458. J. Zou, Y. Dong, S. Zhou, Y. Sum, W. Jun, J. Zhou, T. Huang, S. Yang and H. Zhou, *J. Crystal Growth* **294** (2006) 339.
459. J. Zou and W. Xiang, *J. Crystal Growth* **311** (2009) 3285.
460. P. Kung, A. Saxler, X. Zhang, D. Walker, R. Lavado and M. Razeghi, *Appl. Phys. Lett.* **69** (1996) 2116.
461. J.F. H. Nicholls, H. Gallagher, B. Henderson, C. Trager-Cowan, P.G. Middleton, K.P. O'Donnel, T.S. Cheng, C.T. Foxon and B.H.T. Chai, *Mat. Res. Soc. Symp. Proc.* **395** (1996) 535.
462. T. Ishii, Y. Tazoh and S. Miyazawa, *Jpn. J. Appl. Phys.* **36** (1997) L139.
463. Y. Tazoh, T. Ishii and S. Miyazawa, *Jpn. J. Appl. Phys.* **36** (1997) L746.
464. T. Ishii, Y. Tazoh and S. Miyazawa, *J. Crystal Growth* **189/190** (1998) 208.
465. R.J. Matyl, W.A. Dootlittle and A.S. Brown, *J. Phys. D: Appl. Phys.* **32** (1999) A61.
466. K. Xu, P. Deng, J. Xu, G. Zhou, W. Liu and Y. Tian, *J. Crystal Growth* **216** (2000) 343.
467. A. Pajaczkowska and A. Gloubokov, *Prog. Crystal Growth and Character.* **36** (1998) 123.
468. N. Kimizuka and T. Mohri, *J. Solid State Chem.* **60** (1985) 382.
469. N. Kimizuka and T. Mohri, *J. Solid State Chem.* **78** (1989) 98.
470. H. Takeda, H. Tabata, K. Shimamura and T. Fukuda, *J. Jpn. Assoc. Cryst. Growth (in japanese)* **25** (1998) A11.
471. K. Shimamura, H. Tabata, H. Takeda, V.V. Kochurikhin and T. Fukuda, *J. Crystal Growth* **194** (1998) 209.
472. M. Ito, K. Shimamura, D. A. Pawlak and T. Fukuda, *J. Crystal Growth* **235** (2002) 277.
473. M. Sumiya, T. Shikyow, T. Sasahara, K. Yoshimura, J. Ohta, H. Fujioka, S. Tagaya, H. Ikeya, H. Joinuma and S. Fuke, *Jpn. J. Appl. Phys.* **41** (2002) 5038.
474. N. Futagawa, A. Yoshikawa, K. shimamura and T. Fukuda, *J. Jpn. Assoc. Cryst. Growth (in japanese)* **25** (1998) A10.
475. T. Nishinaga, T. Nakano and S. Zhang, *Jpn. J. Appl. Phys.* **27** (1988) L964.
476. A. Usui, H. Sunakawa, A. Sakai and A. Yamaguchi, *Jpn. J. Appl. Phys.* **36** (1997) L899.
477. Z. Yu, M.A. L. Johnson, J.D. Brown, N.A. El-Masry, J.W. Cook, Jr. and J.F. Schetzina, *J. Crystal Growth* **195** (1998) 333.
478. K. Kobayashi, A. Yamaguchi, S. Kimura, A. Kimura and A. Usui, *Jpn. J. Appl. Phys.* **38** (1999) L611.
479. C. Sasaoka, H. Sunakawa, A. Kumura, M. Nido and A. Usui, *Proc. 2nd Conf. Nitride Semiconductors* (1997) p. 452.
480. K. Linthicum, T. Gehrke, D. thomson, E. Carlson, P. Pajagopa, T. Smith, D. Bathcelor and R. Davis, *Appl. Phys. Lett.* **75** (1999) 196.
481. M. Takeuchi, S. Tanaka, H. Hirayama and Y. Aoyagi, *ICM-OVPE-X* (2000) Th-A5.
482. S. Sakai, T. Wang, Y. Morishima and Y. Naoi, *ICM-OVPE-X* (2000) Tu-P39.
483. M. Iwaya, T. Takeuchi, S. Yamaguchi, C. Wetzel, H. Amano and I. Akasaki, *Jpn. J. Appl. Phys.* **37** (1998) L316.
484. A. Ishibashi, I. Kidoguchi, G. Sugahara and Y. Ban, *IC-MOVPE-X* (2000) Th-A2.
485. M. Ishida, M. Ogawa, K. Orita, O. Imafuji, M. Yuri, T. Sugino and K. Itoh, *IC-MOVPE-X* (2000) Th-A1.
486. A. Sakai, H. Sunakawa, A. Kimura and A. Usui, *J. Electron Microscopy* **49** (2000) 323.
487. K. Hiramatsu, K. Nisiyama, M. Oonishi, H. Mizutani, M. Nakamura, A. Motogaito, H. Miyake, Y. Iyechika and T. Maeda, *J. Crystal Growth* **221** (2000) 316.
488. I. Kidoguchi, A. Ishibashi, G. Sugahara and Y. Ban, *Appl. Phys. Lett.* **76** (2000) 3766.
489. Y. Honda, Y. Iyechika, T. Maeda, H. Miyake and K. Hiramatsu, *Jpn. J. Appl. Phys.* **40** (2001) L309.
490. D.-S. Wuu, H.-W. Wu, S.-T. Chen, T.-Y. Tsai, X. Zheng and R.-H. Horng, *J. Crytal Growth* **311** (2009) 3063.
491. H.-H. Huang, C.-L. Chao, T.-W. Chi, Y.-L. Chang. P.-C Liu, L.-W. Tu, J.-D. Tsay, H.-C. Kuo, S.-J. Cheng and W.-I. Lee, *J. Crystal Growth* **311** (2009) 3029.
492. S. Nagahama, N. Iwasa, M. Senoh, T. Matsushita, Y. Sugimoto, H. Kiyoku, T. Kozaki, M. Sano, H. Matsumura, H. Umemoto, K. Shocho and T. Mukai, *Jpn. J. Appl. Phys.* **38** (2000) L647.
493. W.S. Wong, T. Sands and N.W. Cheung, *Appl. Phys. Lett.* **72** (1998) 599.
494. Y.J. Lee and S.T. Kim, *J. Korean Phys. Soc.* **33** (1998) S330.
495. S.T. Kim, Y. Lee, D.C. Moon, C.H. Hong and T.K. Yoo, *Semicond. Sci. Technol.* **14** (1999) 278.
496. M.K. Kelly, R.P. Vaudo, V.M. Phanse, L. Grogens, O. Ambacher and M. Stutzmann, *Jpn. J. Appl. Phys.* **38** (1999) L217.
497. J.A. Freitas, Jr., G.C.B. Braga, W.J. Moore, J.G. Tishler, J.C. Culbertson, M. Fatemi, S.S. Park, S.K. Park, S.K. Lee and Y. Park, *J. Crystal Growth* **231** (2001) 322.
498. S. Gu, R. Zhang, Y. Shi, Y. Zheng, L. Zhang, F. Dwikusuma and T.F. Kuech, *J. Crystal Growth* **231** (2001) 342.

499. T. Paskova, V. Darakchieva, P.P. Paskov, U. Sodervall and B. Monemar, *J. Crystal Growth* **246** (2002) 207.
500. S. Bohyama, K. Yoshikawa, H. Naoi, H. Miyake, K. Hiamatsu, Y. Iyechika and T. Maeda, *phys. stat. sol.* **194** (2002) 528.
501. Y. Oshima, T. Eri, M. Shibata, H. Sunakawa and A. Usui, *phys. stat. sol. (a)* **194** (2002) 554.
502. Y. Oshima, T. Eri, M. Shibata, H. Sunakawa, K. Kobayashi, T. Ichihashi and A. Usui, *Jpn. J. Appl. Phys.* **42** (2003) L1.
503. Y. Oshima, M. Shibata, T. Eri, T. Yoshida, K. Watanabe and T. Suzuki, *Hitachi Cables (in japanese)* **26** (2007) 31.
504. D. Gogova, H. Larsson, R. Yakimova, Z. Zolnai, I. Ivanov and B. Monemar, *Proc. 5th Int. Conf. Nitride Semicond.* (Nara, Japan, 2003) p. 60.
505. B. Monemar, H. Larsson, C. Hemmingsson, I.G. Ivanov and D. Gogova, *J. Crystal Growth* **281** (2005) 17.
506. M. Hermann, D. Gogova, D. Sche, M. Schmidbauer, B. Monemar, M. Stutzmann and M. Eickhoff, *J. Crystal Growth* **293** (2006) 462.
507. C.E.C. Dam, A.P. Grzegorczyk, P.R. Hageman and P.K. Larsen, *J. Crystal Growth* **290** (2006) 473.
508. A.D. Williams and T.D. Moustkas, *J. Crystal Growth* **300** (2007) 37.
509. H.J. Lee, S.W. Lee, H. Goto, S.-H. Lee, H.-J. Lee, J.S. Ha, T. Goto, M.W. Cho, T. Yao and S.-K. Hong, *Appl. Phys. Lett.* **91** (2007) 192108.
510. S.W. Lee, H. Goto, T. Minegishi, W.H. Lee, J.S. Ha, H.J. Lee, H.J. Lee, S.H. Lee, T. Goto, T. Hanada, M.W. Cho and T. Yao, *phys. stat. sol. (c)* **4** (2007) 2617.
511. Ch. Hennig, E. Richter, M. Weyers and G. Tränkle, *phys. stat. sol. (c)* **4** (2007) 2638.
512. C.M. Lai, W.Y. Liu, J.D. Tsay, P.C. Liu, Y.D. Guo, H.H. Huang, Y.H. Chang and J.C. Yeh, *phys. stat, sol.(c)* **4** (2007) 2231.
513. H. Sasaki, H. Sunakawa, N. Sumi, K. Yamamoto and A. Usui, *J. Crystal Growth* **311** (2009) 2910.
514. K.M. Chen, H.H. Huang, Y.L. Kuo, P.L. Wu, T.L. Chu, H.W. Yu and W.I. Lee, *J. Crystal Growth* **311** (2009) 3037.
515. K. Fujito, S. Kubo, H. Nagaoka, T. Mochizuki, H. Namita and S. Nagao, *J. Crystal Growth* **311** (2009) 3011.
516. K. Kawano, T. Yamamoto, A. Wakahara, A. Yoshida, Y. Seki and O. Oda, *36th Spring Meeting, Jpn. Soc. Appl. Phys. and Related Socs.* (1998) 9a-Zp-1.
517. A. Wakahara, T. Yamamoto, K. Kawano, A. Yoshida, Y. Seki and O. Oda, *Proc. 2nd Int. Symp. on Blue Lasers and LEDs* (Kisarazu, Japan, 1998) p. 190.
518. T. Yamamoto, K. Kawano, A. Wakahara, A. Yoshida, Y. Seki and O. Oda, *17th Electron. Mater. Symp.* (Izu-Nagaoka, 1998) p. 223.
519. A. Wakahara, T. Nishida, K. Kawano, A. Yoshida, Y. Seki and O. Oda, *Proc. 10th Conf. Semicon. Insul. Mater.* (Berkeley, USA, 1998) p. 33.
520. T. Yamamoto, K. Ishio, A. Wakahara, A. Yoshida, Y. Seki and O. Oda, *Ext. Abst. 18th Electron. Mat. Symp.* (Kli-shirahama, Japan, 1999) p. 61.
521. A. Wakahara, T. Yamamoto, K. Ishio, A. Yoshida, Y. Seki, K. Kainosho and O. Oda, *Jpn. J. Appl. Phys.* **39** (2000) 2399.
522. K. Motoki, T. Okahisa, N. Matsumoto, M. Matsushima, H. Kimura, H. Sakai, K. Taketomo, K. Uematsu, T. Hirano, M. Koukitsu and H. Seki, *Jpn. J. Appl. Phys.* **40** (2001) L140.
523. K. Motoki, T. Okahisa, S. Nakahata, N. Matsumoto, H. Kimura, H. Kasai, K. Uematsu, M. Ueno, Y. Kumagai, A. Koukitu and H. Seki, *J. Crystal Growth* **237-239** (2002) 912.
524. Y. Kumagai, H. Murakami, H. Seki and A. Koukitu, *J. Crystal Growth* **246** (2002) 215.
525. Y. Kumagai, F. Satoh, R. Togashi, H. Murakami, K. Takemoto, J. Iihara, K. Yamaguchi and A. Koukitu, *J. Crystal Growth* **296** (2006) 11.
526. Yu Melnick, A. Nikolaev, I. Nikitina, K. Vassilevski and V. Dmitriev, *Mat. Res. Soc. Symp. Proc.* **482** (1998) 269.
527. I. Nikitina, G. Mosina, Yu. ™Melnik, A. Nikovaev and K. Vassilevski, *Mater. Sci. Engineer. B* **61** (1999) 325.
528. N. Sawaki, T. Hikosaka, N. Koide, S. Tanaka, Y. Honda and M. Yamaguchi, *J. Crystal Growth* **311** (2009) 2867.
529. N. Suzuki, T. Uchida, T. Tanikawa, T. Hikosaka, Y. Honda, M. Yamaguchi and N. Sawaki, *J. Crystal Growth* **311** (2009) 2875.
530. T. Tanikawa, Y. Kagohashi, Y. Honda, M. Yamaguchi and N. Sawaki, *J. Crystal Growth* **311** (2009) 2879.
531. D. Iida, M. Iwaya, S. Kamiyama, H. Amano and I. Akasaki, *J. Crystal Growth* **311** (2009) 2887.
532. B. Lucznik, B. Patsuszka, I. Grezegory, M. Bockowski, G. Kamler, E. Litwin-Staszewsk and S. Porowski, *J. Crystal Growth* **281** (2005) 38.

533. I. Grzegroy, B. Lucznik, M. Bockowski and S. Porowski, *J. Crystal Growth* **300** (2007) 17.
534. G. Kamler, B. Lucznik, B. Pastuszka, I. Grzegory and S. Porowski, *J. Crystal Growth* **310** (2008) 3478.
535. O. Kryliouk, M. Reed, T. Dann, T. Anderson and B. Chai, *Mater. Sci. Engineer. B* **56** (1999) 1.
536. O. Kryliouk, M. Reed, T. Dann, T. Anderson and B. Chai, *Mater. Sci. Engineer. B* **59** (1999) 6.
537. O. Kryliouk, M. Reed, T. Dann, T. Anderson and B. Chai, *Mater. Sci. Engineer. B* **66** (1999) 26.
538. O. Kryliouk, M. Reed, T. Dann, T. Anderson and B. Chai, *phys. stat. solidi (a)* **176** (1999) 407.
539. M.D. Reed, O.M. Kryliouk, M.A. Mastro and T.J. Anderson, *J. Crystal Growth* **274** (2005) 14.
540. E. Richer, Ch. Henning, L. Wang, U. Ziemer, M. Weyers and G. Tränkle, *phys. stat. sol. (c)* **4** (2007) 2277.
541. S.V. Novikov, N.M. Stanton, R.P. Campion, R.D. Morris, H.L. Geen, C.T. Foxon and A.J. Kent, *Semicond. Sci. Technol.* **23** (2008) 015018.
542. S.V. Novikov, N.M. Stanton, R.P. Campion, C.T. Foxon and A.J. Kent, *J. Crystal Growth* **310** (2008) 3964.
543. S.V. Novikov, N. Zainal, A.V. Akimov, C.R. Staddon, A.J. Kent and C.T. Foxon, *J. Vacum Sci. Technol. B* **28** (2010) C3B1.
544. S.V. Novikov, C.T. Foxon and A.J. Kent, *phys. status solid (a)* **207** (2010) 1277.
545. S.V. Novikov, C.R. Staddon, C.T. foxon, F. Luckert, P.R. Edwards, R.W. Martin and A.J. Kent, *J. Crystal Growth* **323** (2011) 80.
546. S.V. Novikov, N. Zainal, A.V. Akimov, C.R. Standdon, C.T. Foxon, A.J. Kent, D.P. Nicholls and S.E. Hopper, *Proc. 8th Inter. Conf. Nitride Semicond.* (ICNS-8, 2009) p. 4494.
547. X.L. Chen, Y.G. Cao, Y.C. Lan, X.P. Xu, J.Q. Li, K.Q. Lu, P.Z. Jiang, T. Xu, Z.G. Bai, Y.D. Yu and J.K. Liang, *J. Crystal Growth* **209** (2000) 208.
548. E.G. Villora, K. Shimamura, K. Aoki and N. Ichinose, *J. Crystal Growth* **270** (2004) 462.
549. S. Ohira, M. Yoshioka, T. Sugawara, K. Nakajima and T. Shishido, *Jpn. J. Assoc. Crystal Growth (in japanese)* **32** (2005) 10.
550. A. Mabuchi, Y. Iwase, E. Yasuda, T. Sugiura and H. Minoura, *J. Ceram. Soc. Japan* **113** (2005) 291.
551. T. Hirano, A. Mabuchi, T. Sugiura and H. Minora, *J. Crystal Growth* **311** (2009) 3040.
552. H. Saitoh, W. Utsumi and K. Aoli, *J. Crystal Growth* **310** (2008) 473.
553. A.K. Murali, A.D. Barve, S.H. Risbud, *Mater. Sci. Engineer. B* **96** (2002) 111.
554. B. Schwenzer, L. Loeffler, R. Seshadri, S. Keller, F.F. Lange, S.P. DenBaars and U.K. Mishra, *J. Mater. Chem.* **14**(2004) 637.
555. F.V. Williams, in *Compound Semiconductors, Vol.1. Preparation of III-V Compounds*, eds. by R.K. Willardson and H.L. Goering (Reinhold Publ, Corp., New York, 1962) p. 171.
556. T.L. Chu, M. Gill and R.K. Smeltzer, *J. Crystal Growth* **33** (1976) 53.
557. N. Kato, W. Kammura, M. Iwami and K. Kawabe, *J. Appl. Phys.* **16** (1977) 1623.
558. K. Shohno, H. Ohatake and B. Bloen, *J. Crystal Growth* **45** (1979) 187.
559. Y. Hirai and K. Shono, *J. Crystal Growth* **41** (1977) 124.
560. S. Motojima, Y. Miura, K. Sugiyama and Y. Takahashi, *Bull. Chem. Soc. Japan* **48** (1975) 3161.
561. S. Motojima, S. Yokoe and K. Sugiyama, *J. Crystal Growth* **49** (1980) 1.
562. S.G. Garg and J.H. Nevin, *Mater. Res. Bull.* **16** (1981) 1395.
563. J.O. Schmitt, J.. Edgar, L. Liu, R. Nagarajan, T. Szyszko, S. Podsiadlo, W. Gebicki, H.M. Meyer and S. Speakman, *E-MRS Fall Meeting* Symp. C (2004).
564. Y. Kumashiro, in *"Electric Refractory Materials"* Ch. 22 (Taylor and Francis, 2000).
565. C.E. Whitely, Y. Zhangm Y. Gongm S. Bakalova, A. Mayo, J.H. Edger and M. Kunall, *J. Crystal Growth* **318** (2011) 553.
566. P. Becla, A. F. Witt, *Prog. Rep. DOE/NV*11627-TI (1995).
567. W.P. Allred, B. Paris and M. Genser, *J. Electrochem. Soc.* **105** (1958) 93.
568. H.A. Schell, *Z. Metallk* **46** (1955) 58.
569. R. Linnnebach and K.W. Benz, *J. Crystal Growth* **53** (1981) 579.
570. V.E. Kutny, A.V. Rybka, A.S. Abyzov, L.N. Davydov, V.K. Komar, M.S. Roland, C.F. Smith, *Nucl. Instrum. Methods A* **458** (2001) 448.
571. W.P. Allred, W.L. Mefferd and R.K. Willardson, *J. Electrochem. Soc.* **107** (1960) 117.
572. C.T. Lin, E. Schönherr, H. Bender and C. Busch, *J. Crystal Growth* **94** (1989) 955.
573. C.T. Lin, E. Schönherr and H. Bender, *J. Crystal Growth* **104** (1990) 653.
574. K.B. McAfee, Jr., D.M. Gay, R.S. Hozack, R.A. Laudise and W.A. Sunder, *J. Crystal Growth* **88** (1988) 488.
575. A.F. Witt, P. Becla, C. Counderman, J. DiFrancesco, G. Landahl, K. Morse and J. Sanchez, *Final Report* DOE/NV/11627-2 (1996)
576. V. R. Pino, Y. Ko and P.S. Dutta, *J. Crystal Growth* **290** (2006) 29.
577. G.A. Slack, *Mat. Res. Soc. Symp. Proc.* **512** (1998) 35.

578. R. Dalmau, B. Ragothamacher, M. Dudley, R. Schlesser and Z. Sitar, *Mat. Res. Soc. Symp. Proc.* **798** (2004) Y.2.9.1.
579. M. Bickermann, B.M. Epelbaum and A. Winnnacker, *J. Crystal Growth* **269** (2004) 432.
580. R.A. Youngman and J.H. Harris, *J. Am. Ceram. Soc.***73** (1990) 3238.
581. S.Y. Karpov, A.V. Kulik, A. Segal, M. Ramm and Y. Makarov, *phys. stat. sol. (a)* **188** (2001) 763.
582. G.A. Slack, L.J. Schowalter, D. Morelli and J.A. Freitas, Jr., *J. Crystal Growth* **246** (2002) 287.
583. F.A. Ponce, D. Cherns, W.T. Young and J.W. Steeds, *Appl. Phys. Lett.* **69** (1996) 770.
584. D. Kapolnex, X.H. Wu, B. Heying, S. Kellere, U.K. Mshra, S.P. Denbaars and J.S. Speck, *Appl. Phys. Lett.* **67** (1995) 1541.
585. H. Morkoc, *Mater. Sci. Engineer. R* **33** (2001) 135.
586. J. L. Weyher, P.D. Brown, J.L. Rouviere, T. Wosinski, A.R.A. Zauner and I. Grzegory, *J. Crystal Growth* **210** (2000) 151.
587. J.L. Weyher, L. Macht, G. Kalmer, J. Borysiuk and I. Grzegory, *phys. stat. sol. (c)* **0** (2003) 821.
588. G. Kalmer, J.L. Weyher, I. Grzegory, E. Jezierska and T. Wosinski, *J. Crystal Growth* **246** (2002) 21.
589. C. Youtsey, L.T. Romano and I. Adesida, *Appl. Phys. Lett.* **73** (1998) 797.
590. C. Youtsey, L.T. Romano, R.J. Molnar and I. Adesida, *Appl. Phys. Lett.* **74** (1999) 3537.
591. J.L. Weyher, F.D. Tichelaar, H.W. Zandbergen, L. Macht and P.R. Hageman, *J. Appl. Phys.* **90** (2001) 6105.
592. H. Lu, X.A. Cao, S.F. LeBoeuf, H.C. Hong, E.B. Kaminsky and S.D. Arthur, *J. Crystal Growth* **291** (2006) 82.
593. J.L. Weyher, G. Kamler, G. Nowak, J. Borysiuk, B. Luczink, M. Krysko, I. Grzegory and S. Porowski, *J. Crystal Growth* **281** (2005) 135.
594. D. Zhuang, J.H. Edgar, B. Strojek, J. Chaudhuri and Z. Rek, *J. Crystal Growth* **262** (2004) 89.
595. D. Zhuang and J.H. Edger, *Mater. Sci. Engineer. R* **48** (2005) 1.
596. T. Taniguchi, K. Watanabe and S. Koizumi, *phys. stat. sol. (a)* **201** (2004) 2573.
597. R.T. Bondokov, K.E. Morgan, R. Shety, W. Liu, G.A. Slack, M. Goorsky and L.J. Schowalter, *Mat. Res. Soc. Symp. Proc.* **892** (2006) 763.
598. M. Bickermann, S. Schmidt, B.M. Epelbaum, P. Heimann, S. Nagata and A. Winnacker, *phys. stat. sol. (c)* **4** (2007) 2609.
599. M. Bickermann, S. Schmidt, B.M. Epelbaum, P. Heimann, S. Nagata and A. Winnacker, *J. Crystal Growth* **300** (2007) 299.
600. S.B. Schujman, L.J. Schowalter, R.T. Bondokov, K.E. Morgan, W. Liu, J.A. Smart and T. Bettles, *J. Crystal Growth* **310** (2008) 887.
601. J. Elsner, R. Jones, P.K. Sich, V.D. Porezag, M. Elstner, Th. Frauenhei,. M.I. Heggie, S. Oberg and P.R. Briddon, *Phys. Rev. Lett.* **79** (1997) 3672.
602. N.G. Weimann, L.F. Eastman, D. Doppalapudi, H.M. Ng and T.D. Mouskakas, *J. Appl. Phys.* **83** (1998) 3656.
603. D.C. Look and J.R. Sizelove, *Phys. Rev. Lett.* **82** (1999) 1237.
604. J.M. Baranowski and S. Porowski, *Proc. XXIII Int. Conf. Semicon. Phys.* eds. M. Shcaffen and A. Zimmermeann (World Scientific, Singapore, 1996) p. 497.
605. S.K. Hong, B.J. Kim, H.S. Park, Y. Park, S.Y. Yoon and T.I. Kim, *J. Crystal Growth* **191** (1998) 275.
606. Y. Ono, Y. Iyechika, t. Takada, K. Inui and T. Matuse, *J. Crystal Growth* **189/190** (1998) 133.
607. W. Qian, M. Skowronski, K. Doverspike, L.B. Rowland and D.K. Gaskill, *J. Crystal Growth* **151** (1996) 396.
608. S.K. Hong, T. Yao, B.J. Kim, S.Y. Yoon and T.I. Kim, *Appl. Phys. Lett.* **77** (2000) 82.
609. S. Sakai, *J. Jpn Assoc. Crystal Growth (in japanese)* **27** (2000) 194.
610. I. Grzegory, J.A. Kozubowski, J. Borysiuk, J.L. Weyher, M. Bockowski, B. Lucznik and S. Porowski, *phys. stat. sol. (a)* **216** (1999) 537.
611. S. Nakamura, in *Introduction to Nitride Semiconductor Blue Lasers and Light Emitting Diodes*, eds. S. Nakamura and S. Chichibu (Taylor & Francis, 2000).
612. S.D. Lester, F.A. Ponce, M.G. Craford and D.A. Steigerwald, *Appl. Phys. Lett.* **66** (1995) 1249.
613. Y. Narukawa, Y. Kawakami, M. Funato, S. Fujita, S. Fujita and S. Nakamura, *Appl. Phys. Lett.* **70** (1997) 981.
614. L. Sugiura, *Appl. Phys. Lett.* **70** (1997) 1317.
615. S. Nagahama, T. Kozaki, T. Yanamoto and T. Mukai, *Oyo Buturi (Monthly Publ. Jpn. Soc. Appl. Phys.) (in japanese)* **73** (2004) 210.
616. S. Nakamura, *Science* **281** (1998) 956.
617. P. Kozodoy, J.P. Ibbetson, H. Marchand, P.T. Fini, S. Keller, J.S. Speck, S.P. DenBaars and U.K. Mishra, *Appl. Phys. Lett.* **73** (1998) 975.
618. K. Leung, A.F. Wright and E.B. Stechel, *Appl. Phys. Lett.* **74** (1999) 2495.
619. S.J. Rosner, E.C. Carr, M.J. Ludowise, G. Girolami and H.I. Erikson, *Appl. Phys. Lett.* **70** (1997) 420.
620. B. Monemar and B.E. Sernelius, *Appl. Phys. Lett.* **91** (2007) 181103.

621. X.H. Wu, C.R. Elsass, A. Arare, M. Mack. S. Keller, P.M. Petroff, S.P. DenBaars, J.S. Speck and S.J. Rosner, *Appl. Phys. Lett.* **72** (1998) 692.
622. J.W. Hsu. M.J. Manfra, S.N.G. Chu, C.H. Chen, L.N. Pfelffer and R.J. Molnar, *Appl. Phys. Lett.* **78** (2001) 3980.
623. *Compound Semiconductors* **August** (2001) 23.
624. B. Raghothamachar, J. Bai. M. Dudley, R. Dalmau, D. Zhung, Z. Herro, R. Schlesser, Z. Sitar, B. Wang, M. Callahan, K. Rakes, P. Konkapaka and M. Spencer, *J. Crystal Growth* **287** (2006) 349.
625. J.E. Northrup, J. Neugebaure and L.T. Romano, *Phys. Rev. Lett.* 77 (1996) 103.
626. Z.Z. Bandic and T.C. McGil, *Phys. Rev.* **56** (1997) 3564.
627. W. Götz, L.T. Romano, B.S. Kruser, N.M. Johnson and R.J. Molnar, *Appl, Phys. Lett.* **69** (1996) 252.
628. M. Leszczynski, I. Grzegory, H. Teisseyre, T. Suski, M. Bockowski, J. Jun, J.M. Baranowsku, S. Porowski and J. Domagala, *J. Crystal Growth* **169** (1996) 235.
629. B. Raghothamachar, J. Bai. M. Dudley, R. Dalmau, D. Zhung, Z. Herro, R. Schlesser, Z. Sitar, B. Wang, M. Callahan, K. Rakes, P. Konkapaka and M. Spencer, *J. Crystal Growth* **287** (2006) 349.
630. J. Bai, M. Dudley, R. Raghothamachar, P. Gouma, B.J. Skromme, L. Chen, P.J. Hartlieb, E. Michaels and J.W. Kolis, *Appl. Phys. Lett.* **84** (2004) 3289.
631. E. Goovaerts, S.V. Nstor, D. Ghica and T. Taniguchi, *phys. stat. sol. (a)* **201** (2004) 2591.
632. D.W. Jenkins and J.D. Dow, *Phys. Rev. B* **39** (1989) 3317.
633. N. Nepal, K.B. Nam, M.L. Nakarmi, J.Y. Lin, H.X. Jaing, J.M. Zavada and R.G. Wilson, *Appl. Phys. Lett.* **84** (2004) 1090.
634. F. Tuomisto, J.-M. Mäki, T. Yu. Chemekova, Yu.N. Makarov, O.V. Avdeev, E.N. Mokhov, A.S. Segal, M.G. Ramm, S. Davis, G. Huminic, H. Helava, M. Bickermann and B.M. Epelbaum, *J. Crystal Growth* **310** (2008) 3998.
635. T. Mattila and R. M. Nieminen, *Phys. Rev. B* **55** (1997) 55.
636. D.W. Jenkins and J.D. Dow, *Phys. Rev. B* **39** (1989) 3317.
637. J. Neugebauer and C.G. van de Walle, *Phys. Rev. B* **50** (l994) 8067.
638. J. Neugebauer and C.G. van de Walle, *Appl. Phys. Lett.* **68** (1996) 1829.
639. J. Neugebauer and C.G. van de Walle, *Appl. Phys. Lett.* **69** (1996) 503.
640. R. Boguslawski, E. L. Briggs and J. Bernholc, *Phys. Rev. B* **51** (1995) 17255.
641. R. Boguslawski, E. L. Briggs and J. Bernholc, *Appl. Phys. Lett.* **69** (1996) 233.
642. R. Boguslawski and J. Bernholc, *Phys. Rev. B* **56** (1997) 9496 .
643. S. Limpijumnong and C. G. Van de Walle, *Phys. Rev. B* **69** (2004) 035207.
644. T. Mattila and R. M. Nieminen, *Phys. Rev. B* **55** (1997) 9571.
645. I. Gorczyca, A. Svane, and N. E. Christensen, *Phys. Rev. B* **60** (1999) 8147.
646. J. Neugebauer and C. G. Van de Walle, in *Festkörperprobleme/Advances in Solid State Physics Vol. 35,* ed. R. Helbig (Vieweg, Braunschweig, 1996) p. 25.
647. C.G. Van de Walle and J. Neugebauer, *J. Appl. Phys.* **95** (2004) 3851.
648. I. Gorczyca, N. E. Christensen and A. Svane, *Phys. Rev. B* **66** (2002) 075210.
649. K. Saarinen, P. Seppälä, J. Oila, P. Hautojärvi, C. Corbel, O. Briot and R.L. Aulombard, *Appl. Phys. Lett.* 73 (1998) 3253.
650. K. Saarinenm J. Nissilä, J. Oila, V. Ranki, M. Hakala, M.J. Puska, P. Hautojärvi, J. Likonen, T. Suski, I. Grzegory, B. Lucznik and S. Porowski, *Physica B* **273-274** (1999) 33.
651. A. Uedono, S. Ishibashi, T. Ohdaira and R. Suzuki, *J. Crystal Growth* **311** (2009) 3075.
652. F. Tuomisto, J.-M. Mäki and M. Zajac, *J. Crystal Growth* **312** (2010) 2620.
653. E. Yamaguchi and R.R. Junnarkar, *J. Crystal Growth* **189-190** (1998) 570.
654. S. Chen, A. Uedono, J. Seo, J. Sawahata and K. Akimoto, *J. Crystal Growth* **311** (2009) 3097.
655. K.H. Chow, G.D. Watkins, A. Usui, and M. Mizuta, *Phys. Rev. Lett.* **85** (2000) 2761.
656. T. Mattila, A.P. Seitsonen, and R. Nieminen, *Phys. Rev. B* **54** (1996) 1474.
657. D. J. Chadi and K. J. Chang, *Phys. Rev. Lett.* **60** (1988) 2187.
658. J. Dabrowski and M. Scheffier, *Phys. Rev. Lett.* **60** (1988) 2183.
659. A.F. Wright, *J. Appl. Phys.* **92** (2002) 2575.
660. C.G. Van de Walle, J. Neugebauer, C. Stampel, M. D. McCluskey and N. M. Johnson, *Acta Phys. Pol. A* **96** (1999) 613.
661. I. Gorczyca, A. Svane and N. E. Christensen, *Phys. Rev. B* **61** (2000) 7494.
662. U. Kaufmann, M. Kunzer, M. Maier, H. Obloh, A. Ramakrishan, B. Santic and P. Schlotter, *Appl. Phys. Lett.* **72** (1998) 1326.
663. C.D. Latham, R. Jones, S. Oberg, R.M. Nieminen and P.R. Bridden, *Phys. Rev. B* **68** (2003) 205209.
664. K. Saarinen, V. Ranki, T. Suski, M. Bockowsku and I. Grzegory, *J. Crystal Growth* **246** (2002) 281.
665. S.-G. Lee and K.J. Chang, *Semicond. Sci. Tehcnol.* **14** (1999) 138.

666. C.G. Van de Walle, *Phys. Rev. B* **56** (1997) R10020.
667. C.G. Van de Walle, S. Limpijumnong, and J. Neugebauer, *Phys. Rev. B* **63** (2001) 245205.
668. J. Neugebauer and C. G. Van de Walle, *Phys. Rev. Lett.* **75** (1995) 4452.
669. B. Szucs, A. Gali, Z. Hajnal, P. Deak and C.G. Van de Welle, *Phys. Rev. B* **68** (2003) 085202.
670. C.D. Latham, R. M. Nieminen, C. J. Fall, R. Jones, S. Oberg and P. R. Briddon, *Phys. Rev. B* **67** (2003) 205206.
671. A.Y. Polyakov, N.B. Smironov, A.V. Govorkov, T.G. Yugova, K.D. Scherbatchev, O.A. Avdeev, T. Yu Chemekova, E.N. Morkhov, S.S. Nagaluk, H. Helava and Yu.N. Makarov, *Physica B* **404** (2009) 4939.
672. B. Monemar, P.P. Paskov, J.P. Bergman, A.A. Toporov, T.V. Shubina, *phys. stat. sol. (b)* **244** (2007) 1759.
673. R. Zeisel, M.W. Bayerl, S.T.B. Goennenwein, R. Dimitrov, O. Ambacher, M.S. Brandt and M. Stutzmann, *Phys. Rev. B* **61** (2000) R16283.
674. V.A. Sotamov, I.V. Ilyin, A.A. Solamova, D.O. Tolmachev, E.N. Mokhov and P.G. Baranov, *Diamod Relat. Mater.* **20** (2011) 1085.
675. H. Morkoc, *Handbook of Nitride Semiconductors and Devices (Vol. 1)* (Wiley-Vch, 2008) p. 938.
676. P. Hacke, T. Detchprohm, K. Hiramatsu, N. Sawaki, K. Tadatomo and K. Miyaje, *J. Appl. Phys.* **76** (1994) 304.
677. W. Götz, N.M. Johnson and D.P. Pbour, *Appl. Phys. Lett.* **68** (1996) 3470.
678. C.D. Wang, L.S. Yu, S.S. Lau, E.T310. Yu, W. Kim, A.E. Botchkarev and H. Morkoc, *Appl. Phys. Lett.* **72** (1998) 1211.
679. W. Kim, A.E. Botchkarev, H. Morkoc, Z.-Q. Fang, D.C. Look and D.J. Smith, *J. Appl. Phys.* **84** (1998) 6680.
680. Q.-S. Zhu, H. Nagai, Y. Kawaguchi, K. Hiramatsu and N. Sawaki, *J. Vac. Sci. Technol. A* **18** (2000) 261.
681. H.K. Cho, K.S. Kim, C.-H. Hong and H.J. Lee, *J. Crystal Growth* **223** (2001) 38.
682. L.Lee, F.C. Chang, H.M. Chung, M.C. Lee, W.H. Chen, W.K. Chen and B.R. Huang, *Chin. J. Phys.* **40** (2002) 424.
683. L. Wu. W.E. Meyer, F.D. Auret and M. Hayes, *Physica B* **340-342** (2003) 475.
684. C.B. Soh, S.J. Chua, H.F. Lim, D.Z. Chi, S. Tripathy and W. Liu, *J. Appl. Phys.* **96** (2004) 1341.
685. C.B. Soh, S.J. Chua, H.F. Lim, D.Z. Chi, W. Liu and S. Tripathy, *J. Phys.; Cond. Mater.* **16** (2004) 6305.
686. M.A. Py, Ch. Zellweger, V. Wagner, J.-F. Carlin, H.-J. Buehlmann and M. Ilgems, *phys. stat. sol. (a)* **202** (2005) 572.
687. A. Armstrong, A.R. Arehart, D. Green, U.K. Mishra, J.S. Speck and S.A. Ringle, *J. Appl. Phys.* **98** (2005) 053704.
688. D.C. Look, Z.-Q. Fang and B. Claflin, *J. Crystal Growth* **281** (2005) 143.
689. Y. Tokuda, Y. Matsuoka, H. Ueda, O. Ishiguro, N. Soejima and T. Kachi, *Superlatt. Microstruc.* **40** (2006) 268.
690. J. Nie, E.A. Chevtchenko, J. Xie, X. Ni and H. Morkoc, *Proc. SPIE* **6894** (2008) 689424.
691. A.R. Arehart, A. Corrison, C. Poblenz, J.S. Speck U.K. Mishra and S.A. Ringel, *Appl. Phys. Lett.* **93** (2008) 112101.
692. T. Mitsunaga, Y. Yagishita, M. Kurouchi, S. Kishimoto, J. Osaka and T. Mizutani, *phys. stat. sol. (c)* **5** (2008) 3032.
693. A.Y. Polyakov, N.B. Smirov, A.V. Govorkov, A.V. Markov, E.B. Yakimov, P.S. Vergeles, H. Amano and T. Kawashima, *J. Crystal Growth* **311** (2009) 2923.
694. A. Armstrong, C. Poblenz, U.K. Mishra, J.S. Speck and S.A. Ringle, *phys. stat. sol.* **244** (2007) 1867.
695. U.V. Desnica, M. Pavlovic, Z.-Q. Fang and D.C. Look, *J. Appl. Phys.* **92** (2002) 4126.
696. Y. Gai, J. Li, Q. Hou, X. Wang, H. Xiao, C. Wang and J. Li, *J. Phys. D: Appl. Phys.* **4** (2009) 155403.
697. G.-C. Yi and B.W. Wessels, *Appl. Phys. Lett.* **68** (1996) 3769.
698. Y. Nakano, N. Matsuki, Y. Irokawa and M. Sumiya, *Jpn. J. Appl. Phys.* **50** (2011) 01AD02.
699. T. Ito, Y. Nomura, S.L. Selvaraj and T. Egawa, *J. Crystal Growth* **310** (2008) 4896.
700. C.H. Qiu and J.I. Pankove, *Appl. Phys. Lett.* **70** (1997) 1983.
701. T.W. Kang, Sh.U. Yuldashev, C.S. Park, C.S. Chi, S.H. Park, Y.S. Ryu and T.W. Kim, *Solid State Comm.* **112** (1999) 637
702. S.U. Yuldashev, T.W. Kang, W.H. Jung, M.Y. Kwak, N.H. Kim, Y.S. Ryu, D.Y. Kim, T.W. Kim and K.S. Chung, *Jpn. J. Appl. Phys.* **39** (2000) 5044.
703. M. Palczewska, B. Sychanek, R. Dwilinski, K. Paku, A. Wagner and M. Kaminska, *MRS Int. J. Nitride Semi. Res.* **3** (1998) article 45.
704. J. Bauer, K. Maier, M. Kunzer, U. Kaufmann, J. Schneider, H. Amano, I. Akasaki, T. Dechprohhm and K. Hiramatsu, *Appl. Phys. Lett.* **64** (1994) 857.
705. I. Shalish, L. Kronik, G. Segal, Y. Rosenwaks, Y. Shapira, U. Tisch and J. Salzman, *Phys. Rev. B* **59** (1999) 9748.

706. A. Sedhan, J. Li, J.. Lin and H.X. Jiang, *Appl. Phys. Lett.* **96** (2010) 151902.
707. I. Shalish, C.E.M. de Oliveira and Y. Shapira, *Phys. Rev. B* **64** (2001) 205313.
708. R.X. Wang, S.J. Xu, S.L. Shi, C.D. Beling, S. Fung, D.G. Zhao, H. Yang and X.M. Tao, *Appl. Phys. Lett.* **89** (2006) 143505.
709. R.P. Vaudo, X. Xu, A. Salant, J. Malcarne and G.R. Brandes, *phys. stat. sol. (a)* **200** (2003) 18.
710. D.J. Chadi, *Appl. Phys. Lett.* **71** (1997) 2970.
711. J.L. Lyons, A. Janotti and C.G. Van de Walle, *Appl. Phys. Lett.* **97** (2010) 152108.
712. T. Mattila and A. Zunger, *Phys. Rev. B* **58** (1998) 1367.
713. T. Hashizume, *Jpn. J. Assoc. Crytal Growth (in japanese)* **36** (2009) 59.
714. D.C. Look, *phys. stat. sol.* **228** (2001) 293.
715. D. Johnstone, S. Biyikli, S. Dogan, Y.T. Moon, F. Yun and H. Morkoc, *Proc. SPIE* **5739** (2005) 7.
716. P. Hacke, P. Ramvall, S. Tanaka, Y. Aoyagi, A. Kuramatam K. Horino and H. Munekata, *Appl. Phys. Lett.* **74** (1999) 543.
717. N.M. Schmidt, D.V. Davydov, V.V. Emtsev. I.L. Kresnikov, A.A. Levedev, W.V. Lundin, D.S. Poloskin, A.V. Sakhaov, A.S. Usikov and A.V. Osinsky, *phys. stat. sol. (b)* **216** (2001) 533.
718. P. Muret, A. Phillipe, E. Monroy, E. Munoz, B. Beaumont. F. Omnes and G. Gilbart, *Mater. Sci. Engineer. B* **82** (2001) 91.
719. Q.-S. Zhu, H. Nagai, Y. Kawaguchi, K. Hiramatsu and N. Sawaki, *J. Vac. Sci. Technol. A* **18** (2000) 261.
720. D.J. Kim, D.Y. Thy, N.A. Bojarczuk, J. Karasinski, S. Guha, S.H. Lee and J.H. Lee, *J. Appl. Phys.* **88** (2000) 2564.
721. A. Armstrong, G. Thaler and D.D. Koleske, *J. Appl. Phys.* **105** (2009) 043712.
722. M.A. Reshchikov and H. Morkoc, *J. Appl. Phys.* **97** (2005) 061301.
723. C.H. Seager, A.F. Wright, J. Yu and W. Götz, *J. Appl. Phys.* **92** (2002) 6553.
724. D.S. Green, U.K. Mishra and J.S. Speck, *J. Appl. Phys.* **95** (2004) 8456.
725. A. Armstrong, C. Poblenz, D.S. Green, U.K. Mishra, J.S. Speck and S.A. Ringel, *Appl. Phys. Lett.* **88** (2006) 082114.
726. V. Fiorentini, L.F. Fernardini, A. Bosin and D. Vanderbilt, in *Proc. 23rd Int. Conf. Phys. Semicond. Vol. 4* (World Scientific, 1996) p. 2877.
727. S.N. Mohammad, A.A. Salvador and H. Morcoc, *Proc. IEEE* **83** (1995) 1306.
728. B.E. Foutz, S.K. O'Leary, M.S. Shur and L.F. Eastman, *J. Appl. Phys.* **85** (1999) 7727.
729. V.W. Chin, T.L. Tansley and T. Osotachn, *J. Appl. Phys.* **75** (1994) 7365.
730. T. Taniguchi, K. Watanabe, S. Koizumi, T. Sekiguchi, I. Sakaguchi and S. Yamaoka, *Diamond Relat. Mater.* **12** (2003) 1098.
731. T. Taniguchi, T. Terajim S. Koizumi, K. Waanabe and S. Yamaoka, *Jpn. J. Appl. Phys.* **41** (2002) L109.
732. F. Oba, A. Togo and I. Tanaka, *Phys. Rev. B* **81** (2010) 075125.
733. C. H. Park and D. J. Chadi, *Phys. Rev. B* **55** (1997) 12995.
734. O. Mishima, J. Tanaka, S. Yamaoka and O. Fukunaga, *Science* **238** (1987) 181.
735. M. Kasu and N. Kobayashi, *J. Crystal Growth* **221** (2000) 739.
736. L.X. Yun, L. Riba-Salamanca, M.G. Spencer, K. Wongchigul, P. Zhou, X. Tang, V. Talyansky and T. Venkatesan, *Mat. Res. Soc. Symp. Proc.* **449** (1997) 555.
737. Y. Ohsima, T. Yoshida, T. Eri, M. Shibata and T. Mishima, *phys. stat. sol. (c)* **4** (2007) 2215.
738. S. L. Selvaraj, A. Watanabe and T. Egawa, *Proc. 69th Dev. Res. Conf.* (Santa barbara, 2011) p. 211.
739. S.C. Binari and H.C. Dietrich, in *GaN and Related Materials*, ed. S.J. Pearton (Gordon and Breach, New York, 1997) p. 509.
740. B. Gelmont, K. Kim and M. Shur, *J. Appl. Phys.* **74** (1993) 1818.
741. J.D. Albrecht, R.P. Wang, P.P. Ruden, M. Farahmand and K.F. Brennan, *J. Appl. Phys.* **83** (1998) 4777.
742. M. Ilgemes and H.C. Montgomery, *J. Phys. Chem. Solids* **34** (1973) 885.
743. J.I. Pankove, S. Bloom and G. Harbeke, *RCA Review* **36** (1975) 163.
744. P. Perin, T. Suski, H. Teisseyre, M. Leszczynski, I. Grzegry, J. Jun, S. Porowski, P. Boguslawski, J. Nernholc, J.C. Chervin, A. Polian and T.D. Noustakas, *Phys. Rev. Lett.* **75** (1995) 296.
745. D. Jenkins, J.D. Dow and M. Tsai, *J. Appl. Phys.* **72** (1992) 4130.
746. R. Dingle and M. Ilgems, *Solid State Commun.* **9** (1971) 4438.
747. O. Lagerstedt and B. Monemar, *J. Appl. Phys.* **45** (1974), 2266.
748. V.S. Vavilov, S.I. Makarov, M.V. Chukichev and I.F. Chetverkikova, *Sov. Phys. Semicond.* **13** (1979) 1259.
749. J.J. Song and W. Shan, in *Group III Nitride Semiconductor Compounds,* ed. B. Gil (Clarendon, Oxford, 1998) p. 182.
750. W. Götz, N.M. Johnson, C. Chen, H. Liu, C. Kuo and W. Imler, *Appl. Phys. Lett.* **68** (1996) 3144.
751. S. Nakamura, T. Mukai and M. Senoh, *Jpn. J. Appl. Phys., Part 1* **31** (1992) 2883.
752. H. Wang and A.-B. Chen, *J. Appl. Phys.* **87** (2000) 7859.

753. R. Boguslawski and J. Bernholc, *Phys. Rev. B* **56** (1997) 9496.
754. H. Amano, N. Sawaki, I. Akasaki and Y. Toyoda, *Appl. Phys. Lett.* **48** (1986) 353.
755. H. Amano, I. Akaski, T. Kozawa, K. Hiramatsu, N. Sawaki, K. Ikeda and Y. Ishii, *J. Lumin.* **40&41** (1988) 121.
756. H. Amano, M. Kito, K. Hiramatsu and I. Akasaki, *Jpn. J. Appl. Phys., Part 2* **28** (1989) L2112.
757. S. Nakamura, M. Senoh and T. Mukai, *Jpn. J. Appl. Phys., Part 2* **30** (1991) 1705.
758. S. Nakamura, N. Iwase, M. Senoh and T. Mukai, *Jpn. J. Appl. Phys., Part 1* **31** (1992) 1258.
759. S. Nakamura, T. Mukai, M. Senoh and N. Iwase, *Jpn. J. Appl. Phys., Part 1* **31** (1992) L139.
760. W. Götz, N.M. Johnson, J. Waler, D.P. Bour and R.A. Street, *Appl. Phys. Lett.* 68 (1996) 667.
761. N. Akutsu, H. Tokunaga, I. Waki, A. Yamaguchi and K. Matsumoto, *Mat. Res. Soc. Symp. Proc.* **482** (1998) 113.
762. M.E. Lin, G. Xue, G.L. Zhou, J.E. Greene and H. Morkoc, *Appl. Phys. Lett.* **63** (1993) 932.
763. G. Popovici, H. Morkoc and G.-B. S.N. Mohammad, in *Group III Nitride Semiconductor Com pounds*, ed. B. Gil (Clarendon, Oxford, 1998) p. 19.
764. I. Akasaki, H. Amano, M. Kito and K. Hiramatsu, *J. Lum.* **48&49** (1991) 666.
765. T.D. Moustakas and R.J. Molnar, *Mat. Res. Soc. Symp. Proc.* **281** (1993) 253.
766. T. Tanaka, A. Watanabe, A. Amano, H. Kobayashi, I. Akasaki, S. Yamazaki and M. Koike, *Appl. Phys. Lett.* **65** (1994) 2024.
767. M.A.L. Johnson, Z.Yu.C. Boney, W.C. Hughes, J.W. Cook, Jr., J.F. Schetzina, H. Zhao, B.J. Skromme ad J.A. Edmond, *Mat. Res. Soc. Symp. Proc.* **449** (1997) 215.
768. A.E. Hughes and S.C. Jain, *Adv. Phys.* **28** (1979) 717.
769. O. Brandt, H. Yang, H. Kostial and K.H. Ploog, *Appl. Phys. Lett.* **69** (1996) 2707.
770. H. Yoshida, K. Sato, R. Kato and T. Yamamoto, *J. Jpn Assoc. Crystal Growth (in japanese)* **27** (2000) 233.
771. J. Neugebauer and C.G. Van de Walle, *J. Appl. Phys.* **85** (1999) 3003.
772. F. Mireles and S. E. Ulloa, *Phys. Rev. B* 58(1998) 3879.
773. H. Wang and A.-B. Chen, *Phys. Rev. B* **63** (2001) 125212.
774. B. Pödör, *Semicond. Sci. Technol.* **11** (1996) 827.
775. S. Limpijutnnong, C.G. Van de Walle, and J. Neugebauer, *Mat. Res. Soc. Symp. Proc.* **639** (2001) G4.3.
776. L. E. Ramos, J. Furthmtlller, L.M.R. Scolfaro, J. R. Leite, and F. Bechstedt, *Phys. Rev. B* **66** (2002) 075209.
777. J.I. Pankove, in *GaN and Related Materials*, ed. S.J. Pearton (Gordon and Breach, New York, 1997) p. 1.
778. T. Suski, E. Litwin-Staszewsli, P. Perin, P. Wisniewski, H. Teisseyre, I. Grzegory, M. Bockowski, S. Porowski, K. Saarinen and J. Nissilä , *J. Crystal Growth* **230** (2001) 368.
779. S. Fischer, C. Wetzel, E.E. Haller and B.K. Meyer, *Appl. Phys. Lett.* **67** (1995) 1298.
780. D.J. As and U. Köhler, *J. Phys:Condens. Matter* **13** (2001) 8923.
781. T.A.G. Eberlein, R. Jones, S. O"berg and P.R. Briddon, *Appl. Phys. Lett.* **91** (2007) 132105.
782. L.E. Ramos, J. Furthmüller, J.R. Leite, L.M.R. Scolfaro and F. Bechstedt, *phys.stat. sol. (b)* **234** (2002) 864.
783. C. Poblenz, P. Walterit, S. Rajan, S. Heikman, U.K. Mishra and J.S. Speck, *J. Vac. Sci. Technol. B* **22** (2004) 1145.
784. A. Armstrong A.R. Arehart, B. Moran, S.P. DenBaars, U.K. Mishra, J.S. Speck and S.A. Ringel, *Appl. Phys. Lett.* **84** (2004) 374.
785. R.P. Vaudo, X. Xu, A. Salant, M. Chriss, J. Malcarne and G.R. Brandes, *Proc. 5th Int. Conf. Nitride Semicond.* (Nara, Japan, 2003) p. 99.
786. H. Murakami, *J. Jpn. Assoc. Crystal Growth (in japanese)* **31** (2004) 165.
787. S. Yamaguchi, M. Kariya, S. Nitta, T. Takeuchi C. Wetzel, H. Amano and I. Akasaki, *Appl. Phys. Lett.* **76** (2000) 876.
788. V.Y. Davydov, A.A. Klochikhin, R.P. Seisyan, V.V. Emtsev, S.V. Ivanov, F. Bechstedt, J. Furthmuller, H. Harima, A.V. Mudryi, J. Aderhold, O. Semchinova and J. Graul, *phys. stat. solid. (b)* **229** (2002) R1.
789. T. Matsuoka, H. Okamoto, M. Nakao, H. Harima and E. Kurimoto, *Appl. Phys. Lett.* **81** (2002) 1246.
790. J. Wu. W. Walukiewicz, K.M. Yu, J.W. Arger III, E.E. Haller, H. Lu, W.J. Shaff, Y. Saito and Y. Nanishi, *Appl. Phys. Lett.* **80** (2002) 3967.
791. Y. Saito, H. Harima, E. Kurimoto, T. Yamaguchi, N. Teraguchi, A. Suzuki, T. Araki and Y. Nanishi, *phys. stat. solid (b)* **234** (2002) 796.
792. Y. Nanishi, T. Araki and T. Miyajima, *Oyo Buturi (Monthly Publ. Jpn. Soc. Appl. Phys.) (in japanese)* **72** (2003) 565.
793. Y. Nanishi, T. Araki, A. Suzuki and H. Naoi, *Murata Sci. Found. Ann. Rep.* **18** (2004) 239.

794. H. Morkoc, *Handbook of Nitride Semiconductors and Devices Vol. 2* (Wiley-Vch, 2008) p. 491.
795. K. Watanabe, T. Taniguchi and H. Kanda, *phys. stat. sol. (a)* **201** (2004) 2561.
796. K. Watanabe, T. Taniguchi and S. Kanda, *Nat. Mater.* **3** (2004) 404.
797. Z. Remes, M. Nesladek, K. Haenen, K. Watanabe and T. Taniguchi, *phys. stat. sol. (a)* **202** (2005) 2229.
798. K. Wnanabe, T. Taniguchi, T. Kuroda, O. Tsuda and H. Kanda, *Diamond Relat. Mater.* **17** (2008) 830.
799. K. Watanabe and T. Taniguchi, *Phys. Rev. B* **79** (2009) 193104.
800. U. Vetter, H. Hofsäass and T. Taniguchi, *Appl. Phys. Lett.* **84** (2004) 4286.
801. A. Nakayama, T. Taniguchi, Y. Kubota, K. Watanabe, S. Hishita and H. Kanda, *Appl. Phys. Lett.* **87** (2005) 211913.
802. W.M. Yim, E.J. Stofko, P.J. Zanzucci, J.I. Pankove, M. Ettenberg and S.L. Gilbert, *J. Appl. Phys.* **44** (1973) 292.
803. E. Silveira, J.A. Freitas, S.B. Shujman and L.J. Schowalter, *J. Crystal Growth* **310** (2008) 4007.
804. F. Karel, J. Pastrnak, J. Hejduk and V. Losik, *phys. stat. sol.* **15** (1966) 693.
805. F. Karel and J. Pastrnak, *Czech. J. Phys. B.* **19** (1969) 78.
806. F. Karel and J. Pastrnak, *Czech. J. Phys. B.* **20** (1970) 46.
807. F. Karel and J. Mares, *Czech. J. Phys. B.* **22** (1972) 847.
808. F. Karel and J. Mares, *Czech. J. Phys. B.* **20** (1973) 652.
809. E. Archangelskii, F. Karel, J. Mares, S. Pacesova and J. Pastrnak, *phys. stat. sol. (a)* **69** (1982) 173.
810. A. Akasaki and M. Hashimoto, *Solid State Commun.* **5** (1967) 851.
811. M. Kuball, J.M. Hayes, Y. Shi, J.H. Edgar, A.D. Prins, N.W.A. van Uden and D.J. Dunstan, *J. Crystal Growth* **231** (2001) 391.
812. E. Silveira, J.A. Freitas, Jr., G.A. Slack and L.J. Schowalter, *phys. stat. sol. (c)* 0 (2003) 2618.
813. P. Perlin, C. Jauberthie-Carillon, J.P. Itie, A.S. Miguel, I. Grzegory and A. Polian, *Phys. Rev., B* **45** (1992), 83.
814. M.S. Liu, L.A. Bursill, S. Prawer, K.W. Nugent, Y.Z. Tong and G.Y. Zhang, *Appl. Phys. Lett.* **74** (1999) 3125.
815. M.A. Reshchikov and H. Morkoc, *J. Appl. Phys.* **97** (2005) 061301.
816. W. Shan, T.J. Schmidt, X.H. Yang, S.J. Hwang, J.J. Song and B. Goldenberg, *Appl. Phys. Lett.* **66** (1995) 985.
817. B. Gil, O. Briot and R. Aulombard, *Phys. Rev. B* **52** (1995) R17028.
818. G.D. Chen, M. Smith, J.Y. Lin, H.X. Jiang, S.-H. Wei, M.A. Khan and C.J. Sun, *Appl. Phys. Lett.* **68** (1996) 2784.
819. W. Shan, R.J. Hauenstein, A.J. Fishcer, J.J. Song, W.G. Perry, M.D. Bremser, R.F. Davis and B. Goldenberg, *Phys. Rev. B* **54** (1996) 13460.
820. E. Iliopoulos, D. Doppalapudi, H.M. Ng and T.D. Moustkas, *Appl. Phys. Lett.* **73** (1998) 375.
821. B. Arnaudov, T. Paskova, E.M. Goldys, R. Yakimova, S. Evtimova, I.G. Ivanov, A. Henry and B. Monmar, *J. Appl. Phys.* **85** (1999) 7888.
822. G. Neu, M. Teisseire, P. Lemasson, H. Lahreche, N. Gandjean, F. Semond, B. Beaumont, I. Grzegory, S. Porowski and R. Tribulet, *Physca B* **302-303** (2001) 39.
823. K. Fuji, G. Fujimoto, T. Goto, T. Yao, Y. Kagamitani, N. Hoshino, D. Ehrentraut and T. Fukuda, *phys. stat. sol. (a)* **204** (2007) 4266.
824. J.L. Pankove and J. A. Huchby, *Appl. Phys. Lett.* **24** (1976) 281.
825. J.L. Pankove and J. A. Huchby, *J. Appl. Phys.* **47** (1976) 5387.
826. T. Ogino and M. Aoki, *Jpn. J. Appl. Phys* **19** (1980) 2395.
827. K. Saarinen, T. Laine, S. Kuisma, J. Nissilä, P. Hautojärvi, L. Dobrzynski, J.M. Baranowski, K. Pakula, R. Stepniewski, M. Wojdak, A. Wysmoek, T. Suski, M. Lezszcynski, I. Grzegory and S. Porowski, *Phys. Rev. Lett.* **79** (1997) 3030.
828. R. Armitage, W. Hong, Y. Quing, H. Feick, J. Gebauer, E.R. Weber, S. Hautakangas and K. Saarinen, *Appl. Phys. Lett.* **82** (2003) 3457.
829. G. Li, S.J. Chua, S.J. Xu, W. Wang, P. Li, B. Beaumont and P. Gibart, *Appl. Phys. Lett.* **74** (1999) 2821.
830. J. Kang and T. Ogawa, *Appl. Phys. A: Mater. Sci. Process.* **69** (1999) 631.
831. S.O. Kucheev, M. Toth, M.R. Philips, J.S. Williams, C. Jagadish and G. Li, *J. Appl. Phys.* **91** (2002) 5867.
832. Y.C. Chang, A.E. Obehofer, J.F. Muth, R.M. Kolbas and R.f. Davis, *Appl. Phys. Lett.* **79** (2001) 281.
833. F. Reyrings and F. Tuomisto, *Proc. SPIE* **6473** (2007) 64730M.
834. R. Zhang and T.F. Kuech, *Appl. Phys. Lett.* **72** (1998) 1611.
835. R. Armitage, Q. Yan and E.R. Weber, *J. Appl. Phys.* **97** (2005) 073524.
836. M.W. Bayerl, M.S. Brandt, E.R. Glaser, A.E. Wickenden, D.D. Koleske, R.L. Hengy and M. Stutzmann, *phys. stat. sol. (b)* **216** (1999) 547.
837. D.M. Hofmann, B.K. Meyer, H. Alves, F. Letter, W. Burkhard, N. Romanov, Y. Kim, J. Kruger and E.R. Weber, *phys. stat. sol. (a)* **180** (2000) 261.

838. M. Nyk, W. Strek, J.M. Jablonski, L. Kepinski, R. Kudrawiec and J. Misiewicz, *Mater. Sci. Semicond. Process* **8** (2005) 511.
839. K. Fuji, G. Fujimoto, T. Goto, T. Yao, N. Hoshino,Y. Kagamitani, D. Ehrentraut and T. Fukuda, *phys. stat. sol. (a)* **204** (2007) 3509.
840. J. Bauer, K. Maier, M. Kunzer, U. Kaufmann, J. Schneider, H. Amano, I. Akasaki, T. Dechprohhm and K. Hiramatsu, *Appl. Phys. Lett.* **64** (1994) 857.
841. M. Gowda, J.A. Feiras Jr., R.M. Frazier, B.N. Feigelson and M.V. Rao, *J. Crystal Growth* **310** (2008) 3941.
842. K. Omae, T. Iwahashi, F. Kawahara, M. Yoshimura, Y. Mori, T. Sasaki, *Jpn. J. Appl. Phys.* **45** (2006) L173.
843. C.R. Dean, A.F. Young I. Meric, C. Lee, L. Wang, S. Sorgenfrei, K. Watanabe, T. Taniguchi, P. Kim, K.L. Shepard and J. Hone, *Nature Nanotech.* **22** (2010) 932.
844. J. Xue, J. Sanchez-Yamagishi, D. Bulmash, P. Hacquod, A. Deshpande, K. Watanabe, T. Taniguchi, P. Jarillo-Herrero and B.J. LeRoy, *Nature Mater.* **10** (2011) 282.
845. T. Nishida, T. Makimoto, H. Saito and T. Ban, *Appl. Phys. Lett.* **84** (2004) 1002.
846. Z. Ren, Q. Sun, S.-Y. Kwon, J. Han, K. Davitt, Y.K. Song, A.V. Nurmikko, W. Liu, J. Smart and L. Schowalter, *phys. stat. sol. (c)* **4** (2007) 2482.
847. N. Hashimoto, K. Akita, T. Tanabe, H. Nakahata, K. Takeda and H. Amano, *SEI Tech. Rev.* **177** (2010) 92.
848. M. Hatano, J. Yamazami, N. Yafune, S. Hashimoto, K. Akita, Y. Yamamoto and M. Kuzuhara, *Proc. 9th Top. Workshop Heterostruc. Microelectron.* (TWHM 2011) p. 13.
849. S. Sarayama, *Richo Quat. Rev.* **31** (2009) 49.
850. M. Leszczyski, P. Prystawko, R. Czernecki, J. Lehnert, T. Suski, P. Perlin P. Wisniewski, I. Grzegory, G. Nowak, S. Porowski and M. Albrecht, *J. Crystal Growth* **231** (2001) 352.
851. S. Figge, T. Bottcher, J. Dennemarck, R. Kröger, T. Paskova, B. Monemar and D. Hommel, *J. Crystal Growth* **281** (2005) 101.
852. H. Masui, S. Nakamura, S.P. DenBaars and U.K. Mishra, *IEEE Trans. Electron. Dev.* **57** (2010) b 88.
853.X.A. Cao, S.F. LeBoeuf, M.P. DíEvelyn, S.D. Arthur and J. Kretchmer, *Appl. Phys. Lett.* **84** (2004) 4313.
854. T. Kyono, K. Akita, M. Ueno, T. Nakamura and H. Hideki, *SEI Tech. Rev.* **167** (2005) 102.
855. Y. Enya, Y. Yoshizumi, T. Kyono, K. Akita, M. Ueno, M. Adachi, T. Sumitomo, S. Tokuyama, T. Ikegami, K. Katayama and T. Nakamura, *Appl. Phys. Exp.* **2** (2009) 08210.
856. Y. Yoshizumi, M. Adachi, Y. Enya, T. Kyono, S. Tokuyama, T. Sumitomo, K. Akita, T. Ikegami, M. Ueno, K. katayama and T. Nakamura, *Appl. Phys. Exp.* **2** (2009) 092101.
857. T. Kyono, Y. Shiotani, K. Akita, M. Ueno, M. Adachi, T. Sumitomo, S. Tokuyama, T. Ikegami, H. Katayama and T. Nakamura, *SEI Tech. Rev.* **176** (2010) 88.
858. M. Adachi, T. Kyono, Y. Shiotani, K. Akita, M. Ueno, T. Sumitomo, S. Tokuyama, T. Ikegami, H. Katayama and T. Nakamura, *SEI Tech. Rev.* **176** (2010) 93.
859. P. Perlin, T. Suski, M. Leszcyski, P. Prystawko, T. Swietlik, L. Marona, P. Wisniewski, R. Czernecki, G. Nowak, J.L. Wehyer, G. Lamler, J. Borysiuk, E. Litwin-Staszewska, L. Dmowski, R. Piotrzkowski, G. Franssen, S. Grzanka, I. Grzegory and S. Porowski, *J. Crystal Growth* **281** (2005) 107.
860. P. Perlin, L. Marona, K. Holc, P. Wisniewski, T. Suski, M. Leszcyski, R. Czemercki, S. Najda, M. Zajac and R. Kucharski, *Appl. Phys. Exp.* **4** (2011) 062103.
861. M. Khan, J.W. Yang, W. Knap, E. Frayssinet, X. Hu, G. Simin, P. Prystawko, M. Leszcynski, I. Grzegory, S. Porowski, R. Gaska, M.S. Shur, B. Beaumont, M. Tesseire and G. Neu, *Appl. Phys. Lett.* **76** (2000) 3807.
862. K.K. Chu, P.C. Chao, M.T. Pizzella, R. Actis, D.E. Meharry, K.B. Nicholas, R.P. Vaudo, X. Xu, J.S. Flynn, J. Dion and G.R. Brandes, *IEEE Electron. Dev. Lett.* **25** (2004) 596.
863. T. Horii, T. Miyazaki, Y. Saitoh, N. Hashimoto, T. Tanabe and M. Kiyama, *SEI Tech. Rev.* **174** (2008) 77.
864. T. Kachi, *IEICE Tech. Rep.* **ED2008-72** (2008) 171.
865. M. Kodama, M. sugimoto, E. Hayashi, N. Soejima, O. Ishiguro, M. Kanechika, K. Itoh, H. Ueda, T. Uesugi and T. Kachi, *Appl. Phys. Exp.* **1** (2008) 021104.
866. M. Okdada, Y. Saitoh, M. Yokoyama, T. Nakada, M. Yaegashi, K. Katayama, M. Ueno, M. Kiyama, T. Katsuyama and T. Nakamura, *SEI Tech. Rev.* **178** (2011) 84.
867. J. Heikenfeld, M. Gartner, D.S. Lee, R. Birkhaln and A.J. Steckl, *Appl. Phys. Lett.* 7 (1999) 1189.
868. A.J. Steckl and J.M. Zavada, *MRS Bull.* **Sep.** (1999) 33.
869. B. Damilano, N. Grandjean, F. Semond, J. Messies and M. Leroux, *Appl. Phys. Lett.* **75** (1999) 962.
870. Y. Hori, F. Enjalbert, E. Monroy, D. Jalabert, L.S. Dang, M. Tanaka, O. Oda and B. Daudin, *phys. stat. sol. (c)* **0** (2003) 2695.
871. Y. Fujiwara, Y. Terai and A. Nishikawa, *Oyo Buturi (Monthly Publ. Jpn. Soc. Appl. Phys.) (in japanese)* **79** (2010) 25.
872. T. Mukai, H. Nakamura and S. Nakamura, *Jpn. J. Appl. Phys.* **37** (1995) 479.

873. T. Mukai, *Display (in japanese)* **2** (1996) 24.
874. M. Funato, M. Ueda, Y. Kawakami, Y. Narukawa, T. Kosugi, M. Takayoshi and T. Mukai, *Jpn. J. Appl. Phys.* **45** (2006) L659.
875. C.F. Huang, T.Y. Tang, J.J. Huang, W.Y. Shiao and C.C. Yang, *Appl. Phys. Lett.* **89** (2006) 051913.
876. Z.L. Li and H.W. Choi, *J. Phys.:Conf. Ser.* **209** (2010) 012065.
877. C. Wetzel, T. Salagaj, T. Detchprohm, P. Li and J.S. Nelson, *Appl. Phys. Lett.* **85** (2004) 866.
878. C. Wetzel, M. Zhu, J. Senawiratne, T. Detchprohm and P.D. Persans, *J. Crystal Growth* **310** (2008) 3987.
879. T. Detchprohm, M. Zhu, Y. Li, Y. Xia, C. Wetzel, E.A. Preble, L. Liu, T. Paskova and D. Hanser, *Appl. Phys. Lett.* **92** (2008) 241109.
880. T. Detchprohm, M. Zhu, Y. Li, Y. Xia, L. Liu, D. Hanser and C. Wetzel, *J. Crystal Growth* **311** (2009) 2937.
881. T. Detchprohm, M. Zhu, Y. Li, S. You, C. Wetzel. E.A. Preble, T. Paskova and D. Hanser, *Appl. Phys. Lett.* **96** (2010) 051101.
882. Y. Zhao. S. Tanaka, Q. Yan, C.-Y. Huang, R.B. Chung, C.-C. Pan, K. Fujito, D. Feezell, C.G. Van de Walle, J.S. Speck, S.P. DenBaars and S. Nakamura, *Appl. Phys. Lett.* **99** (2011) 051109.
883. Y. Li, S. You, M. Zhu, L. Zhao, W. Hou, T. Detchprohm, Y. Taniguchi, N. Tamura, S. Tanaka and C. Wetzel, *Appl. Phys. Lett.* **98** (2011) 151102.
884. http://www.imect.be/ScientificReport/SR2008/HTML/1225126.html.
885. T.C. Wen, S.J. Chang, Y.K. Su. L.W. Wu, C.H. Kuo, W.C. Lai, J.K. Sheu and T.Y. Tsai, *J. Electron. Mater.* **32** (2003) 419.
886. M.S. Oh, M.K. Kwon, I.K. Park, S.H. Baek, S.J. Park, S.H. Lee and J.J. Jung, *J. Crysal Growth* **289** (2006) 107.
887. R.D. Dupuis, J.B. Limb, J.P. Liu, *Proc. SPIE* **6894** (2008) 68941D.
888. Y. Inui, *MOST Forum* (Frankfurt, 2010).
889. R. Kucharski, M. Rudzinski, M. Zajac, R. Doradzinski, J. garczynski, L. Sierzputowski, R. Kudrawiec, J. Serafinczuk, W. Strupinski and R. Dwilinski, *Appl. Phys. Lett.* **95** (2009) 131119.
890. J.I. Pankove, E.A. Miler and J.E. Berkeyheiser, *RCA Rev.* **32** (1971) 383.
891. S. Yoshida, S. Misawa and S. Gonda, *Appl. Phys. Lett.* **42** (1983) 427.
892. H.P. Maryska and J.J. Tietjen, *Appl. Phys. Lett.* **15** (1969) 327.
893. S. Nakamura, *Jpn. J. Appl. Phys.* **30** (1991) L1705.
894. S. Nakamura, *Jpn. J. Appl. Phys.* **31** (1992) L189.
895. S. Nakamura, T. Murai and M. Senoh, *Jpn. J. Appl. Phys.* **30** (1991) L1998.
896. S. Nakamura, T. Mukai, and M. Senoh, *Appl. Phys. Lett.* **64** (1994) 1687.
897. S. Nakamura, M. Senoh, N. Iwase and S. Nagahama, *Jpn. J. Appl. Phys.* **34** (1995) L797.
898. S. Nakamura, M. Senoh, N. Iwase, S. Nagahama, Y. Yamada and T. Mukai, *Jpn. J. Appl. Phys.* **34** (1995) L1332.
899. T. Mukai, K. Takakawa and S. Nakamura, *Jpn. J. Appl. Phys.* **37** (1998) L839.
900. S. Nakamura and G. Falso, *The Blue Laser Diodes* (Springer-Verlarg, Berlin, 1997)
901. D. Steigerwald, S. Rudaz, H. Liu, S. Kern, W. Götz and R. Fletcher, *JOM* **Sept.** (1997) 18.
902. K. Bando and T. Kususe, *Oyo Buturi (Monthly Publ. Jpn. Soc. Appl. Phys.) (in japanese)* **71** (2002) 1518.
903. K. Bando, Y. Noguchi, K. Sakano and Y. Shimizu, *Phospor Res. Soc. 264th Meeting Tech. Digest* (Tokyo, Nov. 29, 1996) p. 5.
904. P. Mottler, *LEDs for Lighting Applications* (John Wiley &Sons, 2009)
905. G. Bogner, A. Debray, G. Heidel, K. Höhn, U. Müller and O. Schotter, *Proc. SPIE* **3621** (1999) 143.
906. N. Hirosaki, R.J. Xie and K. Sakura, *Oyo Buturi (Monthly Publ. Jpn. Soc. Appl. Phys.) (in japanese)* **74** (2005) 1451.
907. H. Saito, T. Morioka, O. Kawasaki, T. Yuasa, M. Taneya, S. Tanabe, Y. Kawakami, M. Funato, S. Fujita and M. Harada, *Sharp Tech. Rep. (in japanese)* **91** (2005) 50.
908. F. Orito, *"Next Generation LED Lighting Seminor"* (Denpa Publ. Inc., 2010 August).
909. H. Wu, Z, Zhang, C. Guo, J. Xu, M. Wu and Q. Su, *IEEE Tech. Lett.* **17** (2005) 1041.
910. J.K. Park, M.A. Lim, C.H. Kim, H.D. Park, J.T. Park, *Appl. Phys. Lett.* **82** (2003) 683.
911. Z. Sun, J. Zhang, X. Zhang, S. Lu and X. Wang, *J. Lum.* **122-123** (2007) 955.
912. Y. Miyazaki, N. Shimoda, K. Fujiwara, I. Kuzuhara, R. Yokotani and S. Mori, *Matsushita Elect. Works Tech. Rep.* **55** (2007) 5.
913. T. Mukai, M. Yamada, T. Mitani, Y. Narukawa, S. Shioji, I. Niki, S. Sonobe, K. Izuno and R. Suenaga, *Proc. SPIE* **4996** (2003) 156.
914. T. Taguchi, *IEICE Transactions* **J84-C** (2001) 1040.
915. *Ext. Abst. Int. Symp. The light for the 21st Century* (March 7-8, 2002, Tokyo)

916. K. Tadatomo, H. Okagawa, T. Jyoichi, M. Kato, M. harada and T. Taguchi, *Mistubishi Cable Ind. Bull.* **99** (2002) 35.
917. J.K. Sheu, S.J. chang, C.H. Kuo, Y.K. Su, L.W. Wu, Y.C. Lin, W.C. Lai, J.M. Tsai, G.C. Chi and R.K. Wu, IEEE Photo. *Tech. Lett.* **15** (2003) 18.
918. *"Light Emitting Diodes (LEDs) for General Illumination" OIDA Rep.* (March 2001).
919. K. Bando and T. Kususe, *Appl. Phys.* **71** (2002) 1518.
920. D.A. Steigerwald, *IEEE J. Selec. Top. Quan. Electron.* 8 (2002) 310.
921. T. Taguchi, *O plus E (in japanese)* **26** (2004) 275.
922. Y. Narukawa, S. Nagahama, H. Tamaki and T. Mukai, *Oyo Buturi (Monthly Publ. Jpn. Soc. Appl. Phys.) (in japanese)* **74** (2005) 1423.
923. T. Taguchi, *IEICE Tech. Rep.* **ED2005-27** (2005) 63.
924. Y. Yoshizumi, K. Yoshida, T. Kyono, M. Ueno and T. Nakamura, *SEI Tech. Rev.* **179** (2011) 51.
925. A. Dussainge, J. Braut, B. Damilano and J. Messier, *phys. stat. sol. (c)* **4** (2007) 57.
926. U. Hori, *Doctorate Thesis* (Nagoya Institute of Technology, 2007).
927. H. Amano, *Oyo Buturi (Monthly Publ. Jpn. Soc. Appl. Phys.) (in japanese)* **74** (2005) 1433.
928. D. Morita, S. Masui, S. Nagahama and T. Mukai, *Laser Recearch (in japanese)* **June** (2006) 392.
929. S. Nakamura, *Microelectron. J.* **26** (1994) 651.
930. K. Tadatomo, H. Okagawa, Y. Ohuchi, T. Tsunekawa, Y. Imada, M. Kato and T. Taguchi, *Jpn. J. Appl. Phys.* **40** (2001) L583.
931. K. Tadatomo, Y. Sudo, M. Kato and T. Taguchi, *J. Light & Vis. Env.* **27** (2003) 10.
932. H. Kudo, Y. Ohuchi, T. Jyouichi, T. Tsunekawa, H. Okagawa, K. Tadatomo, Y. Sudo, M. Kato and T. Taguchi, *phys. stat. sol. (a)* **200** (2003) 95.
933. Y. Oba, K. Kaneko, H. Katsuno and M. Kushibe, *Appl. Phys. Exp.* **1** (2008) 101101.
934. H. Amano, *Oyo Buturi (Monthly Publ. Jpn. Soc. Appl. Phys.) (in japanese)* 71 (2002) 1330.
935. A. uasan, R. McClintock, K. Mayers, D. Shiell, L. Gautero, S.R. Darvish, P. Kung and M. Razeghi, *Appl. Phys. Lett.* **83** (2003) 4701.
936. T. Wang, P.J. Parbrook, C.N. Harrison, J.P. Ao and Y. Ohno, *J. Crystal Growth* **267** (2004) 583.
937. A.J. Fischer, A.A. Allerman, M.H. Crawford, K.H.A. Bogart, S.R. Lee, R.J. Kaplar, W.W. Chow, S.R. Kurtz, K.W. Fullmer and J.J. Figiel, *Appl. Phys. Lett.* **26** (2004) 3394.
938. K. Mayers, A. Yasan. R. McClintock, D. Shiell, S.R. Darvish, P. Kung and M. Razeghi, *Appl. Phys. Lett.* **84** (2004) 1046.
939. V. Adivaraha, A. Heidari, B. Zhang, Q. Fareed, M. Isalam, S. Hqang, K. Balakrishanan and A. Khan, *Appl. Phys. Lett.* **2** (2009) 092102.
940. A.S. Usikov, Yu. Melnik, A.I. Pechinikov, V.A. Soukhoveev, O.V. Kovalenkov, E. Shapovalova, S.Yu Karpov and V.A. Dmitriev, UV Solid-State Light Emitters and Detectors, *NATO Science Series* **144** (2004), 15.
941. H. Hirayama, A. Kionoshita, A. Hirata and Y. Aoyagi, *phys. stat. sol. (a)* **188** (2001) 83.
942. H. Hirayama, A. Kinoshita, T. Yamanaka, A. Hirata and Y. Aoyagi, *Mat. Res. Soc. Symp. Proc.* **639** (2001) G.2.8.1.
943. H. Hirayama, A. Kinoshita and Y. Aoyagi, *Oyo Buturi (Monthly Publ. Jpn. Soc. Appl. Phys.) (in japanese)* **71** (2002) 204.
944. S. Hirayama and H. Hirayama, *Appl. Phys. Exp.* **4** (2011) 061002.
945. H. Hirayama, Y. Enomoto, A. Konshita, A. Hirata and Y. Aoyagi, *Mat. Res. Soc. Symp.* **595** (2000) W11.35.1.
946. H. Hirayama, Y. Enomoto, A. Kinoshita, A. Hirata and Y. Aoyagi, *phys. stat. sol. (a)* **180** (2000) 157.
947. H. Hirayama, Y. Enomoto, A. Kinoshita, A. HIrata and Y. Aoyagi, *Appl. Phys. Lett.* **80** (2002) 37.
948. S. Sumiya, Y. Zhu, K. Kosaka, M. Miyoshi, T. Shibata, M. Tanaka and T. Egawa, *Jpn. J. Appl. Phys.* **47** (2008) 43.
949. A. Tyagi, H. Zhong, N.N. Fellows, M. Iza, J.S. Speck, S.P. DenBaars and S. Nakamura, *Jpn. J. Appl. Phys.* **46** (2007) L129.
950. T. Nishida and N. Kobayashi, *phys. stat. sol.(a)* **188** (2001) 113.
951. Y. Taniyasu, M. Kasu and T. Makimoto, *Nature* **441** (2006) 325.
952. Y. Taniyasu and M. Kasu, *Appl. Phys. Lett.* **96** (2010) 221110.
953. Y. Taniyasu and M. Kasu, *NTT Tech. Rev.* **8** (2010) 1.
954. O. Mishima, K. Era and J. Tanaka and S. Yamaoka, *Appl. Phys. Lett.* **53** (1988) 962.
955. Y. Kubota, K. Watanabe, O. Tsuda ad T. Taniguchi, *Science* **317** (2007) 932.
956. K. Watanabe, T. Taniguchi, T. Niiyama, K. Miya and M. Taniguchi, *Nature Photo.* **3** (2009) 591.
957. S. Nakamura, M. Senoh, S. Nagahama, N. Iwasa, T. Yamada, T. Matsushita, H. Kiyoku and Y. Sugimoto, *Jpn. J. Appl. Phys.* **35** (1996) L74.
958. S. Nakamura, M. Senoh, S. Nagahama, N. Iwase, T. Yamada, T. Matsushita, Y. Sugimoto and H. Kiyoku, *Appl. Phys. Lett.* **69** (1996) 4056.

959. S. Nakamura, M. Senoh, S. Nagahama, N. Iwasa, T. Yamada, T. Matsushita, H. Kiyoku, Y. Sugimoto, T. Kozakai, H. Umemoto, M. Sano and K. Chocho, *Jpn. J. Appl. Phys.* **36** (1997) L1568.
960. S. Nakamura, *Semicond. Sci. Technol.* **14** (1999) R27.
961. A. Kuramata. S. Kubota, R. Soejima, K. Domen, K. Horino, P. Hacke and T. Tanahashi, *Oyo Buturi (Monthly Publ. Jpn. Soc. Appl. Phys.) (in japanese)* **68** (1999) 797.
962. S. Nagahama, T. Kozakai, T. Yanamoto and T. Mukai, *Oyo Buturi (Monthly Publ. Jpn. Soc. Appl. Phys.) (in japanese)* **73** (2004) 210.
963. K. Tachibana, S. Nunoue and M. Onomura, *Toshiba Rev.* **99** (2004) 32.
964. C. Skierbiszewski, M. Siekacz, P. Perlin, A. Feduniewicz-Zmuda, G. Cywinski, I. Grzegory, M. Leszynski, Z.R. Wasilewski and S. Prowski, *J. Crystal Growth* **305** (2007) 346.
965. P. Perlin, T. Swietlik, L. Marona, R. Czernecki, T. Suski, M. Leszcynski, I. Grzegory, S. Krukowski, G. Nowak, G. Kamler, A. Czerwinski, M. Plusa, M. Bednarek, J. Rybinsku and S. Porowski, *J. Crystal Growth* **310** (2008) 3979.
966. I. Akasaki, S. Sota, H. Sakai, T. Tanaka, M. Koike and H. Amano, *Electron. Lett.* **32** (1996) 1105.
967. S. Nagahama, T. Yamamoto, M. Sano and T. Mukai, *Jpn. J. Appl. Phys.* **40** (2001) L785.
968. S. Nagahama, T. Yamamoto, M. Sano and T. Mukai, *Jpn. J. Appl. Phys.* **40** (2001) L788.
969. S. Nagahama, T. Yanamoto, M. Sano and T. Mukai, *Jpn. J. Appl. Phys.* **41** (2002) 5.
970. S. Masui, Y. Matsuyama, T. Yanamoto, T. Kozakai, S. Nagahama and T. Mukai, *Jpn. J. Appl. Phys.* **42** (2003) L1318.
971. S. Nagahama, M. Sano, T. Yanamoto, D. Morita, O. Miki, K. Sakamoto, M. Yamamoto, Y. Matsuyama, Y. Kawata, T. Murayama and T. Mukai, *Proc. SPIE* **4995** (2003) 108.
972. S. Kamiyama, K. Iida, T. Kawashima, H. Kasugi, S. Mishima, A. Honshio, Y. Miyake, M. Iwaya, H. Amano and I. Akasaki, *IEEE J. Select. Topics Quan. Electron.* **11** (2005) 1069.
973. H. Yoshida, Y. Takagi, M. Kuwabara, H. Amano and H. Kan, *Jpn. J. Appl. Phys.* **46** (2007) 5782.
974. H. Kawanishi, T. Takano and U. Narita, *Proc. Japan-Germany-Spain Joint Workshop Adv. Semicon. Optoelectron. Dev. Mater.* (Oita, Japan, 2004) p. 13.
975. H. Kawanishi, T. Takano, M. Senuma and T. Nukui, *Oyo Buturi (Monthly Publ. Jpn. Soc. Appl. Phys.) (in japanese)* **74** (2005) 1458.
976. H. Kawanishi, M. Senuma and T. Nukui, *Appl. Phys. Lett.* **89** (2006) 041126.
977. H. Kawanishi, M. Senuma, M> Yamamoto, E. Niikura and T. Nukui, *Appl. Phys. Lett.* **89** (2006) 081121.
978. T. Miyoshi, S. Masui, T. Okada, T. Yanamoto, T. Kozakai, S. Nagahama and T. Mukai, *Appl. Phys. Exp.* **2** (2009) 062201.
979. A. Avramescu, T. Lermer, J. Müller, S. Tautz, D. Queren, S. Lutgen and U. Strass, *Appl. Phys. Lett.* **95** (2009) 071103.
980. M. Razeghi and A. Orgalski, *J. Appl. Phys.* **79** (1996) 7433.
981. Y.A. Goldberg, *Semicond. Sci. Technol.* **14** (1999) R41.
982. M.A. Khan, M. Shatalov, H.P. Maruska, H.M. Wang and E. Kuokstis, *Jpn. J. Appl. Phys.* 44(2005) 7191.
983. E. Monroy, F. Calle, J.L. Pau, E. Munoz, f. Omnes, B, Beaumont and P. Gibart, *J. Crystal Growth* **230** (2001) 537.
984. S. Senda, H. Jiang and T. Egawa, *Appl. Phys. Lett.* **92** (2008) 203507.
985. Z.T. Chen, S.X. Tan, Y. Sakai and T. Egawa, *Appl. Phys. Lett.* **94** (2009) 213504.
986. Z.T. Chen, Y. Sakai, J.C. Zhang, T. Egawa, J.J. Wu, H. Miyake and K. Hiramatsu, *Appl. Phys. Lett.* **95** (2009) 083504.
987. Y. Sakai, P.C. Khai, J. Ichiawa, T. Egawa and T. Jimbo, *J. Appl. Phys.* **109** (2011) 033512.
988. T.H. Tsai, H.I. Chen K.W. Lin, C.W. Hung, C.H. Hsu, T.P. Chen, L.Y. Chen, K.Y. Chu, C.F. Chang and W.C. Liu, *Appl. Phys. Exp.* **1** (2008) 041102.
989. T.H. Tsai, H.I. Chen, I.P. Liu, C.W. Hung, T.P. Chen, L.Y. Chen, Y.J. Liy and W.C. Liu, *IEEE Trans. Electron Dev.* **55** (2008) 3575.
990. M.A. Khan, M.S. Shur, Q. Chen, J.N. Kusnia and C.J. Sun, *Electron. Lett.* **31** (1995) 398.
991. U.K. Mishra, L. Shen, T.E. Kazior and Y.F. Wu, *Proc. IEEE* **96** (2008) 287.
992. V. Kumar, W. Lu, R. Schwindt, A. Kuliev, G. Simin, J. Yang, M.A. Khan and I. Adesida, *IEEE Electron Device Lett.* **23** (2002) 455.
993. H. Xing, S. Keller, Y-F Wu, L. McCarthy, I.P. Smorchkova, D. Buttari, R. Coffie, D.S. Green, G. Parish, S. Heikman, L. Shen, N. Zhang, J.J. Xu, B.P. Keller, S.P. DenBaars and U.K. Mishra, *J. Phys.: Condens. Matter.* **13** (2001) 7139.
994. H. Okumura, *Jpn. J. Appl. Phys.* **45** (2006) 7565.
995. T. Kikkawa, K. Inanishi, N. Hara, H. shigematsu, K. Joshin, *Proc. SPIE* **6894** (2008) 68941.
996. T. Kikkawa, T. Iwai and T. Ohki, *Fujitsu, Sci. Tech. J.* **44** (2008) 333.
997. T. Kikkawa, K. Makiyama, T. Ohki, M. Kanamura, K. Imanishi, N. Hara and K. Joshin, *phys. stat. sol. (a)* **206** (2009) 1135.

998. K. Joshin and T. Kikkawa, *Dev. Res. Conf. Dig. (DRC '05)* **June** (2005) 173.
999. S.L. Selvaraj and T. Egawa, *Appl. Phys. Lett.* **89** (2006) 193508.
1000. S.L. Selvaraj, T. Ito, Y. Terada and T. Egawa, *Appl. Phys. Lett.* **90** (2007) 173506.
1001. S.L. Sevaraj, T. Suzue and T. Egawa, *IEEE Eletrcon. Lett.* **30** (2009) 587.
1002. S.L. Selvaraj, T. Suzue and T. Egawa, *Appl. Phys. Exp.* **2** (2009) 111005.
1003. S. Arulkumaran, S. Vicknesh, N.G. Ing, S.L. Selvaraj and T. Egawa, *Appl. Phys. Exp.* **4** (2011) 084101.
1004. M. Hirota, M. Yanagihara, K. Nakazawa, H. Ueno, Y. Hirose, T. Ueda, Y. Uemoto, T. Tanaka, D. Ueda and T. Egawa, *IEEE IEDM 2004* (Washington DC, 2004) p. 803.
1005. For recent development, see for example *New Markets Expanded by Power Semiconductors* (Denpa Publ. Inc., Feb. 3, 2010)
1006. K. Cheng, M. Leys, S. Degroote, J. Derluyn, B. Sijmus, P. Favia, O. Richard, H. Bender, M. Germain and G. BOrghs, *Jpn.J. Appl. Phys.* **47** (2008) 1553.
1007. http://www2.imec.be/be_en/press/imec-news/imec200mmgan.html
1008. M.C. Benjamin, C. Wang, R.F. Davis and R.J. Nemanich, *Appl. Phys. Lett.* **64** (1994) 3288.
1009. M. Kasu and N. Kobayashi, *Appl. Phys. Lett.* **5**

21. ZnO

21.1 INTRODUCTION

Polycrystalline ZnO is an industrially important material for pigments, phosphors, surge absorbers, piezoelectric transducers, gas sensors, and optical waveguides, transparent electrodes, varistors, surface acoustic wave (SAW) filters, window materials, solar cells and scintillators. Because of its wide bandgap and high exciton binding energy, ZnO has shown recent promise as a compound semiconductor material[1-3] for ultraviolet and blue light emitting diodes[4] and for lasers.[5-8] ZnO is also an attractive new material for applications such as high field electron devices, high power and high temperature electronic devices, future wireless communications systems at frequencies beyond 5 GHz, variable optoelectronic grating and optical power limiting devices.[1] It is also attractive for spin electronic devices.[2] ZnO is also attractive because of its radiation resistance[1,9,10], much higher resistance than Si, GaAs, CdS and GaN, so that it is promising for electronic devices in radioactive environments such as space applications. The growth of ZnO crystals was recently, that is after 2000, largely developed as explained below mainly based on the hydrothermal method. The growth of ZnO cyrstals has been among the most successful, comparable to GaP, GaAs, InP, CdTe, $Hg_{1-x}Cd_xT$ (MCT) and HgI_2.

21.2 PHYSICAL PROPERTIES

ZnO has a large band gap of 3.37 eV and large exciton-binding energy of 60 meV [6,11], so that it is promising for ultra violet (UV) and blue light emitting devices. The bandgap of ZnO can be tuned by the substitution of the cation site with Cd to 2.8 eV and with Mg to 4.0 eV[1,2] and quantum wells can be achieved.[12] ZnO has a wurtzite type crystal structure like GaN and is also expected to be a substrate material not only for homo-epitaxial growth of ZnO layers but also for hetero-epitaxial growth of GaN based layers. The lattice mismatching against GaN is only 1.8 %.[1]

The physical properties of ZnO have been studied[13-20] and are reviewed in Refs. 2, 3 and 21-23. Some of them are summarized in Table 21.1. The phase diagram of ZnO is shown in Fig. 21.1.[20] The melting point of ZnO is determined as 1975°C but the data of solidus lines are few, especially for the O-saturated one as seen in the figure.

Since ZnO amenable to wet chemical etching, device processing is much easier than with competitor materials such as SiC and GaN.[10]

21.3 CRYSTAL GROWTH

In the past, various crystal growth methods have been studied[24] and some of them are summarized in Table 21.2. It was found that the hydrothermal synthesis method is effective in the industrial setting.

Table 21.1 Physical Properties of ZnO

Crystal Structure	wurtzite
Lattice Constant	a = 3.2495 Å
	c = 5.2069 Å
Density	5.6768 g/cm^3
Melting Point	1975 °C
Linear Expansion Coefficient	4.0×10^{-6}/deg
Thermal Conductivity	~ 0.54 W/cm·K
Dielectric Constant	8.84(//), 8.47(⊥)
Refractive Index	2.239
Bandgap at Room Temperature	3.37 eV
Optical Transition Type	direct
Intrinsic Carrier Concentration at R.T.	$<10^6$ cm^{-3}
Electron Mobility at R.T.	200-280 cm^2/V·sec
Hole Mobility at R.T.	5-50 cm^2/V·sec

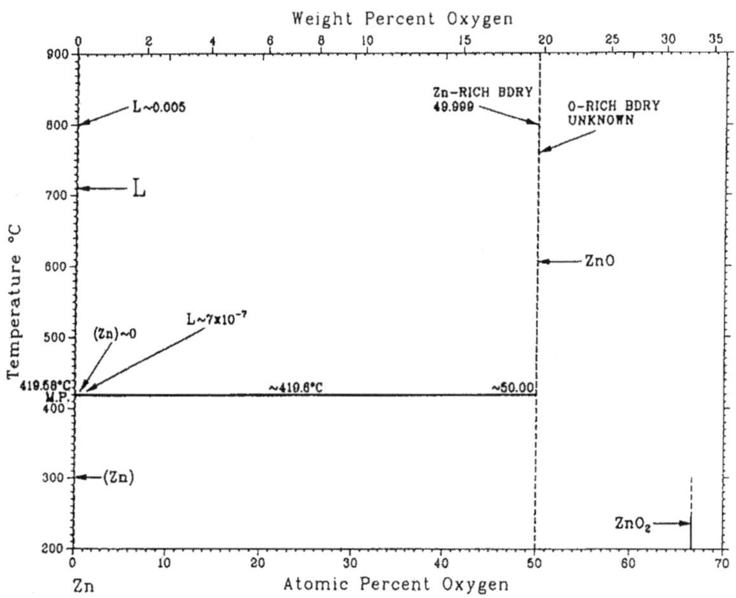

Fig. 21.1 Phase diagram of ZnO (reprinted from Ref. 20 with permission, copyright 1990 ASM International).

21.3.1 Melt Growth

Since the melting point of ZnO is as high as 1975°C, it was thought that it would be difficult to grow single crystals by melt growth.

Nause and Nemeth[25] have used a pressurized melt growth method. This method used an apparatus as shown in Fig. 21.2. The crucible itself is cooled by water during crystal growth and the congruent ZnO material is melted by induction heating. The melt is protected from contamination by the crucible via a thin layer of cooled polycrystalline

Table 21.2 Crystal Growth of ZnO

Method	Contents	Authors	Year	Ref.
HTS	10-20 g ZnO crystals were first grown	Laudise et al.	1964	39, 40
VR	Hydrolysis of zinc fluoride	Kubo	1965	60
Flux	Growth by the THM method using PbF_2 flux	Wolff et al.	1965	29
HTS	Growth from KOH	Kolb et al.	1966	41
VR	Reaction of $ZnCl_2$ with H_2O	Takahashi et al.	1967	61
Flux	Growth from ZnO-PbF_2 flux	Kleber et al.	1966	28
Flux	Growth from ZnO-PbF_2 flux	Chase et al.	1967	30
HTS	Growth of Li-doped ZnO with low carrier concentration	Kolb et al.	1967	42
VR	Reaction of $ZnCl_2$ with H_2O	Weaver	1967	62
VR	Reaction of ZnS or ZnSe vapor with O_2	Park et al.	1967	63
VR	Reaction of Zn vapor with oxygen	Nielsen	1968	64
CVT	Review of CVT growth with various transport agents	Emmenegger et al.	1969	71
Flux	Growth from V_2O_5, P_2O_5 and $(P, V)_2O_5$	Wankyn	1970	31
VR	Oxydation of ZnI_2 and $ZnBr_2$	Hirose et al.	1970-1971	65-67
CVT	Growth using various transport agents	Shiloh et al.	1971	72
VR	$ZnO+H_2=Zn+H_2O$	Helbig	1972	68
CVT	Growth using Br_2 and Cl_2 as transport agents	Piekarczyk et al.	1972	73
VR	Reaction of Zn with oxygen, cyrstals 8 mm thick and 35 mm high	Fishcer	1976	69
HTS	Growth of single crystals measuring 25 × 10 × 5 mm^3	Sakagami et al.	1981	44
HTS	Growth of ZnO, ZnS and Hg_2S by HTS	Kuznetsov et al.	1984	45
CVT	Growth using Zn and $ZnCl_2$ as transport agents	Matsumoto et al.	1985	74
CVT	Growth using (NH_4X), Zn and $ZnCl_2$ as transport agents	Matsumoto et al.	1988	75
CVT	Growth using $HgCl_2$ as the transport agent	Matsumoto et al.	1990	76
Flux	Growth from KOH and NaOH flux	Ushio et al.	1993	32
HTS	Industrial growth of ZnO single crystals, 200 crystals 100-250 g each	Kortounova et al.	1997	54, 55
CVT	Growth using H_2 as the transport agent	Look et al.	1998	77
HTS	Growth using basic mineralizers and characterization	Suscavage et al.	1999	56
CVT	Growth using carbon as the transport agent	Ntep et al.	1999	78
HTS	Growth from a mixture of KOH and LiOH	Sekiguchi et al.	2000-2001	46, 47
HTS	Growth mechanism from the analysis of crystal structure	Demianets et al.	2001	52
Flux	Growth from $V_2O_5+B_2O_3$ and $MoO_3+V_2O_5$ flux by the TSSG and TSFZ methods	Oka et al.	2002	33
HTS	Growth of ZnO cyrstals measuring 50 x 50 x 15 mm^3	Ohshima et al.	2004	48
CVD	Growth of ZnO films by the nearby CVD method	Nishio et al.	2004	83
CVT	Growth using mixtures of H_2+C+H_2O or N_2+C+H_2O as the transport reagent	Mycielski et al.	2004	79
PMG	Growth under pressurized atmosphere upto 100 atm	Nause et al.	2005	25
HTS	50 mm diameter crystals, EPD less than 80 cm^{-2}	Maeda et al.	2005	49
HTS	Production of about 200 crystals, each weighting 100-250 g after 100 days growth	Kortunova	2005	50
CVT	Growth using carbon or CO_2+Zn as the transport agent	Mikami et al.	2005-2007	80-82
HTS	Growth of $Zn_{1-x}Mg_xO$ crystals	Wang et al.	2006	57
Flux	Growth from ZnO-87 mol% PbF_2 flux	Li et al.	2006	34
VB	First growth of ZnO by VB using metal crucibles	Schulz et al.	2006	26

Table 21.2 Crystal Growth of ZnO (continued)

Method	Contents	Authors	Year	Ref.
HTS	Growth of nitrogen-doped ZnO crystals	Wang et al.	2006	57
HTS	Growth of $Zn_{1-x}Mg_xO$ alloy crystals	Wang et al.	2006	58
HTS	Growth of indium-doped conducting ZnO crystals	Wang et al.	2006	59

VB: Vertical Bridgman, HTS: Hydrothermal Synthesis, VR: Vapor Reaction, CVT: Chemical Vapor Transport, CVD: Chemical Vapor Deposition, PMG: Pressurized Melt Growth, THM: Travelling heater Method, TSSG: Top-Seeded Solution Growth, TSFZ: Travelling Solvent Floating Zone.

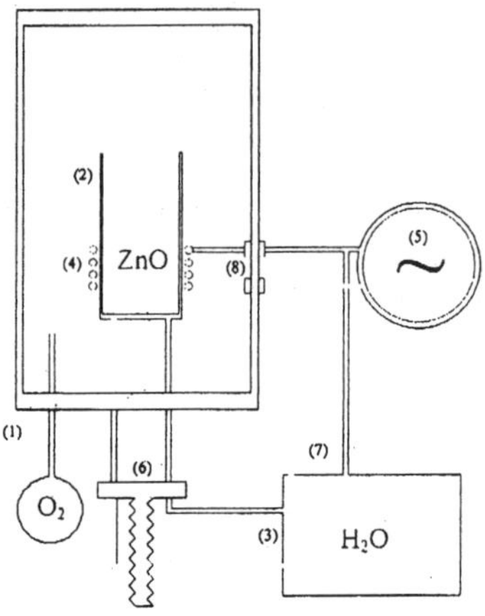

Fig. 21.2 Schematic illustration of the ZnO crystal growth apparatus (reprinted from Ref. 25 with permission, copyright 2005 Institute of Physics).

ZnO covering the crucible surface. ZnO crystal ingots measuring 137 mm diameter, and weighing of the order of a kilogram have been grown. Single crystal ingots up to 50 mm diameter were obtainable from large grown ingots.

Schultz et al.[26] have however shown that crystal growth could be achieved using the vertical Bridgman method with iridium crucibles. Using a [0001] oriented seed crystal in the bottom of the crucible, growth was achieved with 50 bar over pressure using RF coil heating. They were able to grow single crystals of diameter 75 mm and length 30 mm. Full width at half maximum (FWHM) of X-ray rocking curve (XRC) was in the range 22 to 47 arcsec. Figure 21.3 shows the optical micrograph of a zinc oxide wafer of diameter of 62.5 mm.

21.3.2 Solution Growth

(1) Flux method

Nielsen and Dearborn[27] have found that ZnO is highly soluble in PbF_2 as shown in Fig. 21.4[27] and ZnO single crystals have been grown from a ZnO-PbF_2 flux.[27, 28] Wolff and Labelle[29] have grown ZnO crystals by a travelling solvent zone method, using PbF_2 as the solvent zone. Chase and Osmer[30] have grown ZnO crystals from ZnO-80 mol% PbF_2.

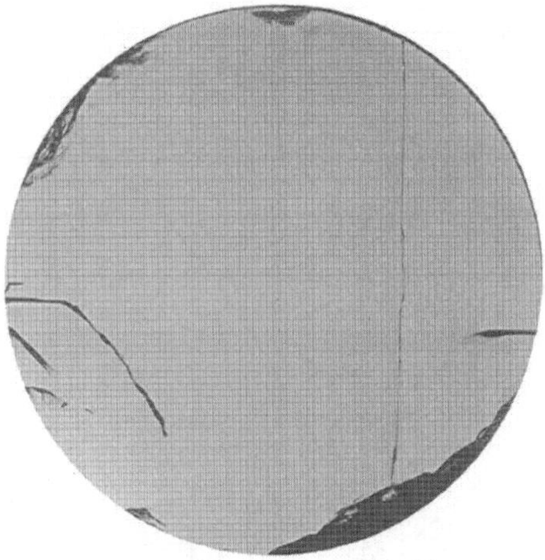

Fig. 21.3 Optical micrograph of an as-grown zinc oxide wafer under crossed polarizers. 62.5 mm diameter {11-20} orientation (reprinted from Ref. 26 with permission, copyright 2006 Elsevier).

Flux growth of ZnO has been extensively studied using various fluxes such as PbF_2, P_2O_5, V_2O_5, KOH, NaOH, MoO_3 and B_2O_3.[31,32] It was found that PbF_2 is a good flux since ZnO is highly solubke in it as shown in Fig. 21.4.

Oka et al.[33] have grown single crystals from $V_2O_5 + B_2O_3$ and $MoO_3 + V_2O_5$ fluxes by top-seeded solution growth (TSSG) and travelling-solvent floating zone (TSFZ) methods using the furnace configuration shown in Fig. 21.5.

Li et al.[34] have grown ZnO single crystals from a PbF_2 flux using the Bridgman configuration. Crystals were grown at a temperature of 1100°C using a starting composition of 78 mol% PbF_2 in a ZnO-PbF_2 system.

(2) Hydrothermal method

This method (Vol. 1, Sec. 2.3.6) which had been successful for the industrial growth of quartz[35,36] was found to be a good method for ithe ndustrial production of ZnO. Ehrentraut et al. have reviewed the application of this method to ZnO.[37,38] This method has been studied in the past by many researchers[18,39-45] and has been developed so that large crystals can be grown.[46-59]

Sekiguchi et al.[46,47] have grown ZnO single crystals measuring about 10 mm using a solvent of KOH and LiOH, and evaluated impurity incorporation and its effect on optical and electrical properties. An autoclave which can be operated at temperatures of 400°C and at pressures of 0.1 GPa was used (Fig. 21.6). Sintered ZnO strings as the nutrient together with a 3M KOH and 1M LiOH aqueous solution were put in a platinum crucible. Seed crystals were suspended by a Pt wire as seen in Fig. 21.6. The temperature of the seed region was set about 10°C lower than that of the nutrient region and the crystal is grown on the seed.

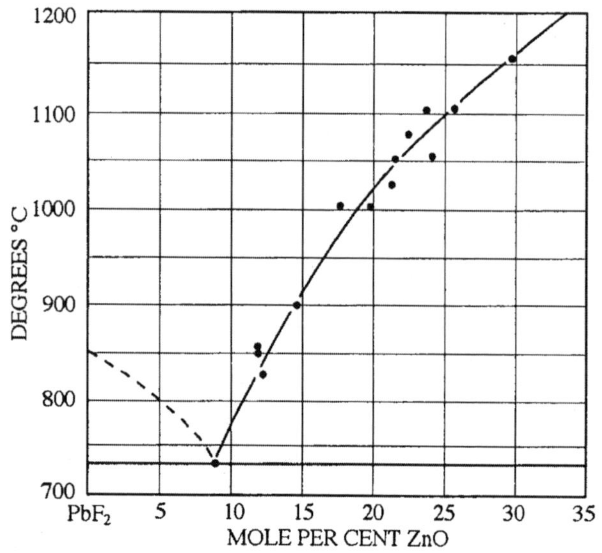

Fig. 21.4 Solubility curve of ZnO in PbF_2 (reprinted from Ref. 27 with permission, copyright 1960 American Chemical Society).

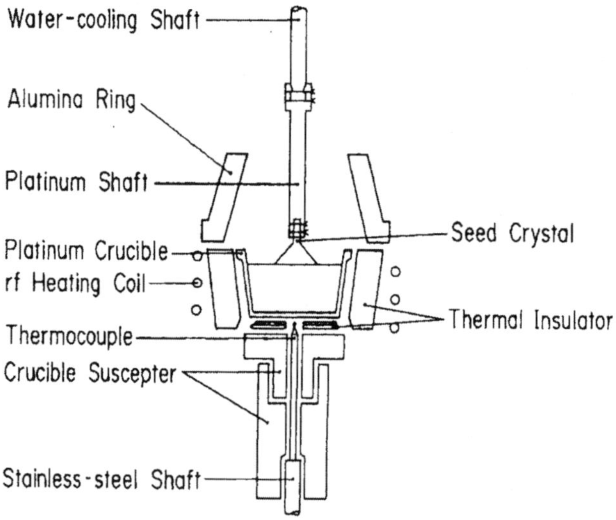

Fig. 21.5 Vertical cross section of the crucible arrangement. TSSG and TSFZ growth of ZnO (reprinted from Ref. 33 with permission, copyright 2002 Elsevier).

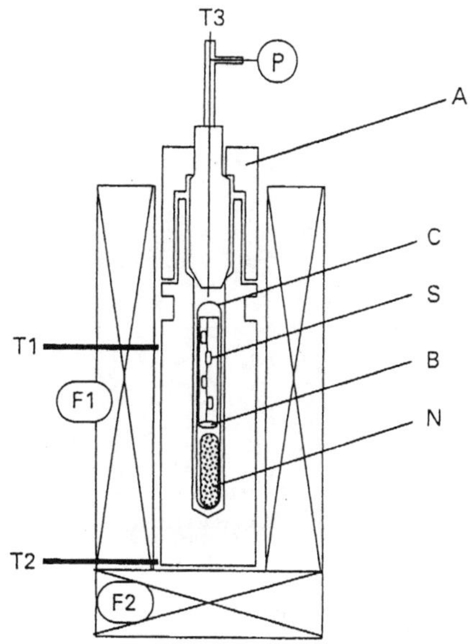

Fig. 21.6 Schematic drawing of a hydrothermal growth system. F: furnaces (1, 2), T: thermocouples for control (1, 2) and monitor (3), P: pressure gauge, A: autoclave, C: Pt crucible, S: seed crystals, N: nutrient, B: baffle (reprinted from Ref. 46 with permission, copyright 2000 Elsevier).

The scaled-up autoclave with the Pt inner container as seen in Fig. 21.7 has been used to grow large size crystals.[48,49] To prevent corrosion, platinum is used as the inner liner. Alkali metal hydroxides are generally used. Pressed and sintered ZnO was used and was placed at the bottom of the autoclave (dissolution zone) with the mineralizer. A mixture of 3M KOH to 1M LiOH is found to be a good mineralizer to give good quality crystals with a reasonable growth rate of 200 μm per day, comparable to 500 μm per day in the case of quartz crystal growth. The upper part of the autoclave (growth zone) was separated by a baffle plate from the dissolution zone to obtain a proper temperature gradient to have a continuous flow of the saturated solution from the dissolution zone to the growth zone. ZnO seed crystals were placed in the growth zone and the crystals were grown along the seed crystal. Crystal growth was performed at temperatures of 300-400°C and under high pressures (80-100 MPa). Large ZnO crystals have been grown using these scaled-up autoclaves[37,38,48-51] with a growth rate of about 200 μm a day and the diameter of the crystals has been increased from 50 mm up to 100 mm.[38] In Fig. 21.8(a), the 50 mm ZnO single crystal is shown with its growth sectors (Fig. 21.8(b)). Figure 21.9 shows 50 mm ZnO single crystals obtained from a single growth run.

The mechanism of crystal growth by this method has been studied by Demianetsa and Kostromarov[52] and the growth anisotropy has been explained from the viewpoint of the charge on the growth surface depending on the ions in the solution. Kortunova et

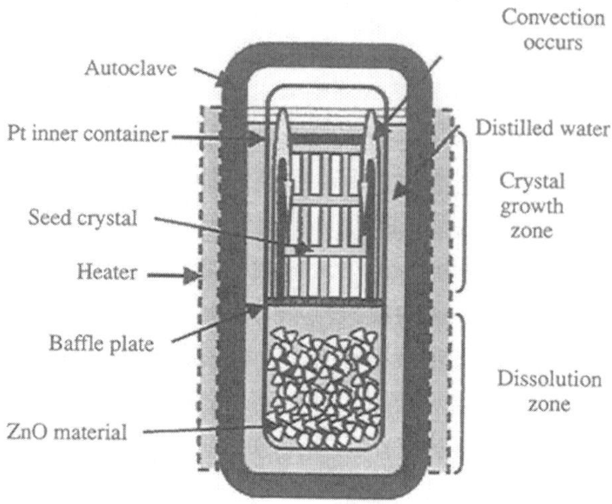

Fig. 21.7 Schematic of the the hydrothermal synthesis system (reprinted from Ref. 49 with permission, copyright 2006 Institute of Physics).

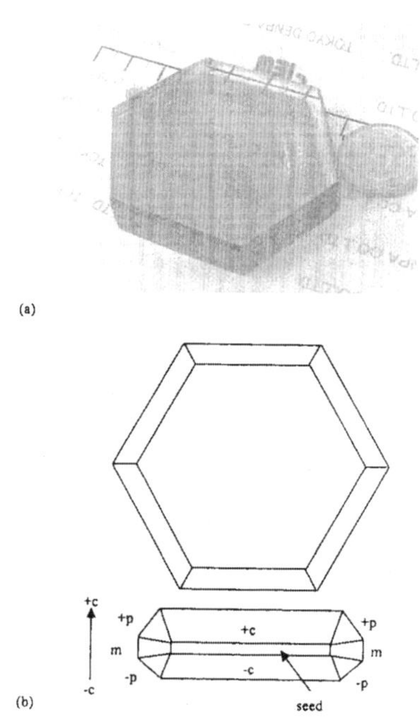

Fig. 21.8 (a) Photographe of a 50 mm ZnO single crystal grown by the hydrothermal method; (b) schematic drawing of growth sectors (reprinted from Ref. 49 with permission, copyright 2006 Institute of Physics).

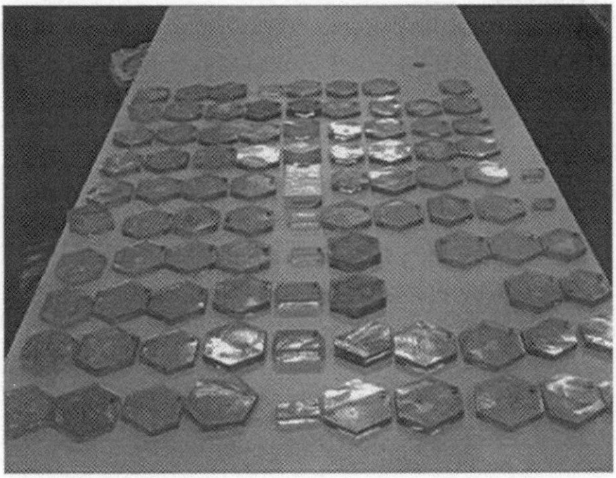

Fig. 21.9 Two inch ZnO crystals from a single growth run (reprinted from Ref. 51 with permission, copyright 2007 Institute of Physics).

al.[53-55] have developed the hydrothermal method, from 24 liter autoclaves in the laboratory to 500 liter autoclaves for the industrial manufacture of ZnO single crystals. Crystals were grown at 330°C at 50 MPa with 3M KOH to 1M LiOH aqueous solutions. They were able to obtain 200 crystals, each of mass 100-250 g.

Suscavage et al.[56] have grown crystals from a solution with a mixture of Li_2CO_3, 4N KOH and 4N NaOH and examined the growth anisotropy. They were able to grow crystals with FWHM-XRC of 37-43 arcsec. In an attmpt to obtain p-type ZnO, nitrogen doped crystals have been grown from alkaline solutions though p-type was not achieved.[57] In order to have good lattice matching with ZnO-based and GaN-based optoelectronic devices, it is preferable to control the lattice constant and $Zn_{1-x}Mg_xO$ crystals have been grown for this purpose. Wang et al.[58] have grown $Zn_{1-x}Mg_xO$ crystals from alkaline solution at 650°C below ~ 22 kpsi. Conductive ZnO doped with In has been grown and the role of In on the crystal growth has been studied. It was found that In has the effect of inhibiting growth in the direction of the c-face while it promotes growth along the m-plane.[59]

21.3.3 Vapor Phase Growth

(1) Vapor reaction

In the past, vapor reaction methods[60-70] have been used to grow ZnO crystals. These are the hydrolysis of zinc fluoride[60], the vapor phase reaction of $ZnCl_2$ with H_2O[61,62], the reaction of ZnS or ZnSe vapor with O_2[63], the reaction of Zn vapor with O_2[64], the reaction of ZnI_2 vapor[65,66] and $ZnBr_2$ vapor[67] with O_2 and the reaction of Zn vapor produced by the reduction of ZnO by hydrogen with O_2.[68,69] ZnO single crystals of 2 mm thickness and 100 mm length and 8 mm thickness and 35 mm length have been obtained.[68] Reynolds et al.[70] have grown ZnO platelets by the vapor phase method and proved that optically pumped ultraviolet lasing could be observed because of the high crystal quality.

(2) Chemical Vapor Transport (CVT)

Emmenneger and Petermann[71] have summarized the chemical vapor transport crystal growth of various oxides including ZnO. Various agents such as HCl, NH_3, NH_4Cl, H_2, Cl_2 and $HgCl_2$ have been examined by Shiloh and Gutman[72] and NH_3, NH_4Cl and $HgCl_2$ have been found to be efficient in growing ZnO single crystals. Various agents such as Br_2 and Cl_2[73], NH_4X, Zn and $ZnCl_2$[74,75], $HgCl_2$[76] have also been investigated.

Look et al.[77] have applied the vapor transport method according to the following reaction,

$$ZnO(s) + H_2(g) = Zn(g) + H_2O(g) \tag{21.1}$$

and were able to grow 50 mm diameter crystals of about 1 cm thickness.

Ntep et al.[78] have shown that ZnO single crystals can be grown at about 970 °C using carbon as a transport agent. Growth occurs according to the following reaction.

$$ZnO(s) + C(s) = Zn(g) + CO(g) \tag{21.2}$$

ZnO and carbon powders were loaded in an ampoule and Zn was vaporized and reacted with CO gas and ZnO crystals were deposited in a lower temperature region.

Mycielski et al.[79] have used $H_2 + C + H_2O$ or $N_2 + C + H_2O$ as the chemical transport agents and evaluated the photoluminescence of the grown crystals. Mikami et al.[80,81] have used CO_2 as the transport agent and have grown crystals according to the following reaction.

$$ZnO(s) + CO(g) = Zn(g) + CO_2(g) \tag{21.3}$$

Figur 21.10 shows the deposition rate as a function of the temperature difference ΔT between source temperature and growth area. Even though carbon was used as the transport agent, there was not significant contamination by carbon as evaluated by secondary ion mass spectroscopy (SIMS) analysis. They also have grown ZnO crystals using an improved CVT ampoule configuration in order to increase the growth rate. By applying this improvement, the kinetic limited transport process was changed to the diffusion-limited transport process leading to an inceased growth rate.[82]

(3) Chemical Vapor Deposition (CVD)

ZnO films have been grown by the nearby CVD (NV-CVD) method[83] The substrate was located over the source material, bis (21,4-pentanedionato) zinc with a separation distance of 2.5 mm as shown in Fig. 21.11. Crystalline ZnO was prepared at low temperatures ranging from 100°C to 300°C.

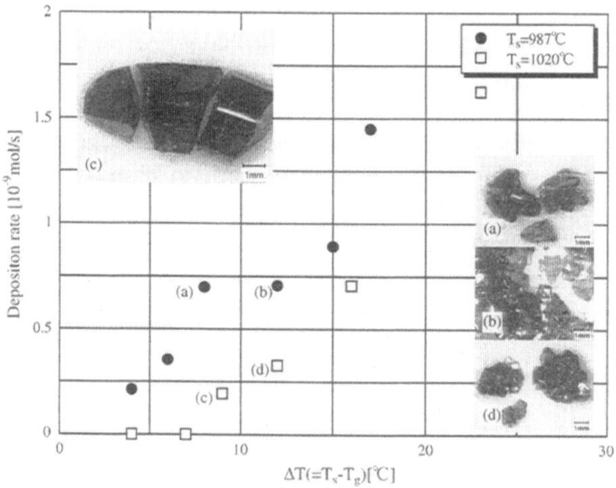

Fig. 21.10 Decomposition rate vs. ΔT at T_s = 987°C and T_s = 1020°C. Insets are photographs of as-grown ZnO crystals. The growth conditions are (a) T_s = 987 °C and ΔT = 8°C, (b) T_s = 987°C and ΔT = 15°C, (c) T_s = 1020°C and ΔT = 9°C and (d) T_s = 1020°C and ΔT = 12°C (reprinted from Ref. 80 with permission, copyright 2005 Elsevier).

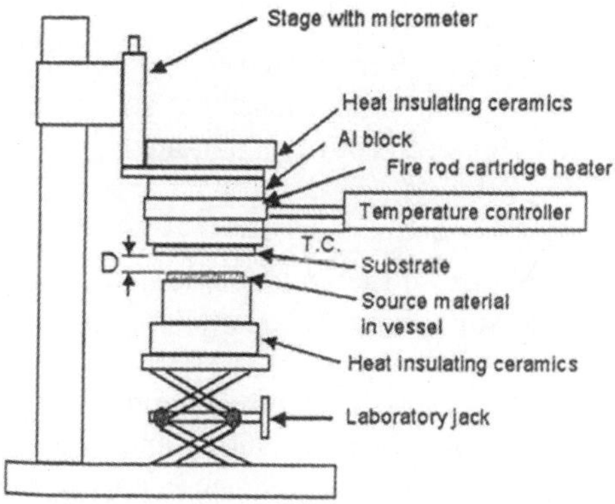

Fig. 21.11 Schematic view of nearby CVD (NV-CVD) apparatus (reprinted from Ref. 83 with permission, copyright 2004 Elsevier).

21.4 CHARACTERIZATION

21.4.1 Purity

It is known that hydrothermally grown crystals contain impurities originating from the mineralizers such as LiOH and KOH at levels of 0.3-20 ppm, though the method has the advantage that most of impurities can be reduced.[1,48,49] Fe and Al are also observed as impurities to which is attributed the pale green color of grown crystals. Kortunova et al.[55] showed detailed analyses of 33 elements and also found that Li and K are important impurities.

Wang et al.[59] have examined the behavior of impurities in crystal growth and have found that Al, Fe and Si show incorporation behavior which supports the growth mechanism that the authors have proposed.

Ehrentraut[38] has shown the behavior of the main impurities such as Li, K, Mg, Al and Fe. It was shown that Al and Fe have two orders of magnitude higher concentrations in (000-1) wafers than (0001) wafers. It was also shown that Li and Al impurities can be reduced by annealing.[37,38]

21.4.2 Defects

(1) Dislocations

Dislocations can be revealed by etching with an aqueous solution of concentrated H_3PO_4 for 5 min. at 25°C. Low dislocation density crystals are etched with 0.7% HCl aqueous solution for 5 min. at 60°C in order to increase the etch rate. The etch pit density of hydrothermally grown crystals is about 300 cm^{-2} and is reduced to less than 80 cm^{-2} by annealing.[49]

(2) Structural defects

Full width at half maximum (FWHM) of X-ray rocking curve (XRC) measurements have been performed for single crystals grown by various methods, and are reported to be 49 arcsec for pressure-melt ZnO[25], 19-30 arcsec for hydrothermal ZnO[37] and 30 arcsec for CVT ZnO.[78]

X-ray topography has also been measured for 50 mm diameter hydrothermal ZnO wafers of thickness 500 μm and it was found that structurally homogeneous crystals whose FWHM-XRC was 19 arcsec could be obtained.[37]

(3) Nonstoichiometry

Sakagami and Shibata[44] evaluated the nonstoichiometry using electrochemical analysis. In order to control the stoichiometry, a small amount of water is contained in H_2 carrier gas during CVT growth.[77]

In hydrothermal synthesis growth, Deminets and Kostomarov [52] have studied the mechanism of impurity incorporation on <0001> and <000-1> faces. They found that Li ions have the effect of reducing the difference in growth rates between the faces and so improving the stoichiometry and crystal quality.

(4) Native defects

Kohan et al.[84] and Janotti et al.[85] have calculated the formation energies and the concentration of various native defects by the first-principle pseudopotential method. An example is shown in Fig. 21.12. The behavior of various defects has been examined as a function of the Fermi level and the Zn partial pressure.

Tuomisto and Look[86] have studied vacancy defects in bulk ZnO single crystals grown by various methods, using positron annihilation spectroscopy. Very low concentrations of vacancy defects were found in crystals grown by the seeded vapor phase methods and melt growth methods, where the dominant defects are Zn vacancies. Rela-

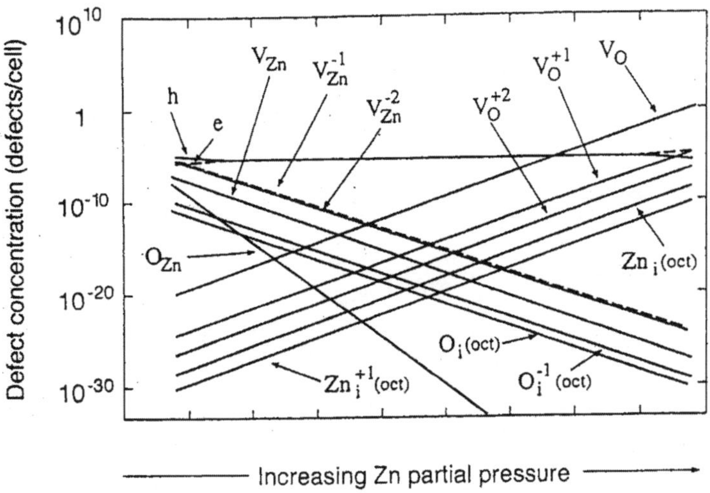

Fig. 21.12 Theoretical intrinsic defect cocentration at 1000 K (reprinted from Ref. 84 with permission, copyright 2000 American Physical Societyt).

tively high Zn and O vacancies have been observed in crystals grown by the hydrothermal and chemical vapor transport methods. Crystals grown by the contactless chemical vapor transport method showed a high concentration of Zn vacancies but not of O vacancies.

21.4.3 Electrical Properties

The electrical properties of ZnO has been extensively studied by Heiland et al.[21]

Nominally undoped ZnO shows n-type conductivity with high electron carrier concentrations between 10^{17} and 10^{21} cm^{-3}. This electron doping in undoped ZnO is attributed to Zn interstitials, oxygen vacancies or hydrogens.[2] The intrinsic defect level for n-type doping is approximately 0.01-0.05 eV below the conduction band. The origin of the n-type doping is however still under discussion and other possibilities such as the impurities Al and Ga are considered candidates.

Electrical properties such as mobility and carrier concentration as a function of temperature have been measured using CVT grown ZnO as shown in Fig. 21.13.[77] Albrecht et al.[87] have made a Monte Carlo simulation of the high field transport properties of ZnO. ZnO showed a saturation velocity of 3.2×10^7 cm/s at an electric field of 270 kV/cm at room temperature, higher than that of GaN. The peak velocity of ZnO is similar to that of GaN, but the peak field for GaN is about 100 kV/cm much lower than that of ZnO. They attributed this difference to the lower satellite valleys in GaN and to the stronger Fröhlich coupling in ZnO.

N-type doping itself is relatively easy with Al, Ga or In doping or via excess Zn. P-

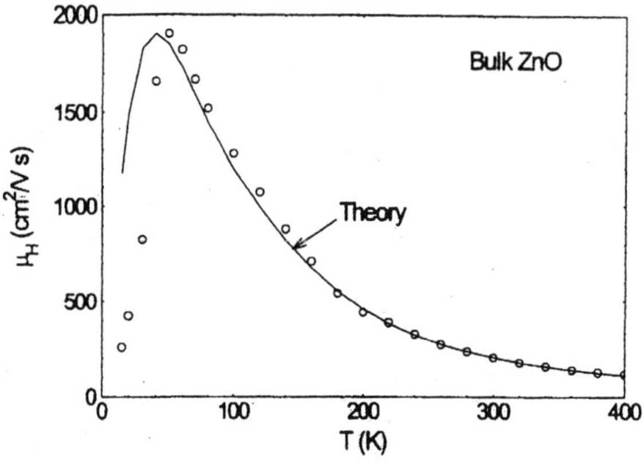

(a)

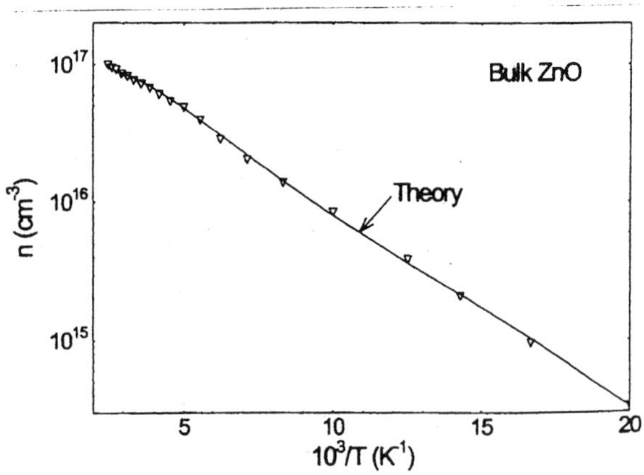

(b)

Fig. 21.13 (a) Experimental (circles) and theoretical (solid line) Hall mobility as a function of temperature and (b) experimental carrier concentration (triangles), corrected for Hall r-factor and theoretical fit (solid line) as a function of inverse temperature in bulk ZnO (reprinted from Ref. 77 with permission, copyright 1998 Elsevier).

type doping is significantly difficult because of the self-compensation effect (Vol. 1, Sec. 4.2.5). ZnO prefers n-type conductivity due to the formation of Zn interstitials or O vacancies. To overcome this problem, co-doping of donors and acceptors has been proposed based on the theoretical calculation of Madelung energies for n-type doping and p-type doping.[88] Tsukazaki et al.[89] concluded that the pn control in co-doping is difficult but they finally realized p-type doping by the high temperature annealed self-buffer layer (HITAB) technique and repeated temperature modulation (RTM).[89,90]

21.4.4 Optical Properties

The optical properties of ZnO are reviewed in Refs. 3, 91 and 92. A strong room temperature near band-edge UV photoluminescence (PL) peak at ~ 3.2 eV is attributed to an exciton state as the exciton binding energy is of the order of 60 meV.[6,11] Table 21.3 contains a summary of various PL emissions reported in the literature.

(1) Edge emission

Reynolds et al.[93] first observed and identified the edge emission lines from I_1 to I_{10}. Neutral-donor-bound-exciton complexes were also studied.[94] They have also studied the PL line between deep band emissions (DBEs) and free excitons (FEs) and discussed it in relation to polaritons.[95] Neutral-accepter-bound-excitons have also been observed[96] though they are still under dispute. Ohashi et al.[97] have examined the cathodoluminescence (CL) emission spectra in the temperature range of 20-300 K, comparing undoped crystals with Al-doped crystals. They found a peak, neither Fe-nLO nor BE-nLO, attributable to a localized state. Hamby et al.[98] have studied the temperature dependence of edge emissions from bulk crystals in the temperature range of 4.2-300 K. The temperature dependence of free-exciton peak energy was analyzed using Mannogian-Wolley equation[99] and it was found that the observed coefficients is in good agreement with the theoretical calculation. In Fig. 21. 14, a typical PL spectrum is shown of the region of the band edge for hydrothermal-grown ZnO. It can be seen that many edge emissions with high resolution are comparable to those of the LPE-grown ZnO.

(2) Impurities

A broad orange-red photoluminescence emission at 1.7-2.8 eV with a peak around 2.3 eV has been attributed to donor-accepter recombination due to Li.[100,101] An emission around 2.4 eV is attributed to copper impurities less than 1 ppm.[102,103] Well defined lines at 1.7-1.8 eV are attributed to Fe^{3+}.[104]

(3) Native defects

Various authors have studied the correlation between photoluminescence and native defects. Vanheysden et al.[105] have found that the green 510 nm emission is related to singly ionized oxygen. Lin et al.[106] have studied the green emission from ZnO films on Si and attributed it to oxide antisite defect O_{Zn}. Garcus et al.[102] have suggested that the green emission at 510 nm is associated with Cu^{2+} ions.

A blue-green emission around 500 nm is explained as transitions involving self-activated centers formed by a doubly ionized zinc vacancy and an ionized interstitial impurity acceptor and/or interstitial O. From the study of the temperature dependence of the green PL emission and the absorption, Leiter et al.[107] concluded that oxygen vacancies are the origin of the emission. Borseth et al.[108] have studied the photoluminescence from single crystalline ZnO annealed in inert, Zn-rich and O-rich atmospheres.

Table 21.3 Photoluminescence Data for ZnO

Peak (eV)	Notation	Assignment	Ref.
Exciton peaks			
3.4318		n=2 state of the exciton associated with the B-valence band	56
3.4295		n=3 state of the Γ5 exciton	56
3.4280		n=3 state of the Γ6 exciton	56
3.4214		n=2 states of the Γ5 and Γ6 exciton	56
3.3885		ground state B-exciton	56
3.3785	FX(B)	Free exciton	56
3.378(4)	FX(A)	Free exciton	98
3.3758	Γ5,A	Γ5 allowed exciton	56
3.374(3.3750)	(LPB_A)	Lower polariton branch of the A valence band	97
3.3740	Γ6,A	Γ5 forbidden exciton	56
3.3725	I_0		93
3.3718	I_1		93
3.368(7)	(D^0,X)	Exciton bound to neutral donor	98
3.3674	I_2	Exciton bound to ionized donor	93
3.3665	I_3	Exciton bound to ionized donor	93
3.364	(D^0,X)	Exciton bound to neutral donor	98
3.365	(D^0,X)	Exciton bound to neutral donor	98
3.3628	I_4	H	92, 93
3.362	(D^0,X)	Exciton bound to neutral donor	98
3.3614	I_5		93
3.361(3.3605)	(D^0,X)	Exciton bound to neutral donor	98
3.3608	I_6	Al	92, 93
3.3600	I_7		93
3.3598	I_8	Ga	92, 93
3.358	(D^0,X)	Exciton bound to neutral donor	98
3.3564	(A^0,X)	Exciton bound to neutral acceptor	
3.3567	I_9	In	92, 93
3.356	(A^0,X)	Exciton bound to neutral acceptor	
3.3531	I_{10}		93
3.353	(A^0,X)	Exciton bound to neutral acceptor	
3.3484	I_{11}		93
3.22	DAP	Li related	92, 101,102
Deep level related peaks			
2.53	DBE	V_O-related defect level	108
2.35	DBE	V_{Zn}-related defect level	108
2.45	DBE	V_O	107
2.38	DBE	Oxygen antisite (O_{Zn})	106

(4) Polishing

Lucca et al.[109] have studied the effect of polishing on the room temperature PL spectra. They showed that the second order exciton-phonon (FX-2LO) emission peak is reduced while that of the first order exciton-phonon (FX-1LO) is increased by chemomechanical polishing. Suscavage et al.[56] have studied the PL of hydrothermally

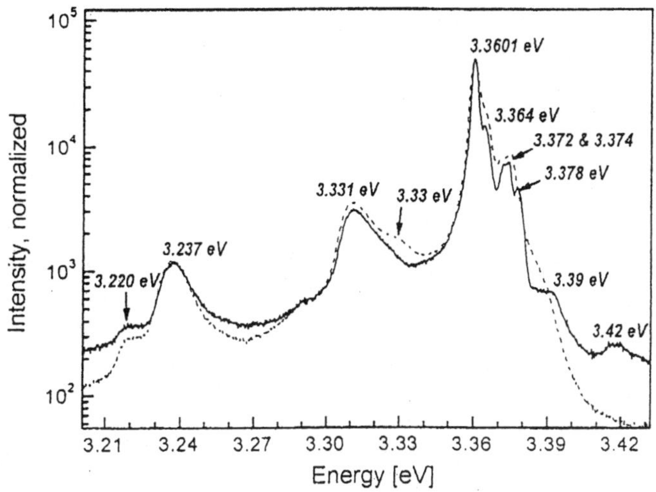

Fig. 21.14 Low temperature PL spectra from the region near the band edge. (a) undoped ZnO subtrate by the hydrothermal method (dashed line), (b) LPE-grown film (solid line) (reprinted from Ref. 37 with permission, copyright 2006 Elsevier).

grown ZnO bulk crystals compared with vapor phase grown crystals and pointed out the effect of surface polishing on the PL spectrum.

(5) Polar surfaces

Sherriff et al.[110] have studied the PL from the two polar faces of ZnO and found difference between the two faces. Phonon replicas of a particular D^0,X transition were dominant on the O face while they were barely observed on the Zn face. PL emissions from free excitons were also more intense on the O face than on the Zn face.

21.5 APPLICATIONS

21.5.1 Substrates

Wenisch et al.[111] have used high resolution X-ray diffraction (HRXRD) ω-scan rocking curve measurement to evaluate the quality of commercial substrates grown by the hydrothermal synthesis method and the seeded physical vapor transport method. Cho et al.[112] have studied the quality of hydrothermal grown ZnO after various surface treatment to remove the damage after chemomechanical polishing.

ZnO substrates are used for homoepitaxial growth of ZnO and for heteroepitaxial growth of GaN. The epitaxial growth of ZnO has been mainly performed on various substrates[113] such as sapphire, $ScMgAlO_4$, $LiNbO_3$, (La, Al)(Sr, Ta)O_3 (LAST), yttrium stabilized zirconia (YSG), Si, $SrTiO_3$, NaCl, since for a long time large single crystal could not be obtained. Large single crystals are now available by the hyrdothermal method, seeded physical transport method and melt growth method, so that homoepitaxial growth on ZnO substrates with no lattice matching is now under investigation. Homoepitaxial growth has been studied by molecular beam epitaxy (MBE)[114] and LPE[115] and heteropitaxial growth of $MgZn_{1-x}O_x$ on ZnO has been studied by LPE.[116]

ZnO has also been studied as the substrate for GaN epitaxial growth since its lattice mismatching is only <1.8%.[1, 117]

21.5.2 Scintillators

In contrast to commonly used ionic scintillators, near-band-edge emitting semiconductors have excited states that extend over many atoms. The emission may arise from band-to-band or free exciton recombination transitions. It is therefore expected that they would make high speed and high radiative efficiency scintillators. In the case of ZnO, Lehman[118,119] intentionally doped Ga and In to enhance fast, near-band-edge emission. It was found that the introduction of ~ 0.3 % Ga introduces a degenerate donor band and increases the conductivity of ZnO. Dorenzo et al.[120] have studied the scintillation from 0.3 % doped ZnO and measured the temperature dependence of the decay time by pulsed X-ray irradiation.

21.5.3 Light Emitting Diodes (LEDs)

ZnO has a high exiciton binding energy with a wide bandgap, so it is a promising material for short wavelength LEDs. In fact, various LED structures have been studied. Most of them are based on p-n heterostructures or metal-insulator-semiconductor (MIS) structures[8] because it is difficult to realize real p-type doping as explained in Sec. 21.4.3.

Alivov et al.[121,122] have fabricated various ZnO-based MIS type, hetero-typed diodes, such as n-ZnO/p-GaN and MIS structures on sapphire substrates. Among various structures, it is claimed that n-ZnO/p-AlGaN/n-SiC[8] is a promising one. They have fabricated n-ZnO doped with Ga by CVD and p-type $Al_{0.12}Ga_{0.88}N$ doped with Mg by HVPE on n-type SiC substrate. It showed a spectral peak at about 387 nm which is attributed to the recombination in the ZnO.

Ohta et al. have fabricated heterostructure pn junction LEDs such as p-$ScCu_2O_2$/n-ZnO[123,124] on yittria-stabilized zirconia (YSZ) substrates. Wang et al.[125] have fabricated n-ZnO/p-diamond on diamond substrates.

Tsukazaki et al.[4,127] have reported a pn junction type LED on $ScAlMgO_4$ substrates. The structure of this LED is shown in Fig. 21.15. p-type ZnO has been realized by nitrogen doping. This work is also reviewed in Ref. 127.

Ryu et al.[128] have fabricated pn-homojunctions on n-SiC substrates. Homo pn junctions were also fabricated on ZnO substrates by Aoki et al.[129] and Guo et al.[130]

21.5.4 Laser Diodes (LDs)

Since ZnO has a bandgap of 3.37 eV and has a high exciton binding energy of 60 meV.[6,11] much higher than that of GaN (24 meV) and than room temperature thermal activation energy (25 meV), it seems likely that laser diodes based on the recombina-

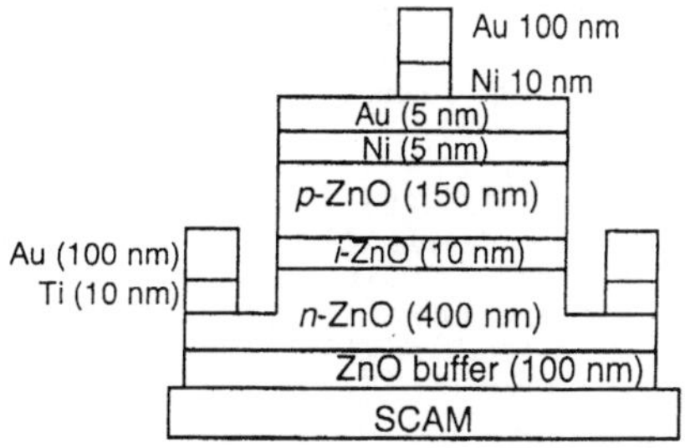

Fig. 21.15 Cross section of ZnO emitting diode (direct current of 5mA at room temperature) (reprinted from Ref. 127 with permission, copyright 1998 The Japan Society of Applied Physics).

tion of free excitons can be realized as predicted by Kayanuma to make possible the production of high-efficiency lasers.[131] This is because the high oscillation frequency is achieved by emission due to exciton recombination. ZnO is therefore a promising material for the fabrication of high efficiency ultraviolet lasers.[132,133]

ZnO lasing has been reported for low temperatures[134,135] and at room temperature.[5,136-138] Ohtomo et al.[139] have observed a sharp stimulated emission at 3.2 eV due to free exicitons through an exciton-exciton collision process at room temperature which was predicted theoretically as mentioned above. The threshold intensity was as low as 24 kW cm^{-2} and the emission intensity was increased by a power of eight as the pumping intensity increased.

21.5.5 Photodiodes

Photodiodes based on heterostructures of n-ZnO on p-Si substrates have been reported.[140-143] These photodiodes have shown high responsivities for UV (310 nm) photons in the depleted n-ZnO and red (650 nm) photons in the depleted p-Si.

Ohta et al. have fabricated heterojunction photodiodes based on p-$ZnRh_2O_4$/n-ZnO[144] and p-NiO/n-ZnO[145] structures on YSZ (111) substrates by pulsed laser deposition (PLD) combined with the solid-phase-epitaxy technique. A photoresponsivity as high as -0.3 AW^{-1} was achieved at 26 V reverse bias under irradiation by 360 nm light.

Alivov et al.[146] have fabricated n-ZnO/p-6H-SiC heterostructure photodiodes by plasma-assisted MBE which had a phtoresponsivity as high as 0.045 A/W at 7.5 V reverse bias with a response peak at 3.283 eV.

REFERENCES

1. R. Triboulet, *Proc. SPIE* **4412** (2001) 1.
2. S.J. Pearton, D.P. Norton, K. Ip, Y.W. Heo and T. Steiner, *Prog. Mat. Sci.* **50** (2005) 293.
3. U. Ozugür, Y.I. Alovov, C. Liu, A. Teke, M.A. Reschikov, S. Dogan, V. Avrutin, S.-J. Cho and H. Morkoç, *J. Appl. Phys.* **98** (2005) 041301.
4. A. Tsukazaki, A. Ohtomo, T. Onuma, M. Ohatani, T. Makino, M. Sumiya, K. Ohtani, S.F. Chichibu, S. Fuke, Y. Segawa, H. Ohno, H. Koinuma and M. Kawasaki, *Nat. Mater.* **4** (2005) 42.
5. P. Zu, Z.K. Tang, G.K.L. Wong, M. Kawasaki, A. Ohtomo, H. Koinuma and Y. Segawa, *Solid State Commu.* **103** (1997) 459.
6. D.M. Bagnall, Y.F. Chen, Z. Zhu, T. Yao, M.Y. Shen and T. Goto, *Appl. Phys. Lett.* **73** (1998) 1038.

7. A.A. Toropov, O.V. Nekrutkina, T.V. Shubina, T. Gruber, C. Kirchner, A. Wang, K.F. Karlsson, P.O. Holtz and B. Monemar, *Phys. Rev. B* **69** (2004) 165205.
8. D.C. Look, B. Claflin, Y.L. Alivov and S.J. Park, *phys. stat. sol. (a)* **201** (2004) 2203.
9. S.O. Kucheyev, J.S. williams, C. Jagadish, J. Zou, C. Evans, A.J. Nelson and A.V. Hamza, *Phys. Rev. B* **67** (2003) 094115.
10. D.C. Look, *Mater. Sci. Eng. B* **80** (2001) 383.
11. D.G. Thomas, *J. Phys. Chem. Solids* **15** (1960) 86.
12. A. Ohotomo, M. Kawasaki, T. Koida, K. Masubuchi, H. Koinuma, Y. Sakurai, Y. Yoshida, Y. Yasuda and T. Segawa, *Appl. Phys. Lett.* **72** (1998) 2466.
13. Y, Li, G.S. Tompa, S. Liange, C. Gorla, C. Lu and J. Doyle, *J. Vac. Sci. Technol. A* **15** (1997) 1063.
14. M. Liu, A.H. Kitai and P. Mascher, *J. Lumin.* **54** (1992) 35.
15. W.A. Roth and P. Chall, *Z. Elektronchem.* **34** (1928) 185.
16. G.F. Huttig, H. Moldner, *Z. Anorg. Allg. Chem.* **211** (1933) 368.
17. E.D. Kolb and R.A. Laudise, *J. Am. Ceram. Soc.* **48** (1965) 342.
18. E.D. Kolb and R.A. Laudise, *J. Am. Ceram. Soc.* **48** (1965) 302
19. D.I. Florescu, L.G. Mourokh, F.H. Pollak, D.C. Look, G. Cantwell and X. Li, *J. Appl. Phys.* **91** (2002) 890.
20. H.A. Wriedt, in *Binary Alloy Phase Diagrams, Sec. Ed. Vol. 3*, eds. by T.B. Massalski, H. Okamoto, P.R. Subramanian and L. Kacprzak, (1990, ASM International) p. 2938.
21. G. Heiland, E. Mollow and F. Srockmann, *Solid State Phys.* **8** (1959) 191.
22. R. Triboulet and J. Perriere, *Prog. Cryst. Growth Char. Mater.* **47** (2005) 65.
23. Landolt Bornstein (Springer, New York, 1982) Vol. 17.
24. I.P. Kuzmina and V.A. Nikitenko, in *Zinc Oxide. Processing and Optical Properties* (Nauka , 1984) p. 166.
25. J. Nause and B. Nemeth, *Semicond. Sci. Technol.* **20** (2005) S45.
26. D.Schulz, S. Ganschow, D. Klimm, M. Neubert, M. Roßberg, M. Schmidbauer and R. Fornari, *J. Crystal Growth* **296** (2006) 27.
27. J.W. Nielsen and E.F. Dearborn, *J. Phys. Chem.* **64** (1960) 1762.
28. W. Kleber, R. Mlodoch, *Krist. Tech.* **1** (1966) 249.
29. G.A. Wolff and H.E. Labelle, *J. Am. Ceram. Soc.* 48 (1965) 441.
30. A.B. Chase and J. Osmer, *J. Am. Ceram. Soc.* **50** (1967) 325.
31. B.M. Wanklyn, *J. Crystal Growth* **7** (1970) 10.
32. M. Ushio and Y. Sumiyoshi, *J. Mater. Sci.* **28** (1993) 218.
33. K. Oka, H. Shibata, S. Kashiwaya, *J. Crystal Growth* **237-239** (2002) 509.
34. X.H. Li, J.Y. Xu, X.M. Li, *Appl. Phys.* **A82** (2006) 373.
35. K. Byrappa, in *Handbook of Crystal Growth, Bulk Crystal Growth* ed. D.T.J. Hurle (North-Holland, Amsterdam, 1994) p. 467.
36. F. Iwasaki and H. Iwasaki, *J. Crystal Growth* **237-239** (2002) 820.
37. D. Ehrentraut, H. Sato, Y. Kagamitani, H. Sato, A. Yoshikawa and T. Fukuda, *Prog. Cryst. Growth Char. Mater.* **52** (2006) 280.
38. D. Ehrentraut, *Jpn. J. Assoc. Crystal Growth (in japanese)* **34** (2007) 3.
39. R.A. Laudise and A.A. Ballman, *J. Phys. Chem.* **64** (1960) 688.
40. R.A. Laudise, E.D. Kolb and A.J. Caporaso, *J. Am. Ceram. Soc.* **47** (1964) 9.
41. E.D. Kolb and R.A. Laudise, *J. Am. Ceram. Soc.* **48** (1965) 302
42. E.D. Kolb, A.S. Coriell, R.A. Laudise, A.R. Hutson, *Mater. Res. Bull.* **2** (1967) 1099.
43. I.L. Khodakovskiy and A.Ye. Yelkin, *Geochemistry* **10** (1975) 1450.
44. N. Sakagami and K. Shibata, *Jpn. J. Appl. Phys.* **20** (1981) 201.
45. V.A. Kuznetsov, I.P. Kuzmina and I.M. Sylvestrova, *Bull. Mater. Sci.* **6** (1984) 2177.
46. T. Sekiguchi, S. Miyashita, K. Obara, T. Shishido and N. Sakagami, *J. Crystal Growth* **214/215** (2000) 72.
47. T. Sakagami, M. Yamasita, T. Sekiguchi, S. Miyashita, K. Obara and T. Shishido, *J. Crystal Growth* **229** (2001) 98.
48. E. Ohshima, M. Sato, I. Niikura, K. Maeda, M. Sato, M. Ito and T. Fukuda, *J. Crystal Growth* **260** (2004) 166
49. K. Maeda, M. Sato, I. Niikura and T. Fukuda, *Semicond. Sci. Technol.* 20 (2005) S49.
50. J. Kou, T. Fukuda, M. Sato and K. Yoshioka, *29th Meeting, Wide Bandgap Semiconductor Opto-electronic dDevices 162 Committee (in japanese)* (2006) p. 16.
51. T. Fukuda, M. Mikawa and D. Ehrentraut, *Proc. SPIE* **6474** (2007) 647412.
52. L.N. Demianets and D.V. Kostomarov, *Ann. Chim. Sci. Mater.* **26** (2001) 193.
53. E.V. Kortounova and V.I. Lyutin, *Ann. Chim. Sci. Mater.* **22** (1997) 647.
54. E.V. Kortounova, V.I. Lyutin, in: *Synthesis of Minerals and Methods of their Study, Proceedings of All Russian Research Institute of Synthesis of Raw Mineral* **24** (1997) 31.

55. E.V. Kortunova, P.P. Chvanski and N.G. Nikolaeva, *J. Phys. IV France* **126** (2005) 39.
56. M. Suscavage, M. Harris, D. Bliss, P. Yip, S.-Q. Wang, d. Schwall, L. Bouthillette, J. Bailey, M. Callahan, D.C. Look, D.C. Reynolds, R.L. Jones and C.W. Litton, *MRS Internet J. Nitride Semicond. Res.* **4S1** (1999) G3.40.
57. B. Wang, M.J. Callahan, L.O. Bouthillette, C. Xu and M.J. Suscavage, *J. Crystal Growth* **287** (2006) 381.
58. B. Wang, M.J. Callahan, L.O. Bouthillette, *Crystal Growth & Des.* **6** (2006) 1256.
59. B. Wang, M.J. Callahan, C. Xu, L.O. Bouthillette, N.C. Giles and D.F. Bliss, *J. Crystal Growth* **304** (2007) 73.
60. J. Kubo, *Jpn. J. Appl. Phys.* **4** (1965) 225.
61. T. Takahashi, A. Ebina and A. Kamiyama, *Jpn. J. Appl. Phys.* **5** (1966) 560.
62. E.A. Weaber, *J. Crystal Growth* **1** (1967) 320.
63. Y.S. Park and D.C. Reynolds, *J. Appl. Phys.* **38** (1967) 756.
64. K.F. Nielsen, *J. Crystal Growth* **3/4** (1968) 141.
65. M. Hirose, Y. Furuya and J. Kubo, *Jpn. J. Appl. Phys.* **9** (1970) 726.
66. M. Hirose, Y. Furuya and J. Kubo, *Jpn. J. Appl. Phys.* **9** (1970) 423.
67. M. Hirose, *Jpn. J. Appl. Phys.* **10** (1971) 401.
68. R. Helbig, *J. Crystal Growth* **15** (1972) 25.
69. K. Fischer, *J. Crystal Growth* **34** (1976) 139.
70. D.D. Reynolds, D.C. Look and B. Jogai, *Solid State Commun.* **99** (1996) 873.
71. F. Emmenegger and A. Petermann, *J. Crystal Growth* **2** (1968) 33.
72. M. Shiloh and J. Gutmann, *J. Crystal Growth* **11** (1971) 105.
73. W. Piekarczyk, S. Gazda and T. Niemyski, *J. Crystal Growth* **12** (1972) 272.
74. K. Matsumoto, K. Konemura and G. Shimaoka, *J. Crystal Growth* **71** (1985) 99.
75. K. Matsumoto and G. Shimaoka, *J. Crystal Growth* **86** (1988) 410.
76. K. Matsumoto and K. Noda, *J. Crystal Growth* **102** (1990) 137.
77. D.C. Look, D.C. Reynolds, J.R. Sizelove, R.L. Jones, C.W. Litton, G. Cantwell and W.C. Harsch, *Solid State Commun.* **105** (1998) 399.
78. J.-M. Ntep, S.S. Hassani, A. Lusson, A. Tromson-Varli, D. Ballutaud, G. Didier and R. Triboulet, *J. Crystal Growth* **207** (1999) 30.
79. A. Mycielski, L. Kowaczyk, A. Stepniewski, J.M. Baranowski, M. Potemski, A. Witowski, R. Jakiela, A. Barcz, B. Witkowska, W. Kaliszek, A. Jedrzejczak, A. Suchocki, E. Lusakowska and E. Kaminska, *J. Alloys Comp.* **371** (2004) 150.
80. M. Mikami, T. Eto, J. Wang, Y. Masa and M. Isshiki, *J. Crystal Growth* **276** (2005) 389.
81. M. Mikami, T. Sato, J. Wang, Y. Masa and M. Isshiki, *J. Crystal Growth* **286** (2006) 213.
82. M. Mikami, S.-H. Hong, T. Sato, S. Abe, J. Wang, K. Matsunoto, Y. Masa and M. Isshiki, *J. Crystal Growth* **304** (2007) 37.
83. J. Nishio and Y. Nosaka, *J. Crystal Growth* **268** (2004) 174.
84. A.F. Kohan, G. Ceder, D. Morgan and C.G. Van de Walle, *Phys. Rev. B* **61** (2000) 15019.
85. A. Janotti and C.G. Van de Walle, *J. Crystal Growth* **287** (2006) 58.
86. F. Tuomisto and D.C. Look, *Proc. SPIE* **6474** (2007) 647413.
87. J.D. Albrecht, P.P. Ruden, S/ Limpijumong, W.R.L. Lambrecht, K.F. Brennan, *J. Appl. Phys.* **86** (1999) 6864.
88. A. Tsukazaki, H. Saito, K. Tamura, M. Ohtani, H. Koinuma, M. Sumiya, S. Fuke, T. Fukumura and M. Kawasaki, *Appl. Phys. Lett.* **81** (2002) 235.
89. A. Tsukazaki, A. Ohotomo, S. Yoshida, M. Kawasaki, C.H. Chia, T. Makino, Y. Segawa, T. Koida, S.F. Chichibu and H. Koinuma, *Appl. Phys. Lett.* **83** (2003) 2784.
90. A. Tsukazaki, A. Ohotomo, T. Ohnuma, M. Ohtani, T. Makino, M. Sumiya, K. Ohtani, S. Fuke, Y. Segawa, H. Ohno, H. Koinuma and M. Kawasaki, *Nature Mater.* **4** (2005) 42.
91. D.M. Kolb and H.-J. Schulz, in *Current Topics in material Science Vol. 7*, ed. by E. Kaldis (North-Holland, Amsterdam, 1981) p. 226.
92. B.K. Meyer, H.R. Alves, D.M. Hofmann, W. Kriegseis, D. Forster, F. Bertram, J. Christen, A. Hoffmann, M. Strassnurg, M. Dworzak, U. Haboeck and A.V. Rodina, *phys. stat. sol. (b)* **241** (2004) 231.
93. D.C. Reynolds, C.W. Litton and T.C. Collins, *Phys. Rev.* **140** (1965) A1726.
94. D.C. Reynolds, D.C. Look, B. Jogai, C.W. Litton, T.C. Collins, W. Harsch and G. Cantwell, *Phys. Rev. B* **57** (1998) 12151.
95. D.C. Reynolds, D.C. Look, B. Jogai and T.C. Collins, *Appl. Phys. Lett.* **79** (2001) 3794.
96. E. Tomzig and H. Helbig, *J. Lumin.* **14** (1976) 403.
97. N. Ohashi, T. Sekiguchi, K. Aoyama, T. Ohgaki, Y. Terada, I. Sakaguchi, T. Tsurumi and H. Haneda, *J. Appl. Phys.* **91** (2002) 3658.

98. D.W. Hamby, D.A. Lucca and M.J. Klopfstein, *J. Appl. Phys.* **93** (2003) 3214.
99. A. Manoogian and J.C. Wooley, *Can. J. Phys.* **62** (1984) 285.
100. O.F. Schimer and D. Zwingel, *Solid State Commun.* **8** (1970) 1559.
101. D. Zwingel, *J. Lumin.* **5** (1972) 385.
102. N.Y. Garces, L. Wang, L. Bai, N.C. Giles, L.E. Halliburton and G. Cantwell, *Appl. Phys. Lett.* **81** (2002) 622.
103. R. Dingle, *Phys. Rev. Lett.* **23** (1969) 579.
104. N.C. Giles, N.Y. Garces, L. Wang, L.E. Haliburton, *Proc. SPIE* **5359** (2004) 267.
105. K. Vanheysden, C.H. Seager, W.L. Warren, D.R. Tallant and J.A. Woight, *Appl. Phys. Lett.* **68** (1996) 403.
106. B. Lin, Z. Fu and Y. Jia, *Appl. Phys. Lett.* **79** (2001) 943.
107. F.H. Leiter, H. Alves, D. Pfisterer, N.G. Romanov, D.M. Hofmann and B.K. Meyer, *Physica B* **340-342** (2003) 201.
108. T.M. Borseth, B.G. Svensson, A. Yu. Kuzznetzov, P. Klason, Q.X. Zhao and M. Willander, *Appl. Phys. Lett.* **89** (2006) 262112.
109. D.A. Lucca, D.W. Hamby, M.J. Klopfstin and G. Cantwell, *phys. stat. sol. (b)* **229** (2002) 845.
110. R.E. Sherriff, D.C. Reynolds, D.C. Look, B. Jogai, J.E. Hoelscher, T.C. Collins, G. Cantwell and W. Harsch, *J. Appl. Phys.* **88** (2000) 3454.
111. H. Wenisch, V. Kirchner, S.K. Hong, Y.F. Chen, H.J. Ko and T. Yao, *J. Crystal Growth* **944** (2001) 227.
112. M.W. Cho, C. Harada, H. Suzuki, t. Minegishi, T. Yao, H. Ko, K. Maeda and I. Nikura, *Supperlatt. Microstruct.* **38** (2005) 349.
113. R. Triboulet and J. Perriere, Prog. *Crystal Growth and Charac. Mater.* **47** (2003) 65.
114. D.C. Look, D.C. Reynolds, C.W. Litton, R.L. Jones, D.B. Eason and G. Cantwell, *Appl. Phys. Lett.* **81** (2002) 1830.
115. D. Ehrentraut, H. Sato, Y. kagamitani, A. Yoshikawa, T. Fukuda, J. Pejchal, K. Polak, M. Nikl, H. Okada, K. Hatanaka and H. Fukumura, *J. Mater. Chem.* **16** (2006) 3369.
116. H. Sato, D. Ehrentraut and T. Fukuda, *Jpn. J. Appl. Phys.* **45** (2006) 190.
117. T. Matsuoka, N. Yoshimoto, T. Sasaki and A. Katsui, *J. Electron. Mater.* **21** (1992) 157.
118. W. Lehmann, *J. Electrochem. Soc.* **112** (1965) 1150.
119. W. Lehmann, *Solid-State Electron.* **9** (1966) 1107.
120. S.E. Dorenzo, M.J. Weber and M.K. Klintenberg, *Nucl. Instrum. Methods Phys. Res.* **A486** (2002) 214.
121. Ya.I. Alivov, D.C. Look, B.M. Ataev, M.V. Chukichev, V.V. Mamedov, V.I. Znenko, Yu.A. Agafonov and A.N. ustovit, *Solid-State Electon.* **48** (2004) 2343.
122. Ya.I. Alivov, J.E. Van Nostand, D.C. Look, M.V. Chukichev and B.M. Ataev, *Appl. Phys. Lett.* **83** (2003) 2943.
123. H. Ohta, M. Orito, M. Hirano and H. Hosono, *J. Appl. Phys.* **89** (2001) 5720.
124. H. Hosono, H. Ohta, K. Hayashi, M. Orita and M. Hirano, *J. Crystal Growth* **237-239** (2002) 496.
125. C.Z. Wang, G.W. Yang, T.C. Zhang, H.W. Lu, Y.H. Han, J.F. Luo, C.X. Gao and G.T. Zou, *Diamond Relat. Mater.* **2** (2003) 1548.
126. H. Ohno, S.F. Chichibu and M. Kawasaki, *Jpn. J. Appl. Phys.* **44** (2005) L643.
127. A. Tsukazaki, M. Kubota, A. Ohtomo, T. Onuma, K. Ohtani, H. Ohno, S.F. Chichibu and M. Kawasaki, *Jpn. J. Appl. Phys.* **44** (2005) L643.
128. Y.R. Ryu, T.S. Lee, J.H. Leem and H.W. Whte, *Appl. Phys. Lett.* **83** (2003) 4032.
129. T. Aoki, Y. Hatanaka and D.C. Look, *Appl. Phys. Lett.* **76** (2000) 3257.
130. X.-L. Guo, J.-H. Choi, H. Tabata and T. Kawai, *Jpn. J. Appl. Phys.* **40** (2001) L177.
131. Y. Kayanuma, *Phys. Rev. B* **38** (1988) 9797.
132. A. Ohtomo and M. Kawasaki, *Parity (in japanese)* **13** (1998) 48.
133. M. Kawasaki and A. Ohtomo, *Solid Physics (in japanese)* **33** (1998) 59.
134. F. H. Nicoll, *Appl. Phys. Lett.* **9** (1966) 13.
135. S.W. Koch, C. Kringshirn et al. *phys. stat. sol. (b)* **89** (1978) 431.
136. D.M. Bagnall, T.F. Chen, Z. Zhu, T. Yao, S. Koyama, M.Y. Shen and T. Goto, *Appl. Phys. Lett.* **70** (1997) 2230.
137. P. Yu, K. Tang, G.K.L. Wong, M. Kawasaki. A. Ohtomo, H. Koinuma and Y. Segawa, *23rd Inter. Conf. Phys. Semicon.* ed. by M. Scheffler and R. Zimmermann (World Scientific, Singapore, 1966) p. 1453.
138. Y. Segawa, A. OHhtomo, M. Kawasaki, H. Koinuma, Z.K. Tang, P. Yu and K.L. Wong, *phys. stat. sol. (b)* **202** (1997) 669.
139. A. Ohtomo, M. Kawasaki, Y. Sakurai, Y. Yoshida, H. Koinuma, P. Yu, Z.K. Tang, G.K.L. Wong and Y. Segawa, *Mater. Sci. Engineer. B* **54** (1998) 24.
140. J.Y. Lee, Y.S. Choi, W.H.Choi, H.W. Yeom, Y.K. Yoon, J.H. Kim and S. Im, *Thin Solid Films* **420-421** (2002) 112.

141. Y.S. Choi, J.Y. Lee, S. Im and S.J. Lee, *J. Vac. Sci. Technol. B* **20** (2002) 2984.
142. C.H. Park, I.S. Jeong, J.H. Kim and S. Im, *Appl. Phys. Lett.* **82** (2003) 3973.
143. I.-S. Jeong, J.H. Kim and S. Im, *Appl. Phys. Lett.* **83** (2003) 2946.
144. H. Ohta, H. Mizoguchi and M. Hirano, *Appl. Phys. Lett.* **82** (2003) 823.
145. H. Ohta, M. Hirano, K. Nakahara, H. Maruta, T. Tanabe, M. Kamiya, T. Kamiya and H. Hosono, *J. Appl. Phys.* **83** (2001) 1029.
146. Ya. Alivov, U. Ozgur, S. Dogan, D. Johnstone, V. Avrutin, N. Onojima, C. Liu, J. Xie, Q. Fan and H. Moroc, *Appl. Phys. Lett.* **86** (2005) 241108.

22. MERCURY COMPOUNDS

22.1 INTRODUCTION

Mercury compounds have a negative bandgap and have some semimetal properties. Among various mercury compounds, the mixed crystal mercury cadmium telluride (MCT; $Hg_{1-x}Cd_xTe$) between HgTe and CdTe is one of the most important compound semiconductors for use in far-infrared detectors.[1-4] This is because it is a direct bandgap semiconducting alloy whose bandgap Eg can be adjusted by varying the alloy composition between $-0.3<Eg<1.6eV$[2,5] and far-infrared detectors in the wavelength of 1-3 μm and 10-14 μm can be fabricated. Because of this importance, the development of bulk crystal growth technology has a long history.

Mercuric iodide (HgI_2) was first discovered by Willig[6,7] and was found to be a good material for X-ray and gamma-ray detectors[8,9] as compared with other materials (Vol. 1, Table 6.5). In fact, Malm[10] has first shown the real gamma spectrum detected by HgI_2. Since then, it has been developed as a detector material for nuclear radiation such as X-ray and gamma-ray measurements. The development of mercury iodide has been reviewed by several authors.[11-13]

In this chapter, crystal growth and characterization of these two main crystals, MCT and HgI_2, are reviewed.

22.2 PHYSICAL PROPERTIES

22.2.1 Mercury Cadmium Telluride (MCT; $Hg_{1-x}Cd_xTe$)

The physical properties of MCT are reviewed in Refs. 1, 14 and 15 and are summarized in Table 22.1.

The physical properties of MCT have been studied extensively. The phase diagram of MCT has been studied by various authors.[16-21] X-T and P-T diagrams are shown in Fig. 22.1. Schmit and Stelzer[22] first measured the energy bandgap of MCT as a function of the composition x. Hansen et al.[23] have summarized all data on the energy bandgap as a function of the composition x of $Hg_{1-x}Cd_xTe$, and determined it as

$$Eg = -0.302 + 1.93x + 5.35 \times 10^{-4}T(1-2x) - 0.810x^2 + 0.832x^3. \qquad (22.1)$$

Glazov and Pavolva[24] have studied the structure of MCT melt and discussed the melt behavior. Andrews et al.[25] have measured the microhardness as a function of the composition x of various mercury ternary alloys including MCT which shows a maximum in the range of $0.6 < x < 0.8$. Self diffusion of Hg in MCT has been studied by Palfrey[26] who showed the possibility of short circuit diffusion paths.

Table 22.1 Physical Properties of Mercury Compounds

	HgTe	$Hg_{0.8}Cd_{0.2}Te$	CdTe
Crystal Structure	zincblende	zincblende	zincblende
Lattice Constant (Å)	6.463	6.4645	6.482
Density (g/cm^{3})	8.08	7.63	5.85
Melting Point (°C)	673	670	1092
Linear Expansion Coefficient ($x10^{-6}$ cm/deg)	4	4.3	5.5
Thermal Conductivity (W/cm·K)	0.04	-	0.07
Dielectric Constant	20.5	18.5	10.5
Refractive index	3.424	3.249	2.675
Bandgap at Room Temperature (eV)	- 0.15	- 0.165	1.505
Optical Transition Type	-	direct	direct
Intrinsic Carrier Concentration at R.T. (cm^{-3})	-	3×10^{16}	-
Electron Mobility at 77K (cm^2/V·sec)	-	2.5×10^5	4×10^4
Hole Mobility at 77K (cm^2/V·sec)	-	7×10^2	3.8×10^4

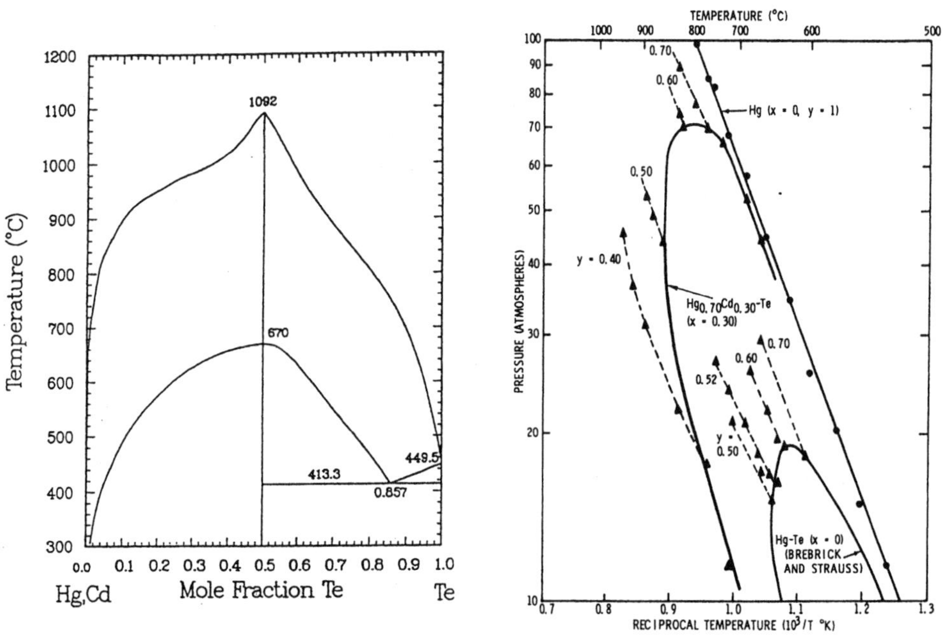

(a) (b)

Fig. 22.1 (a) Liquidus lines for HgTe (lower) and CdTe (upper) and (b) pressure-temperature phase diagram for Hg-Te and $Hg_{0.7}Cd_{0.30}Te$ melts (reprinted from Refs. 18 with permission, copyright 1976 Springer).

22.2.2 HgI_2

Mercuric iodide (HgI_2) has a high atomic number (80 for Hg and 53 for I), a large bandgap of 2.292 eV and a low electron-hole-pair formation energy of 4.22 eV and is one of the best materials for radiation detectors such as X-ray and gamma-ray detectors as compared with other materials as shown in Table 6.5 in Vol. 1. HgI_2 is a quite different structure from zincblende compounds. HgI_2 is an anisotropic layered material consisted of alternating layers of Hg and I. Hg-Hg and Hg-I are covalent bonds while I-I interlayers have Van der Waals bonds as shown in Fig. 22.2.[27] Each Hg atom is tetrahedrally bonded to four I atoms whereas the I atoms display only two bonds. HgI_2 shows a phase transformation at 131.2°C from the tetragonal red structure (α-HgI_2) to the orthorhombic yellow structure (β-HgI_2).[28, 29] Because of this transformation, it is difficult to grow red HgI_2 from the melt even though the melting point is as low as 250°C. Its energy bandgap[30], thermal diffusivity and thermal expansion are summarized by Burger et al.[31] Its microhardness and hardness anisotropy have been studied by Milstein et al.[32] The hardness anisotropy of HgI_2 single crystals have been measured by Basker et al.[33] James and Milstein[34] have measured the plasticity of HgI_2. The main physical properties of HgI_2 are shown in Table 22.2.[8,13]

The phase diagram of HgI_2 has been studied by various authors[35-38] and is shown in Fig. 22.3 [12,36] with a transformation at 70°C.[37,38] The vapor pressure of HgI_2 has been measured by Dishon et al.[39] Photoplastic effect in HgI_2 has been observed by Marchall and Milstein.[40]

22.3 CRYSTAL GROWTH

22.3.1 Mercury Cadmium Telluride (MCT; $Hg_{1-x}Cd_xTe$)

Among various mercury compounds, mixed crystal, MCT was extensively studied as summarized in Table 22.3 because of its importance in far-infrared detectors. The crystal growth of MCT has been reviewed by several authors.[41-46] MCT crystals were grown by the following methods.

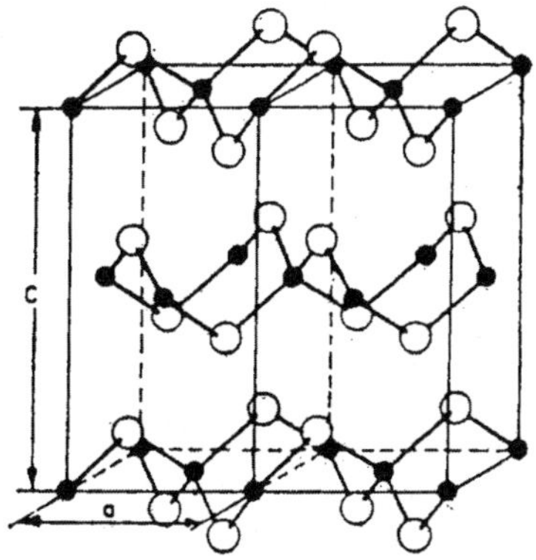

Fig. 22.2 Crystal structure of HgI_2 (reprinted from Refs. 27 with permission, copyright 1975 IEEE).

Table 22.2 Physical Properties of Mercury Iodine

	HgI_2
Crystal Structure	Tetragonal
Lattice Constant	a = 4.361(5) Å c = 12.450(7) Å
Density	6.37 g/cm^3
Melting Point	250°C
Linear Expansion Coefficient	$\alpha_{11} = 11 \times 10^{-6}$ cm/deg $\alpha_{33} = 54 \times 10^{-6}$ cm/deg
Thermal Conductivity at 300K	$k_{11} = 4.08 \times 10^{-3}$ W/cm·K $k_{33} = 1.13 \times 10^{-3}$ W/cm·K
Bandgap at Room Temperature	2.292 eV
$\mu\tau e$ / $\mu\tau h$ at R.T.	10^{-4} - 10^{-5} cm/V
$\mu\tau E$	0.2 cm
Intrinsic Resistivity	10^{17} Ω·cm
Hole-electron formation energy	4.22 eV

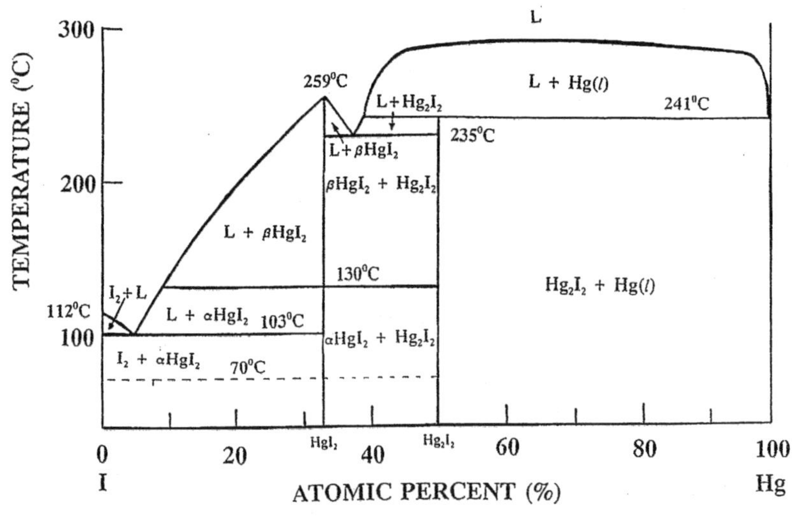

Fig. 22.3 Phase diagram of the Hg-I system with a dashed horizontal line at 70°C indicating an apparent surface transformation of a-HgI_2[37,38] (reprinted from Ref. 36 with permission, copyright 1970 Springer).

(1) Bridgman growth

-Conventional and modified methods

Barlett et al.[47,48] have examined the effect of growth rate on the radial and axial composition uniformity and concluded that it is not possible to grow uniform MCT crystals

Table 22.3 Crystal Growth of $Hg_{1-x}Cd_xTe$ (MCT)

Method	Contents	Authors	Year	Ref.
HZM	First growth of MCT by HZM	Lawson	1959	5
VB	First growth of MCT by VB	Bartlett et al.	1969	47
VZM	First growth of MCT by VZM	Dziuba	1969	71
SSR	First growth of MCT by SSR	Swink et al.	1970	87
VZM	Growth from the Te solution	Ueda et al.	1972	72
SSR	Growth of MCT by slush-recrystallization	Harman	1972	88
ZL	Effect of the growth rate on the S/L interface	Takase	1973	73
Modified VB	Homogeneous growth with a second solid phase	Fiorito et al.	1978	49
VB	Effect of growth rate on compositional variations	Bartlett et al.	1979	48
SSR	Growth by cast-recrystallize-anneal (CRA) method	Bartlett et al.	1979	89
SSR	Growth of MCT by pressurized cast-recrystallization	Vere et al.	1982	90
VB	Effect of growth rate on the interface shape	Capper et al.	1983	53
VB	Effect of growth rate on compositional variations observed versus theoretical	Capper et al.	1983	54
ACRT-VB	Simulation of fluid flow	Capper et al.	1984	61
ACRT-VB	Fluid flow analysis for tall narrow containers	Capper et al.	1984	62
ACRT-VB	Effect of rotational parameters and start compositions	Capper et al.	1986	64
THM	30 mm diameter, 89 mm length, good homogeneity	Triboulet et al.	1985	75
VZM	Growth of MCT without grain boundaries	Yoshikawa	1985	74
SSR	Growth mechanism of recrystallization	Yadava et al.	1985	92
SSR	Incremental quenching of MCT and recrystallization, growth of 20 mm by 50 mm length MCT crystals	Colombo et al.	1985	93
ACRT-VB	Study of solid-liquid interfaces by quenching	Capper et al.	1987	65
ACRT-VB	Interface and flow regimes at various rotation rates	Capper et al.	1988	66
THM	Growth of <111>A oriented crystals	Colombo et al.	1988	76
THM	Growth under Hg pressure control	Mochizuki et al.	1988	77
ACRT-VB	Microprocessor-controlled crystal grower	Coates et al.	1989	67
ACRT-VB	Effect of changes in growth process on crystal quality	Capper et al.	1989	68
THM	Interface shape between grown crystal and CdTe seeds	Shilo et al.	1990	78
THM	Intelligent growth based on capacitance manometry, fiber optic thermometry and nuclear gauge	Witter	1990	80
THM	Inclusion elimination by a first THM pass	Gille et al.	1991	79
Modified VB	Homogeneous growth with second solid phase	Bittner et al.	1991	58
THM	Horizontal THM with crucible rotation	Gille et al.	1991	81
ACRT-THM	High growth rate by ACRT	Bloedner et al.	1993	82
SSR	Horizontal casting to improve uniformity	Sharma et al.	1993	94, 95
PVB	Two-stage pressurized VB method	Wang et al.	2004	96, 97
ACRT-VB	Routine growth of 20 mm diameter, 200 mm length crystals	Capper et al.	2005	69

VB: Vertical Brideman, PVB: Pressurized VB, ACRT-VB: Accelerated Crucible Rotation Technique-VB, HZM: Horizontal Zone Melting, VZM: Vertical Zone Melting, ZL: Zone Leveling, THM: Travelling Heater Method, SSR: Solid State Recrystallizaion.

with controlled electrical properties by the Bridgman method. In order to overcome this problem of the Bridgman method, Fiorito et al.[49] have proposed a modified VB method (Fig. 22.4) in which the melt composition is controlled by means of a second solid phase.

Jones et al.[50,51] have studied the thermal behavior using an electrical analogue model and examined the factors affecting the isotherm shape during Bridgman crystal growth. Liaw et al.[52] have made a numerical study of the Bridgman crystal growth, considering crucible rotation. Capper et al.[53,54] have made quenching studies of Bridgman-grown MCT crystals and examined the solid-liquid interface shape from the viewpoint of interface curvature and boundary layer thickness. They also made a theoretical prediction of the axial and radial compositional uniformity considering no convective mixing and complete mixing. Crystal growth of MCT has been performed varying the growth rate and the initial composition, and the axial and radial compositional uniformity have

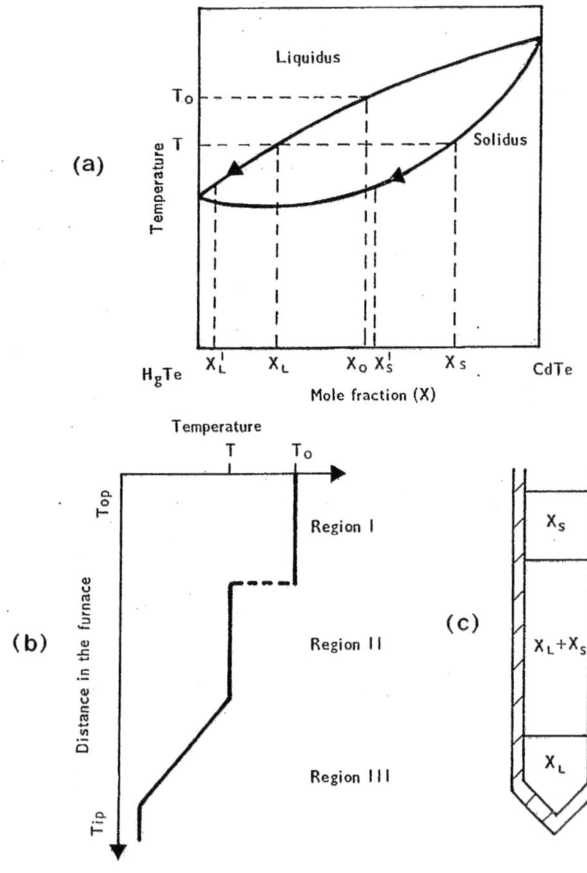

Fig. 22.4 Sketches of a modified vertical Bridgman method: (a) phase diagram of the pseudobinary alloy system HgTe-CdTe; (b) temperature profile of growth apparatus; (c) X_L and X_s phase distribution at the start of growth (reprinted from Refs. 49 with permission, copyright 1978 The Electrochemical Society).

been compared with the predictions.

Segregation in MCT crystal growth has been theoretically discussed by Jones et al.[55,56] and Lehoczky and Szofran.[57] Methods for determining effective segregation coefficients are discussed. The uniformity of the melt during crystal growth has been deduced from the axial composition profile of MCT.

Bittner et al.[58] have grown MCT crystals by the modified Bridgman method by introducing a source of solid material floating in the melt and compared their device properties with those using travelling heater method (THM) crystals.

Wang et al.[59,60] have recently developed a new two-stage pressurized Bridgman method. In the first stage, MCT with $x = 0.04$ is synthesized in a high pressure rocking furnace. In the second stage, synthesized polycrystalline MCT ingot was fed with Hg, Cd, Te in order to obtain MCT melt with $x = 0.21$-0.22 by diffusion of elements. Crystal growth was performed at 710°C with a temperature gradient of about 3-8°C/cm at a velocity of about 2-9 mm/day in the furnace as shown in Figs. 22.5 and 22.6. By this method, they were able to grow 40 mm diameter MCT crystals ($x = 0.210$) as shown in Fig. 22.7.

- *Accelerated Crucible Rotation Technique (ACRT)*

MCT is grown mainly by the Bridgman method using ACRT (Vol. 1, Sec. 2.5.4) in order to prevent the compositional non-uniformity due to segregation. The simulation study of the fluid flow for the application of ACRT to a tall narrow container in the vertical Bridgman method has been performed by Capper et al.[61,62] using mainly water and other fluids like acetone, iso-propanol and iso-buthyl. They analyzed three types of flows, Couette flow, Spiral shearing flow and Ekman flow. The typical flow patterns are as shown in Fig. 22.8. These simulation studies were reviewed by Brice and Capper.[63]

This method was used to grow MCT crystals.[64-69] Capper et al.[65,66] have done a

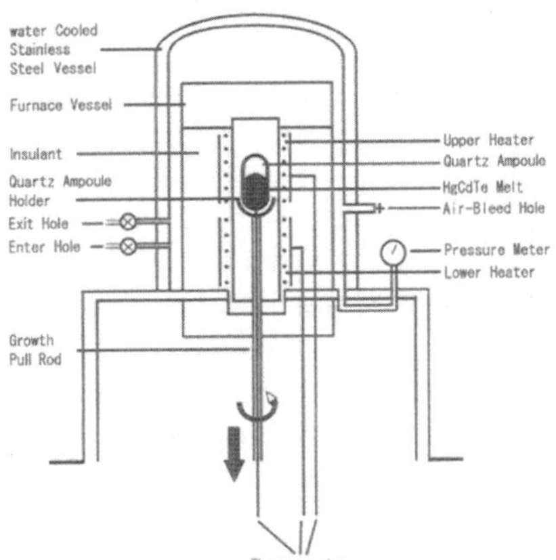

Fig. 22.5 Illustration of the growth furnace (reprinted from Ref. 59 with permission, copyright 2004 Elsevier).

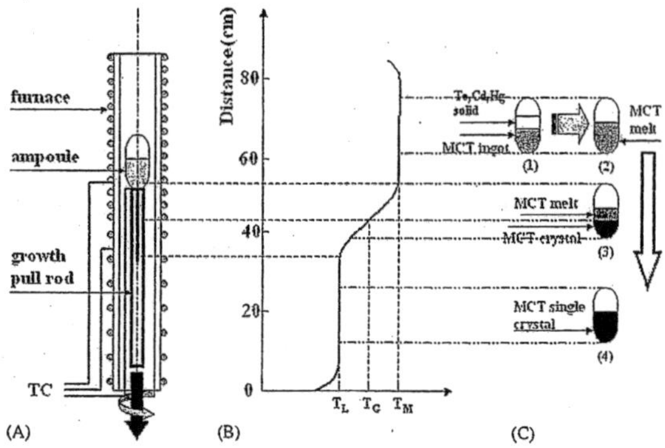

Fig. 22.6 Schematic illustration of the growth procedure by the pressurized-Bridgman method: (A) illustration of furnace, (B) furnace temperature distribution and (C) position of ampoule in the furnace (reprinted from Ref. 60 with permission, copyright 2004 Elsevier).

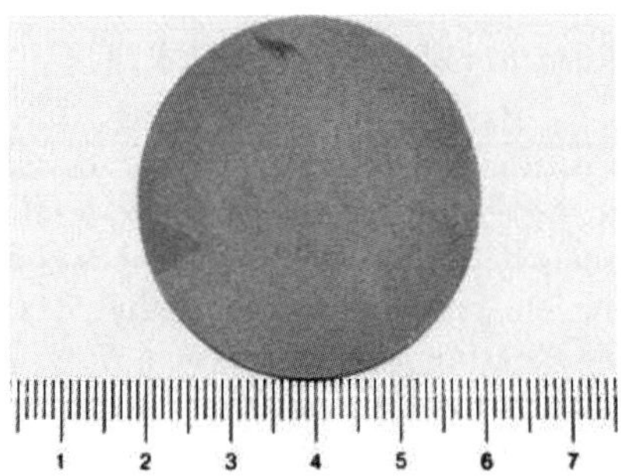

Fig. 22.7 Large $Hg_{1-x}Cd_xTe$ (MCT) crystal wafer of diameter 40 mm grown by the pressurized-Bridgman method (reprinted from Ref. 59 with permission, copyright 2004 Elsevier).

quenching study to examine the solid-liquid interface shape as compared with theoretical predictions. Coates et al.[67] have studied the effect of rotation parameters on the grown MCT crystals. Capper et al.[68] have studied the effect of growth parameters such as growth rates and growth ampoule shape on the grown crystal qualities. Capper et al.[69] have grown 20 mm diameter and 200 mm long crystals weighing 0.5 kg by this method using a furnace arrangement as shown in Fig. 22.9. Liu et al.[70] have made a numerical analysis of the forced convection resulting from the ACRT and its effects on the position and the shape of the melt/crystal interface and the solute redistribution and radial segregation by the finite difference method.

(2) Zone Melting (ZM) and Zone Leveling (ZL)

Even though crystal growth by Bridgman methods has been performed, compositional uniformity remains unattainable because of the segregation effect. Zone melting methods have less segregation effect than the Bridgmann methods.

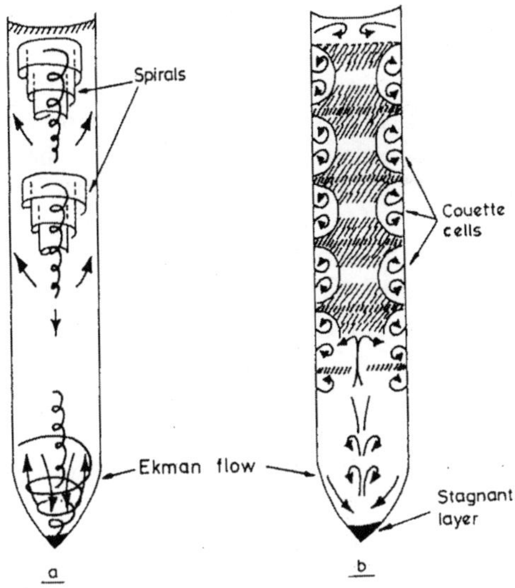

Fig. 22.8 Schematic representation of flows in conical-based container during (a) spin-up and (b) spin-down (reprinted from Ref. 61 with permission, copyright 2004 Elsevier).

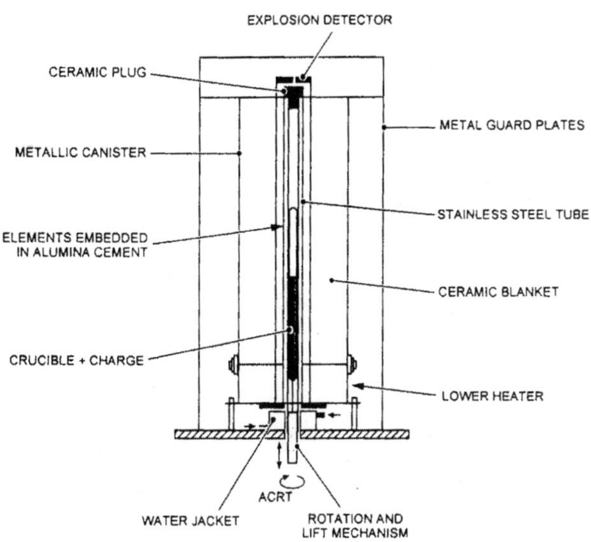

Fig. 22.9 Bridgman/ACRT growth apparatus (reprinted from Ref. 69 with permission, copyright 2005 Elsevier).

Lawson[5] was the first to grow MCT crystals using the horizontal zone-melting method and evaluate their electrical properties. Dziuba[71] has applied the vertical zone melting method to grow homogeneous MCT crystals. He showed the effectiveness of the zone melting method in obtaining crystals with less segregation.

Ueda et al.[72] have applied the zone melting method to obtain homogeneous MCT crystals, using Te as a solvent. Several zone passes have been performed in order to obtain high purity materials. Takase[73] has used a three-zone vertical furnace in the zone leveling method. The solid-liquid interface was studied as a function of growth rate and temperature. He showed that the interface shape is largely affected by the growth rate. Yoshikawa[74] has grown MCT by the modified zone melting method (Fig. 22.10) and evaluated the crystal quality by fabricating infrared sensors.

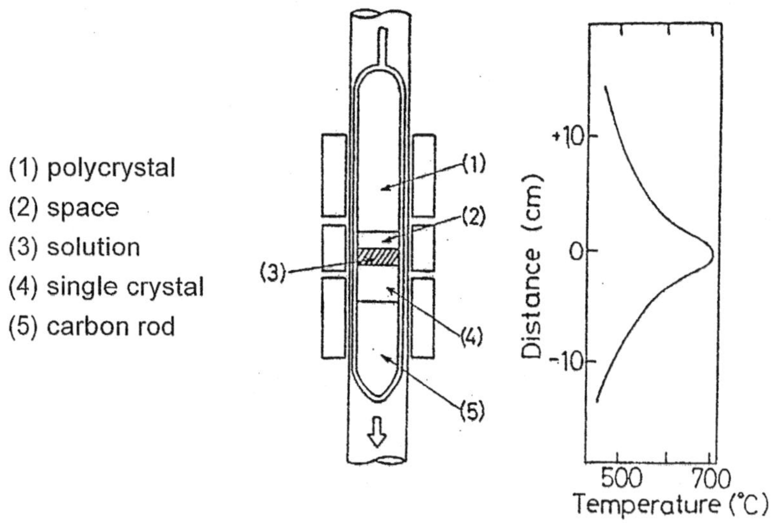

Fig. 22.10 Modified zone-melting method (repreinted from Ref. 74 with permission, copyright 1985 The Institute of Electronics, Information and Cmmuncation Engineers).

(3) Solution growth

- Travelling Heater Method (THM)

MCT crystals have been grown by the THM by various researchers[75-83] and some of them are reviewed by Triboulet.[84] Triboulet et al.[75] were the first to demonstrate that the THM is an effective method for growing MCT crystals using a furnace as shown in Fig. 22.11. They have grown ingots of 30 mm diameter and 80 mm length (Fig. 22.12) with longitudinal homogeneity of ± 0.02 mol. Colombo et al.[76] have grown MCT crystals in the direction of <111>A and <111>B using seed crystals. Shilo et al.[78] have examined the solid-liquid interface in the MCT crystals grown using CdTe seed crystals. Gill et al.[79] have developed a new THM method in which nonstoichiometric source material is used and stoichiometric crystals can be grown using several travelling passes.

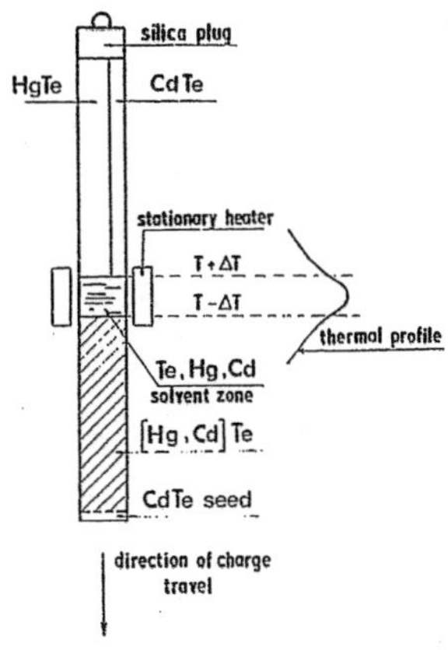

Fig. 22.11 Principle of MCT THM growth using a cylinder consisting of two cylindrical sectors (one HgTe, the other one CdTe)(reprinted from Ref. 75 with permission, copyright 1985 American Vacuum Society).

Fig. 22.12 Photograph of a 30 mm crystal ingot with slices cut from it (reprinted from Ref. 75 with permission, copyright 1985 American Vacuum Society).

Witter et al.[80] have developed an intelligent system in which various measurement systems are combined. They are eddy current analysis for the solvent volume and interface shapes, capacitance manometry for mercury partial pressure, fiber optic thermometry for thermal gradients and nuclear gauges for composition. Gill et al.[81] have developed a horizontal THM with crucible rotation as shown in Fig. 22.13.

Bloedner and Gille[82] have applied an accelerated crucible rotation technique (ACRT) to the THM method to increase the growth rate. Bloedner et al.[83] have grown $Hg_{1-x}Cd_xTe$ and $Cd_{1-x}Zn_xTe$ single crystals applying two different techniques for artificially stirring the solution zone. They are accelerated crucible rotation technique (ACRT) in a vertical arrangement and constant rotation in a horizontal arrangement.

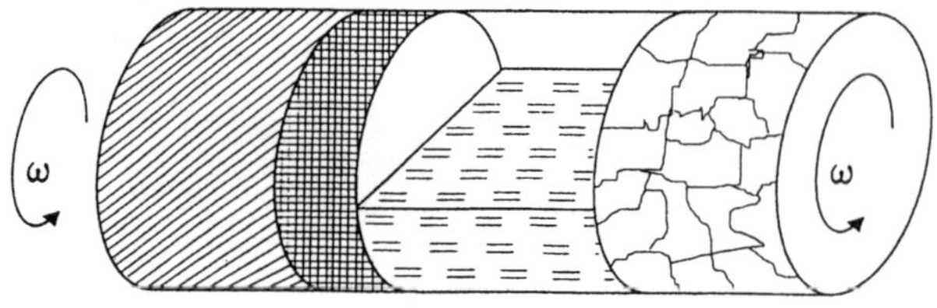

Fig. 22.13 Schematic representation of the horizontal THM crystal growth arrangement (reprinted from Ref. 81 with permission, copyright 1993 Elsevier).

These two techniques have been found to be effective in enhancing convective flow in the solution zone, thus in increasing the crystal growth rate.

Numerical analysis of THM of MCT crystal growth has been performed by Chang et al.[85] and Apanovich and Ljumkis.[86]

(4) Solid State Recrystallization (SSR)

Crystal growth of MCT by solid state recrystallization with modifications like the slush method, the cast-recrytallise-anneal (CRA) method, has been performed by various researchers[87-97] and summarized by Micklethwaite[98] and Tregilgas.[99]

Swink and Brau[87] have grown MCT crystals by recrystallization, quenching the melt and annealing at a temperature 665 °C for at least one week.

Harman[88] has applied a new recrystallization method which is called the slush method (Fig. 22.14). In this method, the melt is quickly moved to a low temperature zone in a vertical furnace, leaving most of melt unfrozen. Under a temperature gradient of approximately 100°C at the solid-liquid interface, crystal growth and annealing are performed for a long period. By this method, high quality single crystals of the size of 3 cm^3 can be grown.

Barlett et al.[89] have grown MCT crystals by casting the raw material, recrystallizing it to obtain single crystals and annealing under mercury vapor to convert the resulting p-type material to n-type. The effects of casting rate, cooling method, and thermal insulation of the crucible on the grown crystals have been examined.

Vere et al.[90] have used a high pressure furnace as shown in Fig. 22.15 to examine various conditions in the CRA process. Boliang et al.[91] have examined the dendrite structures in quenched MCT materials in order to find the best conditions for the subsequent recrystallization process. Yadava et al.[92] have examined the mechanism of the recrystallization growth and found that an excess of Te has the effect of increasing the recrystallization efficiency. Colombo et al.[93] have grown large MCT crystals (20 mm diameter by 50 mm length) by incremental quenching. They have obtained grain sizes

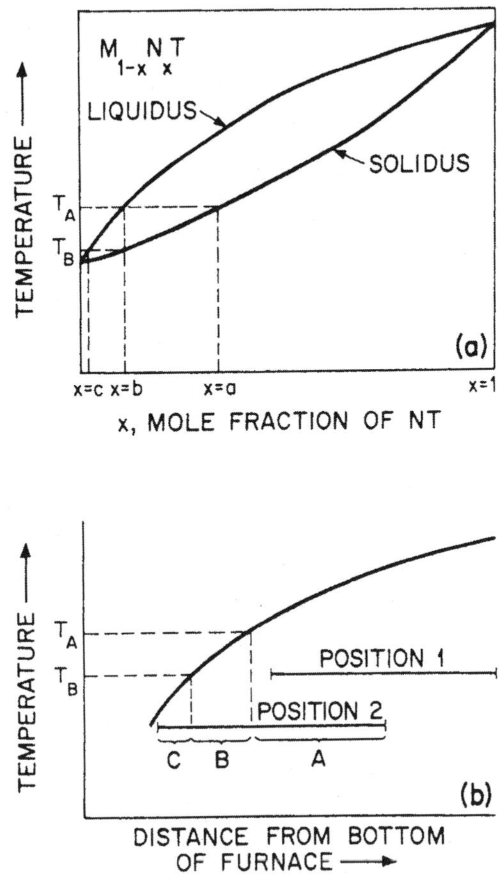

Fig. 22.14. (a) Schematic diagram of a simple pseudobinary M-NT system. (b) Schematic designation of position of charge, liquid fraction (Region A) and solid fractions (Regions B and C) after a steady state condition has been achieved and schematic temperature profile of furnace with initial and final positions of growth ampoule (reprinted from Ref. 88 with permission, copyright 1972 Springer).

from 25 to 35 mm after annealing at 665°C for 28 days.

Sharma et al.[94, 95] have developed a horizontal casting method for SSR as shown in Fig. 22.16 and Fig. 22.17. This method improved the compositional uniformity in the range ± 0.002 for MCT (x = 0.22).

Wang et al.[96,97] have studied various defects introduced in MCT crystals by the SSR method. These defects are point defects, holes, dendrites, second phase, dislocations and subgrain boundaries. The origins of these defects were discussed and it was concluded that they can be reduced by improving the material purity and controlling the stoichiometry.

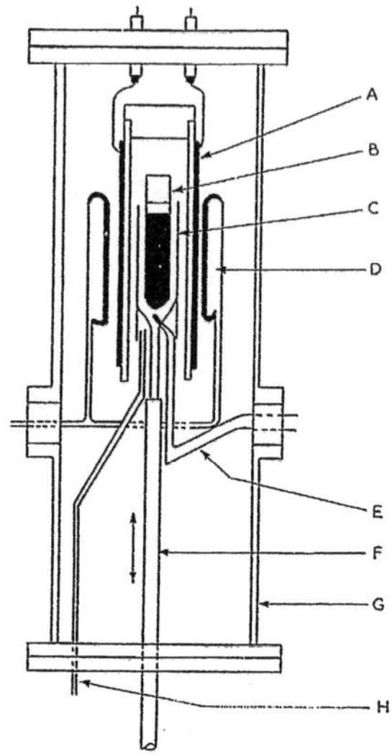

Fig. 22.15 Schematic diagram of the pressurized cast-recrystallise-anneal (CRA) method. (A) strip-element furnace; (B) quartz amouple; (C) quartz ampoule holder; (D) water-cooled induction coils; (E) thermocouple; (F) pull-rod; (G) water cooled stainless steel pressure vessel; (H) high pressure gas jet (reprinted from Ref. 90 with permission, copyright 1982 Elsevier).

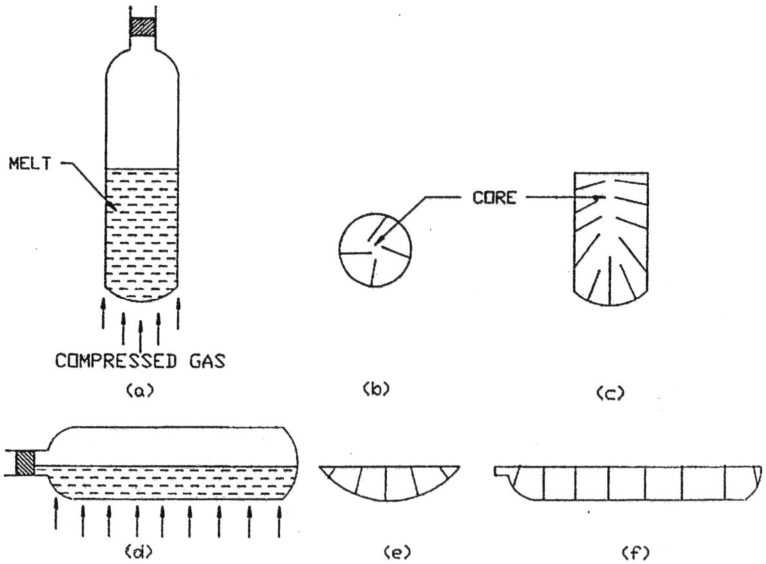

Fig. 22. 16 Arrangement of vertical and horizontal recrystallization (reprinted from Ref. 95 with permission, copyright 1993 Elsevier).

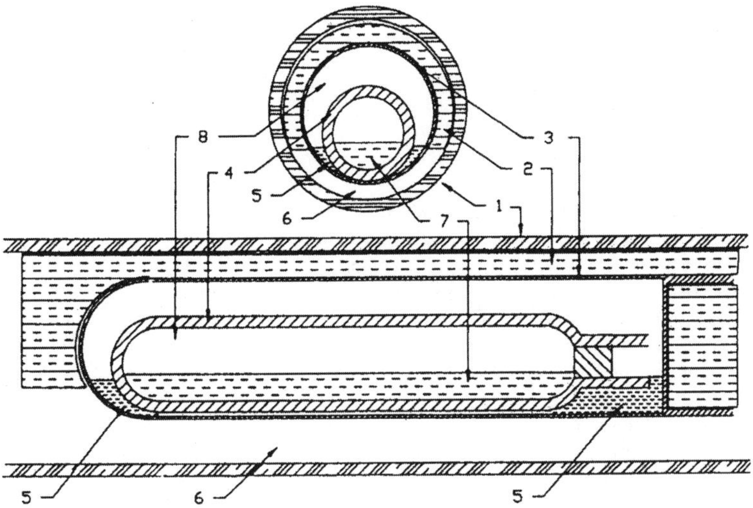

Fig. 22.17. Experimental setup for horizontal casting: (1) furnace, (2) insulation, (3) thin walled quartz ampoule, (4) thick walled quartz ampoule, (5) tin, (6) quenching channel, (7) MCT melt, (8) evacuated and sealed space (reprinted from Ref. 95 with permission, copyright 1993 Elsevier).

(5) Crystal growth of other materials

Su et al.[100] have grown $Hg_{1-x}Zn_xTe$ (MZT) by directional solidification with x ranging from 0.15 to 0.22. The axial and radial composition were studied by various methods such as precision density measurement, infrared (IR) transmission spectra and energy dispersive X-ray spectroscopy (EDX).

Wang and Jie[101] have grown mercury indium telluride $Hg_{1-x}In_xTe$ (MIT) by the vertical Bridgman method. MIT is a crystal promising fast and efficient photodetectors since it is electrically inactive to the introduced impurities and has a high resistance to ionizing radiation. The physical properties of MIT have been measured for the grown crystals.

22.3.2 Mercuric Iodide (HgI_2)

Mercuric iodide can be grown mainly by the vapor phase growth method. This is because HgI_2 has a structural transformation from α-HgI_2 to β-HgI_2 at a low temperature of 131°C. Because of this transformation, it is difficult to grow crystals from the melt. The growth of HgI_2 has been performed extensively as shown in Table 22. 4. They are also reviewed by various authors.[12, 102-104]

(1) Starting material preparation

Since impurities such as hydrocarbons, anions and cations affect the performance of

Table 22.4 Crystal Growth of HgI_2

Method	Contents	Authors	Year	Ref.
PVT	Growth of Hg_2I_2 crystals	Barta	1970	155
SG	Growth from HI solution	Coleman	1970	116
SG	Solvent evaporation using acetone with 4% water	Willig	1971	6
PVT	Platelets 1 mm thick and 3-4 mm diameter	Swierkowski et al.	1973	121
TOM	First application of TOM to the growth of HgI_2 crystals	Scholz	1973	130
PVT	Growth of 3 x 3 x 4 mm crystals	Llacer et al.	1974	122
SG	Solvent evaporation in CH_3I, CH_3OH and CH_2I_2	Schieber et al.	1975	105
PVT	Crystals of about 5 x 8 x 10 mm by a pulling method	Schieber et al.	1975	105
PVT	Growth of about 3 cm length by a pulling method	Ponpon et al.	1975	129
SG	Growth from dimethlsulfoxide (DMSO)	Carlston et al.	1976	116
TOM	TOM for periodic oscillation of the source or crystal temperatures in horizontal and vertical furnaces	Schieber et al.	1976-1978	131-133
TOM	Growth of crystals doped with SbI_3, BiI_3, NH_4I, PbI_2 and TeI_4	van den Berg et al.	1977	134
TOM	Growth in an improved horizontal furnace	Beinglass et al.	1977	135
SG	Growth from dimethlsulfoxide (DMSO) and iodomercurate complexes and growth mechanism	Nicolau et al.	1980	118-120
CVT	Growth using organics as the transport agent	Faile et al.	1980	160
FFM	Principle of the FFM, independent control of mass flow, supersaturation and growth temperature	Omlay et al.	1981	137, 138
CVT	Effect of organics as the transport agent	Burger et al.	1982	161
MG	Growth of HgI_2 doped with SbI_3	Nissenbaum et al.	1983	108
TOM	Growth from Hg_2I_2 source	Gospodinov et al.	1980	136
FFM	Nucleation on the glass surface as a function of supersaturation and temperature	Omlay et al.	1983	139
PVT	Routine growth of about 200 g crystals within 10 weeks	Lamond et al.	1983	104
CVT	The yield of HgI_2 platelets in the presence of polyethlene	Burger et al.	1985	162
PVT	Crystals of about 1 cm using a growth rate of about 500 μm/hr	Kobayashi et al.	1986	123
PVT	Growth of Hg_2Cl_2 crystals of 26 mm diameter and 60 mm length	Singh et al.	1986, 1987	156-158
FFM	Analysis of growth parameters on the FFM	Fournier et al.	1988	140
FFM	Growth by the FFM and characterization of nuclear detection properties	Coupat et al.	1988	141
PVT	Space growth and characterization	van den Berg et al.	1989	143
PVT	Apparatus preparation for space growth	Piechotka et al.	1990	144
PVT	Growth rate as a function of supersaturation	Isshiki et al.	1990	124, 151
PVT	Nucleation control by laser beam observation	Piechotka et al.	1991	125
PVT	Effect of heat conductivity anisotropy on crystal morphology	Zha et al.	1990	126
PVT	Surface kinetics and physical vapor transport in space growth	Coupar et al.	1994	145
MG	Growth of $HgBr_{1.6}I_{0.4}$ crystals by the Bridgman method	Daviti et al.	1996, 1997	110, 111
PVT	Growth in an oil bath furnace and effect of temperature conditions on crystal quality	Zhou et al.	2003	127

Table 22.4 Crystal Growth of HgI_2 (continued)

Method	Contents	Authors	Year	Ref.
MG	Growth of $Hg(Br_xI_{1-x})_2$ crystals by the Bridgeman method	Gospodinov et al.	2005	112
MG	Growth of $Hg_{1-x}Cd_xI_2$($0<x<0.6$) under ~ 20 atm argon pressure by directional solidification	Cummings et al.	2006	113
PVT	Effect of temperature conditions on the formation of a metastable HgI_2 phase, nucleation and growth behaviors	Lin et al.	2006	128
PVT	Growth of large crystals of Hg_2Cl_2 and Hg_2Br_2 (600 g)	Kim et al.	2008	159

MG: Melt Growth, SG: Solution Growth, PVT: Physical Vapor Transport, TOM:Temperature oscillation, FFM; Forced Flux Method, CVT: Chemical Vapor Transport

HgI_2 radiation detectors, it is important to purify the starting materials as much as possible. In order to purify commercially available HgI_2, sublimation cycles[103,105] or melting procedure[103,105] have been used.

Van den Berg and Whited[106] have developed an iodine redressing procedure to achieve a stoichiometric starting material. This procedure consists of sublimation of the purified powders of HgI_2 in an iodine atmosphere and of the direct insertion of the purified material into the growth ampoule (so-called automatic loading). The excess iodine was removed by evacuation where the source material was sublimed into the ampoule.

Skninner et al.[107] have developed an aqueous precipitation process of HgI_2 in baths of 3 kg using $Hg(NO_3)_2$, or $HgCl_2$ and KI. These are based on the following reactions.

$$2KI + Hg(NO_3)_2 \rightarrow 2KNO_3 + HgI_2 \tag{22.2}$$

$$2KI + HgCl_2 \rightarrow 2KCl + HgI_2 \tag{22.3}$$

(2) Melt growth

Since α-HgI_2 has a structural transformation to β-HgI_2 in the vicinity of 131°C, α-HgI_2 crystals are grown at temperatures below 131°C by solution growth or vapor phase growth. It is also known that α-HgI_2 is soft and the degradation of crystals is inevitable during the fabrication of detector devices. It has been however reported that the α to β phase transformation was avoided in the case of HgI_2 crystals doped with SbI_3 grown from the melt.[108] Daviti et al.[109-111] and Gospodinov et al.[112] have grown $Hg(Br_xI_{1-x})_2$ crystals by the Bridgman method without transformation problem. Cummings et al.[113] have grown $Hg_{1-x}Cd_xI_2$ crystals ($0 < x < 0.6$) by the directional solidification process under ~ 20 atm of argon pressure.

(3) Solution growth

Coleman[114] was the first to grow HgI_2 crystals form a HI solution. Willig[6] has grown HgI_2 crystals by a solvent evaporation method using acetone with 4% water. Thereafter, various solvents such as acetone, methanol, NH_4I/H_2O, NH_4I-CH,OH, KI-H_2O, KI-CH,OH, methyl iodide and methylene iodide, and iodomercurate complexes[105, 115-120] have been proposed for solution growth. Kleber et al.[115] have grown HgI_2 single crystals from organic solution such as diiodomethane or 1.2-dibromomethane by slow cooling. Growth of HgI_2 crystals from solutions of iodomercurate complexes has been performed by various authors.[105,115-118]

Dimethylsulfoxide (DMSO) was found to be a good solvent for HgI_2, because HgI_2 is very soluble in it. Carlston et al.[117] have grown HgI_2 crystals from DMSO by slow cooling of the solution and by hydration of the solution. As shown in Fig. 22.18, slow hydration of the DMSO solution reduces the solubility of HgI_2 in the solution which causes the excess HgI_2 to crystallize. Nicolau and Joly[120] have grown HgI_2 single crystals from DMSO solution using equipment as shown in Fig. 22.19. They have also used a screened seeding technique as shown in Fig. 22.20 in order to prevent dislocation propagation from the seed crystal. Applying this technique, single crystals could be obtained as shown in Fig. 22.21 with dislocation density as low as 10-100 cm^{-2}.

(4) Physical Vapor Transport (PVT)

There are several modifications based on static evaporation. These are pulling from the vapor, the temperature oscillation method (TOM) and the forced flux method (FFM).

- Static evaporation

This is a conventional vapor phase growth. It has been however pointed out that the disadvantage is the difficulty in controlling spurious nucleation. This method was first applied to the growth of HgI_2 crystals in order to characterize their material properties[10,121,122] and has been performed using both axial two-zone horizontal and vertical furnaces.[103,105] HgI_2 is now routinely grown using more than 20 horizontal furnaces.[103]

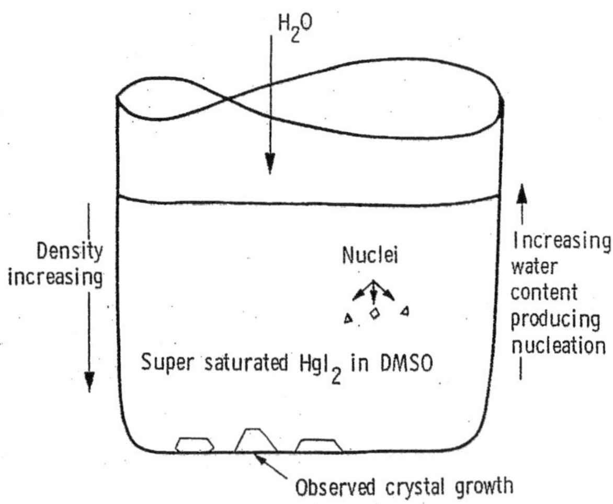

Fig. 22.18 Schematic representation of flows (reprinted from Ref. 117 with permission, copyright 1976 Elsevier).

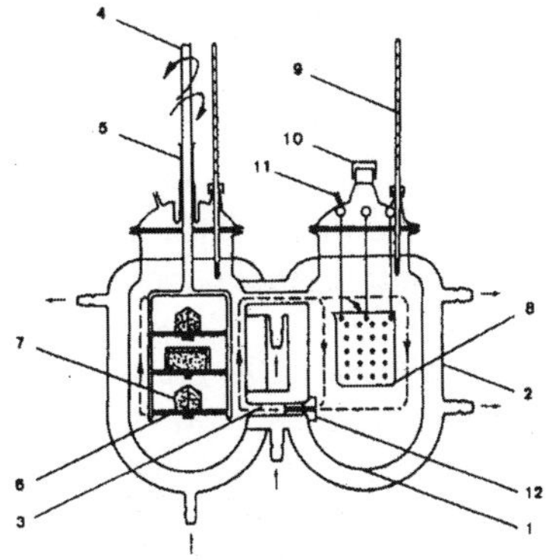

Fig. 22.19 Crystal growth apparatus: 1 quartz reactor vessels, 2 heating jackets, 3 joining tubes, 4 stirrer, 5 guide, 6 seed supports, 8 crystals, 8 Teflon beaker, 9 control thermocouple, 10 SVL stopper, 11 capillary tubes, 12 drilled Teflon stopper (reprinted from Ref. 120 with permission, copyright 1980 Elsevier).

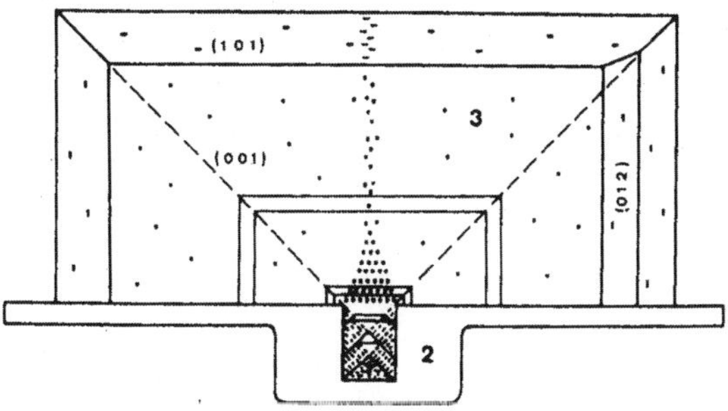

Fig. 22.20 Prevention and elimination of dislocations arising from the seed using screened seeding: 1 seed, 2 Teflon seed support, 3 single crystal (reprinted from Ref. 120 with permission, copyright 1980 Elsevier).

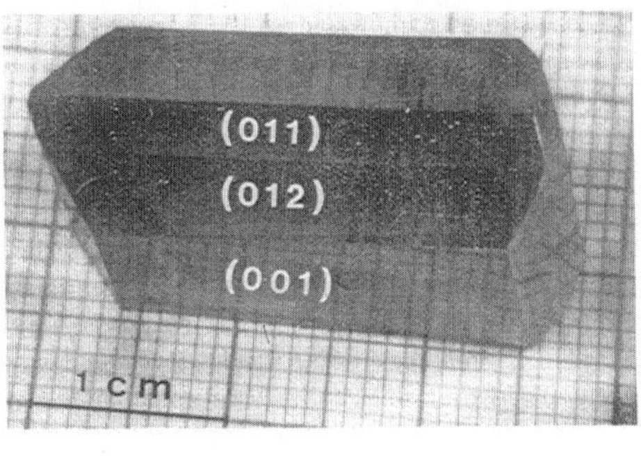

Fig. 22.21 α-HgI_2 single crystal grown by screened seeding in a quartz apparatus (reprinted from Ref. 120 with permission, copyright 1980 Elsevier).

Kobayashi et al.[123] have grown HgI_2 crystals using a furnace as shown in Fig. 22.22. They have examined the effect of impurities on the growth rate and found that the growth rate decreases as a function of growth time. Isshiki et al.[124] have developed a transparent growth apparatus as shown in Fig. 22.23 which allows better control of the crystal growth process. They also have studied the growth rate as a function of growth duration. Piechotka et al.[125] have developed a laser beam method for nucleation control. This method allows the selective evaporation of parasitic nuclei of HgI_2, using an apparatus as shown in Fig. 22.24.

Zha et al.[126] have studied the anisotropy of the kinetics of crystal growth under static evaporation conditions. Zhou et al.[127] and Lin et al.[128] have studied growth conditions such as the source and crystal temperatures and the nucleation conditions to obtain good crystal quality by using an oil-bath furnace. Yang et al.[129] have used the ultrasonic assisted vapor phase deposition (UWAVPD) method for growing crystals.

- *Pulling method*

This requires a triangular-shaped temperature gradient such that the rates of evaporation, condensation and pulling are equal. Ponpon et al. [27] were the first to apply this method using the furnace arrangement shown in Fig. 22.25. Nucleation is accomplished on a conical pointed cold end and the pulling of the crystal is performed by moving the temperature gradient towards the polycrystalline charge. A single crystal boule weighing up to 30 g has been grown by this method.[105] The disadvantage is the sticking of the grown boule to the ampoule walls, which introduces strains in the grown crystal.

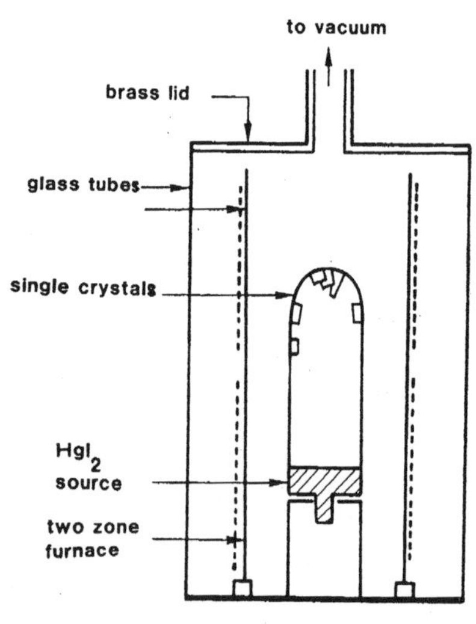

Fig. 22.22 Schematic drawing of the apparatus used for crystal growth of α-HgI_2 by static sublimation (reprinted from Ref. 123 with permission, copyright 1983 The Electrochemical Society).

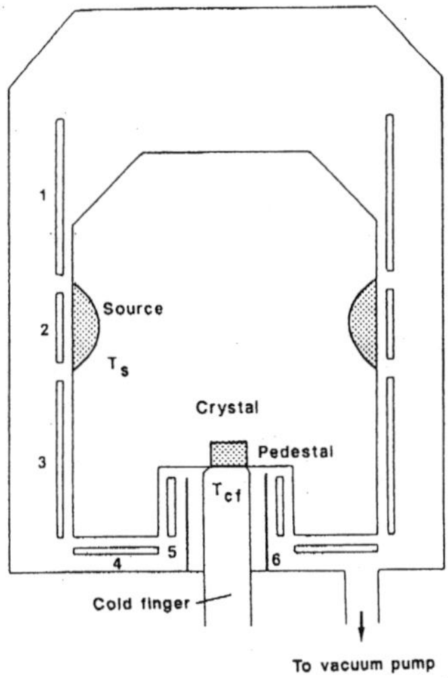

Fig. 22.23 Schematic diagram of the growth apparatus for mercuric iodide crystals: (1) upper heater; (2) source heater; (3) lower heater; (4) bottom heater; (5) pedestral heater; (6) radiation shield (reprinted from Ref. 124 with permission, copyright 1990 Elsevier).

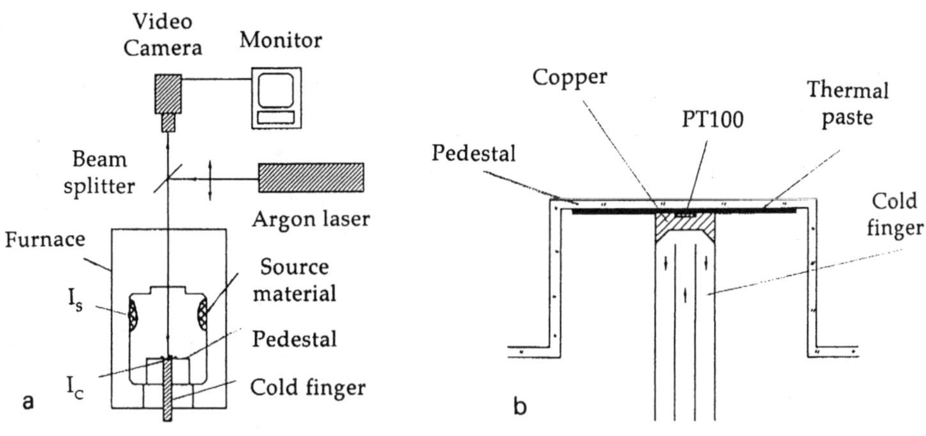

Fig. 22.24 (a) Schematic diagram of the experimental setup for the control of mercuric iodide nucleation by means of a laser beam. (b) Close-up of the growth chamber showing the nucleation plate (pedestral) and the cooling finger (reprinted from Ref. 125 with permission, copyright 1991 Elsevier).

- Temperature Oscillation Method (TOM)

This is a method first applied by Scholtz[130] in order to prevent spurious nucleation by applying oscillated temperature, consisting of a periodic reversal of the temperature

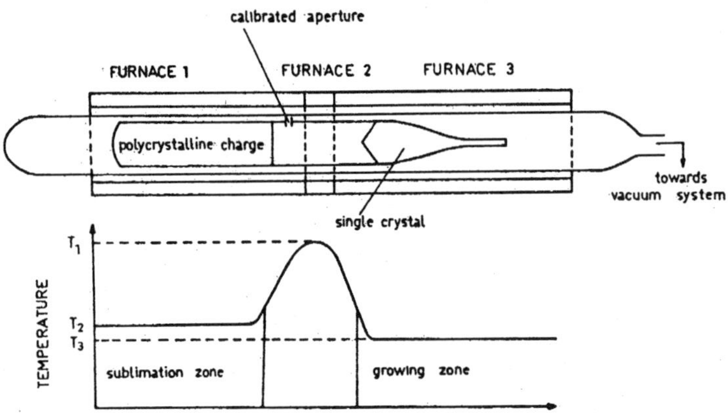

Fig. 22.25 The growing system and the temperature gradient applied to the furnace (reprinted from Ref. 27 with permission, copyright 1975 IEEE).

gradient between the growing crystal and the source material. This method causes alternating growth and reevaporation of the crystal. The growth has been performed in vertical and horizontal furnaces.[102,131-136]

Crystal growth is performed under periodic oscillation of the source temperature (POST) or that of the crystal temperature (POCT).[131,132] The POST method is applied in horizontal or vertical furnace configurations while the POCT method is applied only in a vertical configuration. In Fig. 22. 26, the horizontal configuration and in Fig. 22.27, the vertical configuration are shown. The grown crystals with POST and POCT are shown in Fig. 22.28.

TOM crystal growth has been performed with an excess of mercury or iodine and in an atmosphere containing a foreign element such as SbI_3, BiI_3, NH_4I, PbI_2 and TeI_4.[134] Beinglass et al.[135] have grown HgI_2 single crystals using an improved horizontal furnace from purified and iodine treated starting materials.

Gospodinov and Sveshtarov[136] have grown TOM crystals in excess vapor pressure of one constituent and evaluated the crystal quality. It was found that the highest quality crystal could be obtained with an excess of iodine.

- *Forced Flux Method (FFM)*

The furnace arrangement of this method is shown in Fig. 22.29. This method was invented theoretically by Omlay et al.[137] and was applied to the growth of HgI_2.[138,139] This method is based on the independent control of the transport and kinetic parameters. In a three zone furnace, the source is placed at one end of a closed pyrex tube at tempera-

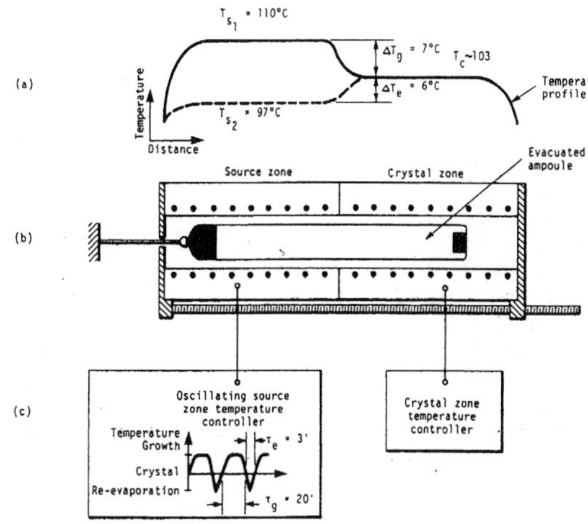

Fig. 22.26 Schematic representation of a two-zone horizontal furnace operating with the POST method: (a) temperature profile or source temperature oscillation between T_{S1} and T_{S2} at constant crystal temperature T_C; (b) sketch of furnace with control arrangements; (c) temperature-time variation (reprinted from Ref. 131 with permission, copyright 1976 Elsevier).

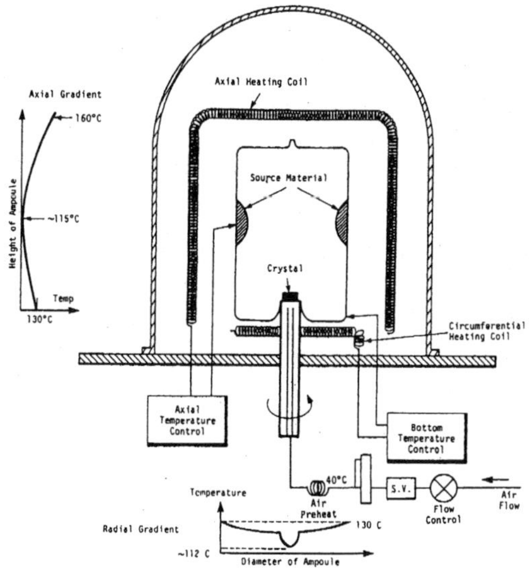

Fig. 22.27 Schematic representation of the vertical furnace, showing an approximate axial and radial temperature distribution at the outside wall of the ampoule. Temperatures indicated correspond to the growth cycle of the POCT method (reprinted from Ref. 131 with permission, copyright 1976 Elsevier).

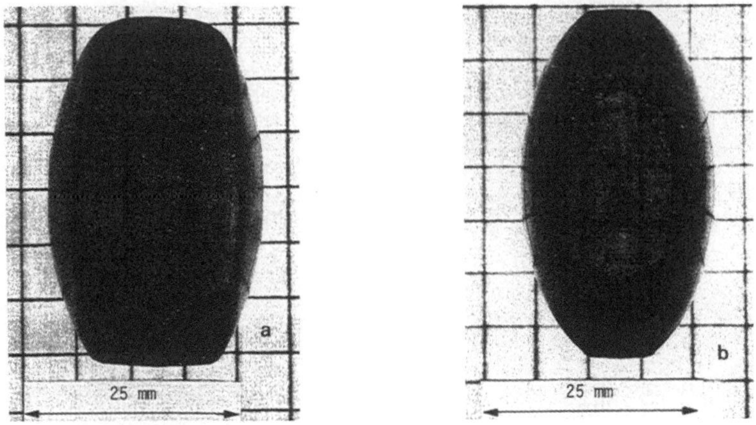

Fig. 22.28 100 g HgI_2 crystals grown using the TOM with (a) POCT and (b) POST variations (reprinted from Ref. 105 with permission, copyright 1976 Elsevier).

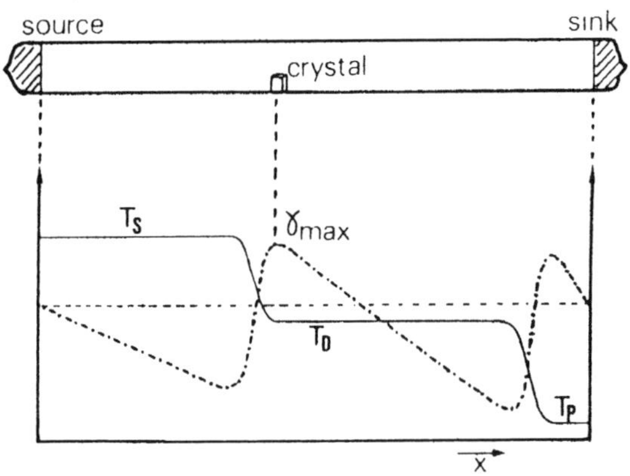

Fig. 22.29. Thermal profile (-------) and supersaturation profile (- - - -) in the forced flux method (FEM) (reprinted from Ref. 141 with permission, copyright 1988 Elsevier).

ture T_s. The crystal is placed in the middle at temperature T_c. The other end at the lowest temperature acts as the sink. Evaporated HgI_2 is partly deposited on the crystal and partly at the sink. This method is called the FFM since HgI_2 is forced to pass the crystallization zone on the way to the sink. By this method, the supersaturation and the average flux can be controlled independently.

Fournier et al.[140] have studied the effect of the thermal gradients on supersaturation, impurity incorporation and diffusion transport. Coupat et al.[141] have grown HgI_2 crystals by the FFM and fabricated radiation detectors and evaluated the crystal quality.

- *Space growth*

Van den Berg et al.[142,143] have grown HgI_2 crystals using a furnace as shown in Fig. 22.23 aboard the Spacelab III. The quality of the grown crystals was compared with those grown on the ground and it was found that the stability of the crystal surfaces is much higher than on the ground where convection is present. It was also found that the gamma ray rocking curve is superior and the electronic transport properties are improved. Piechotka and Kaldis[144] have studied the transport rate prior to crystal growth in a space experiment. Coupat et al.[145] have performed crystal growth in space and on the earth and studied the influence of microgravity on the growth rate distribution as a function of the flow, the orientation and the location of the wall. They have concluded that the effect of gravity appears as a nonlinear disturbance on the concentration distribution around the growing crystals.

- *Vaporization*

The vaporization of HgI_2 has been extensively studied in order to clarify the vaporization mechanism.[146-150] The vapor pressure of HgI_2 was first reported by Stranski and Totomanov.[146] Dishon et al.[147] have found that the vapor pressure varied with the impurity content and the stoichiometry. Impure and nonstoichiometric HgI_2 showed higher vapor pressures than those of purified and more stoichiometric materials. Pitchotka and Kaldis[148-150] have systematically studied the effect of impurities such as hydrocarbons and water and stoichiometry on vaporization by using mass spectroscopy. It was found that during vaporization, nonstoichiometric vapors remain until the end of evaporation if they contain hydrocarbons. When the material is sufficiently pure, excess mercury or iodine disappear during evaporation leaving stoichiometric solids.

- *Reduction of growth rate*

The reduction of growth rate during the PVT crystal growth as a function of time is well known in the case of the vapor phase growth of HgI_2, especially when the crystal reaches about 100 g in size.[103] This reduction is attributed to impurities gathered on the surface of grown crystals and poisoning the surface steps[123] and on the non-ideal temperature distribution in the vapor phase around the growing crystal.[124] Isshiki et al.[124,151,152] have made a systematic study of the growth rate as a function of supersaturation. They proposed a model of the reduction of the growth rate attributed to microfacetting of the growing crystal facet surface because of the reduction of the temperature difference between the crystal and the source.

The behavior of the residual gas in the crystal growth ampoule was systematically studied by Laskowski et al.[153] using mass spectroscopy. Residual gas, molecules such as H_2, H_2O, CH_4, CO, CO_2 and NO have been detected. It was found that these residual gases decrease the growth rate so that the elimination of these impurities is important.

-*Mercury halides*

Monovalent mercury halides such as Hg_2X_2 (X = Cl, Br, I) are promising materials for acousto-optic and optoelectronic devices because of their high transmittance for long wavelengths up to 40 μm, high double refraction and high diffraction efficiency.[154, 155] The PVT method has been applied to grow these Hg_2Cl_2[155-159] and Hg_2Br_2[159] crystals for these applications. Singh et al.[158] have grown Hg_2I_2 single crystals using a conventional vertical PVT furnace as shown in Fig. 22.30. Fig. 22.31 shows the Hg_2I_2 crystal grown by this method. Kim et al.[159] have recently grown Hg_2I_2 and Hg_2Br_2 crystals up to 48 mm in diameter and weighing 400-600 g by the unseeded PVT method.

(5) Chemical Vapor Transport (CVT)

Faile et al.[160] have developed a novel CVT method which uses organic monomers or polymers such as styrene or polyethylene as the transport agent. This method gives platelet HgI_2 crystals which can be directly used for radiation detectors. Burger et al.[161] have grown HgI_2 crystals with and without polyethylene using the furnace arrangement shown in Fig. 22.32. It was found that the existence of polyethylene is not essential for the growth of platelet crystals but is important to grow stoichiometric HgI_2 crystals with better radiation detector performances. They also reported that the yield of grown HgI_2 platelets is dependent on the ampoule geometry.[162] The quality of grown crystals was compared with TOM crystals and it was found that these crystals have good nuclear spectroscopic results and lower dislocation densities but the uniformity is worse than that of TOM crystals.[162]

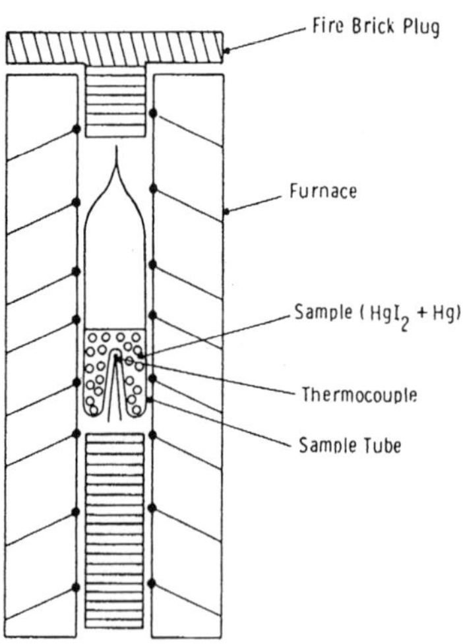

Fig. 22.30 Tube used for freezing point determination (reprinted from Ref. 158 with permission, copyright 1987 Elsevier).

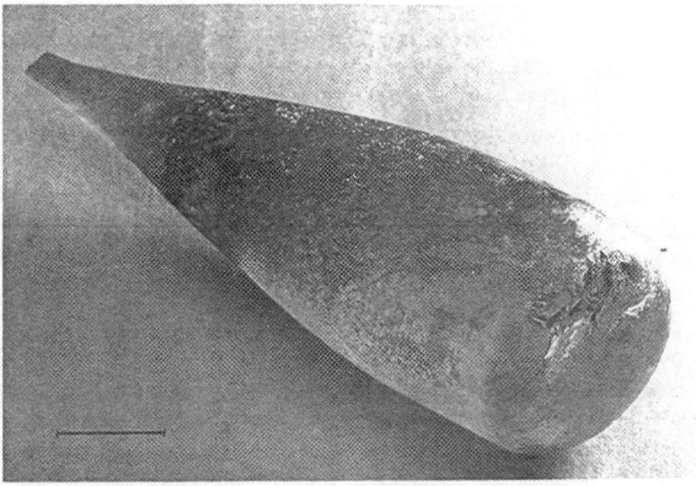

Fig. 22.31 Single crystal of Hg_2I_2. Marker represents 1 cm (reprinted from Ref. 158 with permission, copyright 1987 Elsevier).

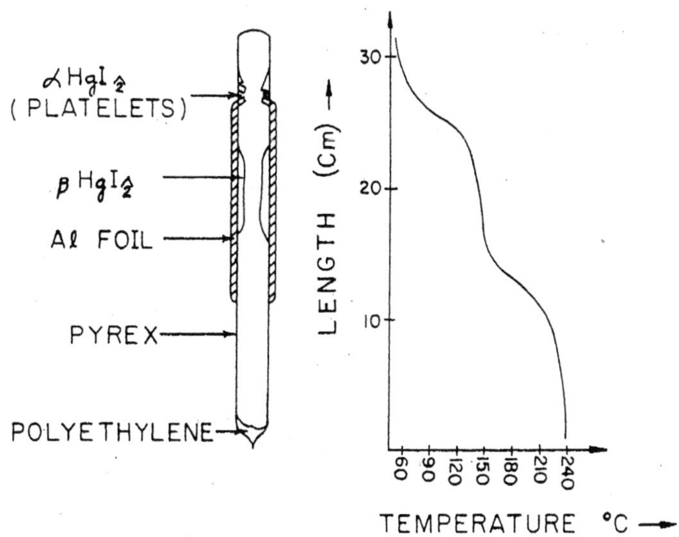

Fig. 22.32. Experimental arrangement of HgI_2 platelet growth (reprinted from Ref. 161 with permission, copyright 1982 Elsevier).

22.4 CHARACTERIZATION

22.4.1 Purity

(1) $Hg_{1-x}Cd_xTe$ (MCT)

Impurity behavior has been studied extensively by Capper et al.[163-168] Segregation coefficients of various impurities have been determined for the normal Bridgman method[163, 164] and for the accelerated crucible rotation technique (ACRT)-vertical Bridgman (VB) method[165] by intentional doping of various impurities to the starting materials and by chemical analysis and Hall measurement. They are summarized together with THM data and epitaxial growth data.[166, 167] These segregation coefficients are shown in Table 22.5.[166] Wang et al.[97] have studied the effect of impurities on the various defect formation during solid state recrystallization (SSR).

(2) HgI_2

Hole traps due to impurities limit the energy resolution so purification is an important matter for HgI_2 radiation detectors.[8] Pitchotka[13] has reviewed the purification methods and analytical methods, together with the behavior of inorganic and organic impurities.

Muheim et al.[169] analyzed HgI_2 crystals from different laboratories. They contained various elemental (C, N, O, Na, Al, K, Ca, Cr, Mn, Fe) and hydrocarbon (CH_3, C_2H_3, C_3H_3, C_3H_4, C_4H_4, C_4H_7, C_4H_8) impurities of the order of a few hundred ppm. This is attributed to the layer structure of HgI_2 and the resulting binding properties.

Nicolau[170] has prepared high-purity HgI_2 by recrystallization, electrolysis, sublimation and zone melting and concluded that first precipitation by recrystallization or electrolysis and final purification by distillation with vapor superheating is effective.

Kobayashi et al.[123] have grown HgI_2 crystals by a static sublimation method from various materials, such as HgI_2 prepared by synthesis from elements, commercial HgI_2 powders and HgI_2 purified by repeated sublimation. They found that the growth rate decreases as a function of time by one order of magnitude when unpurified HgI_2 material is used. They speculated that this is due to step-poisoning by impurities.

The origin of hydrocarbon contamination is believed to be from the iodine sources[171] and contamination from vacuum oil. The mechanism of hydrocarbon accommodation into HgI_2 is discussed, not only from the viewpoint of intercalation but also from that of mosaic structure.[172] Purification of iodine itself was proposed by the synthesis of a closed packed structure with no place available for long chains of hydrocarbons and by the decomposition of this compound back to iodine.[171] Purification of HgI_2 was also performed by heating with oxygen.[173] Coupat et al.[141] have measured many kinds of impurities by spark mass spectroscopy and found that the elimination of hydrocarbons and certain mineral elements improved the performance of radiation detectors. Skinner et al.[107] have shown that the yield of radiation detectors could be greatly improved by reducing the carbon impurities using their own purification process. James et al.[174] have measured water and hydrocarbons in HgI_2 by Fourier transform infrared (FTIR) spectroscopy and found that water could be reduced by storing the material in a desiccator containing P_2O_5. They also speculated small amounts of impurities such as mercurous iodide, calcium iodate, sodium iodate and compounds containing a carbonate

Table 22.5 Segregation Coefficients of Various Impurities of MCT (reprinted from Ref. 166 with permission, copyright 1991 AIP).

Element	Bridgman	ACRT	THM
Li	0.4[a]	0.7	
Na	1.0	0.04	
K	2.1	<0.01	
Cu	0.2[a]	0.5[a]	
Ag	0.1[a]	0.3[a]	
Au	0.02[a]	0.03[a]	0.03[a]
Mg	2.2		
Ca	2.0		
Al	2.8[a]	1.0[a]	
Ga	1.8	0.5	
In	2.0	1.6	
C	0.5	0.06	
Si	2.4	2.3	
Sn	<0.01		
Pb	<0.01		
P	1.0[a]		
As	0.1[a]	0.01[a]	
Sb	<0.01[a]	<0.01[a]	
Bi	<0.01	<0.01[a]	
Cr	0.15	<0.01	
Mo		1.0	
O	0.01	0.04	
S	1.0		
Mn	1.6		
Cl	0.4[a]	0.02[a]	
Br	<0.01	0.5	
Fe	0.5[a]	0.5	
Co	0.15		

[a] From singly doped crystals

anion. It is also pointed out that water impurity incorporated in HgI_2 as H_2O^+ is not easily removed by repeated sublimation.[12]

22.4.2 Defects

(1) Dislocations

- $Hg_{1-x}Cd_xTe$ (MCT)

Parker and Pinnell[175] have applied the solution of 15 conc. HNO_3 + 15 H_2O + 1 conc. HCl + 15 % bromine in methanol as the dislocation etchant for MCT. When Cd concentration is increased, it is necessary to dilute the etch solution in order to obtain well-defined patterns. Boinnykh et al.[176] used this etchant for evaluating Bridgman grown MCT crystals and found it does not include dislocations in the subgrain boundaries. Polisar et al.[177] have used solutions with HF + HCl + H_2O +Br_2 in methanol (No. 1) or Br_2 in CH_3COOH (No.2). Brown and Willoughby[178] have shown that Plisar etch No. 1

developed pits with 25° of the (111) plane and Polisar etch No.2 revealed etch pits on all planes of MCT. A mixture of HCl, HF and chromic acid was proposed for (110) plane of MCT.[179] Hädnert and Schenk[180] have developed a new etchant, a dilution of the mixture of 1 conc. HF + 1 chromic acid (50 wt.% CrO_3 in H_2O) and 1 conc. HCl. This etchant is effective to reveal etch pits for (111)A, $(\bar{1}\bar{1}\bar{1})$B and (110) planes of MCT. Etching with a variable mole fraction of MCT ($0 \leq x \leq 0.75$) is also possible by varying the dilution of the etchant. Sharma et al.[95] have used a mixture of 1 HCl + 21% Br_2 in CH_3OH + 10 CH_3OH + 30 HNO to reveal the dislocation density of SSR crystals.

The reported dislocation densities are 1×10^4-2×10^5 cm^{-2} for THM crystals[76], 10^5-10^6 cm^{-2} for Bridgman crystals[59], and 4×10^4 cm^{-2} - 10^7 cm^{-2} for SSR crystals.[75, 93, 95] Colombo et al. have reported a lowest dislocation density of 4 x 10^4 cm^{-2} [95] for the central region of a SSR crystal in a horizontal arrangement as shown in Fig. 22.17.

-HgI_2

The dislocation etch pits can be revealed on the (001) surfaces simply by using various alcohol etchants.[27, 120,123] James and Milstein[181] have used an etchant consisting of 5% alcohol in trichloroethylene (TCE). It was found that this etchant can demonstrate etch pits on the (001) surface that are associated with dislocations based on their motion under deformation.

Ponpon et al.[27] has grown HgI_2 crystals by the pulling technique and measured the dislocation etch pit density around 10^4 cm^{-2} by using methanol or ethanol alcohol as the etchant. Schieber et al.[131,132] used 0.2 % bromine in ethanol to reveal etch pits on both (110) and (001) faces of vapor-grown crystals. The dislocation densities were in the range of 10^3 to 10^7 cm^{-2}. Kobayashi et al.[123] have studied the dislocation etch pits using a 0.2 % Br_2 solution in methanol for PVT grown crystals under static evaporation and reported dislocation density of 10^3 cm^{-2}. The dislocation densities of HgI_2 crystals grown by the temperature oscillation method (TOM) were measured by etching with 0.2 % bromine in methanol and were reported to be around 10^5 cm^{-2} [131] and 10^3-10^4 cm^{-2}.[132]

Niclau and Joly[120] have shown using a methanol etching, that the dislocation densities of HgI_2 crystals grown by solution growth were of the order of 10 cm^{-2} on the {001} faces and 10-100 cm^{-2} on the {012} faces and 100 cm^{-2} on the {110} faces. These low dislocation densities were achieved using a screened seeding procedure as shown in Fig. 22.20. Some screw dislocations emerged on these faces but they did not propagate because of the screened seeding.

Burger et al.[162] have evaluated the dislocation density of HgI_2 crystals grown by the chemical vapor transport method using polyethylene as the reagent. It was evaluated using the etchant of James and Milstein[181] and was found to be 10^2-10^3 cm^{-2}. This low density was attributed to the growth of platelet crystals which may have less distortion compared with bulk grown crystals.

Randtke and Ortale[182] have studied the effect of dislocations on the performance of radiation detectors in respect of charge collection and electric field uniformity. It was found that dislocation density levels of 10^3-10^4 cm^{-2} show acceptable performance, but higher dislocation densities or lineage structure degrade them.

(2) Structural defects

- $Hg_{1-x}Cd_xTe$ (MCT)

Genzel et al.[183] have examined the effect of the ratio of growth and solidus temperature T_g/T_s for THM crystals using X-ray double crystal topography to investigate crystal quality. They found that lowering T_g/T_s can result in obtaining higher quality crystals with larger subgrains and lower dislocation densities.

Shaake and Tregilgas[184] have studied the kinetics of Te precipitation in MCT. For a starting concentration of 2×10^{18} cm^{-3} excess tellurium, precipitation occurs very quickly, within 15 min, while for a starting concentration of 10^{17} cm^{-3} excess tellurium, precipitation is extremely slow, with a time constant of the order of hours. Kiessling and Leipner[185] have studied the precipitates in THM crystals and found that mercury vacancy loops occur in as-grown crystals at a density of about 1×10^{15} cm^{-3}. These defects could be completely eliminated by annealing under saturated mercury vapor conditions. Neubert et al.[186] have studied the kinetics of Te precipitation in THM grown MCT using the positron annihilation technique and found that the kinetics can be quatitatively explained based on the Hg vacancy diffusion coefficient.

Bartlett et al.[89] have studied structural defects such as dendrites in cast-recrystallise-anneal (CRA) crystals. It was found that the casting rate strongly influences the formation of structural defects. Wang et al.[96, 97] have studied defects in MCT crystals grown by the solid state recrystallization (SSR) method. In SSR crystals various defects occur, such as subgrain boundaries, second phases, dendrites, stacking faults and stresses due to these defects. Since these defects are due to impurities, deviation from stoichiometry and unsuitable rapid quenching, better management of these issues is necessary.

-HgI_2

HgI_2 crystals have various metastable phases apart from the main α and β phases. They are white phases[115,187] and orange phases.[188,189] It is also reported that the α-phase transforms slowly to the β-phase.[190] It is also pointed out that the surface of as-grown crystals exhibits a much higher concentration of defects than the bulk, probably due to this surface transformation. For this surface transformation, a transformation line at 70°C is drawn in the Hg-I phase diagram.[12]

Shieber et al.[191] have studied crystal perfection by various methods and determined the correlation between crystal perfection and detector performances. Leon-Gits and Fries[192] have characterized HgI_2 crystals grown by physical vapor transport on the ground and in a microgravity environment by X-ray topography.[29] Remy et al.[193] have used synchrotron radiation white-beam X-ray transmission topography to examine the origin of observed defects.[194] Gamma-ray transmission rocking-curves have also been performed to evaluate crystal quality.[195]

(3) Nonstoichiometry

- $Hg_{1-x}Cd_xTe$ (MCT)

Deviation from stoichiometry is caused by a number of native defects such as antisite defects, interstitials and vacancies. Berding et al. have calculated the defect energies of various native defects.[196] Fig. 22.33 (a) shows the pressure-temperature diagram[69] in which the stoichiometric region[197] is shown where p-type conductivity is converted to n-type. In Fig. 22.33 (b), the concentration of V_{Hg} and Hg_i are shown as a function of temperature.[229] Control of stoichiometry can be achieved not only during crystal growth but also by annealing as explained in Sec. 22.4.3 (1).

- HgI_2

Since the resistivity of HgI_2 is high, it is difficult to measure its nonstoichiometry electrically. Chemical analysis[198,199] has been performed with a precision of 10^{-4} and excess of mercury or iodine has been observed depending on the material. I-rich and Hg-rich nonstoichiometry of HgI_2 have both been proved by quadrupole mass spectroscopy analysis of the vapors diffused from a Knudsen cell.[150] It was found that contamination by hydrocarbons prevents the preparation of stoichiometric material by the sublimation procedure. In fact, the excess of Hg could not be improved by sublimation.[12,150]

Nonstoichiometry of HgI_2 has been extensively reviewed by Piechotka and Kaldis[12] and it was concluded that the synthesis of HgI_2 from the elements under pure condi-

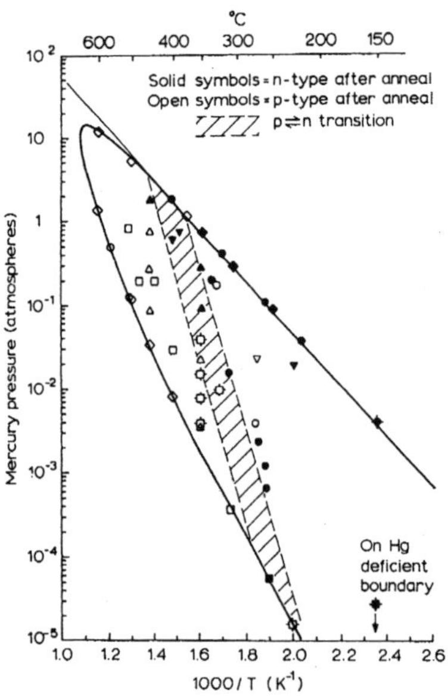

(a)

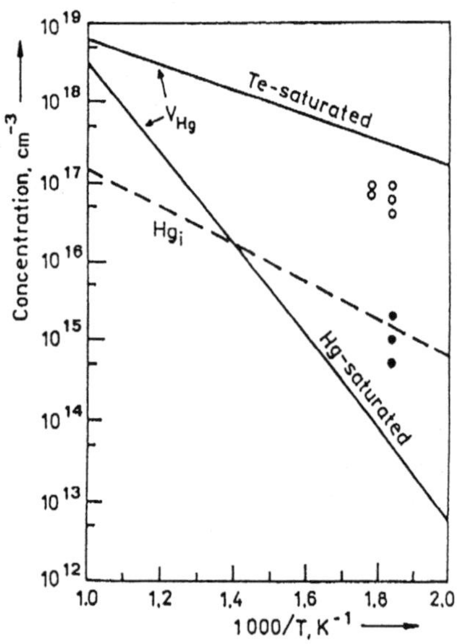

(b)

Fig. 22.33 (a) Mercury pressure vs temerature diagram for x = 0.2 (reprinted from Ref. 69 with permission, copyright 1989 Elsevier). Solid symbols = n-type after anneal, open symbols = p-type after annealing, hatching region = p-n transition region proposed by Farrar et al.[197] (b) Total concentration of V_{Hg} and Hg_i vs temperature under Te and Hg saturated conditions (reprinted from Ref. 229 with permission, copyright 1992 Elsevier).

tions can effectively obtain stoichiometric materials. The deviation from stoichiometry has been determined by measuring lattice unit cell dimensions and densities of various solution- and vapor-grown crystals. The stoichiometric composition of I/Hg = 2.000 was taken to give a unit cell of maximum volume which corresponds to a density of 6.3698 g/cm³ as shown in Fig. 22.34 (a).[103,200] The stoichiometry was also determined by measuring the transition temperature T_{tr} from the red to the yellow phase of HgI_2. By plotting T_{tr} as a function of the I/Hg ratio, maximum T_{tr} has been found to correspond to the stoichiometric ratio as shown in Fig. 22.34 (b).[103]

Tadjine et al.[194] have correlated the chemical stoichiometry measurements with detector performance and concluded that the stoichiometry ratio closest to I/Hg = 2 showed the least polarization and the best spectra at higher energies.[171] Stoichiometry was measured by Raman spectroscopy for excess mercury[201] and by spectrophotometric methods for excess iodine.[202]

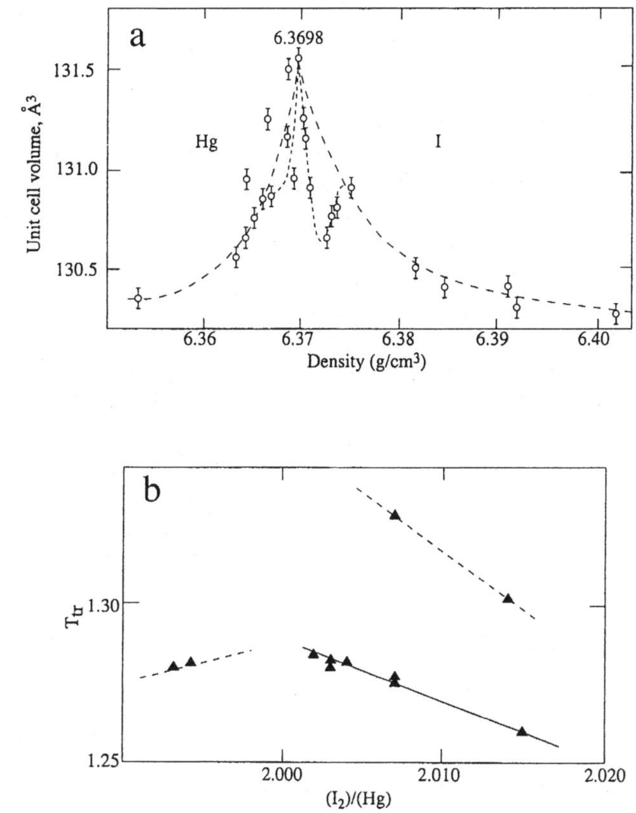

Fig. 22.34 Stoichiometry study. (a) Indirect method by volume of unit cell versus density, (b) direct method by red to yellow phase transformation T_{tr} versus I/Hg ratio for various HgI_2 samples (reprinted from Ref. 103 with permission, copyright 1983 Elsevier).

(4) Native defects

- $Hg_{1-x}Cd_xTe$ (MCT)

Berding et al.[203] have calculated the vacancy formation energies for CdTe, HgTe and MCT using Harrison's tight-binding theory, extended to a cluster embedded in an extended crystal and the variation of the energies with the alloy composition of MCT. They have also calculated the total energies of various impurities and defects in HgTe and CdTe using a self-consistent linear muffin-tin orbital (LMTO) method within the local density and atomic spheres approximation.[196] The variation with alloying was predicted to be a linear variation of the substitution energy with the local concentration.

Myles et al.[204] have extended the deep level theory in MCT to include (vacancy, impurity) nearest neighbor pairs. They have measured the energy levels of various defects as a function of the composition of MCT, and compared the theoretical predictions with the experimental results of deep level transient spectroscopy (DLTS) and photoluminescence (PL).

- HgI_2

The performance of HgI_2 radiation detectors is affected by charge trapping centers which create nonuniform internal electric fields degrading the charge collection. The main source of traps are native defects regarding nonstoichiometry, impurity-related defects, impurities themselves and mechanical defects like plastic deformation. It is therefore important to evaluate these defects but since HgI_2 is highly resistive, various techniques such as mobility-lifetime ($\mu\tau$) measurement[194,205,206], thermally stimulated current (TSC)[194,205-211], optical DLTS including photoinduced current transient spectroscopy (PITS)[206,212], photon-stimulated current (PSC)[213,214], transient charge technique (TCT)[215], and photoluminescence (PL)[216,217] have been used for evaluating these defects. These data are summarized in Table 22.6. It can be known that there are many traps in HgI_2.

The theoretical calculation of native defects in HgI_2 has been performed by Ayres et al.[218,219] using the total energy spin-polarized augmented plane wave method considering full atomic relaxation. In Fig. 22.35, the formation energies of V_I and V_{Hg} are shown as a function of Fermi level. They claim that iodine vacancies are the most important intrinsic defects in HgI_2.

22.4.3 Electrical Properties

(1) $Hg_{1-x}Cd_xTe$ (MCT)

MCT is a defect semiconductor in which the electrical properties are determined mainly by native defects, such as Hg vacancies and Hg interstitials, and impurities related to these native defects so that annealing after crystal growth makes a big difference to the electrical properties.

- Conductivity mechanism

The intrinsic carrier concentration n_i is an important electrical parameter of MCT since it varies strongly with temperature and the mole fraction of cadmium. The intrinsic

Table 22.6 Comparison of Energy Levels in HgI_2

Energy level (eV)	TSC[194,207-211]	DLTS[206]	PITS[212]	TCT[215]	PL[216,217]
0.01-0.11	0.01(Pd) 0.10-0.13 0.16-0.19	0.074			0.08(Ag)
0.1-0.2	0.19(Ni) 0.16	0.15	0.19	0.13(Hg)h 0.18(Hg)e	
0.2-0.3	0.20(Au) 0.28 0.21-0.26 0.25-0.30	0.22			
0.3-0.4	0.32 0.34 0.38	0.31	0.32	0.37($C_{10}H_8$)h 0.39(Hg)h	
0.4-0.5	0.48(Cu) 0.49(C) 0.48(Ni) 0.40 0.43 0.40-0.44 0.45	0.46	0.44 0.49	0.40(C_{23})h 0.42(I_2)h 0.43(pure)e 0.45(C_{23})e 0.46(pure)h 0.47(I_2)e 0.51($C_{10}H_8$)e	
0.5-0,6	0.54(Pd) 0.57(Al) 0.50 0.59 0.50-0.59				
0.6-0.7	0.61(Al) 0.64(Al)				
0.7-1.0	0.92(Ag) 0.77	0.79			

carrier concentration of MCT as a function of temperature and composition has been calculated by Hansen and Schmit[220] using the Kane nonparabolic approximation for band as shown in Fig. 22.36.

The electrical parameters and/or conduction type of MCT stored near 300 K for 3-13 years are known to have changed. Micklethwaite and Redden[221] have studied the mechanism of this change and explained it by the cross diffusion of Hg in the presence of an interstitial-vacancy concentration gradient and/or surface exchange with the ambient.

The electron transport properties of MCT have been calculated by a Monte Carlo simulation by Palermo et al.[222] and it was found that hot-electron transport and impact ionization processes occur for relatively weak electric fields of around 100 V/cm.

- *Impurities*

Capper[163] has summarized the effect of various impurities on the electrical properties by measuring them using crystals doped with various impurities grown by the cast-

(a)

(b)

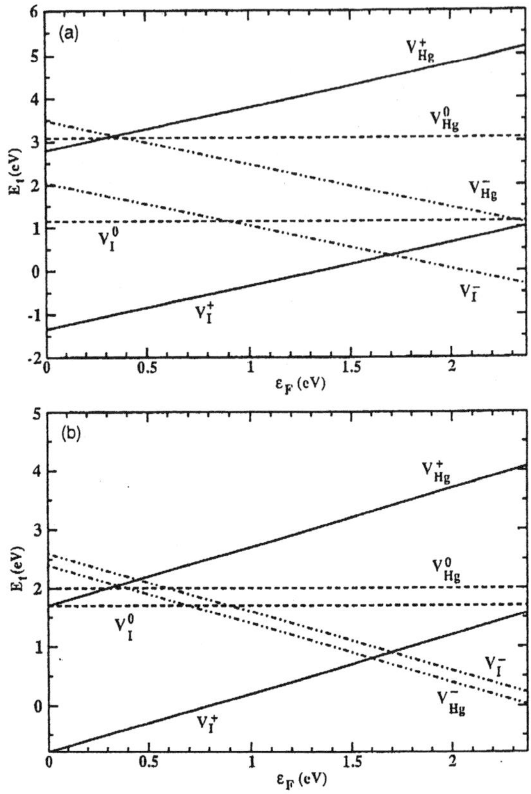

Fig. 22.35 Formation energies of V_I and V_{Hg} at several charge states as a function of the Fermi level for (a) mercury-rich and (b) iodine-rich conditions. Full, dashed and dotted-dashed lines represent, respectively, the vacancies in positive, neutral and negative charge states (reprinted from Ref. 219 with permission, copyright 2006 American Institute of Physics).

recrystallise-anneal (CRA) process. Various impurities were found to act as acceptors or donors as shown in Table 22.7. Jones et al.[223] have studied the electrical behavior of gold in Bridgman grown MCT crystals and found that it does not act as an acceptor to convert the conductivity type to p-type while it lowers the Hall mobility indicating mixed conduction.

Various impurities have been added to MCT crystals grown by the accelerated crucible rotation technique (ACRT), and the influence of stoichiometry on the electrical activity of impurities was studied by annealing under Hg vapor pressure.[224] It was found that group IB and IIIB elements are acceptors and donors, respectively, on metal sites irrespective of stoichiometry. Group VB and VIIB elements are inactive when a high Hg vacancy concentration is present (under the Te-rich condition).

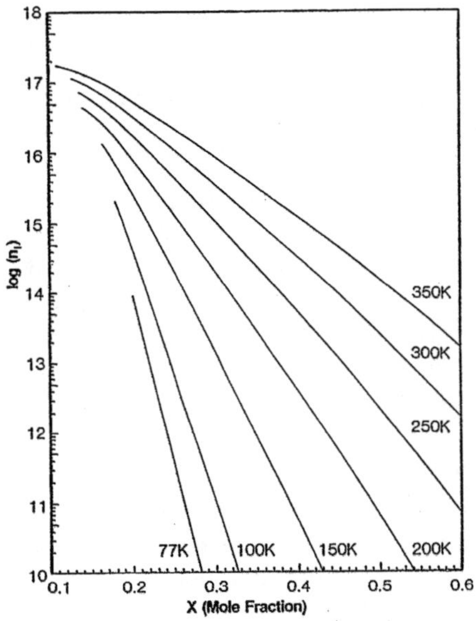

Fig. 22.36 Intrinsic carrier concentration of MCT as a function of the mole fraction x and temperature (reprinted from Ref. 220 with permission, copyright 1983 American Institute of Physics).

Kenworthy et al.[225] have determined the ionization energy as a function of the mole fraction of impurity acceptors of the group I elements such as Ag, Cu, As and Sb which were doped in Bridgmann and ACRT grown MCT crystals. Before the electrical measurement, ACRT grown crystals were annealed under a Hg atmosphere in order to eliminate the effect of Hg vacancies. They found the following empirical equation for the ionization energy.

Table 22.7 Expected Role of Impurities (reprinted from Ref. 163 with permission, copyright 1982 Elsevier)

Group	I	II	III	IV	V	VI	VII	VIII
Expected role	Acceptor		Donor	Donor Acceptor	Acceptor		Donor	Donor
Element	Li Na K Cu Ag Au	Cd Hg	B Al Ga In	C Si Sn Ti Pb	N P As Sb Bi	Te	F Cl Br I	Fe Ni

$$E_a = 42x + 1.36 - 1.40 \times 10^{-5} N_a^{1/3} \tag{22.4}$$

- *Annealing*

Since the electrical properties of MCT are dominated by Hg vacancies, V_{Hg}, native donors such as Hg interstitial atoms, Hg_i, and foreign impurities, annealing has a large effect on these properties. Annealing experiments have been performed by several researchers, and the main message is that p-type conductivity is converted to n-type conductivity by appropriate annealing.[167,226-228] It is also reported that when the material has a large amount of Te precipitates, it is difficult to achieve the conversion.[229] Kalish et al.[230] have shown a simple annealing method to accelerate the process of conductivity conversion by stabilizing Hg saturated conditions. Savitzky and Srotchum[231] have studied rapid thermal annealing such as by laser irradiation and by electric current, which can allow mercury interchanges with the surrounding atmosphere.

- *Minority carrier lifetime*

The lifetime (τ) of minority carriers in MCT is important since it has a large effect on photodetector performance. Lifetime has been measured for crystals prepared by solid state recrystallization (SSR), Bridgman and Bridgman ACRT methods, and for undoped n-type and p-type crystals. The references to these measurements can be seen in papers by Pratt et al.[232,233] and Laklison and Capper.[234] The lifetime is determined by a combination of band-to-band and Schockley-Read recombination processes. In Fig. 22.37, lifetime is shown as a function of carrier concentrations.

(2) HgI_2

Bao et al.[235] have reviewed the electrical properties of HgI_2. Since HgI_2 is a high resistivity material with resistivity up to $\sim 10^{13}$ Ω.cm, HgI_2 detectors can be operated at room temperature. It is however difficult to measure the electrical properties by the conventional Hall measurement.

Since the dark resistivity determines the leakage current in HgI_2 detectors and affects the energy resolution of the detector, the higher dark resistivity is useful in obtaining high performance detectors. Factors affecting the leakage current in HgI_2 have been studied[236] and it was associated with the accumulation of electric charges which modifies the electric fields in the bulk crystal.

Mobility, lifetime and mobility-lifetime product are measured by various methods such as time of light measurements (current and charge transient techniques), photocurrent, photo-Hall effect and photomagentoelectric effect (PME) measurements and X-ray spectra and photocurrent methods.[235] Mobilities for electrons and holes have been evaluated[121, 122, 237-239] and typical data are shown in Fig. 22.38 [237].

Mobility-lifetime ($\mu\tau$) is an important measurement in predicting the performance of radiation detectors. Swierkowski et al.[121] have evaluated $\mu\tau$ as 8×10^{-5} cm^2 /V from the detector spectra. Llacer et al.[122] measured the carrier mobility in HgI_2 as 3 cm^2/V.sec by analysis of the transient waveform from alpha particle radiation. Blacjak et al.[240]

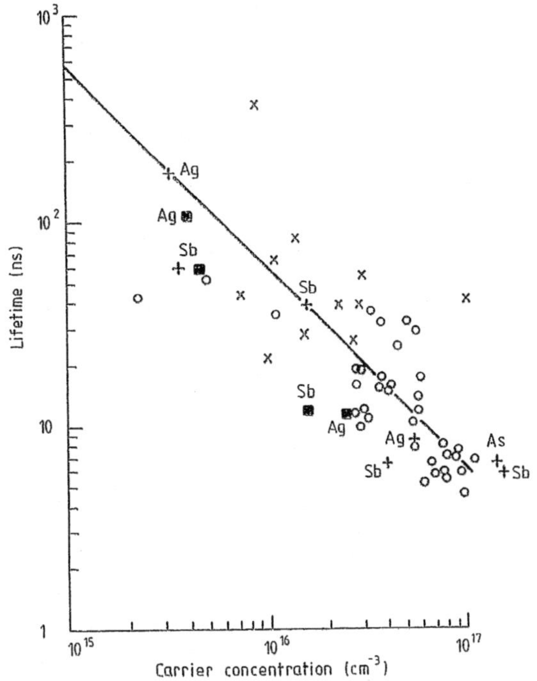

Fig. 22.37 Lifetime versus carrier concentration for various types of p-type material at 77 K (reprinted from Ref. 234 with permission, copyright 1987 Institute of Physics).

have fabricated gamma-ray detectors and evaluated the $\mu\tau$ product as 9×10^{-4} - 7×10^{-3} cm^2/V. Alexiev et al.[241] have measured the $\mu\tau$ value and reported $\mu\tau$ of 4×10^5 cm^2/V for electrons and 3×10^5 cm^2/V for holes.

Electron and hole transport properties in HgI_2 are strongly affected by any deviation from stoichiometry. White and van den Berg[205] have measured $\mu\tau$ products for HgI_2 crystals, both Hg-rich and I-rich and have shown that $\mu\tau$ products vary from $3x10^{-6}$ to 2×10^{-4} cm^2/V for electrons and from $<10^{-8}$ to 1.5×10^{-5} cm^2/V for holes as shown in Table 22.8. Hermon et al.[242] have measured the lifetime of electrons and holes in crystals deviating from stoichiometry. Ponpon et al.[236] have studied current instability in HgI_2 devices and have shown that Che variation with time of the current is due to positive charges on trapping centers. Cho et al.[243] have observed space-charge-limited current and space-charge free currents for electron and holes and compared the results with the previous data.

Marinova et al.[244] have evaluated the electron and hole mobility of $Hg(Br_xI_{1-x})_2$ crystals with compositions x = 0.25, 0.50, 0.75 and 1.00, grown by the PVT method. The electron mobility and hole mobility diminished from 100 and 4 $cm^2/V{\cdot}sec$ respectively to immeasurable levels.

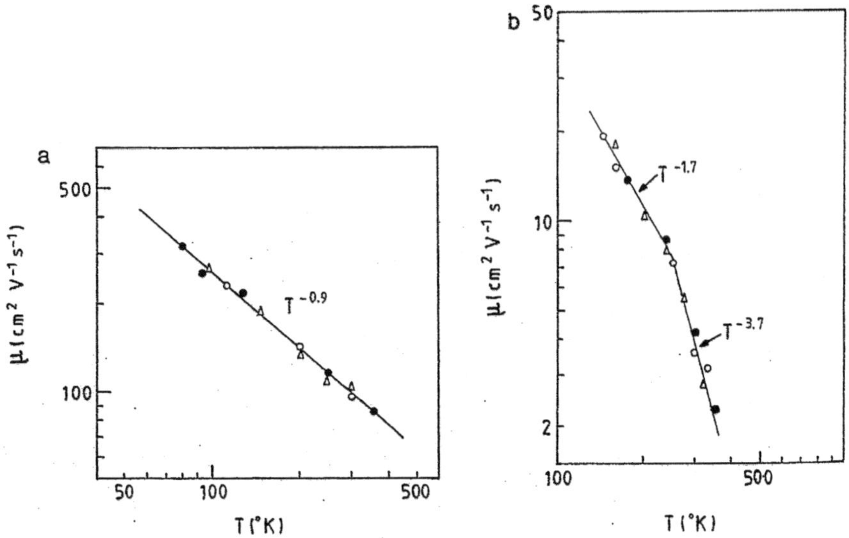

Fig. 22.38 Temperature dependence of the (a) electron and (b) hole mobility parallel to the c-axis (reprinted from Ref. 237 with permission, copyright 1974 American Institute of Physics).

Table. 22.8 HgI_2 Mobility and Lifetime(reprinted from Ref. 205 with permission, copyright 1977 IEEE)

Stoichiometry	$\mu_n\tau_n$ (cm^2/V)	$\mu_p\tau_p$ (cm^2/V)
Heavily I_2	3×10^{-6}	3×10^{-7}
Ligthly I_2	4×10^{-6}	8×10^{-7}
Typical	7×10^{-5}	4×10^{-7}
Lightly Hg	3×10^{-4}	$<10^{-7}$
Lightly Hg	5×10^{-4}	$<10^{-8}$
Best	2×10^{-4}	1.5×10^{-5}
High Purity	1×10^{-4}	2.6×10^{-6}

22.4.4 Optical Properties

(1) $Hg_{1-x}Cd_xTe$ (MCT)

Photoluminescence (PL) spectra of MCT have been studied by several authors.[245-249] The observed PL peak varies as a function of the composition, according to the change in the bandgap. They have observed several lines in bulk and LPE materials, line A which is related to band-to-band and bound exciton recombination luminescence, B which is related to band-to-acceptor and donor-acceptor luminescence and line C whose origin was not identified. They have also evaluated the ionization energies of donors and acceptors from PL data, as $E_d = 1.0 \pm 1.0$ meV ($x = 0.32$) and 4.5 ± 2.0 meV (x = 0.48) and as E_a = 14.0 ± 1.5 meV (x = 0.32) and 15.5 ± 2.0 meV (x = 0.48).

Miles et al.[204] have evaluated the PL of MCT and compared the results with their prediction on native defects and identified the origin of these defects. They suggested the origin of line C as a vacancy-impurity complex where the impurity occupies the anion (Te) site.

Kruse[250] has measured the photoconductive and photovoltaic effects of infrared detectors fabricated from MCT and discussed the response time of the detectors. Capper et al.[251] have developed the wavelength mapping method using an automatic double-beam IR spectrophotometer in order to evaluate the compositional uniformity.

(2) HgI_2

Photoluminescence (PL) of HgI_2 has been extensively studied by various researchers.[27, 252-260]

Edge emission peaks at low temperatures were studied from as far back as 1957[252] and the identification and the assignments have been systematically made by Novikov and Pimonenko[253, 254] and Merz et al.[255] Typical PL peaks for edge emission[253,255] are summarized in Table. 22.9. Bao et al.[256] have measured more emission lines by measurement at higher resolution.

Wu and Merz[257] have compared various crystals with different sublimation and doping conditions and discussed the origin of several edge emission peaks. Akopyan et al.[258] have compared HgI_2 crystals grown by various methods. They also made assignments for 560, 620 and 635 nm peaks.[259]

Anderson[260] has studied time-resolved PL and analyzed the origin of a broad band peak related to native defects. It has been concluded that the broad band peak must be small to obtain good material for radiation detectors. Wen and Ohno[261] have measured the temperature dependence of the time-resolved PL and discussed the origin of near band-gap exciton lines in 530 nm and a broad band at 630 nm. A typical PL spectrum is shown in Fig. 22.39.[260]

The effect of various device processes such as chemical etching and electrode formation on PL has been studied.[262-267]

As electrodes for radiation detectors, palladium [262], indium, tin and indium-tin-oxide (ITO)[263, 264] have been deposited on HgI_2 crystals and the PL has been measured to examine the effect of the electrode deposition. It was found that the beneath of the palladium and ITO electrodes, the PL spectrum is altered and that is attributed to the interaction of metals with HgI_2. This indicates that the selection of the electrode material is important.

The effect of chemical etching with KI[265-267] and HNO_3[266] has been studied and it was found that these chemical etchants introduce carrier traps in the near-surface region. Heating and vacuum exposure also affects the defect structure of HgI_2 but not as greatly as chemical etching.[267]

Bao et al.[268] have clarified the relationship between detector performance and PL spectra. It was found that the intensity of the broad peak at 6200 Å is weak in crystals that produce high-quality detectors. It was also found that strong emission in the exciton region is required for high-quality detectors.

Table 22.9 Low Temperature Photoluminescence Data for HgI_2 (reprinted from Ref. 255 with permission, copyright 1983 Elsevier)

Energy (eV)	Line	Assignment and coments
2.3402	1A*	Free exciton excited state
2.3379	1A	doublet; free exciton ground state; increases with temperature
2.3300	1B	Dominant bound exciton; increases with band 2 (5600 Å) decreases with increasing T
2.3286	1C	Dominant bound exciton; decreases as band 2 increases; deceases with T
2.3262	1A*-LO	
2.3214	1A-LO	
2.3198		
2.3162	1B-LO	
2.3144	1C-LO	
2.3120	1A*-2LO	
2.3105	1A-2LO	
2,3083		
2.3022	1B-2LO	
2.3012	1C-2LO	
2.2977	1A*-3LO	
2.2959	1A*-3LO-TO	

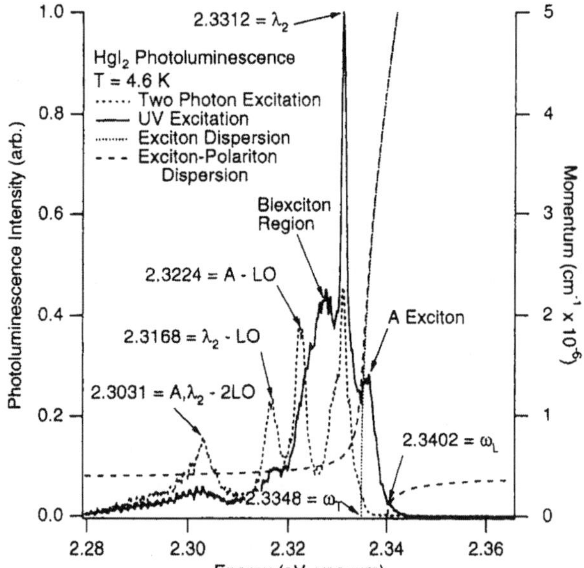

Fig. 22.39 Typical photoluminescence of HgI_2 (reprinted from Ref. 260 with permission, copyright 1996 Elsevier).

The bandgap[269], dielectric constant[270], refractive index[271], reflectivity and absorption[272], and the Urbach rule for luminescence[273] have been studied extensively.

22.5 APPLICATIONS

22.5.1 Far Infrared Detectors

Since the bandgap of MCT can be varied from negative values to 1.6 eV by varying the composition x from 0 to 1[4], MCT is the most widely used material for detecting infrared radiation. It is especially useful for the high transmission wavelength region through atmospheric windows (Fig. 22.40[274]), one is of 3-5 μm (mid-wave infrared (MWIR)) ($x = 0.3$) and the another is 10-14 μm (very long-wave infrared (VLWIR))($x = 0.22$), since through these atmospheric windows objects can be detected over long distances.

MCT has been widely used in various kinds of infrared detectors, especially in thermal imaging, not only in military applications but also homeland defence applications, for detecting fire/earthquake victims in buildings and for intruder detection in security, as well as medical imaging, chemical sensing and astronomy. There are various types of detector structures as reviewed by various authors.[1,42,69,275-282] The first commercial device was the signal processing in the element (SPRITE) photoconductive detector with several element photoconductive devices.[283-291] High performance SPRITE detectors were fabricated using Bridgman and ACRT crystals.[281, 288] Charlton[277] described discrete IR detectors with photoconductive and photovoltaic devices for various space applications. Ashley et. al.[289] described photoconductive room-temperature IR detectors. In Fig. 22.41, the principle of SPRITE detectors is shown.[287] There are single element types and linear arrays of up to 180 elements in commercial use. Linear arrays with a maximum of 1024 or 2000 elements have also been reported.[291]

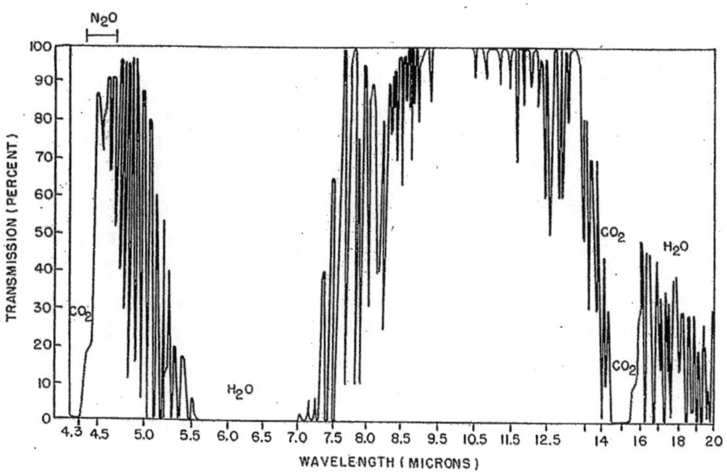

Fig. 22.40 Atmospheric window where high transmission can be obtained for 1-3 μm and 4-14 μm (reprinted from Ref. 274 with permission, copyright 1969 IEEE).

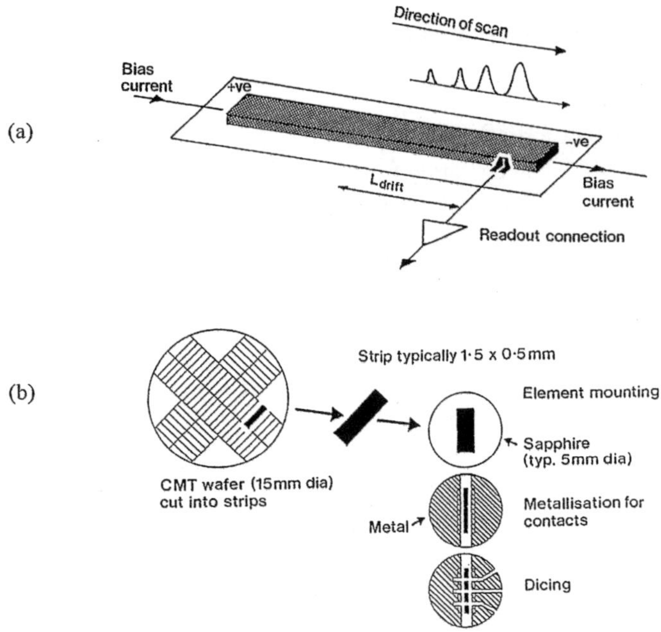

Fig. 22.41 (a) Principle of operation of SPRITE, (b) Fabrication technology used for SPRITE (reprinted from Ref. 287 with permission, copyright 1984 International Society for Optics and Photonics).

Much effort has been put into the development of infrared focal plane arrays (IRFPA).[279,292-298] They are two-dimensional arrays of infrared detectors on the focal plane with multiplexed read-outs. Photovoltaic devices are necessary for IRFPAs, and arrays with 256 × 256 elements have been fabricated.[298] There are detectors with metal-insulate-semiconductor (MIS) structures which can be used as a charged coupled device (CCD) and a charge injection device (CID). The applications of high-performance infrared focal plane arrays are described in Refs. 1, 4, 275, 276, 278 and 280. A staring array image sensor in the 8-14 um wavelength has been demonstrated.[296] Using ACRT materials, 32-element linear arrays, 32 × 32 arrays and 64 × 64 arrays have been fabricated.[293] A typical focal plane array architecture is shown in Fig. 22.42. As the third generation of MCT detectors, photovoltaic devices with two-dimensional focal plane arrays containing thousands of detector elements are under research.

In order to realize IR detectors of greater area, epitaxial growth of MCT on CdTe or $Cd_{1-x}Ze_xTe$ (CZT) substrates (Vol. 1, Ch. 15) has been extensively studied[1,280], using liquid phase epitaxy (LPE)[20,299], metal organic chemical vapor deposition (MOCVD)[300], molecular beam epiraxy (MBE)[301, 302] and other technologies. The MCT detectors fabricated using these epitaxial layers have been reported.[1,303,304] MCT bulk materials are however still in production for making detectors since it is much easier to fabricate devices without using the epitaxial growth process.[69]

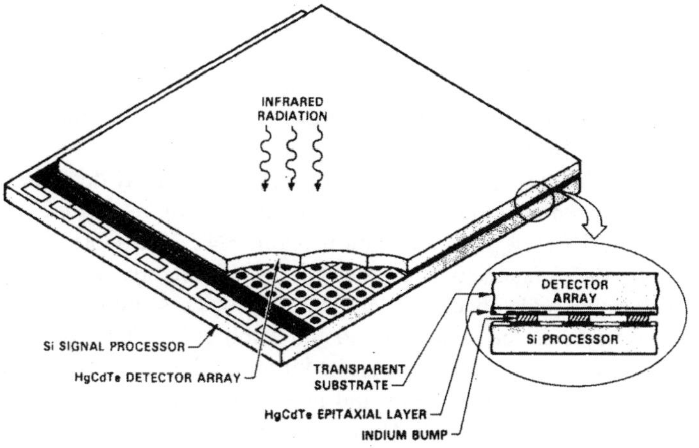

Fig. 22.42 Hybrid focal plane architecture (reprinted from Ref. 298 with permission, copyright 1991 Elsevier).

Ashley et al.[305,306] have studied a non-equilibrium mode of operation for infrared photodetectors in order to reduce the operating temperature by holding the carrier densities below their equilibrium, near-intrinsic levels. Wilson et al.[307] have developed a heterodyne detector of CO_2 laser radiation for rangefinding, velocity measurement and anemometry. Recently, MCT detectors coupled with fibers have been developed[308] for applications to real-time non-contact optical metering systems for characterizing various objects whose direct access is difficult. Resonant-cavity-enhanced MCT detectors have also been developed aimed at higher temperature operation.[309]

22.5.2 Radiation Detectors Using HgI_2

Because of its large average atomic number and stopping power for radiation, HgI_2 is the most promising of various materials for radiation detectors. It is summarized in Sec. 6.5.1 of Vol. 1 and also compared with other materials in Refs. 8, 9, 104, 310 and 311. Based on its reported performance, HgI_2 has been studied for applications such as X-ray counters, X-ray and gamma-ray spectrometers, photodetectors and position sensitive/imaging devices for nuclear medicine, astronomy and radiography. The development of HgI_2 for gamma-ray and X-ray detectors is summarized in Ref. 312.

The main application of HgI_2 is in nuclear radiation detectors, such as gamma-ray detectors because of its good performance. Gamma-ray detectors actually have been fabricated by various researchers.[8,9,108,313-318] Nissenbaum et al.[108] have compared the nuclear response of undoped HgI_2 and SbI_3-doped HgI_2 grown by slow solidification from the melt and found that SbI_3-doped material could detect the 60 keV radiation from ^{241}Am and 122 keV radiation from ^{57}Co, while undoped HgI_2 did not have any nuclear response. The nuclear response of vapor-phase-grown HgI_2 of thickness from 0.4 to 4.7 cm has been evaluated and a spectral resolution of less than 10 % for gamma-ray energies above 1 MeV was found.[313]

Wang et al.[314] have fabricated gama-ray spectrometers using HgI_2 photodetectors combined with scintillators and obtained an energy resolution of 4.58 % FWHM for 662 keV ^{137}Cs. Nelson[317] has compared the performance of room temperature HgI_2 detectors with that of cooled high purity germanium detectors for measuring the ^{239}Pu/^{240}Pu ratio in order to identify nuclear materials.

Bolotnik et al.[318] have fabricated Frisch-ring type gamma-ray detectors with 3 × 3 × 6 mm^3 HgI_2 crystals and obtained an energy resolution of less than 5 % full width at half maximum (FWHM) of 511 keV photons from ^{22}Na. The poorer energy resolution was explained by the presence of macroscopic crystal defects.

X-ray detectors have also been fabricated by various researchers.[8, 9, 319-324] Flat panel X-ray imaging sensors are being developed and Kabir et al.[320] have analyzed the charge carrier transport and the sensitivity of HgI_2. Street et al.[321] fabricated flat panel X-ray imaging sensors using HgI_2 thin films and found that single-photon detection is possible which yields better resolution. Hartsughe et al.[322] have also fabricated flat panel X-ray image sensors measuring up to 6.5 x 6.5 cm^2 and have evaluated the dark current as 0.5 nA/cm^2 and the X-ray response as 20 $\mu C/R\ cm^2$. Zentai et al.[323] compared HgI_2 with PbI_2, for fabricating flat panel X-ray image sensors and have shown that HgI_2 is better than PbI_2 because of the long image lag.

Jinwel et al.[324] have fabricated an array of seven HgI_2 detectors for detecting X-ray radiation and the energy spectrum in the range between 15-150 keV. It was installed in the HL-1M Tokamak and proved effective in measuring the spatial and temporal distribution of the X-ray energy spectrum.

REFERENCES

1. A. Rogalski and J. Pitrowski, *Prog. Quant. Electron.* **12** (1988) 87.
2. *Infrared Detectors and Emitters, Materials and Devices*, eds. by P. Capper and C.T. Elliot (Kluwer Academic Publ., Boston, 2000) Chs. 11 and 12.
3. M.B. Reine, A.K. Sood, T.J. Tredwell, *Semicon. Semimet. Vol.18*, eds. by R.K. Willardson and A.C. Beer (Academic Press, New York, 1981) Ch. 6, p. 201.
4. P.W. Kruse, *Semicon. Semimet., Vol.18*, ed. R.K. Willardson and A.C. Beer (Academic Press, New York, 1981) p. 1.
5. W.D. Lawson, S. Nielson, E.H. Putley and A.S. Young, *J. Phys. Chem. Solids* **9** (1959) 325.
6. W.R. Willig, *Nucl. Instrum. Methods* **96** (1971) 615.
7. W.R. Willig, *Nucl. Instrum. Methods* **101** (1972) 23.
8. R.C. Whited and Schiber, *Nucl. Instr. Methods* **162** (1979) 113.
9. *Proc. 5 th International Workshop on Mercuric Iodide Nuclear Radiation Detectors*, ed. M. Schieber (Jerusalem, 1982), *Nucler. Instr. Methods* **213** (1983).
10. H.L. Malm, *IEEE Trans. Nucl. Sci.* **NS-19** (1972) 263.
11. R.C. Whited and Schiber, *Nucl. Instr. Methods* **162** (1979) 119.
12. M. Piechotka and E. Kaldis, *Nucl. Instr. Methods Phys. Res. A* **283** (1989) 111.
13. M. Piechotka, *Mater. Sci. Engineer. R* **18** (1997) 1.
14. P. Capper, *Properties of Mercury Cadmium Telluride*, ed. by J.C. Brice and P. Capper (INSPEC/IEE, London, 1987).
15. *Properties of Narrow-gap Cadmium Based Compounds*, ed. by P. Capper (INSPEC/IEE, London, 1994).
16. R.F. Brebrick and A.J. Strauss, *J. Phys. Chem. Solids* **26** (1965) 989.
17. J.L. Schmit and C.J. Speerschneider, *Infrared Phys.* **8** (1968) 247.
18. J. Steininger, *J. Electron. Mater.* **5** (1976) 299.
19. T.-C. Yu and R.F. Brebrick, in *Properties of Narrow-gap Cadmium Based Compounds*, ed. by P. Capper (INSPEC/IEE, London, 1994) p. 55.
20. T.C. Harman, *J. Electron. Mater.* **9** (1980) 945.
21. P. Moskvin, V. Khodakovsky, L. Rashkovetskyi and A. Stronski, *J. Crystal Growth* **310** (2008) 2617.

22. J.L. Schmit and E.L. Stelzer, *J. Appl. Phys.* **40** (1969) 4865.
23. G.L. Hansen, J.L. Schmit and T.N. Casselman, *J. Appl. Phys.* **53** (1982) 7099.
24. V.M. Glazov and L.M. Pavlova, *J. Crystal Growth* **184/185** (1998) 1253.
25. R.N. Andrews, S.D. Walck, M.W. Price, F.R. Szofran, C.-H. Su and S.L. Lehoczky, *J. Crystal Growth* **99** (1990) 717.
26. H.D. Palfrey, *J. Crystal Growth* **94** (1989) 778.
27. J.P. Ponpon, R. Stuck, P. Siffert, B. Meyer and C. Schwab, *IEEE Trans. Nucl. Sci.*, **NS-22** (1975) 182.
28. J.B. Newkirk, *Acta Metall.* **4** (1956) 316.
29. A. Burger, S. Morgan, C. He, E. Silberman, L. van den Berg, C. Ortale, L. Franks and M. Schieber, *J. Crystal Growth* **99** (1990) 988.
30. A. Burger and D. Nason, *J. Appl. Phys.* **71** (1992) 2717.
31. A. Burger, S.H. Morgan, E. Silberman, D. Nason and A.Y. Cheng, Nucl. *Instr. Methods Phys. Res. A* **322** (1992) 427.
32. F. Milstein, T.W. James and G. Georgeson, *Nucl. Instr. Methods Phys. Res. A* **285** (1989) 500.
33. K. Baskar, K. Thangaraj and R. Gobinathan, *J. Mat. Sci. Lett.* **10** (1990) 85.
34. T.W. James and F. Milstein, *J. Mater. Sci.* **18** (1983) 3249.
35. R. Dworsky, K.L. Komarek, *Monatshefte Chem.* **101** (1970) 976.
36. R. Dworsky, K.L. Komarek, *Monatshefte Chem.* **101** (1970) 984.
37. M. Piechotka and E. Kaldis, *J. Less Common Metals* **115** (1985) 315.
38. M. Piechotka and E. Kaldis, *Helv. Phys. Acta* **58** (1985) 797.
39. G. Dishon, M. Schieber and L. Ben-Dor, *J. Crystal Growth* **69** (1984) 47.
40. J. Marschall and F. Milstein, *Appl. Phys. Lett.* **62** (1993) 2784.
41. D.A. Nelson, W.M. Higgins and R.A. Lancaster, *SPIE* **225** (1980) 48.
42. W.F.H. Micklethwaite, *Semicon. Semimet. Vol.18* (New York, Academic Press, 1981) p. 47.
43. J.L. Schmit, *J. Crystal Growth* **65** (1983) 249.
44. P. Capper, Prog. *Crystal Growth and Charact.* **19** (1989) 259.
45. C.F. Freeman, *J. Vac. Sci. Technol.* **B9** (1991) 1613.
46. P. Capper, *Prog. Crystal Growth and Charact.* **28** (1994) 1.
47. B.E. Bartlett, J. Deans and P.C. Ellen, *J. Mater. Sci.* **4** (1969) 266.
48. B.E. Bartlett, P. Capper, J.E. Harris and M.J.T. Quelch, *J. Crystal Growth* **46** (1979), 623.
49. G. Fiorito, G. Fasparrini and D. Passoni, *J. Electrochem. Soc.* **125** (1978) 315.
50. C.L. Jones, P.Capper and J.J.G. Gosney, *J. Crystal Growth* **56** (1982) 581.
51. C.L. Jones, P. Capper, J.J.G. Gosney and I. Kenworthy, *J. Crystal Growth* **69** (1984) 281.
52. G.S. Liaw, J.D. Mo, S.M. Li and L.C. Chou, *Proc. 22nd Inter. SAMPE Tech. Conf.* (November, 6-8, 1990) p. 638.
53. P. Capper, J.J. Gosney and C.L. Jones and M.J.T. Quelch, *J. Crystal Growth* **63** (1983), 154.
54. P. Capper, C.L. Jones, E.J. Pearce and M.J.T. Quelch, *J. Crystal Growth* **62** (1983) 487.
55. C.L. Jones, P.Capper, J.J.G. Gosney, C.K. Ard and I. Kenworthy, *J. Crystal Growth* **64** (1983) 403.
56. C.L. Jones and P. Capper, *J. Crystal Growth* **72** (1985) 733.
57. S.L. Lehoczky and F.R. Szofran, *J. Crystal Growth* **69** (1984) 201.
58. H. Bittner, P. Hoschl and B. Schubert, *Cryst. Res. Technol.* **26** (1991) 667.
59. Y. Wang, Q. Li, Q. Han, Q. Ma, R. Ji, B. Song, W. Jie, Y. Zhou and Y. Inatomi, *J. Crystal Growth* **273** (2004) 54.
60. Y. Wang, Q. Li, Q. Han, B. Song, W. Jie, Y. Zhou and Y. Inatomi, *J. Crystal Growth* **263** (2004) 273.
61. P. Capper, J.J. Gosney and C.L. Jones, *J. Crystal Growth* **70** (1984) 356.
62. P. Capper, J.J. Gosney, C.L. Jones and E.J. Pearce, *J. Electron Mater.* **15** (1986) 361.
63. J.C. Brice, P. Capper, C.L. Jones and J.J.G. Gosney, *Prog. Crystal Growth Character.* **13** (1986) 197.
64. P. Capper, J.J. Gosney, C.L. Jones and I. Kenworthy, *J. Electron Mater.* **15** (1986) 371.
65. P. Capper, W.G. Coates, C.L. Jones, J.J. Gosney, C.K. Ard and I. Kenworthy, *J. Crystal Growth* **83** (1987) 69.
66. P. Capper, J.C. Brice, C.L. Jones, W.G. Coates, J.G. Gosney, C. Ard and I. Kenworthy, *J. Crystal Growth* **89** (1988) 171.
67. W.G. Coates, P. Capper, C.L. Jones, J.J. Gosney, C.K. Ard, I. Kenworthy and A. Clark, *J. Crystal Growth* **94** (1989) 959.
68. P. Capper, W.G. Coates, C.L. Jones, J.J. Gosney, I. Kenworthy and C.K. Ard , *J. Crystal Growth* **102** (1990) 848.
69. P. Capper, C. Maxey, C. Butler, M. Grist and J. Price, *J. Crystal Growth* **275** (2005) 259.
70. Z. Liu, W. Jie and Y. Zhou, *J. Crystal Growth* **209** (2000) 751.
71. E.Z. Dziuba, *J. Electrochem. Soc.* **116** (1969) 104.
72. R. Ueda, O. Ohtsuki, K. Shinohara and Y. Ueda, *J. Crystal Growth* **13/14** (1972) 668.
73. Y. Takase, *Jpn. J. Appl. Phys.* **13** (1973) 539.
74. M. Yoshikawa, *IEICE Tech. Rep.*, **CPM85-65**(1985) 35.

75. R. Triboulet, T. Nguyen Duy and A. Durand, *J. Vaccum Sci. Technol. A* **3** (1985) 95.
76. L. Colombo, R.R. Chang, C.J. Chang and B.A. Baird, *J. Vaccum Sci. Technol. A* **6** (1988) 2795.
77. K. Mochizuki, H. Iwanaga, T. Masumoto and Y. Furukawa, *Jpn. J. Assoc. Crystal Growth* **15** (1988) 67.
78. I. Shilo, E. Kedar and D. Szafranek, *Mater. Res. Soc. Symp. Proc.* **214** (1990) 47.
79. P. Gille, F.M. Kiessling and M. Burkert, *J. Crystal Growth* **114** (1991) 77.
80. D.E. Witter, in *Intelligent Processing Materials*, ed. by H.N.G. Wadley and W.E. Eckhart, Jr. (The Minerals, Metals & Materials Socierty, 1990) p. 91.
81. P. Gille, M. Presia, R.U. Bloedner and N. Puhlmann, *J. Crystal Growth* **130** (1993) 188.
82. R.U. Bloedner and P. Gille, *J. Crystal Growth* **130** (1993) 181.
83. R.U. Bloedner, M. Presia and P. Gille, *Adv. Mater. Opt. Electron.* **3** (1994) 233.
84. R. Triboulet, *Prog. Crystal Growth and Charact.* **28** (1994) 85.
85. C.J. Chang, B. Baird, P.K. Liao, R. Chang and L. Colombo, *J. Crystal Growth* **98** (1989) 595.
86. Yu.V. Apanovich and E.D. Ljumkis, *J. Crystal Growth* **110** (1991) 839.
87. L.N. Swink and M.J. Brau, *Metall. Trans.* **1** (1970) 629.
88. T.C. Harman, *J. Electron. Mater.* **1** (1972) 230.
89. B.E. Bartlett, P. Capper, J.E. Harris and M.J.T. Quelch, *J. Crystal Growth* **47** (1979) 341.
90. A.W. Vere, B.W. Straughan, D.J. Williams, N. Shaw, A. Royle, J.S. Gough and J.B. Mullin, *J. Crystal Growth* **59** (1982) 121.
91. C. Boliang, S. Jie, D. Shuzheng, *J. Electon. Mater.* **13** (1984) 47.
92. R.D.S. Yadava, S. Sinha, B.B. Sharma and A.V.R. Warrier, *J. Crystal Growth* **73** (1985) 343.
93. L. Colombo, A.J. Syllaos, R.W. Perlaky and M.J. Brau, *J. Vaccum Sci. Technol. A* **3** (1985) 100.
94. B.B. Sharma, R.K. Sharma, V.K. Singh and N.K. Nayyar, *Proc. Natl. Seminar on Recent Trends in IR Devices and Their Defense Applications* (Solid State Phys. Lab., Delhi, Mrch 1989) p. 4.
95. R.K. Sharma, V.K. Singh, N.K. Nayyar, S.R. Gupta and B.B. Sharma, *J. Crystal Growth* **131** (1993) 565.
96. Y. Wang, Q. Li, Q. Han and B. Song, *SPIE* **1519** (1991) 428.
97. Y. Wang, Q. Li, Q. Han and B. Song, *SPIE* **3553** (1998) 1.
98. W.F.H. Micklethwaite, *Semicon. Semimet. Vol.18* (Academic Press, New York, 1981) p. 63.
99. J.H. Tregilgas, *Prog. Crystal Growth and Charact.* **28** (1994) 57.
100. C.-H. Su, S.L. Lehoczky and F.R. Szofran, *J. Crystal Growth* **86** (1988) 87.
101. L. Wang and W. Jie, *J. Crystal Growth* **290** (2006) 203.
102. M. Schieber, *Nucl. Instr. Meth.* **144** (1977) 469.
103. M. Schieber, M. Roth and W.F. Schnepple, *J. Crystal Growth* **65** (1983) 353.
104. H.A. Lamonds, *Nucl. Instrum. Methods Phys. Res. A* **213** (1983) 5.
105. M. Schieber, R.C. Carlton, H.A. Lamonds, P. Randtke, W.F. Schnepple and J. Llacer, *J. Crystal Growth* **24/25** (1976) 205.
106. L. van den Berg and R.C. Whited, *IEEE Trans. Nucl. Sci.* **NS-25** (1987) 395.
107. N.L. Skinner, C. Ortale, M.M. Schieber and L. Van Den Berg, *Nucl. Instr. Methods Phys. Res. A* **283** (1989) 119.
108. J. Nissenbaum, I. Shilo, A. Burger, A. Levi, M. Shieber, L. Keller and C.N.J. Wagner, *Nucl. Instrum. Methods Phys. Res. A* **213** (1983) 27.
109. M. Daviti, K. Paraskevopoulos, A.N.Anagnostopoulos, E.K. Polychroniadis, *Crys. Res. Technol.* **31** (1996) 11.
110. M. Daviti, K. Paraskevopoulos, A.N. Anagnostopoulos, E.K. Polychroniadis, *Nucl. Instr. Meth. Phys. Res. A* **380** (1996) 62.
111. M. Daviti, K. Paraskevopoulos, A.N. Anagnostopoulos, E.K. Polychroniadis and M.M. Gospodinov, *Inst. Phys. Conf. Ser.* **155** (1997) 937.
112. M.M. Gospodinov, D. Petrova, I.Y. Yanchev, M. Daviti, M. Manolopoulou, K.M. Paraskevopoulos, A.N. Anagnostopoulos and E.K. Polychroniadis, *J. Alloys Comp.* **400** (2005) 249.
113. T. Cummings, C. Marin, A.G. Ostrogorsky, A. Burger and M. Bliss, *J. Crystal Growth* **297** (2006) 334.
114. *Workshop on HgI_2 (June 11, 1975) Informal Proc.* ed. P. Siffert (University Strasbourg, 1975).
115. W. Kleber, H. Raidt and K.O. Leupold, *Kristall Tech.* **3** (1968) 65.
116. C.C. Coleman, *J. Crystal Growth* **6** (1970) 203.
117. R.C. Carlston, M.M. Schieber and W.F. Schnepple, *Mater. Res. Bull.* **11** (1976) 959.
118. I.F. Nicolau, *J. Crystal Growth* **48** (1980) 45.
119. I.F. Nicolau, *J. Crystal Growth* **48** (1980) 51.
120. L.F. Nicolau and J.P. Joly, *J. Crystal Growth* **48** (1980) 61.
121. S.P. Swierkowski, G.A. Armantrout and R. Wichner, *Appl. Phys. Lett.* **23** (1973) 281.
122. J. Llacer, K. Watt, M. Schieber, R. Carlston and W. Schnepple, *IEEE Trans. Nucl. Sci.* **NS-21** (1974) 305.
123. T. Kobayashi, J.T. Muheim, P. Waegli and E. Kaldis, *J. Electrochem. Soc.* **130** (1986) 1183.

124. M. Isshiki, M. Piechotka and E. Kaldis, *J. Crystal Growth* **102** (1990) 344.
125. M. Piechotka, M. Zha and E. Kaldis, *J. Crystal Growth* **113** (1991) 251.
126. M. Zha, M. Piechotrka and E. Kaldis, *J. Crystal Growth* **115** (1991) 43.
127 H. Zhou, C. Lee, J. Chung, C. Shin, K. Chiu and S. Lan, *Mater. Res. Bull.* **38** (2003) 1987.
128. J. Lin, C. Shih, H. Zou, Y. Luo, J. Liao and K. Chiu, *J. Crystal Growth* **290** (2006) 111.
129. W. Yang, L. Nie, D. Li, Y. Wang, J. Zhou, L. Ma and K.C. Chie, *J. Crystal Growth* **324** (2006) 149.
130. H. Schotz, *Acta Electron.* **17** (1974) 69.
131. M. Schieber, W.F. Schnepple and L. van den Berg, *J. Crystal Growth* **33** (1976) 125.
132. M. Schieber, I. Beinglass, G. Dishon and A. Holzer, in *Crystal Growth and Materials, Current Topics in Materials Science, Vol.2*, eds. E. Kaldis and H.J. Scheel (North-Holland, Amsterdam, 1977) p. 279.
133. M. Schieber, I. Beinglass, G. Dishon, A. Holzer and G. Yaron, *Nucl. Instrum. Methods* **150** (1978) 71.
134. L. van den Berg, W.F. Schnepple and C. Ortale, *J. Crystal Growth* **42** (1977) 160.
135. I. Beinglass, G. Dishon, A. Holzer and M. Schieber, *J. Crystal Growth* **42** (1977), 166.
136. M. Gospodinov and P. Sveshtarov, *J. Crystal Growth* **65** (1980) 365.
137. J. Omlay, M. Robert and R. Cadoret, *Mater. Res. Bull.* **16** (1981) 785.
138. J. Omlay, M. Robert and R. Cadoret, *Mater. Res. Bull.* **16** (1981) 1261.
139. J. Omlay, M. Robert and R. Cadoret, *Nucl. Instrum. Methods* **213** (1983) 19.
140. J.P. Fournier, M. Silga, B. Coupat et J. Omlay, *Annales de Chimie-Sciences de Materiaux* **13** (1988) 85.
141. B. Coupat, J.P. Fournier, M. Silga and J. Omaly, *Nucl. Instrum. Methods Phys. Res. A* **269** (1988) 213.
142. L. van den Berg and W.F. Schnepple, in *Materiasl Processing in the Reduced Gravity Environment in Space*, ed. G.E. Rindone (Elsevier-North Holland, 1982) p. 439.
143. L. van den Berg and W.F. Schnepple, *Nucl. Instrum. Methods A* **283** (1989) 335.
144. M. Piechotka and E. Kaldis, *J. Crystal Growth* **99** (1990) 1300.
145. B. Coupat, J.P. Badaud, J.P. Fournier, S. Er-Raji, R. Cadoret and A. Magnan, *J. Crystal Growth* **141** (1994) 465.
146. I.N. Stranski and D. Totomanov, *Z. Phys. Chem.* **A163** (1933) 399.
147. G. Dishon, M. Schieber and L. Ben-Dor, *J. Crystal Growth* **69** (1984), 47.
148. M. Piechotka and E. Kaldis, *J. Crystal Growth* **79** (1986) 469.
149. M. Piechotka and E. Kaldis, *J. Less Common Metals 115* (1986) 315.
150. M. Piechotka and E. Kaldis, *J. Electrochem. Soc.* **133** (1986) 200.
151. M. Isshiki, M. Piechotka and E. Kaldis, *Proc. 7th European Symp. on Materials Science and Fluid Science in Microgravity* (Oxford, September, 1989) **ESASP-295** (1990) 13.
152. M. Isshiki, *Boundary (in japanese)* **March** (1993) 42.
153. J. Laskowski, M. Qledzka and J. Przyluski, *J. Mater. Sci.* **25** (1990) 589.
154. I.M. Silverstrova, C. Barta, G.F. Dobrzhanskii, L.M. Belayaev and Yu.V. Pisarevskii, *Sov. Phys. Cryst.* **20** (1976) 649.
155. C. Barta, *Kristall Tech.* **5** (1970), 541.
156. N.B. Singh, R.H. Hopkins, R. Mazelsky and J.J. Conroy, *J. Crystal Growth* **75** (1986) 173.
157. N.B. Singh, R.H. Hopkins, R. Mazelsky and M. Gottlieb, *J. Crystal Growth* **74** (1986) 667.
158. N.B. Singh, R.H. Hopkins, R. Mazelsky and M. Gottlieb, *J. Crystal Growth* **85** (1987) 240.
159. J. Kim, S.B. Trivedi, J. Soos, N. gupta and W. Palosz, *J. Crystal Growth* **310** (2008) 2457.
160. S.P. Faile, A.J. Dabrowski, G.C. Huth and J.D. Iwanczyk, *J. Crystal Growth* **50** (1980) 752.
161. A. Burger, M. Roth and M. Schieber, *J. Crystal Growth* **56** (1982) 526.
162. A. Burger, A. Levi, J. Nissenbaum, M. Roth and M. Schieber, *J. Crystal Growth* **72** (1985) 643.
163. P. Capper, *J. Crystal Growth* **57** (1982) 280.
164. P. Capper, J.J.G. Gosney, C.L. Jones and I. Kenworthy, *J. Crystal Growth* **71** (1985) 57.
165. P. Capper, I.G. Gale, F. Grainger, J.A. Roberts, C.L. Jones, J.J. G. Gosney and I. Kenworthy, C.K. Ard and W.G. Coates, *J. Crystal Growth* **92** (1988), 1.
166. P. Capper, *J. Vac. Sci. Technol. B* **9** (1991) 1667.
167. P. Capper, E.S. O'Keefe, C. Maxey, D. Dutton, P. Mackett, C. Butler and I. Gale, *J. Crystal Growth* **161** (1996) 104.
168. B.E. Bartlett, P. Capper, J.E. Harris and M.J.T. Quelch, *J. Crystal Growth* **49** (1980) 600.
169. J.T. Muheim, T. Kobayashi and E. Kaldis, *Nucl. Instrum. Methods Phys. Res. A* **213** (1983) 39.
170. J.F. Nicolau, *Nucl. Instrum. Methods Phys. Res. A* **213** (1983) 13.
171. M. Piechotka and E. Kaldis, *Monatsh Chem.* **118** (1987) 137.
172. W.B. Yelon, R.W. Alkire, M.M. Schieber, L. van den Berg, S.E. Rasmussen, H. Cristensen and J.R. Schneider, *J. Appl. Phys.* **52** (1981) 4604.
173. J.M. Whelan, US Pat. 4,554,150 (Nov. 19, 1985).
174. R.B. James, D.K. Ottesen, D. Wong, T.E. Schlesinger, W.F. Schnepple, C. Ortale and L. van den Berg, *Nucl. Instrum. Methods Phys. Res. A* **283** (1989) 188.

175. S.G. Parker and J.E. Pinnell, *J. Electrochem. Soc.* **18** (1971) 1868.
176. N.M. Boinykh, A.M. Sokolov, G.B. Indenbaum and A.B. Banukov, *Russ. Metall.* **5** (1974) 195.
177. E.L. Polisar, N.M. Boinykh G.V. Indenbaum, A.V. Vanyukov and V.P. Schastlivyi, *J. Sov. Phys.* **11** (1968) 48.
178. M. Brown and A.F.W. Willoughby, *J. Phys. Suppl.* **C6 40** (1979) 151.
179. J.I. Zvadanovich, L.J. Kusnezova, A.M. Sokolov and J.G. Puchov, SU Patent HO1L 21/465.
180. I. Hähnert and M. Schenk, *J. Crystal Growth* **101** (1990) 251.
181. T.W. James and F. Milstein, *J. Mater. Sci.* **16** (1981) 1167.
182. P.T. Randtke and C. Ortale, *IEEE Trans. Nucl. Sci.* **NS-24** (1977) 129.
183. C. Genzel, P. Gille, I. Hahert, F.M. Kiessling and P. Rudolph, *J. Crystal Growth* **101** (1990) 232.
184. H.F. Schaake and J.H. Tregilgas, *J. Vac. Sci. Technol.* **A4** (1986) 2181.
185. F.M. Kiessling and H.S. Leipner, *J. Crystal Growth* **128** (1993) 599.
186. M. Neubert, F.M. Klessiling, B. Barz and K. Jacobs, *Semicon. Sci. Technol.* **8** (1993) 2151.
187. G. Tamman, *Z. Anorg. Chem.* **109** (1920) 213.
188. H.W. Kohlshuetter, Koloidchem. *Beihefte* **24** (1927) 319.
189. D. Schwazenbach, *Z. Krist.* **128** (1969) 87.
190. N.V. Long, K. Kleinstueck, J. Tobisch, P. Klinger, K. Prokert and V. Schuricht, *Cryst. Res. Technol.* **18** (1983) K93.
191. M. Schieber, C. Ortale, L. van den Berg, W. Schnepple, L. Keller, C.N.J. Wagner, W. Yelon, F. Ross, G. Georgeson and F. Milstein, *Nucl. Instrum. Methods Phys. Res. A* **283** (1989) 172.
192. S. Leon-Gits and E. Fries, *Nucl. Instrum. Methods Phys. Res. A* **285** (1989) 513.
193. F. Remy, J. Gastaldi, G. Grange, C. Jourdan and G. Le Lay, *J. Crystal Growth* **121** (1992) 243.
194. A. Tadjine, D. Gosselin, J.M. Koebel and P. Siffert, *Nucl. Instrum. Methods Phys. Res. A* **213** (1983) 77.
195. W.B. Yellon, R.W. Alkire, M. Schieber, L. van den Berg, S. Rasmussen and G. Schneider, *J. Appl. Phys.* **57** (1981) 4606.
196. M.A. Berding, M. van Schifgaarde, A.T. Paxton and A. Sher, *J. Vac. Sci. Technol. A* **8** (1990) 1103.
197. R.A. Farrar, G.J. Gilham, B. Bartlett and M. Quelch, *J. Mater. Sci.* **12** (1977) 836.
198. G. Dishon, M. Schieber, L. Ben-Dor and L. Halitsz, *Mater. Res. Bull.* **16** (1981) 565.
199. M.C. Delong and R. Rosenberger, *Mater. Res. Bull.* **16** (1981) 1445.
200. I.F. Nicolau and G. Rolland, *Mater. Res. Bull.* **16** (1981) 759.
201. H. Hermon, M. Roth, M. Schieber and J. Shamur, *Vibrat. Spec.* **2** (1991) 155.
202. H. Hermon, M. Schieber, M. Roth and J. Shamur, *Anl. Chim. Acta* **259** (1992) 181.
203. M.A. Berding, A. Sher and A.-B. Chen, *J. Vac. Sci. Technol. A* **5** (1987) 3009.
204. C.W. Myles, P.F. Williams, R.A. Chapman and E.G. Nylander, *J. Appl. Phys.* **57** (1985) 5279.
205. R.C. Whited and L.van den Berg, *IEEE Trans. Nucl. Sci.* **NS-24** (1977) 165.
206. D. Alexiev, N. Dylewski, M.I. Reinhard and L. Mo, *Nucl. Instrum. Methods Phys. Res. A* **517** (2004) 226.
207. U. Gelbert, Y. Yacoby, I. Beinglass and A. Holzer, *IEEE Trans. Nucl. Sci.* **NS-24** (1977) 135.
208. T. Mohammed-Brahim, A. Friant and J. Mellet, *phys. stat. sol.* **79** (1983) 71.
209. T.E. Schlesinger, X.J. Bao, R.B. James, A.Y. Cheng, C. Ortale and L. van den Berg, *Nucl. Instrum. Methods Phys. Res. A* **322** (1992) 414.
210. X.J. Bao, T.E. Schlesinger, R.B. James, A.Y. Cheng, *Mat. Res. Soc. Symp. Proc.* **242** (1992) 767.
211. J.M. Van Scyoc, T.E. Schlesinger, R.B. James, A.Y. Cheng, C. Ortale and L. van den Berg, *Mater. Res. Soc. Symp. Proc.* **302** (1993) 115.
212. F. Alvarez and J. Saura, *J. Mater. Sci.* **9** (1990) 569.
213. S.L. Sharma, T. Pal and H.N. Acharya, *J. Appl. Phys.* **78** (1994) 7884.
214. T. Pal. S.L. Sharma and H.N. Acharya, *J. Phys, D: Appl. Phys.* **28** (1995) 1439.
215. H. Hermon, M. Schieber and M. Roth, *Nucl. Instrum. Methods Phys. Res. A* **380** (1996) 10.
216. R.B. James, X.J. Bao, T.E. Schlesinger, J.M. Markakis, A.Y. Cheng and C. Ortale, *J. Appl. Phys.* **66** (1989) 2578.
217. R.B. James, X.J. Bao, T.E. Schlesinger, A.Y. Cheng and C. Ortale, *J. Appl. Phys.* **67** (1990) 2571.
218. F. Ayres, W.V.M. Machado , J.F. Justo and L.V.C. Assali, *Physica B* **340-342** (2003) 918.
219. F. Ayres, L.V.C. Assali, W.V.M. Machado and J.F. Justo, *Appl. Phys. Lett.* **88** (2006) 011918.
220. G.L. Hansen and J.L. Schmit, *J. Appl. Phys.* **54** (1983) 1639.
221. W.F.H. Micklethwaite and R.F Redden, *Appl. Phys. Lett.* **36** (1980) 379.
222. C. Palermo, L. Varani, J.C. valisseire, E. Starikov, P. Shiktorov, V. Gruzinskis and B. Azais, *Solid-State Electron.* **53** (2009) 70.
223. C.L. Jones, P. Capper, M.J.T. Quelch and M. Brown, *J. Crystal Growth* **64** (1983) 417.
224. P. Capper, J.A. Roberts, I. Kenworthy, C.L. Jones, C.K. Ard and W.G. Coates, *J. Appl. Phys.* **64** (1988) 6227.

225. I. Kenworthy, P. Capper, C.L. Jones, J. Gosney and W.G. Coates, *Semicond. Sci. Technol.* **5** (1990) 854.
226. H.F. Schaake, J.H. Tregilgas, J.D. Beck and M.A. Kinch, *Solid State Commun.* **50** (1984) 133.
227. H.F. Schaake, J.H. Tregilgas, J.D. Beck, M.A. Kinch and E. Gnade, *J. Vac. Sci. Technol. A* **3** (1985) 143.
228. C.L. Jones, M.J.T. Quelch, P. Capper and J.J. Gosney, *J. Appl. Phys.* **53** (1982) 9080.
229. E. Belas, P. Hoschl, P. Moravec and J. Franc, in *Non-Stoichiometry in Semiconductors*, eds. K.J. Bachmann, H.-L. Hwang and C. Schwab (Elsevier Sci., 1992) p. 81.
230. R. Kalish, R. Fastow, V. Richter and M. Shaanan, *Appl. Phys. Lett.* **51** (1987) 1158.
231.V.G. Savistky and P.E. Storchum, *Mater. Sci. Engineer. B* **44** (1997) 297.
232. R.G. Pratt, J. Hewett, P. Capper, C.L. Jones and M.J. Quelch, *J. Appl. Phys.* **54** (1983) 5152.
233. R.G. Pratt, J. Hewett, P. Capper, C.L. Jones and N. Judd, *J. Appl. Phys.* **60** (1986) 2377.
234. D.E. Lacklison and P. Capper, *Semicond. Sci. Technol.* **2** (1987) 33.
235. X.J. Bao, T.E. Schlesinger and R.B. James, *Semicon. Semimet.* **43** *(*1995) 111.
236. K.P. Ponpon, R. Stuck and M. Amann, *Solid-State Electron.* **44** (2000) 29.
237. R. Minder, G. Majni, C. Canali, G. Ottaviani, R. Stuck, J.P. Ponpon, C. Schwab and P. Shiffert, *J. Appl. Phys.* **45** (1974) 5074.
238. H.L. Mulm, *IEEE Trans. Nucl. Sci.* **NS-19** (1972) 263.
239. H.B. Serreze and G. entine, *IEEE Trans. Nucl. Sci.* **22** (1975) 1995.
240. J. E. Baciak, Z. He and R.P. DeVito, *IEEE Trans. Nucl. Sci.* **49** (2002) 1264.
241. D. Alexiev, N. Dytlewski, M.I. Reinhard and L. Mo, *Nucl. Instrum. Methods Phys. Res. A* **517** (2004) 226.
242. H. Hermon, M. Roth, J. Nissenbaum and M. Schieber, *J. Crystal Gowth* **109** (1991) 376.
243. Z.H. Cho, M.K. Watt, M. Slapa, P.A. Tove, M. Schieber, T. Davies, W. Schnepple, P.T. Randtke, P. Carlston and d. Sarid, *IEEE Trans. Nucl. Sci.* **NS-22** (1975) 229.
244. V. Marinova, I. Yanchev, M. Daviti, K. Kyritsi and A.N. Anagnostopoulos, *Mater. Res. Bull.* **37** (2002) 1991.
245. C.T. Elliot, I. Melnagailes, T.C. Harman and A.G. Foyt, *J. Phys. Chem. Solids* **33** (1972) 1527.
246. V.I. Ivanov-Omskii, V.A. Maltseva, A.D. Britov and S.D. Sivachenko, *phys. stat. sol. (a)* **46** (1978) 77.
247. M.G. Andrukhiv, V.A. Maltseva, V.I. Ivanov-Omskii, V.K. Ogorodnikov and T.Ts. Totievo, *Sov. Phys. Semicond.* **13** (1979) 210.
248. A.T. Hunter and T.C. McGill, *J. Appl. Phys.* **52** (1981) 5779.
249. B.J. Feldman, J. Bajaj and S.H. Shin, *J. Appl. Phys.* **55** (1984) 3873.
250. P.W. Kruse, *Appl. Opt.* **0** (1965) 687.
251. P. Capper, C.L. Jones, I. Kenworthy, M.R. Bennett, D. Davidson and J. McIntosh, *Infrared Phys.* **24** (1984) 473.
252. R.H. Bube, *Phys. Rev.* **106** (1957) 703.
253. B.V. Novikov and M.M. Pimonenko, *Sov. Phys. Semicon. 64* (1971) 1785.
254. B.V. Novikov and M.M. Pimonenko, *Sov. Phys. Semicon.* **6** (1972) 671.
255. J.L. Merz, Z.L. Wu, L. van den Berg and W.F. Schnepple, *Nucl. Instrum. Methods* **213** (1983) 51.
256. X.J. Bao, T.E. Schlesinger, R.B. James, C. Ortale and L. van den Berg, *J. Appl. Phys.* **68** (1990) 2951.
257. Z.L. Wu and J.L. Merz, *J. Lum.* **24/25** (1981) 197.
258. I. Kh. Akopyan, B.V. Bondarenko, B.A. Kazennov and B.V. Novikov, *Sov. Phys. Solid States* **29** (1987) 238.
259. I. Kh. Akopyan, B.V. Bondarenko, O.N. Volkova, B.V. Novikov and T.A. Pavlova, *Phys. Solid State* **39** (1997) 58.
260. R.J.M. Anderson, *Nucl. Instrum. Methods Phys. Res. A* **380** (1996) 88.
261. Z-M. Wen and N. Ohno, *Jpn. J. Appl. Phys.* **40** (2001) 6464.
262. D. Wong, X.J. Bao, T.E. Schlesinger, R.B. James, A. Cheng, C. Ortale and L. van den Berg, *Appl. Phys. Lett.* **53** (1988) 1636.
263. R.B. James, X.J. Bao, T.E. Schlesinger, J.M. Markakis, A.Y. Cheng and C. Ortale, *J. Appl. Phys.* **66** (1989) 2578.
264. R.B. James, X.J. Bao, T.E. Schlesinger, A.Y. Cheng and C. Ortale, *J. Appl. Phys.* **67** (1990) 2571.
265. X.J. Bao, T.E. Schlesinger, R.B. James, R.H. Stulen, C. Ortale abd A.Y. Cheng, *J. Appl. Phys.* **68** (1990) 86.
266. R.B. James, T.E. Schlesinger, X.J. Bao, A.Y. Cheng, C. Ortale and L. van den Berg, *Nucl. Instrum. Methods Phys. Res. A* **322** (1992) 435.
267. C.-Y. Hung, X.J. Bao, T.E. Schlesinger, R.B. James, A.Y. Cheng, C. Ortale and L. van den Berg, *J. Appl. Phys.* **73** (1993) 4591.

268. X.J. Bao, T.E. Schlesinger, R.B. James, S.J. Harvey, A.Y. Cheng, V. Gerrish and C. Ortale, *Nucl. Instrum. Methods Phys. Res. A* **317** (1992) 194.
269. A. Berger and D. Nason, *J. Appl. Phys.* **71** (1992) 2717.
270. M. Veleva, M.M. Gospodinov, M. Daviti, A.N. Anagnostopoulos, K. Paraskevopoulos, E.K. Polychronidadis and I.Y. Yanchev, *J. Mater. Sci. Lett.* **19** (2000) 1019.
271. V. Marinova, St. Shurulinkov, M. Daviti, K. Paraskevopoulos and A. Anagnostopoulos, *Opt. Mater.* **14** (2000) 95.
272. K. Kanzaki and I. Imai, *J. Phys. Soc. Jap.* **32** (1972) 1003.
273. T. Goto, J. Takeda, T. Ishihara and M. Matsuoka, *J. Lum.* **38** (1987) 55.
274. E.L. Stelzer, J.L. Schmit and O.N. Tufte, *IEEE Trans. Electron. Devices* **ED-16** (1969) 880.
275. C.T. Elliott, in *Handbook on Semiconductors Vol. 4*, ed. C. Hilsum (North-Holland, Amsterdam, 1981) p. 727.
276. R.M. Broudy and V.J. Mazurczyk, *Semicon. Semimet.* **18** (1981) 157.
277. D.E. Charlton, *J. Crystal Growth* **59** (1982) 98.
278. M.B. Reine, *Proc. SPIE* **443** (1984) 2.
279. W.E. Tennant, C.A. Cockrum, J.B. Gilpin, M.A. Kinch, M.B. Reing and R.P. Ruth, *J. Vac. Sci. Technol. B* **10** (1992) 1359.
280. S.C. Shen, *Microelecton. J.* **25** (1994) 713.
281. C.T. Elliott, in *Properties of Narrow-gap Cadmium Based Compounds*, ed. by P. Capper (INSPEC/IEE, London, 1994) p. 311.
282. T.J. Phillips, *III-Vs Review* **15** (2002) 32.
283. C.T. Elliott, *Electron Lett.* **17** (1982) 312.
284. C.T. Elliott, D. Day and D.J. Wilson, *Infared Phys.* **22** (1982) 31.
285. A. Blackburn, M.V. Blackman, D.E. Charlton, W.A.E. Dunn, M.D. Jenner, K.J. Oliver and J.T.M. Wotherspoon, *Infared Phys.* **22** (1982) 57.
286. T. Ashley, C.T. Elliott, A.M. White, J.T.M. Wortherspoon and M.D. Johns, *Infrared Phys.* **24** (1984) 25.
287. J.T.M. Wotherspoon, R.J. Dean and M.D. Johns, *SPIE* **510** (1984) 102.
288. M.F. Kimmitt, G.C. Lopez, J.C. Giles, M. Takai, H.P. Roser, B.T. McGuckin and A. Black, *Infared Phys.* **25** (1985) 767.
289. T. Ashley, C.T. Elliott and A.T. Harker, *Infared Phys.* **26** (1986) 303.
290. A. Campbell, C.J. Elliott and A.M. White, *Infrared Phys.* **27** (1987) 125.
291. I.M. Baker, G. J. Crimes, C.K. Ard, MS. Jenner and J.W. Parsons, *4th Int. Conf. Advanced IR Detectors ans Systems* (IEE, London, 1990) p. 78.
292. I.M. Baker and R.A. Ballingall, *SPIE* **510** (1984) 121.
293. I.M. Baker, J.E. Parsons, J.H. W. Lewis, R.A. Lockett and J.T.M. Wotherspoon, *SPIE* **588** (1985) 16.
294. G. Finger, M. Meyer and A.F.M. Moorwood, *SPIE* **865** (1987) 94.
295. R.A. Ballingual, *Electron. Lett.* **18** (1982) 285.
296. R.A. Ballingall, *SPIE* **1310** (1990) 70.
297. K. Vural and L. Kozlowski, *SPIE* **1310** (1990) 107.
298. R.B. Baily, L.J. Kozlowski, J. Chen, D.Q. Bui, K. Vural, D.D. Edwall, R.V. Gil, A,B. Vanderwyck, E.R. Gertner and M.B. Gubala, *IEEE Trans. Electron. Devices* **38** (1991) 1104.
299. J.L. Schmit, R.J. Hager and R.A. Wood, *J. Crystal Growth* **56** (1982) 485.
300. K.A. Porter, J.E. Hails, J.A.J. Pardoe, D.J. Cole-Hamilton and N.J. Blacker, *J. Crystal Growth* **272** (2004) 829.
301. J.P. Faurie, S. Sivananthan and P.S. Wijewarnasuriya, *SPIE* **1735** (1992) 141.
302. V.S. Varavin, S.A. Dvoretsky, V.I. Liberman, N.N. Mikkhilov and Yu.G. Sidorov, *Thin Solid Films* **267** (1995) 121.
303. D.J. Hall, L. Buckle, N.T. Gordonm J.E. Hailsm C. Matlby and T. Ashley, *Appl. Phys. Lett.* **85** (2004) 2113.
304. C.D. Maxey, J.P. Camplin, I.T. Guilfoy, J. Gardner, R.A. Locket, C.L. Jones, P. Capper, M. Houltona and N.T. Grodon, *J. Electron. Mater.* **32** (2003) 656.
305. T. Ashley, C.T. Elliott and A.M. White, *SPIE* **572** (1985) 123.
306. T. Ashley and C.T. Elliott, *Electron. Lett.* **21** (1985) 451.
307. M.C. Wilson, W.A.E. Dunn and D.J. Wilson, *SPIE* **588** (1985) 29.
308. I.Vu. Lartsev, M.S. Nikitin, G.V. Chekanova, V.G. Artiouchenko and T.V. Sakharova, *J. Alloys Compounds* **371** (2004) 37.
309. J.G.A. Whener, C.A. Musca, R.H. Sewell, J.M. Dell and L. Faraone, *Solid-State Electron.* **50** (2006) 1640.
310. V.M. Zaletin, *Atomic Energy* **97** (2004) 773.
311. A. Owens, *J. Synchrotron Rad.* **13** (2006) 143.

312. J.P. Ponpon and M. Sieskind, *Nucl. Instrum. Methods Phys. Res. A* **380** (1996) 173.
313. K. Hull, A. Beyerkem, B. Lopez and J. Markak, *IEEE Trans. Nucl. Sci.* **NS-30** (1983) 402.
314. Y.J. Wang, B.E. Patt and J.S. Iwanzyk, *Nucl. Instrum. Methods Phys. Res. A* **353** (1994) 50.
315. M. Schieber, A. Zuck, M. Braiman, J. Nissenbaum, R. Turchetta, W. Dulinski, D. Husson and J.L. Riester, *Nucl. Phys, B (Proc. Suppl.)* **61B** (1998) 321.
316. M. Schieber, A. Zuck, L. Melekhov, J. Nissenbaum, R. Turchetta, W. Dulinski, D. Husson and J.L. Riester, *Nucl. Instrum. Methods Phys. Res. A* **410** (1998) 107.
317. M.B. Nelson, Thesis, *Air Force Instituet Technology 2003* AFIT.GNE/ENP/03-07.
318. A.E. Bolotnikov, J. Baker, R. DeVito, J. Sandoval and L. Szubart, IEEE Trans. Nucl. Sci. **52** (2005) 468.
319. M. Schieber, A. Zuck, L. Melekhov, R. Shatunovsky, H. Hermon and R. Turchetta, *SPIE* **3768** (1999) 296.
320. M. Z. Kabir and S.O. Kasap, *Appl. Phys. Lett.* **80** (2002) 1664.
321. R.A. Street, J.-P. Lu and S.R. Ready, *IEE Prc.-Circuits Devices Syst.* **150** (2003) 250.
322. N.E. Hartsough, J.S. Iwanczyk, B.E. Patt and N. L. Skinner, *IEEE Trans. Nucl. Sci.* **51** (2004) 1812.
323. G. Znetai, L. Partain, R. Pavlyuchkova, C. Proano, M. Schieber, K. Shah, P. Bennett, L. Melekhov and H. Gilboa, *IEEE Trans. Nucl. Sci.* **53** (2006) 2506.
324. Y. Jinwei, Z. Wei and W. Shiqing, *Rev. Sci. Instr.* **75** (2004) 2513.

23. SiC

23.1 INTRODUCTION

SiC is mechanically and thermally a very stable material and has long been used in varistors and lightning arresters, abrasive powders and refractory bricks.[1] SiC has a tetrahedrally coordinated structure similar to silicon and diamond mainly with covalent bonding due to sp^3 hybrid orbits but has 12% ionicity due to electronegativity between Si and C. SiC has a large bandgap, as much as 2.2-3.2 eV and impurity doping achieves easy p-n control. It was therefore a promising material for visible short wavelength light emitting diodes (LEDs)[2-4] but GaN based LEDs were soon developed (Ch. 20) and displaced SiC in this function. SiC however has uses in other color LEDs[5], UV photodetectors [6,7], gas sensors[8] and radiation detectors.[9] SiC is also thermally stable, chemically inert and mechanically strong and radiation-resistive with high saturation drift velocity and a high breakdown voltage. SiC is therefore an electronic material with promise for high power devices [10,11], high frequency devices[12] and radiation-resistive devices[13] and for devices in harsh environments such as high temperatures up to 500-600°C. SiC was also predicted to be a promising substrate for GaN epitaxial growth because of its good lattice matching with GaN[14] and is now widely used for blue and white LEDs.[15]

Because of the interest in this material, SiC was very actively studied for a while.[16-20] This activity however did not continue due to the difficulty of satisfactory crystal growth and the development was abandoned for a time. The pause in its development was swept aside by the arrival of a new crystal growth method based on seeded sublimation which is called the modified Lely method.[21] This method allowed the growth of large diameter SiC single crystals and the development of SiC crystals and devices was thereby strongly boosted[22-29] and various electronic devices are now going to be at their industralization stage as reviewed in several books.[1,30-33]

23.2 PHYSICAL PROPERTIES

SiC is a very stable material and does not melt at atmospheric pressure as seen from the phase diagram (Fig. 23.1).[34,35] SiC exists as a melt only at high temperatures and pressures (higher than 2830°C and several tens of atmospheres) but melt growth under such extreme conditions is not practicable. Since Si has a solubility for carbon as seen in Fig. 23.2[34, 35], this raises a possibility of solution crystal growth. SiC however sublimes as C and Si rich vapors at temperatures higher than 2000°C as seen in Fig. 23.3.[17,36,37] Sublimed SiC takes various forms such as Si, Si_2C and SiC_2.[37] These sublimation properties become the basis of the sublimation method.

SiC has various polytypes such as 3C (cubic), 4H, 6H (hexagonal) and 15R (rhombohedral), due to the stacking period of atoms perpendicular to the mostly closed

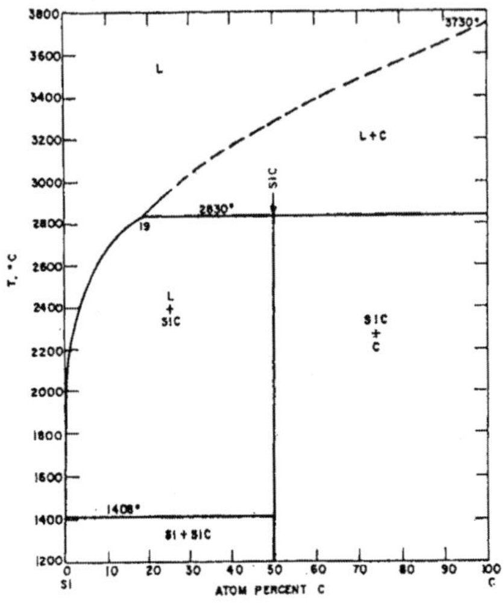

Fig. 23.1 Phase diagram of SiC (reprinted from Ref. 34 with permission, copyright 1959 American Institute of Physics).

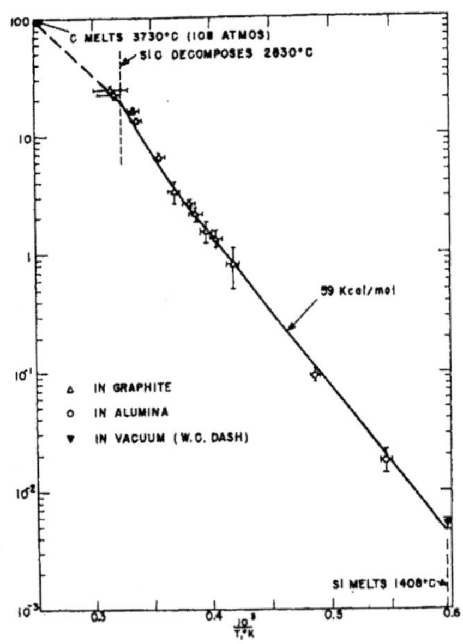

Fig. 23.2 The solubility of carbon in silicon as a function of reciprocal temperature (reprinted from Ref. 34 with permission, copyright 1959 American Institute of Physics).

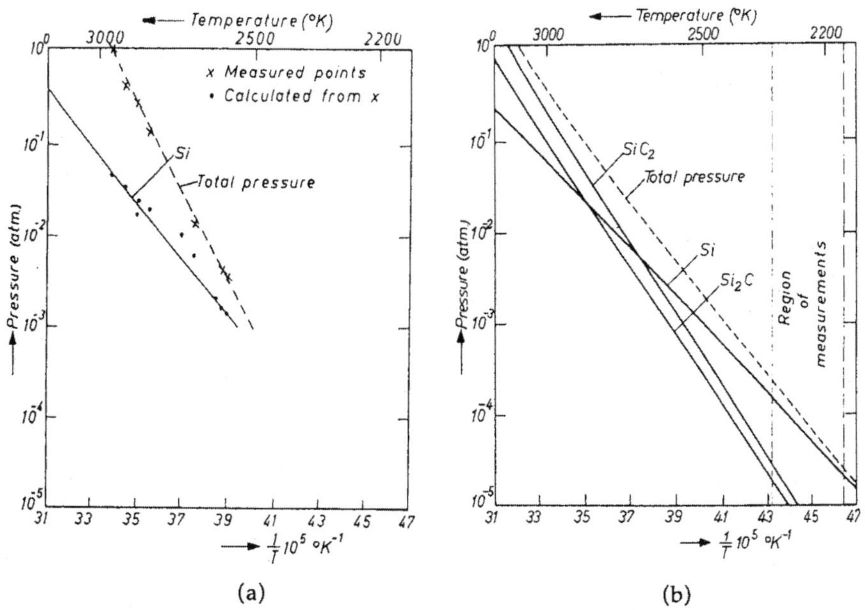

Fig. 23.3 (a) Total pressure and partial pressure curves (reprinted from Ref. 37 with permission, copyright 1958 AIP), (b) total pressure and partial pressures of Si, SiC_2 and Si_2C in the system SiC + C[17, 37] (reprinted from Ref. 17 with permission, copyright 1963 Koninklijke Philips Electronics N.V.).

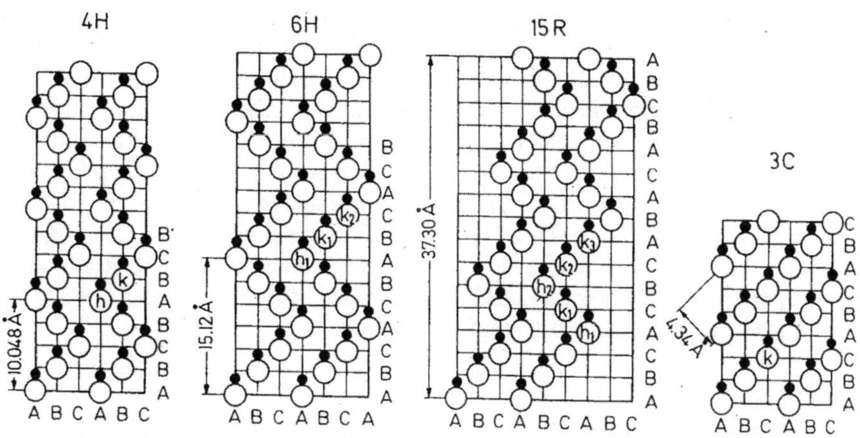

Fig. 23.4 Arrangement of Si and C atoms in plane (1120) in 4H-, 6H-, 15R-, 3C-polytypes of SiC (reprinted from Ref. 38 with permission, copyright 1977 Springer).

packed {0001} plane as shown in Fig. 23.4.[38] Here, the first number means the atomic unit layer in a crystal unit cell and the alphabetical letter designates the crystal system. 3C-SiC is called β-SiC and 4H- and 6H-polytypes are called α-SiC. 200 polytypes of SiC are reported and each has its own physical properties. In SiC crystal growth, the

Table 23.1 Physical Properties of SiC

	3C-SiC	4H-SiC	6H-SiC
Crystal Structure	zincblende	wurzite	wurzite
Lattice Constant (Å)	a=4.358	a =3.08 c=10.05	a=3.0817 c=15.117
Density (g/cm³)	3.2	3.2	3.2
Linear Expansion Coefficient (cm/deg)	2.9×10^{-6} (300 K)		4.7×10^{-6} 4.2×10^{-6} (700 K)
Thermal Conductivity (W/cm•K)	3.2	3.7	3.6
Dielectric Constant	9.72	9.66 (c-axis) 10.03 (a-axis)	9.66 (c-axis) 10.03 (a-axis)
Refractive index	2.7	2.7	2.7
Bandgap at RT (eV)	2.39	3.26	3.02
Optical Transition Type	indirect	indirect	indirect
Electron Mobility at R.T. (cm²/V•sec)	1000	1200	460
Hole Mobility at R.T. (cm²/V•sec)	70	120	90
Saturated Drift Velocity (cm/s)	2.7×10^7	2.2×10^7	2.0×10^7

control of polytypes which depends on growth temperature, growth rate, and doped impurities becomes very important as will be explained later.

As seen in Table 23. 1 [17,39-41], SiC has high thermal conductivity [42], large saturated drift velocity[43], high breakdown voltage[44] and large bandgap.[45] Among various polytypes, 6H-SiC and 4H-SiC single crystals are now industrially produced by the modified Lely method. The main applications for 6H-SiC are as substrates for GaN LEDs and 4H-SiC which is reported to have higher electron mobilities with less anisotropy and shallower levels for donors and acceptors, is used in various electronic devices[46], especially power devices.[47] 3C-SiC is recently receiving attention for power devices because of its isotropic electrical properties and electron mobility as high as 1000 cm²/Vs, high saturation drift velocity (2.7×10^7 cm/s at 2×10^5 V/cm) and high intrinsic carrier concentrations, five orders of magnitude larger than 6H or 4H-SiC.[48] 3C-SiC is especially promising for n-type inversion channel MOSFETs because the near interface traps are localized in the conduction band and have little effect on the transport properties in the inversion channel.[48] Because of this interest, growth of 3C-SiC crystals is under development by various methods.

23.3 CRYSTAL GROWTH

Since SiC decomposes without melting at atmospheric pressure, conventional melt growth methods such as Czochralski and Bridgman methods can not be applied. For preparing SiC source material, the Acheson method is used industrially. For single crystal growth, the solution growth method and the vapor phase transport method, that is the sublimation method (the Lely method and the modified Lely method) have been developed. Of these methods, the vapor phase growth method was believed to be the most practicable[17] and in fact the modified Lely method based on physical vapor transport (PVT) has been applied industrially. In addition to these methods, several other methods such as solution growth, chemical vapor deposition (CVD), high temperature chemical vapor deposition (HT-CVD), smart cut (SC) and closed vapor transport (CVT) methods have also been examined. These crystal growth methods have been reviewed by various authors.[17,23,49-60]

23.3.1 Crystal Growth Methods

(1) Acheson method

SiC has been used in resistors, heaters, and abrasives and is industrially fabricated by the Acheson method in which silica (SiO_2) is reacted with cokes in an electric furnace at temperatures higher than 2000°C.[17,61] SiC single crystals are grown as by-products, most of them as 6H-SiC and some as 15R-SiC. By this method, only small crystals of dimensions 10 mm square are grown and it is not appropriate for industrial production of single crystals. These small single crystals are however used as the starting seed crystals in various crystal growth methods.

(2) Solution growth method

This method is under development and various methods have been applied as summarized in Table 23.2. At the melting point of Si (1420°C), carbon dissolves slightly.[34,35] as shown in Fig. 23.2 so that SiC can be precipitated at lower temperatures under supersaturation so that this method confers the ability to control the polytype.

In the past, the solution growth of SiC from Si melt and other binary and ternary alloy melts using Fe, Cr, Sc, Tb and Pr has been examined not only as the method of bulk crystal growth[38,62-66] but also as that of epitaxial growth.[67-74] These binary and ternary alloys were known to increase the growth rate from only 7 μm/h[72] in Si melt to 150 μm/h in Si-Sc melt.[74] Liquid phase epitaxy was thereafter developed by various authors[75-81], not only growing 6H-, 4H-SiC but even 2H-SiC.[80,81] The solution growth method has been reexamined as a bulk crystal growth method since 1996[82-85] because of its advantage of giving high crystal quality and the possibility of eliminating micropipe defects. The solubility of carbon in various melts is summarized as shown in Fig. 23.5.

When Si is melted in a graphite crucible under 1 atm Ar atmosphere, carbon is dissolved from the higher temperature part of the crucible and SiC is precipitated at the lower temperature part. By this method, the growth of 3C, 4H and 6H polytypes has been reported, but large crystals can not be grown even if seed crystals are used. Only 2-3 mm crystals with developed {0001} planes were obtainable in the past.[82] Müller et al.[83] however made a theoretical study of the possibility of increasing the growth rate considering various growth parameters and concluded that the growth must be performed at temperatures higher than 2000°C with forced convection of the solution.

When the solution temperature is increased in order to increase the solubility and the growth rate, silicon starts to evaporate so that crystal growth becomes difficult. To solve this problem, Hofman and Müller[84] performed the crystal growth under high pressure of 200 bar at 2000°C using the top-seeded solution growth (TSSG) method as seen in Fig. 23.6. They obtained 25 mm diameter single crystals at a growth rate of 0.2 mm/h. Ogura et al.[85] have grown 2H-SiC using Li as a flux at a temperature below 850°C.

Table 23.2 Solution Growth of Crystals

Method	Contents	Authors	Year	Ref.
SG	Growth of b-SiC from metal alloy solution	Halden	1960	62
SG	Growth of a-SiC from Cr solution	Griffith et al.	1964	63
SG	Growth of a-SiC from Si-Sc-C solution	Yakimova et al.	1975	65
SG	Growth from Si-Tb-SiC system	Tairov et al.	1986	66
SG	Growth from Si + transition metal solution	Ivantsov et al.	1998	82
SG	Calculation of solution growth rate	Müller et al.	1998	83
TSSG	Growth from Si-C solution with growth rate 0.05-0.2 mm/h	Hofmann et al.	1999	84
TSSG	Growth of 6H-SiC from Si-Ti-C solution with and without ACRT	Kusunoki et al.	2005 2010	88-91
TZM	Growth of 3C-SiC from a 4H-SiC seed crystal	Eid et al.	2006	111
VLS	Growth of 3C-SiC on 6H- and 4H-SiC seeds from Ge-Si solution fed by propane	Soudeidan et al.	2005-2008	103-107
GF-TSSG	Growth of 4H-SiC from Co-Si solution fed by propane	Caussende et al.	2006	110
TSSG	Growth from Si-(Co, Fe)-C solution	Yashiro et al.	2006	92
TSSG	Growth from Si-Ti and Si-Ge-Ti solution with ACRT	Tanaka et al.	2009	93
TSSG	Growth of 3C-SiC from Si-C, Si-Ti-C solution under N_2-H_2 atmosphere, with a cold crucible	Kusunoki et al.	2009	94-95
TSSG	Growth of 3C-SiC on bulk and epitaxial 3H-SiC seed with C face and Si face	Fujihara et al.	2009	96-98
TSSG	Growth of 3C-SiC with a high growth rate	Mercier et al.	2009	99
TSSG	Heavily p-doped 3C- and 6H-SiC from Al-Si-C solution	Mercier et al.	2009	100
TSSG	Growth of 4H-SiC from Si-Cr solution	Danno et al.	2010	101
TSSG	Polytype transition from 4H to 6H	Alexander et al.	2010	102

SG: Solution Growth, TSSG: Top-Seeded Solution Growth, GF-TSSG: Gas-Fed TSSG, VLS: Vapor-Liquid-Solid, TZM: Travelling Zone Melting, ACRT: Accelerated Crucible Rotation Technique

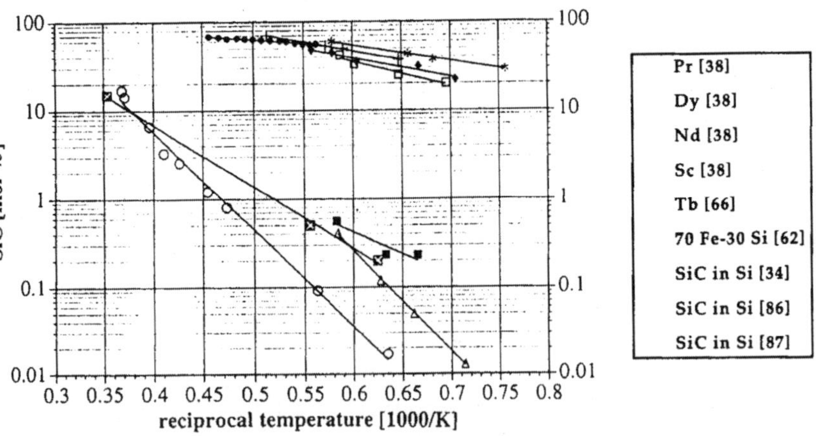

Fig. 23.5 Solubility of C in Si melt and in different other liquid Si-containing compounds versus temperature (reprinted from Ref. 84 with permission, copyright 1999 Elsevier).

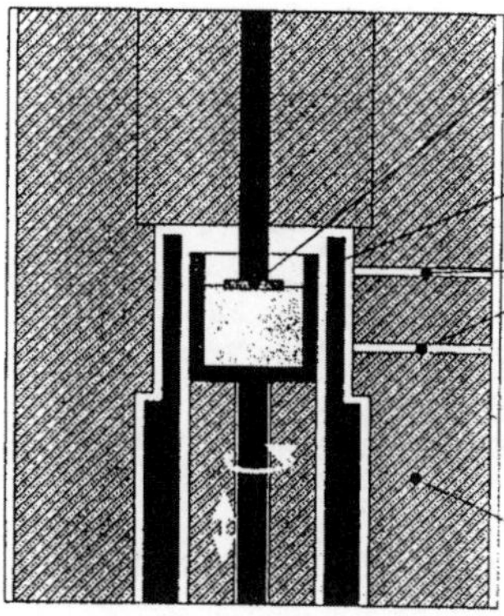

Fig. 23.6 Scheme of the experimental setup for SiC growth by the top-seeded solution growth (TSSG) technique (reprinted from Ref. 84 with permission, copyright 1999 Elsevier).

Kusunoki et al.[88-91] have grown 6H-SiC single crystals using Si-Ti-C ternary solution by the top-seeded solution growth (TSSG) method together with the accelerated crucible rotation technique (ACRT) and have obtained single crystals of diameter 50 mm and thickness 10 mm. The possibility of other solutions such as Si-(Co, Fe)-C [92] and Si-Ti-Ge-C[93] has also been studied.

The growth of 3C-SiC crystals on 6H-SiC seed crystals has been studied by adding Ge to the solution [93], by growing crystals under N_2-He atmosphere[94] and by using the cold crucible technique[95] as shown in Fig. 23.7. It was found that 3C-SiC single crystals can be grown and this is because nitrogen dissolved in the solution favors the 3C-SiC polytype formation or because the forced convection enhances the growth of 3C-SiC polytype. The growth of 3C-SiC crystals on 3H-SiC seed crystals (bulk grown crystal and CVD grown epitaxial crystal) with C face and Si face have been studied.[96-98] Mercier et al.[99] have grown 3C-SiC crystals at a growth rate of a few tens of μm/hr using a modified furnace. Heavily p-doped 6H- and 3C-SiC crystals have been grown using the Si-Al solution, where Al is incorporated as the p-type dopant.[100]

Danno et al.[101] have used a Si-Cr based metal solution and have grown 4H-SiC(0001) oriented single crystals 25 mm in diameter by the TSSG method at 2000°C. The maximum growth rate was 1120 μm/hr. Alexander et al.[102] have examined the polytype transition from 4H- to 6H-SiC during solution growth.

Soueidan et al.[103-107] have developed a method based on the vapor-liquid-solid (VLS) mechanism and have grown 3C-SiC single crystals on 6H-SiC seed crystal in the Ge-Si solution, providing carbon as a propane gas feed as shown in Fig. 23.8. They found that high-quality 3C-SiC single crystals can be obtained with reasonably low stacking fault (SF) densities. The mechanism whereby high quality 3C crystals can be grown even on a 6H seed crystal was explained as due to step flow crystal growth. They have also shown that 3C-SiC crystals can be grown on on-axis 4H-SiC seed crystals. These high quality 3C-SiC crystals have been used as the seed crystals for the continuous-fed PVT

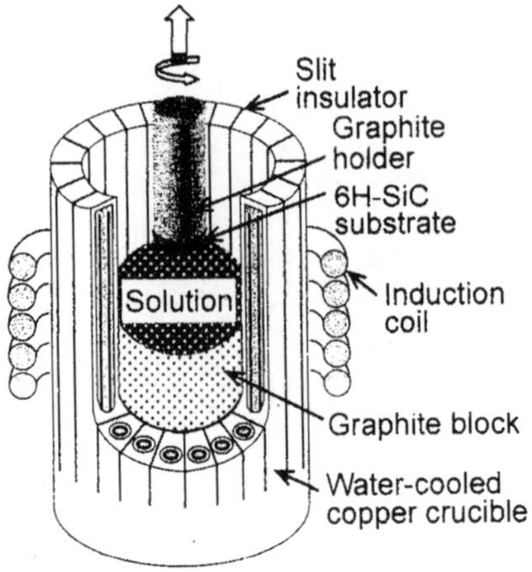

Fig. 23.7 Cold crucible technique for SiC growth by the top seeded solution growth (TSSG) technique (reprinted from Ref. 95 with permission, copyright 2009 Trans Tech Publlications).

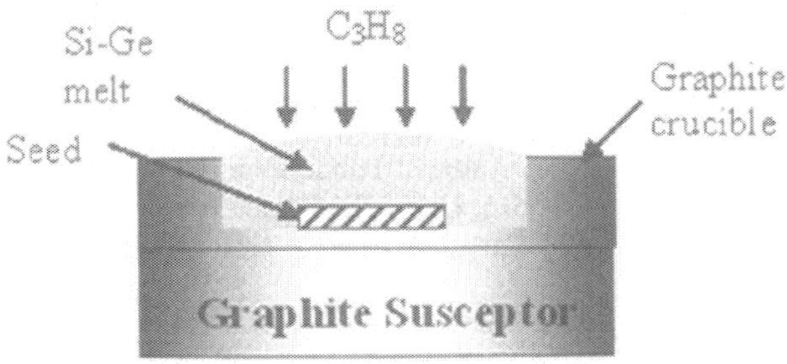

Fig. 23.8 Schematic drawing of the crucible used for the growth of 3C-SiC by the vapor-liquid-solid (VLS) method in Ge-based melts (reprinted from Ref. 104 with permission, copyright 2006 John Wiley & Sons).

(CF-PVT) method[108] as described later and for solution growth.[109]

Based on the VLS method, Chaussende et al.[110] have developed a novel method, gas fed top-seeded solution growth (GF-TSSG) as shown in Fig. 23.9. In a Co-Si solution, propane gas is fed in at 1350 °C and thick layers of 4H-SiC have been grown at 28 μm/h.

Eid et al.[111] have grown 3C-SiC single crystals by the travelling zone melting (TZM) method as shown in Fig. 23.10 using various seed crystals and have grown single crystals as shown in Fig. 23.11. The crystal quality was dependent on the polytype of seed crystal.

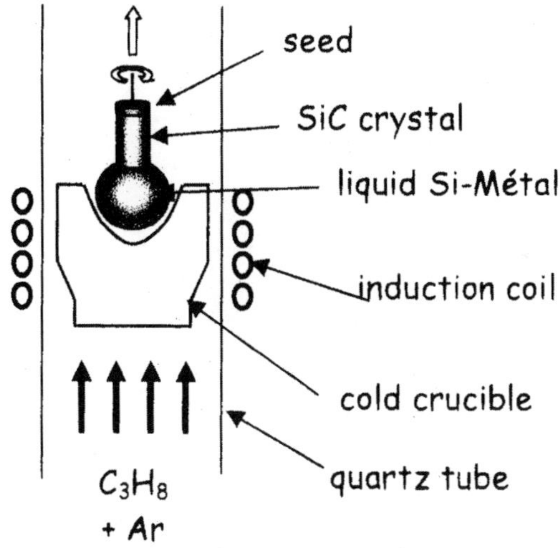

Fig. 23.9 Schematic view of the gas fed-top seeded solution growth (GF-TSSG) process (reprinted from Ref. 110 with permission, copyright 2006 Trans Tech Publlications).

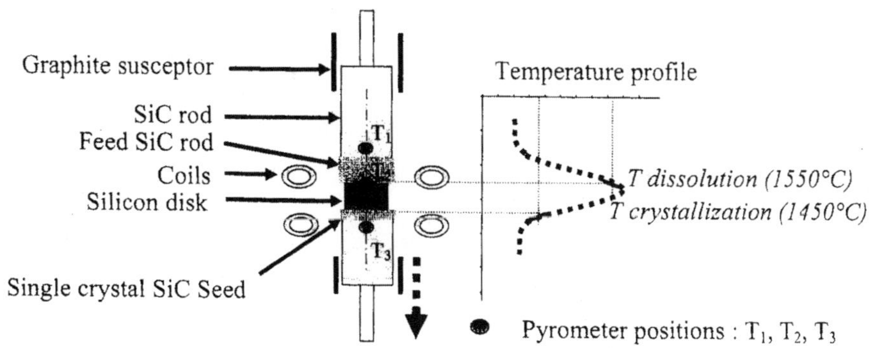

Fig. 23.10 Schematics of the travelling zone melting (TZM) set up (reprinted from Ref. 111 with permission, copyright 2006 Trans Tech Publlications).

(3) Sublimation method (Lely method)

Sublimation methods include the Lely method and the modified Lely method (often called as the modified PVT (MPVT) method or simply the PVT method) and many studies have been performed as shown in Table 23.3.

Sublimation of SiC becomes prominent at 2000°C under reduced pressure and at 2500°C at atmospheric pressure. Single crystals are fabricated by sublimation by the Lely method[112-118] in a furnace as shown in Fig. 23.12. SiC granules are charged in the furnace as a sintered cylinder or set directly between the partition and the graphite

Fig. 23.11 Typical as-grown travelling zone melting (TZM) system(reprinted from Ref. 111 with permission, copyright 2006 Trans Tech Publlications).

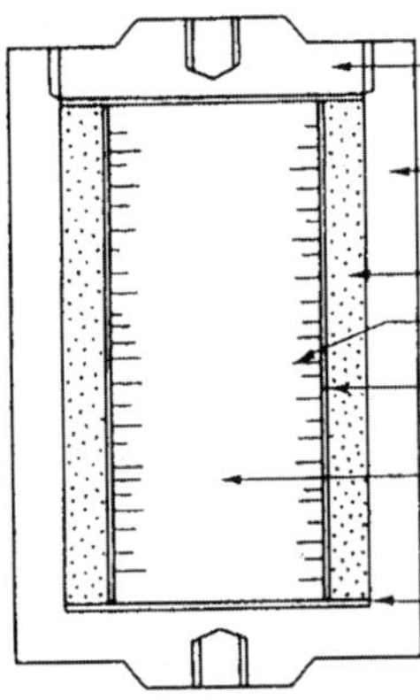

Fig. 23.12 Lely method (reprinted from Ref. 115 with permission, copyright 1968 The Japan Society of Applied Physics).

container. When the container is heated in Ar gas at 2200-2600°C, SiC sublimates through the partition and recrystallizes at the inner wall.

By this method, platelet 6H single crystals with a maximum length of 5-10 mm were obtainable, but larger crystals could not be grown.[113] Since they grow on a spontaneous nucleus, they show irregular shapes and their size is limited. Inomata et al.[114, 115] have grown 6H- and 15R-SiC crystals which are stable above 2000°C. The relative amount of 15R-SiC increased as the temperature decreased and 6H-SiC showed the opposite tendency. Lilov[116, 117] made a theoretical examination of the effect of the diameter of the holes in the graphite crystallization sleeve on the growth rate. He also examined the evaporation mechanism.[118]

Table 23.3 Crystal Growth by the Sublimation Method

Method	Contents	Authors	Year	Ref.
Lely	First Lely method	Lely	1955	112
Lely	Growth of SiC and the stability of polytypes	Inomata et al.	1968	114
MPVT	Growth using (0001) seed crystals 14 mm in diameter and 18 mm in length	Tairov et al.	1974 1981	21, 119- 121
MPVT	Growth of SiC from [11-20] and [-1100] seed crystals by sublimation	Shushlenina et al.	1977	122
MPVT	Growth of 6H-SiC crystals with 22 mmϕ × 24 mm	Ziegler et al.	1983	124
MPVT	Growth of 6H-SiC crystals with 33 mmϕ × 14 mm, study of growth kinetics	Koga et al.	1985 -1991	123, 125 -129
PVT	Crystal growth of 6H-SiC on 3C-SiC(100) substrates, polytype control by substrate transformation	Yoo et al.	1990- 1991	130, 131
MPVT	Polytype control between 6H and 4H-SiC	Kanaya et al.	1991	133
MPVT	Growth of 6H-SiC crystals with 25 mmϕ × 18 mm	Barrett et al.	1991	147
Lely	Growth mechanism by the Lely method	Lilov et al	1992 -1995	116- 118
MPVT	Effect of seed polarity on polytype formation	Stein et. al.	1992- 1993	294, 295
PVT	Crystal growth of 3C-SiC on 6H-SiC(0001) and 3C-SiC(111)	Furukawa et al.	1992- 1993	206- 207
PVT	Growth of 3C-SiC on 6H- and 3C-SiC seeds	Yang et al.	1994	208
MPVT	Growth of 6H-SiC with 60 mm diameter	Barrett et al.	1993	148
MPVT	Growth of 6H-SiC, 50 mm diameter, 75 mm length c-axis and a-axis direction	Hobgood et al.	1993- 1994	149- -151
MPVT	Defect reduction by the growth on (1-100) and (11-20) faces for 6H- and 4H-SiC growth	Takahashi et al.	1994- 1997	134- 137
PVT	Crystal growth of 3C-SiC on 6H-SiC(0001) and 3C-SiC(100)	Yang et al.	1994	79
MPVT	Growth of 75 mm diameter SiC	Brandt et al.	1994	147
MPVT	Vanadium-doped SI SiC	Hobgood et al.	1995	152
MPVT	Influence of face polarity of the seed crystal	Takahashi et al.	1995	141
MPVT	Study of the initial stage growth on the structure of SiC	Avramenko et al.	1996- 1999	191, 192
MPVT	Effect of various growth conditions on the defect formation	Hofmann et al.	1998- 1999	158, 159
MPVT	Near-equilibrium growth for micropipe-free 6H-SiC	Shulze et al.	1998	160
MPVT	Influence of the growth conditions and the crucible geometry on the growth front shape	Ankin et al.	1999	195
MPVT	Growth of 100 mm diameter 6H-SiC	Hobgood et al.	2000	153
MPVT	In-situ visualization of crystal growth by X-ray imaging	Wellmann et al	2000	161
MPVT	50 mm-diameter 4H-SiC, vanadium doped SI crystal	Augustine et al.	2000	166, 167
MPVT	In-situ observation of crystal growth by X-ray topography	Yamaguchi et al.	2000	168
MPVT	Enlargment by the cone-shaped platform	Bahng et al.	2000	171
MPVT	High quality 6H-SiC with no micropipe and with the EPD $4x10^3$ cm^{-2}	Oyanagi et al.	2002	176
MPVT	Surface morphology of 4H-SiC, effect of seed crystal orientation	Schulz et al.	2002	162
MPVT	Growth of 75 mm diameter SiC, reduction of micropipes and low angle grain boundaries	Ohtani et al	2002	142, 143
MPVT	Growth of 4H-SiC in the <03-38> directions	Nakayama et al.	2002	179

Table 23.3 Crytstal Growth by the Sublimation Method (continued)

Method	Content	Authors	Year	Ref
CF-PVT	Growth of 3C-SiC on 6H-, 4H- seeds by feeding TMS gas	Chausssemde et al.	2003-2016	213-218
MPVT	Reduction of SFs by the growth on off-angled [11-20] seed crystals	Ohtani et al.	2003-2005	138-140
MPVT	Increase of single crystal yield by temperature gradient tailoring	Herro et al.	2004	163
PVT	Growth of 3C-SiC on CVD-grown 3C-SiC on Si substrate and on 3C-SiC platelet seed	Sommerlroth et al.	2004-2007	209-211
MPVT	Reduction of dislocation density by repeated a-face (RAF) growth process	Nakamura et al.	2004-2006	181, 182
MPVT	Radial expansion growth and microcrack defects	Fujimoto	2004	198
PVT	Growth of 3C-SiC on 6H-SiC seeds	Freudenberg et al.	2005	212
CF-PVT	Growth by feeding the source as gases (SiH_4, C_3H_8)	Wellmann et al.	2005-2006	164, 165
MPVT	Growth of 100 mm diameter 6H- and 4H-SiC crystals	Gupta et al.	2006-	202, 203
MPVT	Growth of MP-free 75 mm diameter 4H-SiC crystals	Basceri et al.	2006	201
MPVT	Growth of 4H-SiC in the <03-38> directions	Furusho et al.	2006	180
MPVT	Optimization of the geometry of the source material	Grasza et al.	2006	187
MPVT	Addition of H_2 to the growth ambient	Fanton et al.	2006	188
MPVT	Heat treatment of the source SiC powder	Kim et al.	2006	189
MPVT	Crucible with a guide tube covered by a Ta foil	Ku et al.	2006	193
MPVT	Growth of low B concentration crystals	Fanton et al.	2006	196
MPVT	Transformation of 6H-SiC to 4H-SiC	Tupitsyn et al.	2007	297
CF-PVT	Growth of 3C-SiC on a VLS grown seed by feeding TMS	Chaussende et al.	2008	108, 219
MPVT	Growth of 100 mm diameter micropipe-free 4H-SiC	Leonard et al.	2009	154
MPVT	Growth of heavily nitrogen doped 4H-SiC	Ohtani et al.	2009	142
MVPT	Growth of 100 mm diameter 4H-SiC and chracterization of various defects	Nakabayashi et al.	2009, 2010	145, 146
MPVT	Growth of 6H- and 4H-crystals along rhombohedral [01-1n] directions	Flip et al.	2009	183
MPVT	Improvement of the furnace design	Tupitsyn et al.	2009	190
MPVT	Growth under quasi-equilibrium conditions	Tymicki et al.	2009	194
MPVT	Growth of 75 mm 6H-SiC	Makarov et al.	2010	204
CF-PVT	Spontaneous seeding of 3C-SiC	Galuben et al.	2010	223
MPVT	Growth of nonpolar SiC with m-plane and a-plane	Yeo et al.	2010	184
MPVT	Gowth of 4H-SiC on off-angled 6H-SiC seed crystals	Tymicki et al.	2010	185
MPVT	Growth of 100 mm diameter 4H-SiC	Straubinger et al.	2010	205

PVT: Physical Vapor Transport, MPVT: Modified PVT (Modified Lely), CF-PVT: Continuous Feed PVT

(4) Sublimation method (modified Lely method)
- Growth of 6H- and 4H-SiC single crystals

For growing larger crystals, the modified Lely method in which a seed crystal is set in a graphite crucible was first developed by Tairov et al.[21,119-121] and Shushlebina et al.[122] This method was subsequently developed by many researchers as described below, first for blue LEDs and later for substrates for GaN based blue and white LEDs and for various electronic devices. This method is presently used for industrial production and the diameter has risen to 150 mm.

This method is one of the open PVT methods with seed crystals (Fig. 23.13) so that the modified Lely method is often called the modified PVT (MPVT) method or simply the PVT method. Conventionally, a (0001) plane SiC seed crystal is used in this method. SiC powders are set in the furnace and heated by a RF heater or a resistive heater under Ar gas atmosphere of 100-10000 Pa. The raw materials are heated at 2200-2500°C and the seed crystal is heated at 2100-2400°C. Since 6H and 4H polytypes are grown spontaneously, it is important to control the polytype during crystal growth. Crystals are grown at the rate of 1-4 mm/h depending on the inner pressure. From the data on growth rate and temperature, an activation energy of 124-170 kcal/mol for the sublimation of SiC is obtained.[123] The sublimation process of this method is explained schematically in Fig. 23.14. SiC sublimes as Si, Si_2C and SiC_2 according to the following equations.[52]

$$P_{Si_2C} = 2.85 \times 10^{2}\, e^{(-1.79 \times 10^{4}/T)}\, P_{Si} \tag{23.1}$$

$$P_{SiC_2} = 9.41 \times 10^{28}\, e^{(-14.5 \times 10^{4}/T)} / P_{si} \tag{23.2}$$

C also sublimes from the crucible and these species react and transport to the seed crystal where SiC is grown under supersaturation due to the pressure difference between the sublimed species and the the equilibrium vapor pressure at the seed temperature.

Ziegler et al.[68] used a seed crystal in the lower part of the furnace as shown in Fig. 23.13(a) and succeeded in growing 6H-SiC single crystals with a yield greater than 85%. During heating up, the Ar overpressure was held at 1 atm and it was lowered to 2 mbar

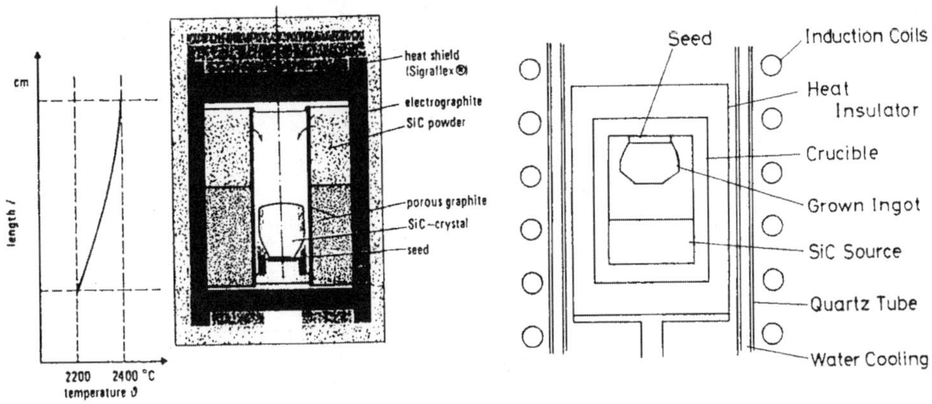

(a) (b)

Fig. 23.13 Crucible assembly for the growth of SiC crystals by the modified Lely method. (a) the seed crystal at the bottom of the furnace cavity (reprinted from Ref. 147 with permission, copyright 1991 Elsevier), (b) the seed crystal at the bottom of the furnace cavity (reprinted from Ref. 135 with permission, copyright 1994 Elsevier).

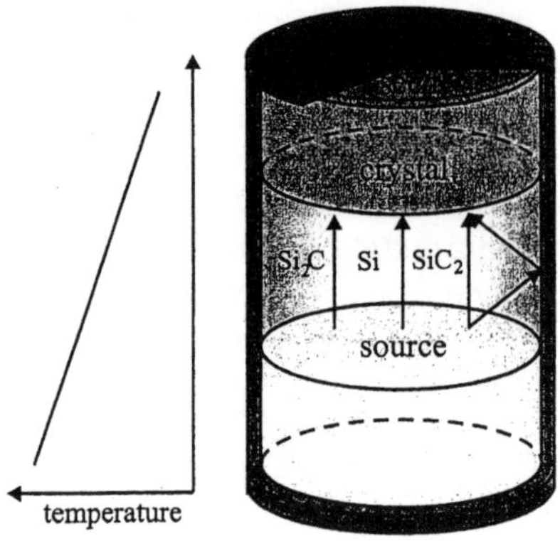

Fig. 23.14 Schematic representation of a crucible configuration for seeded sublimation growth and the mass transfer processes of the dominant gaseous Si-C species inside the crucible, including possible chemical reactions with the crucible wall (reprinted from Ref. 54 with permission, copyright 2000 Elsevier).

after the growth temperature was reached to start the crystal growth. They have grown diameter 20 mm, length 24 mm single crystals on (0001) seed crystals.

Koga et al.[123,124-129] have grown 6H-SiC single crystals with a view to fabricating blue LEDs. They have grown single crystals with of diameter 30-33 mm and thickness 14 mm with the seed crystal at the top of the crucible. They found that when crystals are grown on the (0001) seed substrate, the growth proceeds not only in the direction of the c-axis but also in that of the a-axis so that the grown diameter is enlarged 1.4 times (Fig. 23.15).

Yoo et al.[130,131] have developed a method of growing 6H-SiC by the phase transformation of 3C-SiC crystals grown by the sublimation method on 3C-SiC(100) substrates

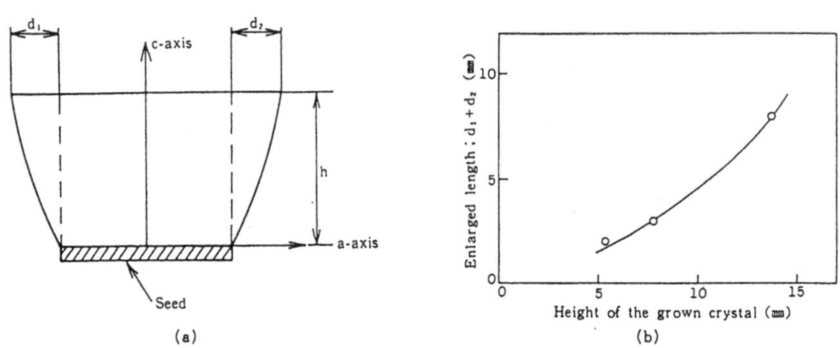

Fig. 23.15 Degree of extension in the a-axis direction of the grown crystal. (a) Schematic cross section of the grown ingot crystal in the c-axis direction (d1 + d2: the extension length in the crystal), (b) relation between the height of the grown crystal and the enlarged length (d1 + d2) of the diameter (reprinted from Ref. 125 with permission, copyright 1985 The Japan Society of Applied Physics).

prepared by the CVD method[132] and by successive high temperature annealing. They were able to grow 6H-SiC (01-14) bulk substrates up to 15 mm in diameter and 6 mm in length. The orientation relationship between 3C-SiC and transformed 6H-SiC was analyzed and the transformation mechanism was discussed.

The Nippon Steel group has conducted much work on the development of SiC crystal growth. Kanaya et al.[133] have grown selectively 6H- or 4H-SiC single crystals by optimizing the seed temperature and the ambient atmosphere pressure to control the occurrence of polytypes. They have used a furnace in which the seed crystal is set at the top of the crucible as shown in Fig. 23.13(b).

Takahashi et al.[134-137] have studied the relationship between the orientation of crystal growth and the occurrence of micropipes. They found that micropipes (cored screw dislocations as explained in Sec. 23.3.3(1)) can be eliminated when the crystal is grown perpendicular to the SiC basal plane (0001), that is oriented to [1-100] and [11-20]. This is because growth in these orientations does not allow the screw dislocation of the seed crystal to propagate to the growing crystal. Growth in these orientations however does promote the occurrence of stacking faults (SFs). These SFs could however be reduced when SiC crystals were grown using off-oriented [11-20] seed crystals.[138-140]

Takahashi et al.[141] have also studied the effect of the seed face polarity on the electrical properties of 6H-SiC crystals. It was found that undoped crystals grown on the C-face were n-type while those on the Si face were highly resistive p-type. This is attributed to the fact that the Si-face preferentially incorporates boron and inhibits nitrogen incorporation. Heavily nitrogen-doped 4H-SiC crystals have been grown by lowering the growth temperature. A nitrogen concentration of 5.3×10^{19} - 1.3×10^{20} cm^{-3} was achieved and almost no indication of 3C inclusions or stacking faults was detected.[142]

They also developed the growth of large diameter crystals up to 100 mm, together with the reduction of various defects, such as micropipes, low angle grain boundaries and SFs.[143-146]

The Westinghouse and Cree group have also been leaders in the development of SiC crystal growth. Barret et al.[147] began the growth of 6H-SiC with a furnace similar to that of Ziegler et al. as shown in Fig. 23.13(a) using a Lely grown platelet measuring 5 x 8 mm as the seed crystal. They were able to grow 6H-SiC single crystals measuring 33 mm in diameter and 18 mm in length as shown in Fig. 23.16. They also have shown that the growth rate is a function of the growth temperature, the system pressure and the distance between the sublimation source and the surface of the growing crystal (Figs. 23.17, 18).[148] Hobgood et al.[149-154] increased the growth size up to 40 mm diameter using a 25 mm diameter wafer seed and then increased the crystal size step by step from 50 mm diameter to 75 mm and 100 mm for 6H- and 4H-SiC. A diameter of 150 mm was recently achieved.[155, 156] This is preferable for power devices since they need large diameter substrates of at least 150 mm from the viewpoint of production costs to compete with Si power devices.

The group at Erlangen University has done a lot of work on the fundamentals of SiC crystal growth. Hofmann et al.[158] have made a theoretical study of the growth rate and its possible correlation with defect formation and compared it with the experimental result. They found that the growth rate is limited by diffusion, kinetics and crystal

Fig. 23.16 SiC boule grown by seeded sublimation vapor transport (reprinted from Ref. 147 with permission, copyright 1991 Elsevier).

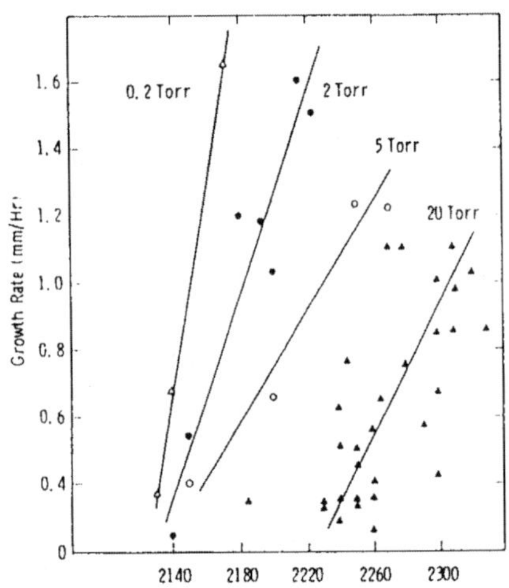

Fig. 23.17 Crystal growth rate increases with temperature (at constant source to seed ΔT) and decreased with system pressure. Below 20 Torr, vapor transport is diffusive, while scatter in the data at 20 Torr indicate gaseous convective effects (reprinted from Ref. 148 with permission, copyright 1991 Elsevier).

length. They have found that the main origin of micropipes is related to second-phase formation, especially the occurrence of C inclusions.[159,204,205] They also found that the micropipe density depends on the axial temperature gradient.

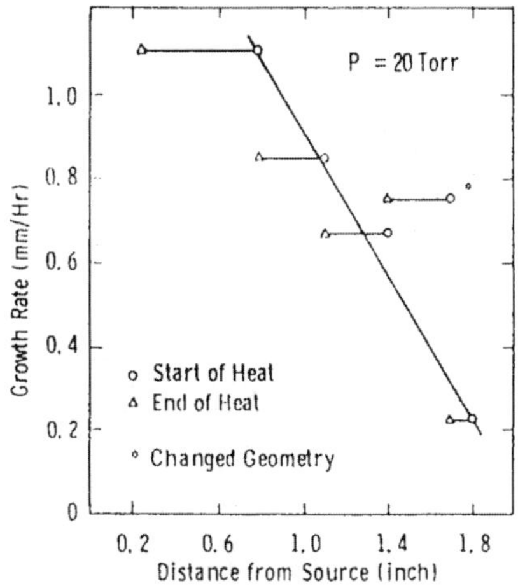

Fig. 23.18 The crystal growth rate is a strong function of distance between the sublimation source and the surface of the growing crystal (reprinted from Ref. 148 with permission, copyright 1991 Elsevier).

Schulze et al.[160] have studied the near-equilibrium growth of micropipe free 6H-SiC single crystals. They found that the maximum average thermal gradient inside the growth furnace must be lower than 5 K/cm. A gradient of 7.5 K/cm resulted in heavy defect formation. Micropipe-free 6H-SiC single crystals were grown using the C face and the Si face of 6H-SiC Lely platelets without micropipes as the seed crystals.

Wellmann et al.[161] first tried in situ observation of crystal growth by transparent X-ray imaging. The crystal shape and the SiC source material degradation were monitored on-line during growth and trends in the the growth rate were analyzed.

Schulz et al.[162] have studied the mass transport and the morphology of 4H-SiC crystal growth. They studied especially the very early stages of crystal growth and examined the influence of the seed thickness and the seed orientation on the morphology of grown surfaces and the formation of defects.

Herro et al.[163] examined the effect of the thermal field on the enlargement of crystals and on defect formation. They have shown that tailoring the temperature gradient is effective in growing large diameter crystals. They have also shown that a conical-shaped seed holder is more effective than a cylindrical one in terms of single crystal growth yield.

Wellmann et al.[164,165] have developed an additional gas pipe line for the modified Lely method as shown in Fig. 23.19, in order to control the gas phase composition. Small amounts of the Si- and C-containing gases silane (SiH_4:H_2-1:10) and propane (C_3H_8) as well as doping gases were introduced in the furnace using the gas pipe. Wellmann et al.[165] have studied the influence of the additional gases (He, Ar. N_2 and propane) on the growth system, experimentally and theoretically using numerical modelling.

The Northrop Graumman group[166, 167] have grown 50 mm diameter 4H-SiC crystals

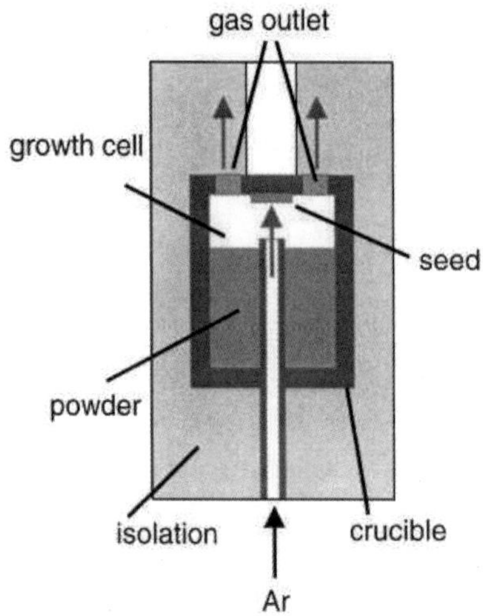

Fig. 23.19 Sketch of the modified physical vapor transport (MPVT) growth configuration. A growth pipe is used for additional species feeding and for fine tuning of the gas phase composition in front of the SiC crystal growth interface (reprinted from Ref. 165 with permission, copyright 2002 Elsevier).

using the same furnace arrangement as Ref. 150, with micropipe density less than 100 cm^{-2}, based on the numerical simulation of the temperature distribution in the furnace. They have also achieved semi-insulating 4H-SiC by vanadium doping similarly to Hobgood et al.[152]

The Advanced Industrial Science and Technology (AIST) group has done unique work on SiC crystal growth. Yamaguchi et al.[168,169] have developed an in-situ X-ray topography method (Fig. 23.20). Nishizawa et. al.[170] have controlled the thermal field to grow high quality single crystals and Bahng et al.[171] have studied the mechanism of the enlargement of SiC single crystals during growth. Kato et al.[172] applied this method to the in-situ observation of SiC crystal growth. They found that large defects such as micropipes and macrodefects originate in the accumulation of dislocations in the initial stage.

Kato et al.[173] have invented an in-process etching method as shown in Fig. 23.21 by which the seed crystal is self-etched by sublimation and the seed crystal shape is controlled in a way that the temperature distribution is automatically accommodated in an isothermal profile where crystal growth starts with a lower temperature gradient. Nishizawa et al.[174] have performed a numerical simulation of heat and mass transfer and invented a method to control the gas flux by using an oblique inner guide tube[175] as seen in Fig. 23.22. They have found that the gap between the seed crystal and the guide-tube was the important parameter for growing single crystals uncontaminated by polycrystals. Oyanagi et al.[176] reported the growth of high quality 6H-SiC single crystals which contained no micropipes and had an EPD of 4×10^3 cm^{-2}. These developments are summarized in Refs. 177 and 178.

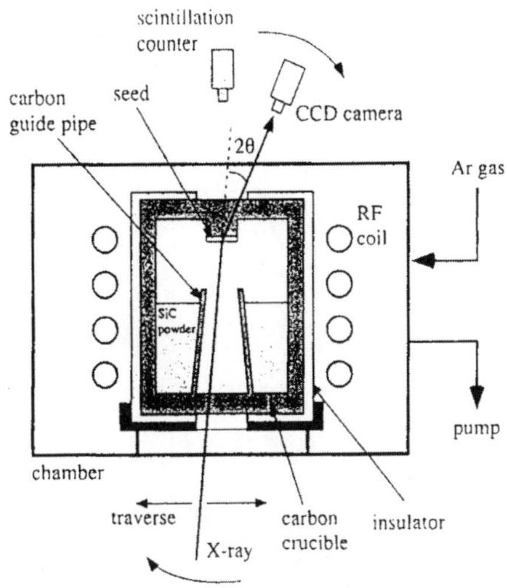

Fig. 23.20 X-ray topographic system for in-situ observation of SiC single-crystal growth by the modified Lely method. X-ray beam irradiates the grown crystal from below through a carbon guide pipe located in the crucible and is detected by a CCD camera and a scintillation counter (reprinted from Ref. 172 with permission, copyright 2001 Elsevier).

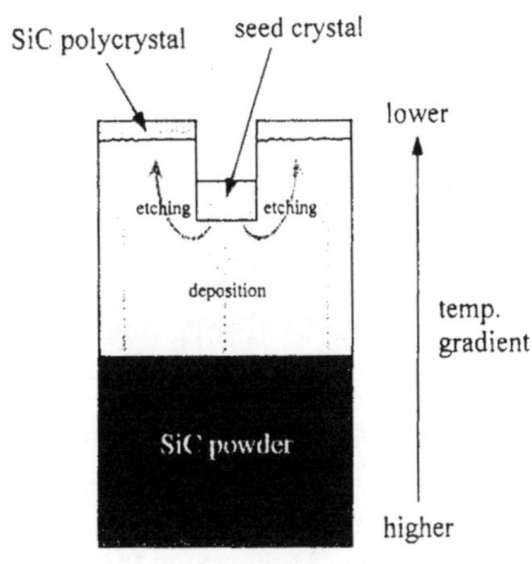

Fig. 23.21 Schematic figure of sublimed gas flow inside the crucible for SiC growth. The gas flow is carried from higher to lower temperature region following the supersaturation gradient. As the seed crystal is located between the highest and lowest temperature portions in the crucible, the gas not only deposits on the seed crystal but also sublimes from its surface to the other part at lower temperature. The balance of deposition and sublimation decides the growth or etching rate of the crystal (reprinted from Ref. 173 with permission, copyright 2001 Elsevier).

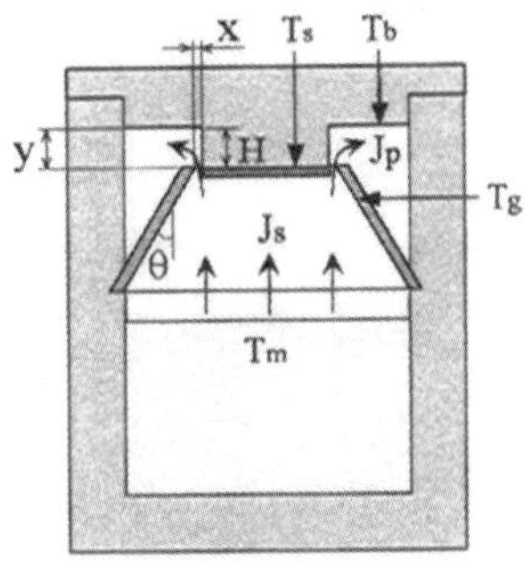

(a) Crucible design

(b) SiC grown crystal

Fig. 23.22 (a) Taper wall crucible structure for single crystal/polycrystal separation and (b) the grown crystal (reprinted from Ref. 177 with permission, copyright 2002 The Japanese Association for Crystal Growth).

- *Growth on various orientations*

As mentioned already, Takahashi et al.[134-137] have grown crystals with orientations perpendicular to the SiC basal plane (0001), such as [1-100] and [11-20] and found it to be effective in reducing the density of micropipes.

Nakayama et al.[179] and Furusho et al.[180] have grown <03-38> oriented 4H-SiC using a <03-38> seed crystal cut from a <0001> 4H-SiC single crystal. They obtained single crystals free from micropipes and stacking faults since these defects did not propagate in the <03-38> direction.

For the reduction of the densities of defects such as micropipes, dislocations and stacking faults, the repeated a-face (RAF) method was invented by Nakamura et al.[181, 182] In this process as shown in Fig. 23.23, in a first step, a-face {11-20} or {1-100} growth is performed on the a-face seed crystal. At the second step, a-face {11-20} or {1-100} growth perpendicular to the first a-face growth direction is performed using the {11-20} or {1-100} plane seed crystal cut from the crystal grown by the first step. At the third step, c-face {0001} growth is performed using the {0001} seed crystal cut with off-angle of 10° from the crystal grown by the second step. By this method, it has been proved that the defect density can be greatly reduced as shown in Fig. 23.24.

The effectiveness of growth on other planes than (0001) has also been shown by several authors. Filip et al.[183] have grown 6H- and 4H-crystals along rhombohedral [01-1n]

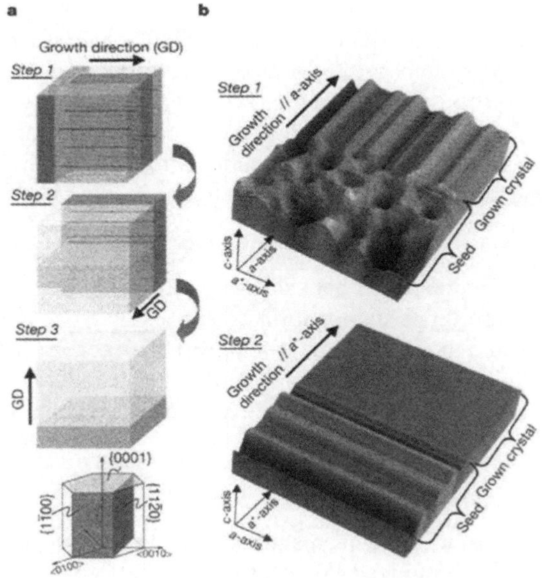

Fig. 23.23 Growth sequences of repeat a-face (RAF) process (reprinted from Ref. 181 with permission, copyright 2004 Nature).

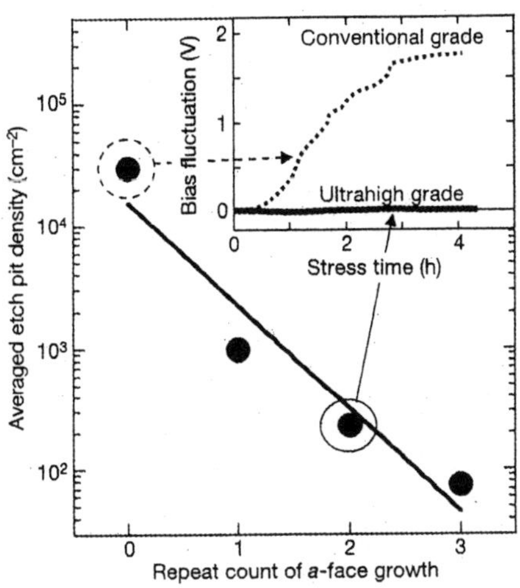

Fig. 23.24 Etch pit density (EPD) decay curves versus repeat count of a-face growth and result of current stress testing. The insert shows forward bias fluctuation of pin diodes under the stress with a constant current density (broken line; conventional substrate, solid line; repeat a-face (RAF) substrate) (reprinted from Ref. 181 with permission, copyright 2004 Nature).

directions and found that the formation of natural rhombohedral faces in the SiC habit gives rise to lattice relaxation via the formation of basal plane dislocations (BPDs) which makes it possible to grow micropipe (MP) free crystals.

50 mm diameter crystal growth using m-plane (1-100) and a-plane seeds (11-20) has been preformed by Yeo et al.[184] and their crystal qualities were compared. It was found that the quality of SiC single crystals grown on a-plane seed was better than on m-plane seed.

Tymicki et al.[185] have grown 4H-SiC using 8° off-axis C-face 6H-SiC seeds inclined toward [11-20] with the open seed backside and also examined the effect of cerium doping on the 4H-SiC polytype stabilization.

Jones[186] has calculated the orientation relationship between nH hexagonal structures from the viewpoint of low surface energy, closed packed, conducive to growth hexagonal semi-polar planes. The nH$(3\text{-}302n)_H$ and $(3\text{-}30\text{-}n)_H$ hexagonal planes were identified.

- *Improvement of growth conditions*

Various other growth factors have been studied to improve the quality of grown crystals. Grasza et al.[187] have optimized the geometry of the source material which led to the termination of micropipes on microfacets formed on the crystallization front during growth. Fanton et al.[188] have added H_2 to the inert growth atmosphere, which enhances the carbon transport resulting in reduced N incorporation and electron trap concentrations. Kim et al.[189] have studied the thermal treatment of SiC powder source materials prior to crystal growth which led to the lowest micropipe densities. This is attributed to fewer Si droplets in the source powder which raises the Si vapor pressure in the initial growth stage.

It is known that growth parameters especially temperatures in the furnace change significantly thorough the growth period. Tupitsyn et al.[190] have designed a new resistive furnace aimed at achieving better controllability of the growth process during long-term growth. The crucible design was also improved, for example a double wall or coaxially arranged cylinder type crucible[191,192] and a crucible using a guide tube and a tantalum foil[193] which were effective in reducing micropipes and dislocations.

Ankin et al.[194] improved the growth conditions in such a way that a flat growth surface can be achieved under quasi-equilibrium conditions which resulted in reduced low grain boundary (LGB) density and in the enlargement of crystals. Tymicki et al.[195] have applied an extremely low growth rate of 0.05-0.2 mm/h in order to cause growth to occur under quasi-equilibrium. They were able to achieve the growth of 6H-SiC single crystals in which a micropipe free area could be realized in the center part of grown crystals by a micropipe bonding phenomenon.

- *Purity control*

Since the growth temperature is extremely high in the modified Lely method, impurities such as sulphur, boron and metallic elements come from carbon materials and also from the source materials, silicon and carbon. However, the use of purifying carbon/

graphite materials and high purity source materials can reduce these impurities in the grown crystals down to 0.1 ppm.[54,58]

It is known that residual nitrogen, acting a shallow donor, stems from the source material and from the reactor material.[123,148] The reduction of residual nitrogen is important in order to achieve high purity semi-insulating SiC (HPSI-SiC) in which the nitrogen concentration must be lower than the concentration of the intrinsic defects acting as deep acceptor levels. Purification has been achieved to the extent that nitrogen concentration is as low as 3×10^{14} cm^{-3} [58] and HPSI-SiC crystals were able to be obtained (Sec. 23.3.4(3)).

The origin of boron contamination is the SiC material, the graphite furnace component and the carbon insulation. Fanton et al.[196] have reduced the boron concentration down to 2.4×10^{15} cm^{-3}, by using a high purity SiC source, refractory metal carbide coatings to commercial graphite and high purity pyrolytic graphite. Reducing residual boron is also important in achieving HPSI-SiC crystals since the boron concentration must be lower than residual donor and deep donor concentrations (Sec. 23.3.4(3)).

It has also been realized that the graphite crucible itself sublimes carbon as was proved experimentally using a carbon isotope. Rohmfeld et al.[197] have used a source material with Si and ^{13}C isotope with a stoichiometric charge and examined the carbon concentration in the grown crystal. It was found that ^{13}C increases up to about 40 % in the initial stage of the crystal growth and reaches the equilibrium concentration of 27 % after 1.5 mm growth.

-Enlargement of crystal diameter

As mentioned already, the crystal diameter is enlarged during crystal growth radially from the seed crystal (Fig. 23.25). Using this process, the diameter can be increased lot by lot in growth from small Acheson or Lely seed crystals.

There were however various pitfalls, such as the increase of various defects such as micropipes, dislocations, pinholes [198], and small angle grain boundaries (SAGBs)[195] during the enlargement. The increase of these defects is especially enhanced at the periphery of the growing crystals and polycrystallization takes place.[31, 175] Fujimoto et al.[199] have found that the radial expansion growth of 6H-SiC crystals induces crevasse-like defects in the expansion region of the crystal and claimed that subgrain boundaries existing in the seed crystals trigger the subsequent formation of minor facet planes, giving rise to the crevasse defects.[68, 120] In order to solve this problem, various solutions such as growth under quasi equilibrium conditions [194,195], in process etching [173], single crystal/polycrystal separation by crucible structure [175], and conical-shaped seed holder structure[163] have been developed.

Large diameter single crystals with low defect densities are crucial in the industrial production of SiC from the viewpoint of the production cost of LED fabrications and especially for power devices where SiC has to compete with Si devices whose cost is already low. Applying the above mentioned technologies in the modified Lely method, the diameter has been increased year by year, from the 1 cm diameter Acheson seed crystal to 75 mm and 100 mm. The growth of 75 mm and 100 mm 6H- and 4H-SiC single crystals has been reported from various affiliations and 75 mm and 100 mm substrates are now commercially available from various manufacturers. The progress in the diameter of 6H-SiC crystal is shown in Fig. 23.25. Even 150 mm diameter crystals have been successfully grown, as recently announced.[155,156]

Hobgood et al.[153] reported the growth of 100 mm diameter 6H-SiC single crystals with residual impurities of 10^{15} cm^{-3}. Powell et al.[200] and Sumaeris et al.[58] have reported 100 mm 4H-SiC single crystals with micropipe densities coming down from 30 to 16 cm^{-2}. Basceri et al.[201] have succeeded in growing MP-free conducting 50 mm and 75 mm diameter 4H-SiC single crystals, by optimizing process instabilities which resulted in the elimination of polytype inclusions. Leonard et al.[154] have reported the growth of MP-free 100 mm 4H-SiC singe crystals based on the technology described in Ref. 201. When this technology was applied to 75 mm 4H-SiC single crystal growth, a density of screw dislocations of 175 cm^{-2} was achieved, which is much lower than the conventional 2-4 $\times$ 10^3 cm^{-2}.

Ohtani et al.[143] have achieved the growth of crystals 100 mm in diameter, by paying attention to crucible design and optimizing the thermal profile in the growth cavity thus minimizing thermal stress. Crack-free and low micropipe density 4H-SiC single crystals 100 mm in diameter have been grown by optimizing the temperature distribution thereby reducing the tensile stress during growth.[144]

Gupta et al.[202,203] have reported the growth of 100 mm diameter V-doped semi-insulating 6H-SI SiC single crystals and n-type 4H-SiC single crystals. n-type 75 mm 4H-SiC crystals are grown with nitrogen doping with resistivity 0.018 Ωcm and with micropipe density about 1 cm^{-2}. 100 mm diameter 6H-SI SiC single crystals have micropipe density 2-8 cm^{-2}, and dislocation density of 3-4 $\times$ 10^4 cm^{-2} and a FWHM of the X-ray rocking curve of 50-100 arcsecond.

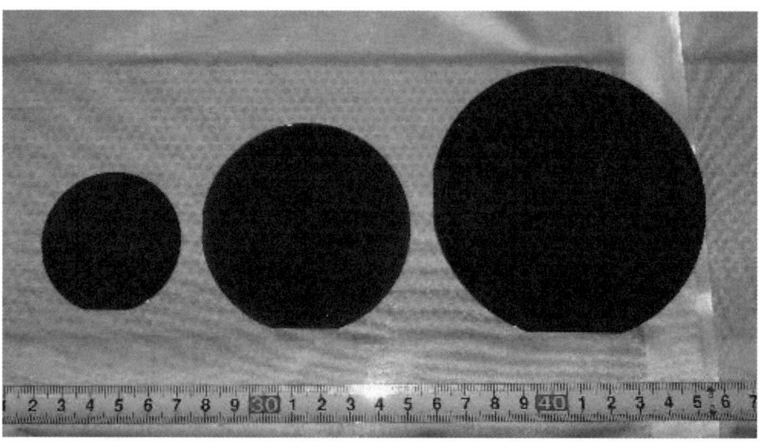

Fig. 23.25 Commercially available 100 mm diameter 4H-SiC substrate together with smaller diameter (50 mm, 75 mm) 4H-SiC substrates (courtesy of Prof. J. Suda at Kyoto University).

Marakov et al.[204] reported the growth of 75 mm diameter 6H SiC single crystals with an MP density of 10-40 cm^{-2}, dislocation density of less than 5 x 10^3 cm^{-2}, and FWHM of the X-ray rocking curve of 25-50 arcsecond and a carrier concentration of 1-5 × 10^{18} cm^{-3}.

Straubinger et al.[205] have reported 100 mm diameter 4H-SiC single crystals without micropipes, with the dislocation density in the range of 10^4 cm^{-2}, and FWHM of the X-ray rocking curve of 20 arcsecond. The resistivity of substrates was down to 16 mΩ·cm without risk of stacking fault formation during annealing at 1150°C.

The results described above are as published each year and the quality has been improving year by year, including the diameter and the production cost. The price was reported to have dropped substantially from 650 US$/$cm^2$ (1994) to 16 US$/$cm^2$ (2004).[58]

- *Growth of 3C-SiC single crystals*

Furukawa et al.[206,207] have grown 3C-SiC(111) single crystals on 6H-SiC(0001) or 3C-SiC(111) seeds. The seed temperature was 1750-1800°C and the source temperature was 2250-2300°C. The growth rate was 0.8 μm/h and 3C-SiC single crystals of thickness 4 mm could be grown. The grown crystals on the 6H-SiC(0001) seed crystal did not show any double positioning boundaries (DPBs). Yang et al.[208] have also studied the growth of 3C-SiC using 6H-SiC(0001) or 3C-SiC(100) seed crystals, though the grown crystal was polycrystalline.

The PVT method was used to grow 3C-SiC crystals on 3C-SiC layers grown on Si substrates by the CVD method[209] and on 3C-SiC platelets.[209,210] It was found that 6H-SiC inclusions were generated in the 3C-SiC grown crystals and this polytype conversion was studied by micro Raman spectroscopy.[211]

Freudenberg et al.[212] have grown 3C-SiC on 6H-SiC seed crystals and double positioning twins and double positioning boundaries (DPBs) have been observed. By optimizing the seed temperature, silicon excess stream and supersaturation, growth in one orientation was made to dominate.

Chaussende et al.[213,214] have developed the continuous feed physical vapor transport (CF-PVT) method, in which Si could be fed using a gas flow of trimethylsilane (TMS) diluted in argon as shown in Fig. 23.26. The supersaturation surrounding of the seed crystal can be precisely controlled. In order to grow DPB-free 3C-SiC single crystals, this method was applied to grow (111)3C-SiC crystals on (0001)6H-SiC seed crystals[215], (0001)4H-SiC seed crystals[214] and (100)3C-SiC grown on Si substrates. Mass transfer during CF-PVT growth has been observed by in situ X-ay imaging.[216] The grown crystals were evaluated by transmission electron microscopy (TEM)[216,217], and electron back scattering diffraction (EBSD) to examine grown-in defects such as microtwins, stacking faults, dislocations and DPBs.[218,219] The growth of 3C-SiC by CF-PVT has been reviewed and the growth conditions of 3C-SiC on 3C-SiC seed crystals and the stability of the growth process have been discussed.[220]

Chaussende et al.[108] also have used a 3C-SiC crystal grown by VLS on a 6H-SiC substrate as the seed crystal to grow 3C-SiC crystals by the CF-PVT method. By this two-step process, twin boundaries (TB) could be eliminated and a growth rate of 0.2

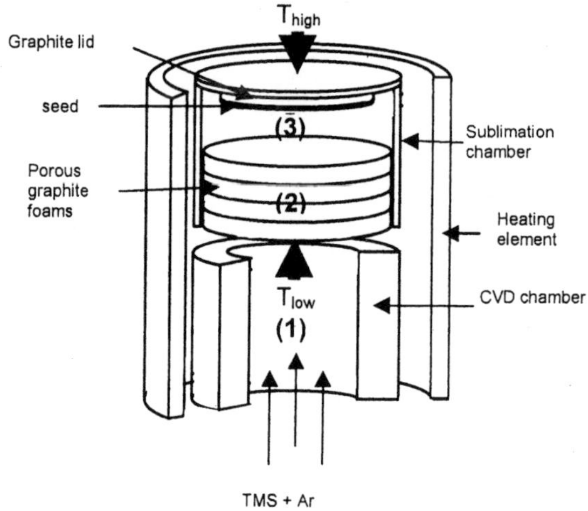

Fig. 23.26 Continuous feed physical vapor transport (CF-PVT) crucible setup. Pyrometric measurement. (1) Feeding zone, (2) transfer zone and (3) sublimation zone (reprinted from Ref. 216 with permission, copyright 2006 Trans Tech Publlications).

mm/h has been achieved. The VLS process itself is explained in Sec. 23.3.1(2). The grown crystals were evaluated by high resolution X-ray reciprocal space mapping (RSM)[221] and birefringence microscopy.[222]

Spontaneous nucleation of 3C-SiC single crystals was studied using the CF-PVT method by Galben-Sandulache et al.[223] The necking procedure was applied during PVT growth and a single 3C-SiC crystal seed could be obtained as shown in Fig. 23.27.

Sun et al.[224] have further improved the growth conditions of the CF-PVT method and have succeeded in growing 3C-SiC crystals of diameter 7.0 mm and height 3.3 mm and a high growth rate of 0.8 mm/hr. The possibility of the growth of 3C-SiC single crystals was reviewed by Ferro.[225]

- *Modelling and simulation*

Various modeling and simulation techniques have been studied by various researchers in order to understand the physical phenomena occurring especially during crystal growth of SiC by the modified Lely method. Various factors affecting the crystal growth phenomena have been considered, such as the electromagnetic field produced by RF heating, heat generated in the graphite susceptor radiation heat transfer, temperature distribution and growth kinetics.

Pearson et al.[226] theoretically examined the surface process from the viewpoint of crystal growth from the vapor. Hofmann et al.[159, 227] have modelled the temperature distribution at the growth temperature of 2573 K and the system pressure up to 3500 Pa.

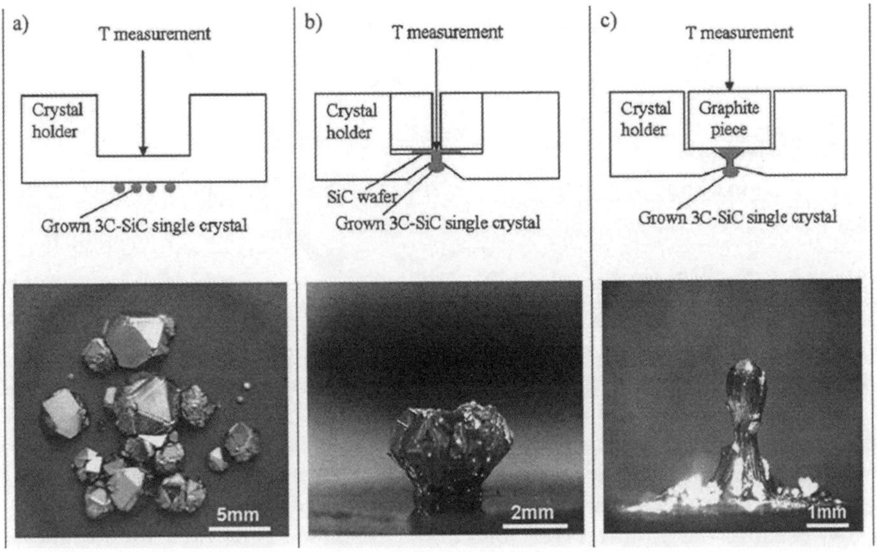

Fig. 23.27 3C-SiC grown by self-nucleation: a) on the standard crystal holder at 2070°C for 5 hrs of growth; b) with the cone-shaped cavity at 2000°C for 2 hrs of growth ; c) with the double cone-shaped cavity at 1960°C for 5 hrs of growth (reprinted from Ref. 223 with permission, copyright 2010 Trans Tech Publlications).

Raback et al.[228] have presented a model for the growth rate and estimated the parametric dependencies of the growth rate. Egorov et al.[229] modeled the global heat transfer inside the system in a tantalum container. Pons et al.[230] calculated the electromagnetic field and temperature distribution, and found that the predicted temperatures for the seed and the powder surface are much larger than the external temperatures measured at the top and bottom of the crucible. Müller et al.[54] have calculated the temperature distributions in inductively heated SiC growth reactors and found that the temperatures in the source powder are highly nonuniform. Karpov et al.[231] predicted the growth rate on a tantalum container. Selder et al.[232] have introduced a modeling approach for the simulation of heat and mass transfer. Klein et al.[233] have made a transient numerical simulation of heat transfer, investigating the respective influence of radiative and convective contributions and of the semi-transparency of the growing crystal. Chen et al.[234, 235] proposed a kinetic model for predicting the growth rate as a function of temperature, temperature gradient and inert gas pressure. Ma et al.[236, 237] have modeled the thermal stresses in the growing crystals. Cherendinichenko et al.[238] have examined theoretically the vapor transport phenomenon and found that it is a self-congruent process. Chen et al.[239] have calculated the effects of induction heating on the temperature distribution and the growth rate. Drachev et al.[240] have described the formation of a faceted step surface on low off-axis angle seed substrates in terms of Mullins theory of thermal etching. Matukov et al.[241] suggested a model of facet formation based on microscopic consideration of the crystal growth by a step flow mechanism, using a growth kinetics model extending the Burton-Cabrera-Frank approach to binary crystals and multicomponent vapor. Drachev et al.[242] have studied the temperature distributions for a furnace design with two plane resistive heaters.

(5) Other Methods

Various other methods of growing SiC crystals are shown in Table 23.4 and explained below.

- Chemical Vapor Deposition (CVD)

Matsunami et al. have developed the growth of 3C-SiC on Si substrates using a buffer layer[243,244] or by the carbonization of the substrate prior to growth[245] using step-controlled epitaxy.[246,247] Based on this surface carbonization, the epitaxial growth method known as step controlled epitaxy has been extensively developed. It was however found that anti-phase-boundaries (APBs) (plane defects) are formed during epitaxial growth because of large lattice mismatching. The mechanism of the formation of APBs is explained in Fig. 23.28. When the step height is an odd number of layers, APBs are formed because the atomic layer stacking is changed. They then found that using

Table 23.4 Crystal Growth by Other Methods

Method	Contents	Authors	Year	Ref.
CSVT	First application of CSVT to SiC growth and study of growth kinetics	Tairov et al.	1976-1979	276, 277
CSVT	Growth rate higher than 10^3 μm/h	Vodakov et al.	1979	278
CSVT	Study of growth mechanism	Lilov et al.	1979	279
CVD	Growth of 3C-SiC on Si substrates with carbonization and step flow epitaxial growth	Matsunami et al.	1981-1987	243-250
CSVT	Growth of 3C-SiC with the growth rate of 30 μm/hr	Sgields et al.	1993	280
SC	SiC wafer bonding to Si substrates	Tong et al.	1995	288
SC	SiC wafer bonding to Si substrates using ion implantation	Cioccio et al.	1996	289
HTCVD	High growth rate at temperatures of 1800-2300°C	Kordina et al.	1996	266
TD	Review of thermal decomposition process	Gorin et al.	1997	284
CVD	Growth of 3C-SiC on an undulated Si substrate and reduction of planar defects	Nagasawa et al.	1997-2008	48, 251-257
HTCVD	Increase of growth rate upto 0.8 mm/hr	Ellison et al.	1998-1999	267, 268
CSVT	Growth rate and the substrate temperature	Matsunami	2003	281
HTCVD	Growth of HPSI 4H-SiC of resistivity higher than 10^{11} Ω·cm	Ellison et al.	2003-2004	269, 270
HCVD	Growth rate upto 100-300 μm/h	Nigam et al.	2006	271
HCVD	Growth rate upto 250-300 μm/h, growth of HP-SI 4H- and 6H-SiC	Polyakov et al.	2006	272
SC	SiC wafer bonding to polycrystalline SiC	Chichignoud et al.	2006	292

CSVT: Closed Space Vapor Transport, CVD: Chemical Vapor Deposition, HTCVD, High Temperature CVD, HCVD: Halide CVD, TD: Thermal Decomposition, SC: Smart Cut

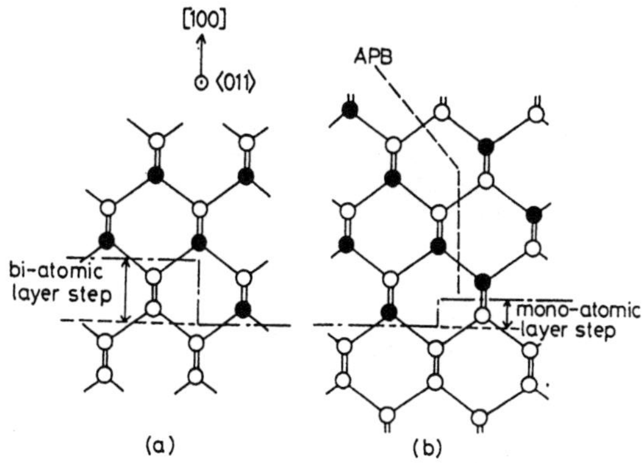

Fig. 23.28 A model of the origin of anti-phase-boundary (APB). On the assumption that the first grown layer consists of C atoms only. (a) Steps with bi-atomic layer height result in APB-free growth. (b) Steps with mono-atomic layer height result in formation of APBs (reprinted from Ref. 248 with permission, copyright 1986 Elsevier).

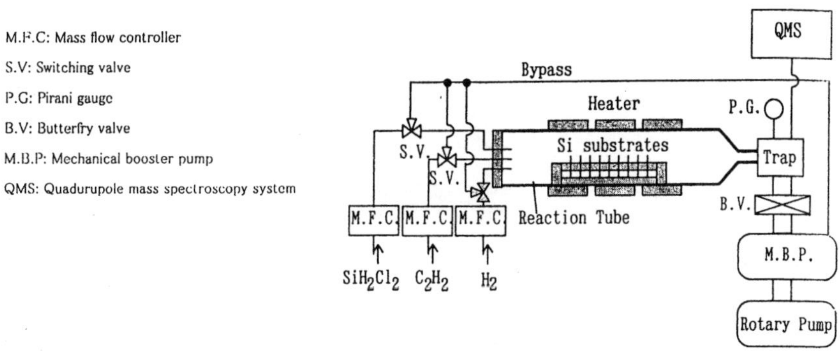

Fig. 23.29 Hot-wall type reactor for vapor-phase epitaxy (reprinted from Ref. 251 with permission, copyright 1997 The Japanese Association for Crystal Growth).

Si(100) substrates off-angled towards (011), these APBs can be eliminated because of step flow growth.[248-250] Twin boundaries (TBs) however remain.

Nagasawa et al.[48, 251-257] developed a CVD method in which 3C-SiC thick epitaxial layers are grown on Si (100) substrates to eliminate both APBs and TBs. In an open tube furnace as shown in Fig. 23.29 , SiH_2Cl_2 and C_2H_4 gases are fed in with H_2 to grow SiC on Si substrates. In order to eliminate the APBs, the Si substrates are off-angled and to eliminate the TBs, undulated Si substrates as shown in Fig. 23.30 are used. After growing thick SiC layers, TBs with different orientations coalesce and are eliminated as shown in Fig. 23.31. Si substrates are removed after the 3C-SiC epitaxial growth and free standing 3C-SiC substrates of diameter up to 150 mm have been obtained.

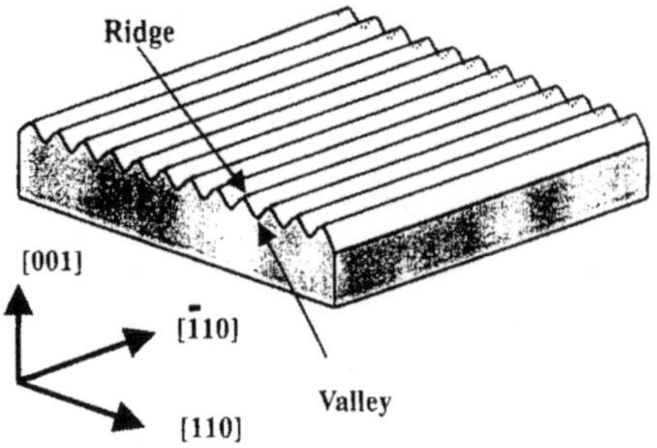

Fig. 23.30 Schematic structure of the surface of "Undulant-Si" (reprinted from Ref. 253 with permission, copyright 2002 Elsevier).

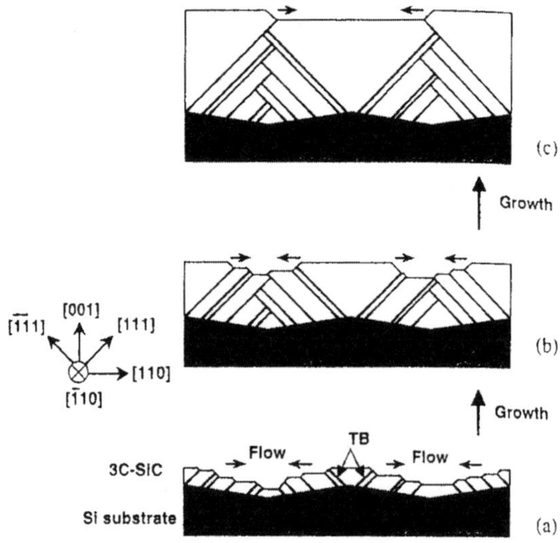

Fig. 23.31 Model of the elimination of twin boundaries (TBs) in the 3C-SiC layer on "Undulant-Si" with 3C-SiC growth. The (-111) cross-section structure of 3C-SiC changes from (a) to (c) as growth proceeds (reprinted from Ref. 253 with permission, copyright 2002 Elsevier).

Other methods, such as tuning the carbonization step conditions[258], using silicon-on insulaor (SOI) substrates[259] or Ge deposited Si substrates[260] and using the "checker-board carbonization" process[261] have been applied. Neudeck et al.[262] have grown 3C-SiC on 4H/6H mesa substrates. The bow is a problem for SiC on Si substrates and the origin and reduction of the bow were studied.[263-265]

- *High Temperature CVD (HT-CVD)*

This is a CVD method which is performed at high temperatures in order to increase the

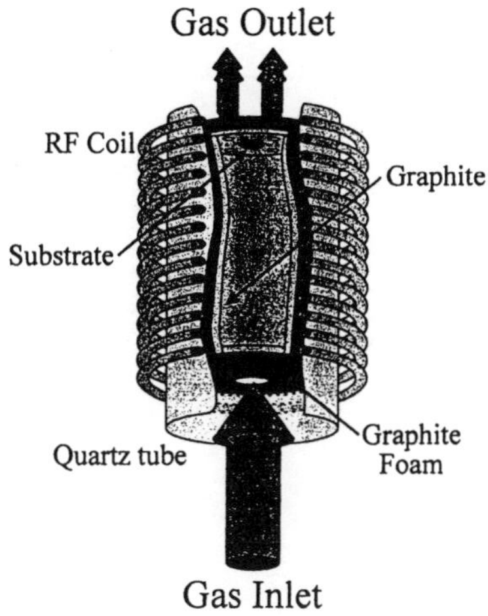

Fig. 23.32 HT-CVD Schematic drawing of the graphite crusible used of the high temperature CVD (HTCVD) method (reprinted from Ref. 266 with permission, copyright 1996 Elsevier).

growth rate. Crystal growth is performed in an open tube furnace as shown in Fig. 23.32 with silane (SiH_4) and hydrocarbons such as methane (CH_4), ethylene (C_2H_4) or propane (C_3H_8) gases flowing using H_2, Ar, and/or He as carrier gases. A growth rate of up to 0.3-0.8 mm/h was achieved at 1700-2300°C using 6H- and 4H(0001) or 6H Lely platelets as the seed crystal.[266-270] 6H- and 4H-SiC thick layers up to a thickness of 7 mm have been grown. The growth rate as a function of the growth temperature is shown in Fig. 23.33.

In order to solve the problem of the instability of SiH_4, the halide CVD (HCVD) method has been developed.[271, 272] Simulations of the growth of SiC for the reaction systems, $SiH_4 + C_3H_8 + H_2$[273] and $SiCl_4 + C_3H_8 + H_2$[274, 275] have been performed and some data have been compared with the experimental results.

- Closed Space Vapor Transport (CSVT)

In this method, the SiC substrate is placed close to the source SiC materials, at a distance of a few mm as seen in Fig. 23.34. Since the substrate and the source are semi-packed, growth occurs under quasi-equilibrium conditions. This method has been used in the past to prepare SiC materials for fundamental studies[276-279] and was known as the sublimation sandwich method. This method has been applied to the growth of 3C-SiC(100) crystals by Sgields et al.[280] 3C-SiC(100) seed crystals were prepared by removing Si substrates from 3C-SiC epitaxial layers prepared by the CVD method. Growth was carried out at 1860 and 1910°C in a 0.1 Torr argon atmosphere. A growth rate of 84 mm/h was achieved and 3C-SiC layers of thickness up to 750 μm could be

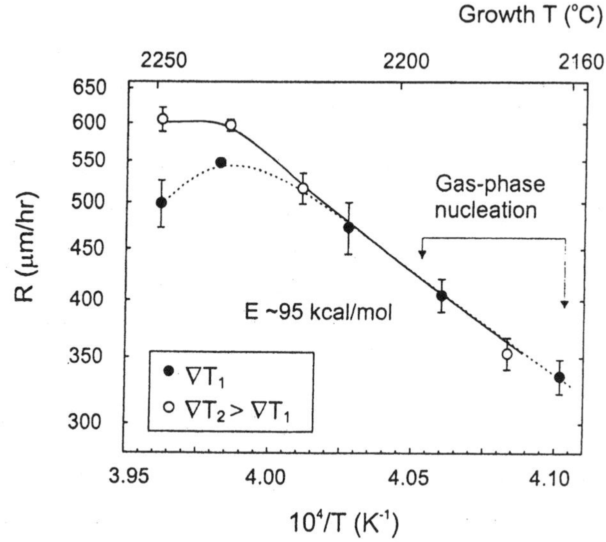

Fig. 23.33 Temperature dependence of growth rates for off-axis 4H-SiC C-face substrates, for two seed-holder configurations differing in the temperature gradient (∇T) applied to the seed (reprinted from Ref. 268 with permission, copyright 1999 Elsevier).

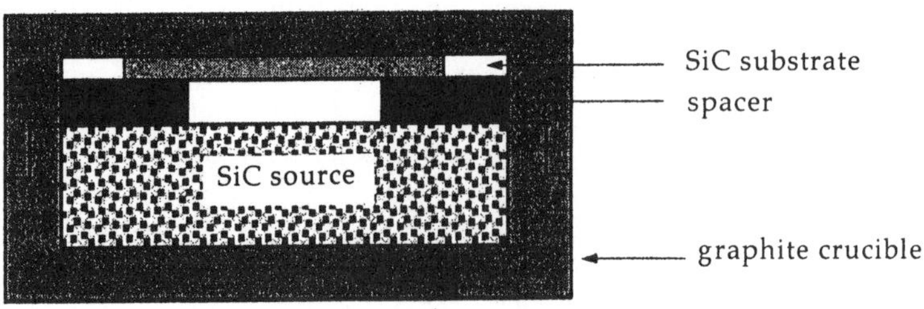

Fig. 23.34 Schematic set-up of closed space vapor transport (CSVT) method (reprinted from Ref. 30 with permission, copyright 2003 Nikkan Kogyo Co.).

obtained as shown in Fig. 23.35.

Matsunami[281] has applied this method to the growth of SiC crystals using 4H-SiC and 6H-SiC seed crystals. Growth was performed in the temperature range 1930-2070°C under 10-100 Torr for 4H-SiC and 1830-2040°C under 400 Torr for 6H-SiC. The growth rate of 4H-SiC is shown in Fig. 23.36. A growth rate of about 200 μm/h has been achieved at 2070°C, with a distance of 1 mm between the source and the substrate and a pressure of 10 Torr.

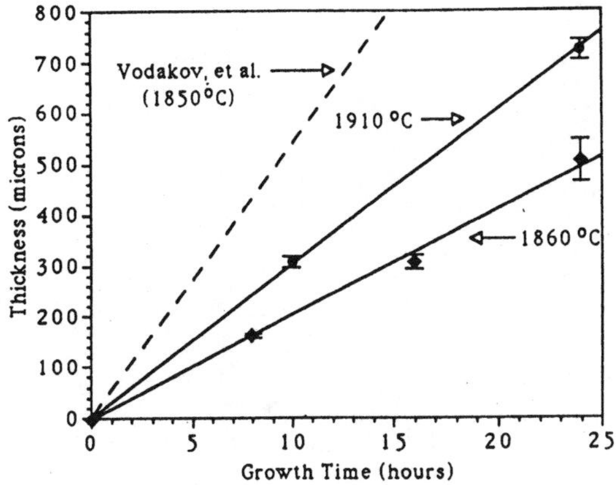

Fig. 23.35 The average 3C-SiC growth rates at 1860°C and 1910°C together with the data of Volokov et al.[278] at 1850 °C (reprinted from Ref. 280 with permission, copyright 1969 American Institute of Physics).

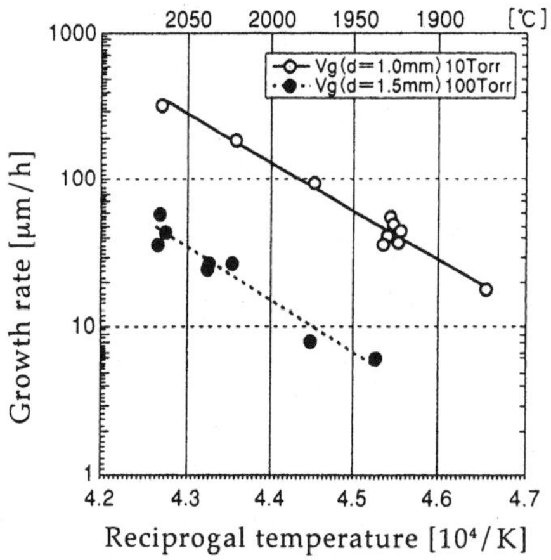

Fig. 23.36 The dependence of the growth rate on the substrate temperature (reprinted from Ref. 30 with permission, copyright 2003 Nikkan Kogyo Co.).

Thermal Decomposition (Barikov Process)

3C-SiC single crystals have been grown by the thermal decomposition of methyl trichlorosilane (MTCS) in hydrogen at a temperature of 1500 to 2100 K. This method which is known as the Van Arkel process[282] was first applied to SiC growth by Pring and Fielding.[283] This method was then developed by many Russian researchers as reviewed by Gorin and Ivanov.[284] This method is also called the Barikov process since it

Fig. 23.37 Photograph of plate-like crystals of 3C-SiC without {100} faces. The trapezoid faces are Si{111}; the bootlike faces are Si{211} (reprinted from Ref. 284 with permission, copyright 1997 John Wiley & Sons).

was developed in Baikov Institute of Metallurgy.

The SiC crystals can be grown in the form of platelets, prisms and skeletal crystals with sizes up to 5 x 5 mm^2 according to the following reaction.

$$(CH_3)_2SiCl_2 = SiC + C + 2HCl + 2H_2 \qquad (23.3)$$

Typical crystals grown by thermal decomposition are shown in Fig. 23.37. The grown crystals have been evaluated by synchrotron white beam X-ray topography and included defects have been identified.[285]

- *Smart Cut*

This method was developed for the production of silicon on insulator (SOI), called as the Smart Cut process [286,287], which is widely used in the silicon device industry. This method was also applied for SiC[288-292] as seen in Fig. 23.38 and the material structure is called as the SiC on insulator (SiCOI) structure.

The smart cut process is as follows.

Step 1: H^+ ions are implanted on SiC substrates (wafer A) capped with oxide layers.
Step2: This wafer A is then chemomechanically polished, cleaned and bonded hydrophilically at room temperature with Si or polycrystalline SiC substrates (wafer B).
Step3: The bonded wafer is heat-treated at first below 1100°C to split the thin SiC layer from wafer A, and then above 1100°C to strengthen the chemical bond between two oxide layers on wafer B.
Step4: Wafer B on which SiC layers are bonded is finally polished in order to remove

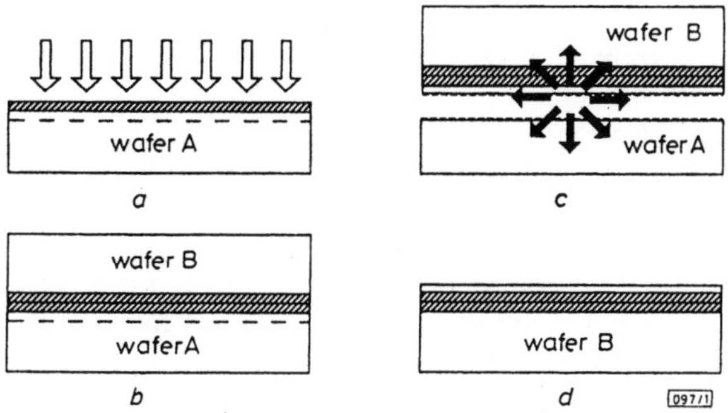

Fig. 23.38 Principle of Smart Cut process. (a) Step 1: H^+ implantation, (b) Step 2: chemomechanical polishing (CMP) + cleaning + bonding, (c) Step 3: heat treatment, (d) Step 4: polishing (reprinted from Ref. 289 with permission, copyright 1996 The Institution of Engineering and Technology).

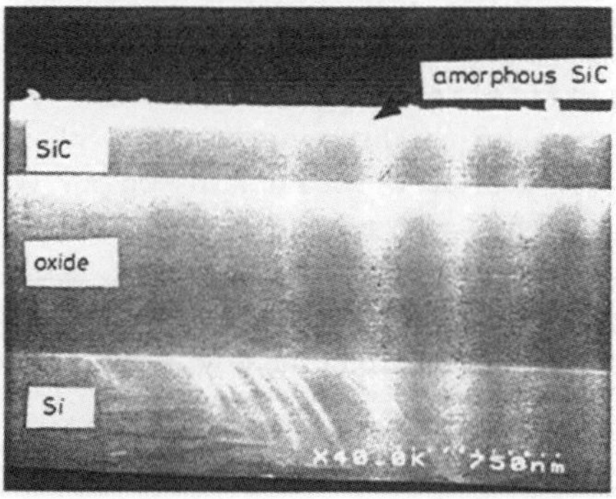

Fig. 23.39 Scanning elecron microscopy (SEM) observation of structure, after transfer of film and before final polishing. Transferred SiC film (350 nm thick) and the interface between two bonded oxide layers are clearly visible (reprinted from Ref. 289 with permission, copyright 1996 The Institution of Engineering and Technology).

the damaged layer and to improve the microroughness of the surface of the split thin SiC layers.

The cross section of the SiC grown by this method is shown in Fig. 23.39. The electrical properties of the transferred thin SiC layers were evaluated and it was found that there was no alteration of the properties.[290] It should be noted that Wafer A (in this case, expensive SiC material) can be reclaimed and reused several times. Industrial silicon production lines can be used as they are designed for this method so that this method has a big advantage in cost reduction.

23.3.2 Control of Polytypes

As mentioned already, several polytypes grow simultaneously even under similar growth conditions.[17,21,68] The probability of occurrence of various polytypes as a function of temperature is shown in Fig. 23.40. It is therefore necessary to control polytypes to grow one-phase SiC crystals.

Various growth parameters in the modified Lely method such as seed polarity [30,49,127,166,293-295], growth temperature [133,293,296], supersaturation[50,133,296,297], vapor phase stoichiometry [53,296-299], impurities [121,145,300-302], seed off-cut angle [303,304], inclusions [305,306], facets [201,307] and polytype transformation [130,131,201,297,308] have been studied.

Koga et al.[127] first tried the control of polytypes by studying the effect of the polarity of the seed substrate face and growth conditions. It was found that at high temperatures, 4H-SiC can be grown on the C face of a 6H-SiC seed crystal. Yoo et al.[130, 131] have obtained 6H-SiC crystals by the phase transformation by high temperature annealing of 3C-SiC crystals grown by the sublimation method on CVD-grown 3C-SiC(100) substrates.

Kanaya et al.[133] studied the effect of pressure and growth temperature on the occurrence frequency of 6H-SiC and 4H-SiC (Fig. 23.41). They found that when the 6H-SiC(000-1)C face is used as the seed crystal, with lower seed temperature, lower pressure and a greater temperature difference between the source and the seed the result is a higher yield of 4H-SiC crystal growth. They claimed that when supersaturation is high, growth rate is high and the 4H polytype can be grown.[50] On the other hand, when growth was performed on the 6H-SiC(0001)Si face, no 4H-SiC growth was observed.[49] Stein et al.[294, 295] have studied the effect of seed polarity on polytype formation from the viewpoint of surface energy. They reported that independently of the polytype of the seed crystal, 6H-SiC is grown on (0001)Si face and 4H-SiC is grown on (000-1)C face. 4H-SiC is also known to grow on the 4H-SiC (000-1) C face.[30,166]

It is pointed out that impurities do affect the polytype selection. It is reported that

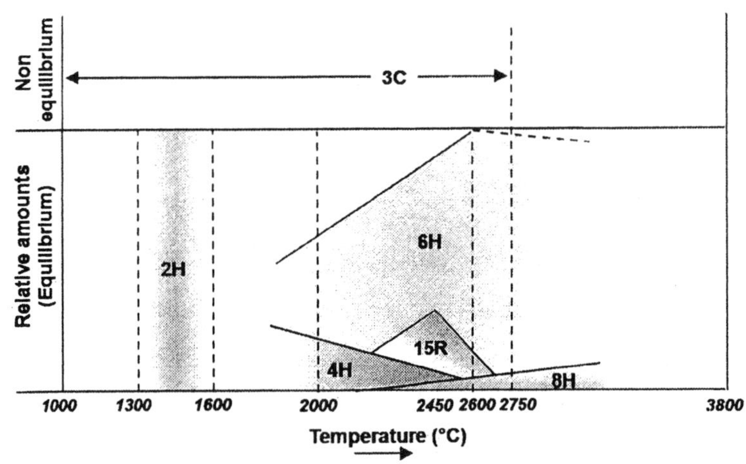

Fig. 23.40 Relation between structure and temperature of occurrence of polytypes. The relative amounts of the different hexagonal structures are shown (reprinted from Ref. 17 with permission, copyright 1963 Koninklijke Philips Electronics N.V.).

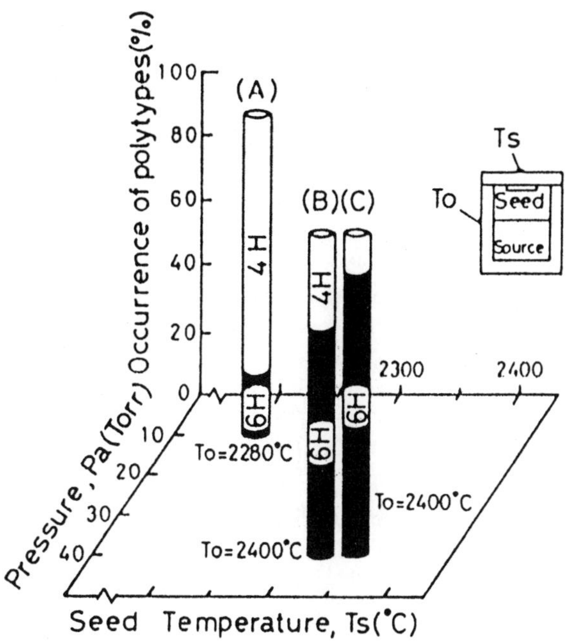

Fig. 23.41 Occurence of polytypes of single crystals grown on the 6H-SiC(000-1) plane under the conditions of (A) T_s = 2200°C, T_0 = 2280°C, P_a = 10 Torr; (B) T_s = 2320°C, T_0 = 2400°C, P_a = 40 Torr; (C) T_s = 2350°C, T_0 = 2400°C, P_a = 40 Torr (reprinted from Ref. 133 with permission, copyright 1969 American Institute of Physics).

Sc[121], Ce[300] and N[50] contribute to the selective growth of 4H-SiC. Rost et al.[301] have studied the effect of nitrogen doping below 2×10^{19} cm^{-3}. The onset of polytype transformation to 15R and 6H was observed after the interruption of nitrogen doping. Chen et al.[302] have shown that when nitrogen is highly doped in 6H-SiC growth, it promotes the transformation to 4H- and 15R-polytypes. Nakabayashi et al.[145] have shown that nitrogen incorporation into SiC crystals and the optimization of a convex thermal distribution environment can enhance the stability of 4H polytype growth, resulting in 100 mm-diameter crystals with extremely low micropipe densities.

Basceri et al.[201] have found that 3C polytype nucleation occurs at the initial and subsequent growth stages in the presence of the (0001) facet. Schmitt et al.[298,299] have studied the occurrence of foreign polytype inclusions in 4H-SiC growth and found that to reduce them moderate growth conditions with a reduced temperature gradient is one prerequisite but also stoichiometry in the gas phase and its modulation by nitrogen doping must be considered.

Fujimoto et al.[305] have examined 4H-SiC grains existing in 6H-SiC single crystals and suggested the these grains are formed from SiC inclusions which act as their embryos in the crystals. Li et al.[306] have shown that 15R-SiC inclusions occur easily in 6H-SiC single crystal growth when the growth interface is not appropriate and when the growth rate exceeds the critical rate of 300 μm/hr.

Tupitsyn et al.[297] have shown the transition from 6H-SiC seeding substrates to 4H-SiC growth by controlling the supersaturation and the vapor stoichiometry. Aoki et al.[308] have observed the sequence of the polytype transformation of 4H-6H-4H and also observed 8H-SiC at the interface between 6H- and 4H-SiC.

Kusunoki et al.[94] have shown that during seeded solution growth from Si solution, a N_2-He atmosphere favors the growth of 3C-SiC on the 6H-SiC seed crystal. Tanaka et al.[95] have shown that in the solution growth by the cold crucible technique, 3C-SiC (111) can be grown on 6H-SiC(0001) but when the dip depth in the solution is increased, 6H-SiC is grown instead of 3C-SiC. The stability of 3C-SiC in solution growth has been discussed from the viewpoint of the seed polarity and the growth rate and the growth temperature.[96,99] Alexander[102] discussed the polytype transition from 4H- to 6H-SiC in solution growth from the viewpoint of growth mode.

23.3.3 Reduction of the Defect Densities

There are various defects in SiC crystals such as micropipes (MPs) (cored screw dislocations), stacking faults (SFs), various dislocations (basal plane dislocations (BPDs), noncored screw dislocations, edge dislocations, and mixed dislocations) and the reduction of these defects is a key technology in the industrial application of SiC.

(1) Micropipes (MPs)

In SiC, specific defects called micropipes are generated. This micropipe defect was first predicted by Frank.[309] He predicted that when the Burgers vector b is large enough, the core of the screw dislocation becomes hollow as seen in Fig. 23.42 and the diameter D is expressed as follows.

$$D = \mu b^2 / 4\pi\gamma \tag{23.4}$$

Here, μ is the shear modulus and γ is the specific surface energy. This prediction is actually observed in the case of SiC[310,311] and Eq. 23.4 was validated by direct observation of the micropipe diameter and the Burgers vector.[312] It is observed in PVT SiC crystals as hollow tubes of a diameter exceeding 1 μm at the center of the growth spirals, with step height in the range from a few to a few tens of nanometers.[313] The relationship between the micropipe diameter and the Burgers vector was then found to be in good agreement with Frank's theory modified by Cabrera and Levine[314] considering kinetic effects.[315,316] Maximenko et al.[317] have deduced the surface free energy from the analysis of the micropipe diameter and found that it was in agreement with other surface energy data.

The origin and the mechanism of micropipe formation have been studied by various authors[51,159,315-317] and models such as carbon inclusions[159], silicon droplets in the source material[189], mixed dislocations[315], surface depression[318], twist-type misorientation[319], stacking fault clusters[320] and surface steps[321,322] have been proposed.

Since it is known that micropipes give rise to the deterioration in breakdown voltage

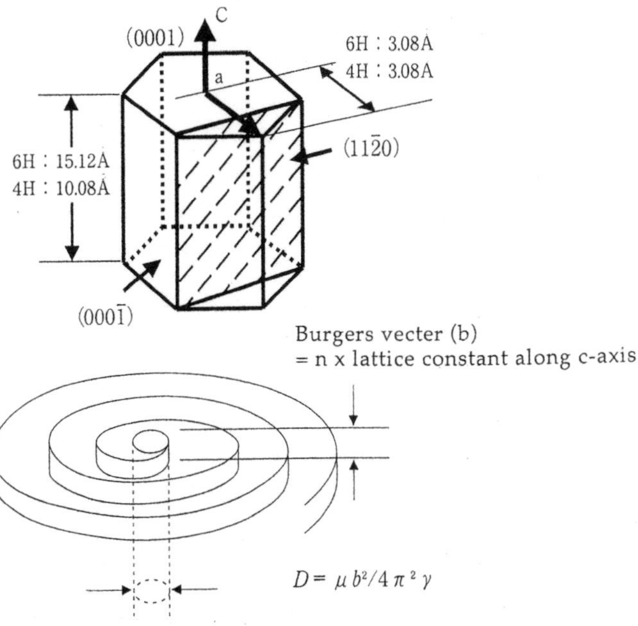

Fig. 23.42 (a) Hexagonal SiC crystal structure with plane indexes and (b) schematic micropipe structure with the relationship betweein the micropipe diameter and the Burgers vector (reprinted from Ref. 31 with permission, copyright 2003 Ohmsha).

of diodes and to increased reverse leak current[323] because of the generation of microplasma inside their hollow core, these defects become decisive in electronic devices which use SiC substrates as the current flow medium.

- *Reduction by the modified Lely method*

In the modified Lely method, micropipes are mainly formed as the propagation of screw dislocations from the (0001) seed crystal. The density of micropipes varied from 10^3 cm^{-2} to 10^2 cm^{-2} depending on the crystal growth conditions and the seed treatment prior to growth.[325,326] It was also reported to be dependent on the purity.[324]

As mentioned in Sec. 23.3.1(4), growth in various orientations except <0001> is effective since micropipes in the seed crystal propagate only in the <0001> direction for growth on (0001) planes. Actually, it was shown that by growing in the orientations other than <0001>, the density of micropipes could be greatly decreased even down to zero.[134-137,179-182,325]

Sculze et al.[160] have shown that MP-free crystals can be grown by optimizing the growth conditions so as to be as near equilibrium as possible. The maximum average thermal gradient inside the growth furnace was set below 5 K/cm. Kato et al.[173] showed that the etching back of the seed crystal prior to the crystal growth by the modified Lely method (Fig. 23.21) was effective in decreasing the density of micropipes down to 1/10.

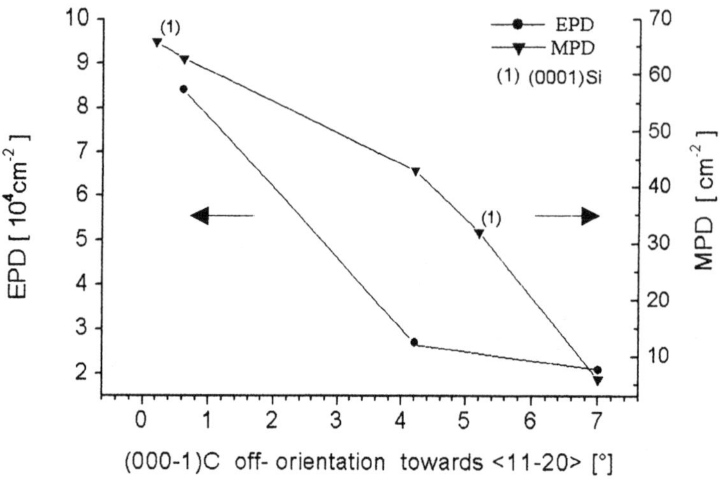

Fig. 23.43 Etch pit density (EPD) and micropipe density (MPD) averaged across a wafer from the same crystal position as a function of the off-orientation for growth in c-direction (riprinted from Ref. 326 with permission, copyright 2006 Trans Tech Publlications).

Rost et al.[303,326] have systematically studied the effect of polarity and off-orientation on the crystal perfection of 4H-SiC single crystals grown by the modified Lely method. In the case of standard c-plane growth on the (000-1)C face, it was found that MPs and etch pit densities (EPDs) are reduced as shown in Fig. 23.43. The stacking faults (SFs) however increase as off-orientation increases. The difference of the SF formation between a-plane and m-plane has been discussed in relation to stress relaxation.

Basceri et al.[201] have shown that polytype inclusions are the origin of the generation of MPs and 75 mm diameter MP free 4H-SiC crystals can be obtained by controlling the formation of polytypes under appropriate growth conditions realizing near equilibrium thermal conditions using (0001) seed crystals by the modified Lely method. Kim et al.[189] have also shown that appropriate thermal treatment of the SiC source powders can reduce the MP density.

Graza et al.[327] have studied micropipe reduction by the 6H-4H transformation technique. When off-angled 6H crystal is used for growing SiC by the modified Lely method, 4H single crystals are grown due to the 6H-4H transformation and micropipe free 4H crystals can be grown.

The density of micropipes in PVT-grown crystals has been decreasing year by year[58,200,201] as seen in Fig. 23.44.[154] In fact, 100 mm diameter micropipe free 4H-SiC single crystals are being realized.[154,205]

- *Closing by epitaxial growth*

It was first found that micropipes in substrates prepared by a modified Lely method could be closed by thick rapid epitaxial growth reapplied by the modified Lely

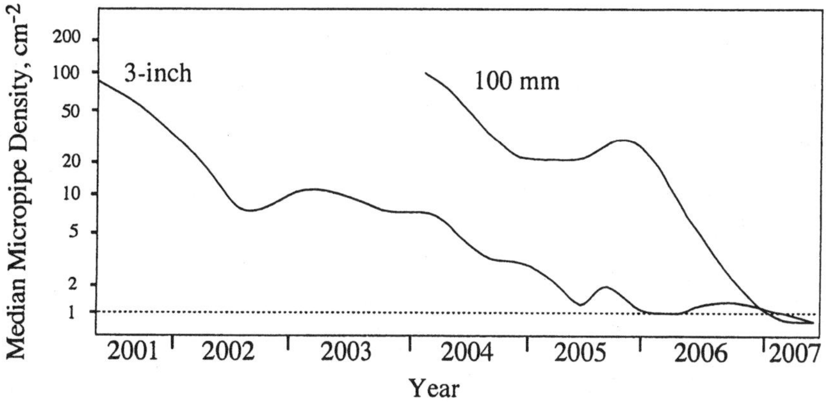

Fig. 23.44 Median micropipe density for production 4H-SiC vs. time for 75 mm and 100 mm diameter crystals (riprinted from Ref. 154 with permission, copyright 2009 Trans Tech Publlications).

method[328] filling up the existing micropipes. Very large micropipes of diameter ~ 70 μm could be overgrown.

The liquid phase epitaxy (LPE) method was found to be effective in closing micropipes[83,84,329-331] in SiC substrates prepared by the modified Lely method using various melts such as Si-Sc[329,330] and Si-Ge.[331] LPE growth allows the growth conditions to be close to thermodynamic equilibrium so that the driving force for growth is small. Epitaxial growth proceeds at low supersaturation compared with the PVT method where supersaturation is much higher and leads to micropipe formation. Flip et al.[331] found that the micropipe closing efficiency is remarkably improved up to 80 % by using Si-Ge melts and off-oriented substrates. The mechanism of micropipe closing was discussed and micropipes were found to decompose into a number of non-hollow-core dislocations with smaller Burgers vectors[331, 332] as shown in Fig. 23.45.

The same phenomenon has been observed in the case of homoepitaxial growth by the CVD method. The CVD method itself has been well developed by Matsunami et al.[246] as step controlled epitaxy. The CVD method has been used to grow SiC epitaxial wafers and it was found that micropipes can be closed during epitaxial growth.[333-335] It was found that micropipes are decomposed to several screw dislocations with smaller Burgers vectors by the step flow epitaxial growth as seen in Fig. 23.46. On the surface of the epitaxial layer, several etch pits were found instead of a micropipe in the substrate. The micropipe closing efficiency reached about 78%, close to that of the LPE method.

Micropipe closing has also been reported for the HT-CVD method and up to 90 % of micropipes existing in the PVT grown substrates could be closed under favorable conditions and the density of micropipes could be reduced to 0.3 cm^{-2} for 50 mm diameter 4H-SiC epitaxial wafers.[269, 270]

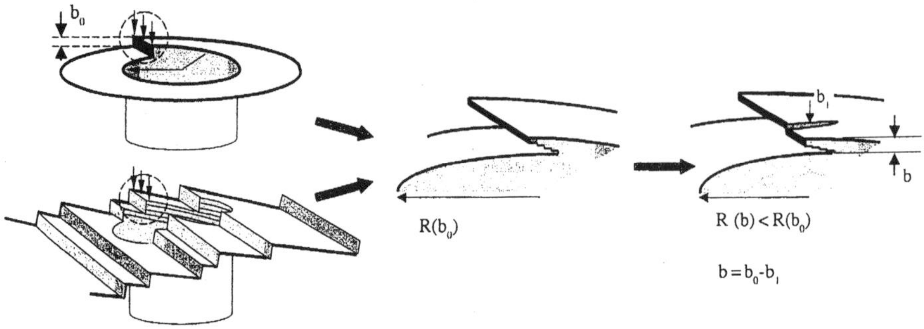

Fig. 23.45 Sketch which illustrates the decomposition process of a hollow-core dislocation (MP) (reprinted from Ref. 331 with permission, copyright 2004 Elsevier).

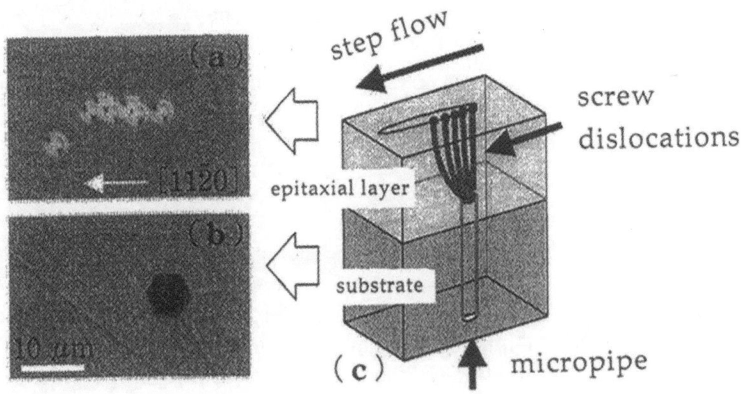

Fig. 23.46 The surface of the epitaxial layer after micropipe closing. (a) The etched surface of the substrate, (c) sketch explaining micropipe closing (reprinted from Ref. 335 with permission, copyright 2002 The Japan Society of Applied Physics).

(2) Dislocations

Various dislocations other than the micropipes described above such as basal plane dislocations (BPDs), screw dislocations and threading dislocations also affect device performance. BPDs are known to degrade the forward bias characteristics of bipolar devices[336] and affect the reliability of the gate oxide for MOSFETs.[337] The reduction of dislocations is therefore an important issue for the industrial application of SiC substrates.

Theoretical considerations of various dislocations in SiC have been extensively studied[338-341] in order to understand the electrical role of dislocations. The energy level of various types of dislocations are calculated using the cluster approach, the supercell approach and the supercell-cluster hybrid approach.

The behavior of dislocations in PVT crystal growth was systematically studied by Ohtani et al.[342-346] It was found that most threading dislocations in PVT crystals are grown-in type dislocations and introduced at the initial nucleation stage. They are also caused by secondary phase and/or foreign polytype inclusions. On the other hand, BPDs are caused by thermoelastic stresses during growth and/or post-growth cooling, since the basal plane (0001)<11-20> glide system is easily activated at high temperatures.

BPDs interact with threading screw dislocations (TSDs) and/or micropipes (MPs). When BPDs on the (0001) plane interact with TSDs threading from the seed crystal, there occurs a dislocation jog of the length of the Burgers vector of the TSDs. This dislocation jog does not move as easily as BPDs, especially when the jog length is large as in the case of MPs, when it acts as a Frank-Read type dislocation multiplication source and acts to increase the dislocation density.[342] When BPDs interact with threading edge dislocations (TEDs) and/or low angle grain boundaries (LAGBs), the situation is different. BPDs and TEDs/LAGBs lie in slip planes orthogonal to each other. However, when BPDs have dislocation jogs parallel to the c-axis, they interact with TEDs.

Nishiguchi et al.[347] have observed the generation of BPD pairs in the growth of 4H-SiC on (1-100) 4H-SiC substrates. It was suggested that the mechanism was the interaction between stacking faults and threading screw dislocations and it is important that the growth keeps an atomically flat surface to prevent the generation of threading screw dislocations and so to suppress the generation of BPD pairs.

Dement et al.[348] performed deformation experiments below the brittle to ductile transition and found that Si-core and C-core partials are evidenced in the basal plane and perfect dislocations are observed on other crystallographic planes. Dislocations can be nucleated down to room temperature and other slip planes than the basal one are active so that the dislocation behavior is complex in SiC.

The effect of dopants at high concentrations has been studied. Eberlin et al.[349] have reported the possibility of the pinning effect of nitrogen and boron but it was not observed in the case of nitrogen up to a concentration of 10^9 cm^{-3}.[350] Wellmann et al.[351] have compared the generation behavior of BPDs in heavily doped p-type and n-type 6H-, 4H-SiC crystals grown by the modified Lely method and found that the density of BPDs is much higher for n-type SiC than for p-type SiC.

Various procedures have been used to reduce these dislocations as follows.

(i) Repeated a-face (RAF) crystal growth[181, 182]
(ii) Etch back process of the seed crystal[173]
(iii) Low growth rate[176,194,195]
(iv) Quasi equilibrium conditions[194,195,201]
(v) Use of off-cut angle seed crystal[303,304]

Wu et al.[352] have found the dislocation density of PVT crystals is proportional to $w^{-0.5}$, where w is the axial position in the crystal. The dislocation density has a tendency decrease as the growth proceeds which is attributed to the annihilation of dislocations of opposite sign during growth.

Dislocation densities of conventional PVT crystals are in the range 10^3-10^4 cm^{-2}, but extremely low density down to 175 cm^{-2} for 1c screw dislocations in PVT crystals has been reported.[154]

(3) Stacking faults (SFs)

SFs are a kind of plane defect due to the misordering of the stacked planes (Vol. 1, Sec. 4.4) and are known to be generated very easily during crystal growth in orientations other than <0001>. As already explained above, micropipes can be eliminated when crystal growth is performed in orientations other than <0001> but that enhances the generation of stacking faults. Another problem is that stacking faults can be generated during the operation of the device and affect its performance. This is especially serious because of the forward bias voltage drop of pn bipolar devices (Sec. 23.4.4.(3)).

The SF formation energy has been calculated and it is known that it is rather low compared with other semiconductor materials.[353-356] It has even been reported that the SF formation energy is negative.[355] It has also been found that (34) and (44) Schokley-type SFs have very low formation energies of around a few mJ/m^2 and are expected to form and expand very easily.[356] Iwata et al.[357] have calculated the energy levels of stacking faults in various SiC polytypes and found that they are 0.2 eV below the conduction band. This may be the main reason why stacking faults are generated during the operation of the pn device.

As explained in Sec. 23.3.1(4), the density of micropipes can be greatly reduced when crystals are grown along [1-100] and [11-20] since micropipes in the seed crystal do not propagate in these directions but only in the [0001] direction. It was however found that SFs are generated during growth in these directions. 6H-SiC growth in the direction of [1-100] showed a density of SFs in the range 10^3-10^4 cm^{-1} while 6H-SiC growth in the direction of [11-20] and 4H-SiC growth in the direction of [1-100] and [11-20] showed a density of SFs in the range 10^2-10^3 cm^{-1}.[358] It was however found that [11-20] seed crystals off-angled toward the <0001> direction were effective in reducing the density of SFs from 100-150 cm^{-1} to about 10 cm^{-1} as the degree of off-orientation is increased from zero to 10°.[138-142, 358]

Katsuno et al.[359] have shown that SF formation in highly nitrogen doped 4H-SiC substrates prepared from modified Lely grown crystals after annealing at 1100 °C depends greatly on the surface roughness of the substrate left after polishing preparation. The SF density was lower than 10 cm^{-1} when the roughness was less than 0.1 nm while it was more than 1000 cm^{-1} when the roughness was more than 10 nm.

Nakabayashi et al.[360] have studied the SF formation mechanism and the stacking sequence in SFs generated in 4H-SiC is 6H (3, 3) stacking sequence for c-face growth while it was different to those for a-face growth.

Seki et al.[361] have shown that SF formation during the growth of 6H-SiC on a 3C-SiC seed crystal by solution growth arose from the interface between 6H-SiC and 3C-SiC and the density of SF decreases with increasing distance from the interface. They indicated that the formation of SFs is caused by relaxation of the lattice-mismatch strain.

Siche et al.[304] have found that the density of SFs is greatly increased from 200 cm^{-2} to 4500 cm^{-2} when the off-angle is increased from 4.2° to 7.7° for (0001) 4H-SiC crystal growth while the dislocation density decreases because SFs have a pinning effect. They also suggested that nitrogen doping has an effect to stabilize the 4H polytype.[362]

(4) Other defects

Apart from micropipes, dislocations and stacking faults, other defects include inclusions, macrodefects, low angle grain boundaries (LAGBs) and pinholes. These defects are often related to each other.

Polytype inclusions which are generated at the interface between the seed and the grown crystal at the initial stage, are known to be the origin of the generation of micropipes.

During the radical enlargement of crystals growing from the seed crystal, various defects such as macrodefects and LAGBs are generated in the periphery of the grown crystals.[163,199] The generation of these defects can be suppressed using conical shaped seed holders[163,171,175] and sublimation etching.[173,198]

Ankin et al.[194] have shown that controlling the growth front shape so that it is as flat as possible by optimizing the growth conditions and the crucible geometry is crucial in decreasing the low grain boundary densities and enlarging the crystal diameter.

Jokubavicius et al.[363] have studied the effect of the Si/C ratio on void formation at the initial stage of 3C-SiC grown on 6H-SiC seed by the sublimation method.

23.3.4 n-Type, p-Type and Semi-Insulating (SI) SiC Crystals

(1) n-Type crystals

- Nitrogen doping

It is known that nitrogen is an unintended shallow donor and its concentration is in the level of 10^{15}-10^{16} cm^{-3} depending on purity control. During PVT crystal growth, nitrogen comes from the source material and/or residual nitrogen in the reactor.[68,123] Undoped SiC therefore shows as n-type except when the nitrogen concentration is less than the residual shallow acceptor impurity, boron, which comes from the source SiC powders.

Nitrogen is the main n-type dopant used to obtain low resistivity n-type crystals, during crystal growth by the modified Lely method. Nitrogen gas is added into the growth atmosphere[126], or nitrogen is added into the source powders and a carrier concentration in the range 2×10^{17} to 1×10^{18} cm^{-3} could be obtained.[126] The kinetics of nitrogen incorporation in 6H-, 4H-SiC has been studied by varying various crystal growth conditions such as growth rate, surface polarity, polytype, seed orientation[364-368], N_2/(N_2 + Ar) ratio in the atmosphere [369], the nitrogen partial pressure [370], the addition of Si powder to the SiC powder [371] and the growth temperature.[372]

There is an increase in nitrogen incorporation from (0001)Si to (11-20) to (1100) to (0001)C while boron is preferentially incorporated on (0001)C.[367] Resistivity as low as 7.6×10^{-3} Ω.cm for 6H-SiC and 5.3×10^{-3} Ω.cm for 4H-SiC has been obtained.[366] Kim et al.[371] found that nitrogen incorporation is increased by the addition of Si to the SiC powders. Ohtani et al.[372] have achieved resistivity as low as 1.5 Ω·cm for 4H-SiC without an increase in structural defects such as stacking faults.

The maximum nitrogen concentration is reported as 2-5 × 10^{19} cm^{-3}, one order of magnitude lower than the solubility limit. This deactivation of nitrogen has been explained theoretically, from the viewpoint of point defect formation.[373] Rurali et al.[374] have also theoretically discussed the activation limit of nitrogen from the viewpoint of nitrogen cluster formation. The solubility of nitrogen has been theoretically calculated and compared with the experimental results.[375]

Henry et al.[376] have used photoluminescence (PL) to determine the nitrogen doping concentration in 6H-SiC. Smith et al.[377] have developed a method of measurement of nitrogen concentration based on secondary ion mass spectroscopy (SIMS) data. Puche et al.[378] have found a series of peaks in the Raman spectra of nitrogen doped crystals and discussed the origin of these peaks.

- *Phosphorus doping*

Phosphorus, which has an ionization energy very close to that of nitrogen[379,380] has recently been studied as an efficient alternative n-type dopant to increase the carrier concentration. Phosphorus occupies Si lattice sites while nitrogen occupies C lattice sites.[381] Phosphorus however has low diffusivity into the material during crystal growth and the free carrier concentration was limited in the range 10^{16} cm^{-3} [382] but a possible way of increasing the phosphorus incorporation is discussed in terms of the surface polarity and growth orientation.[375,383,384] Rauls et al.[385] have shown that neutron transmutation of ^{30}Si-enriched 6H-SiC and subsequent high temperature annealing can achieve efficient n-type doping with free carrier concentrations of 3 x 10^{18} cm^{-3}. The solubility of phosphorus has been theoretically studied and the activation mechanism has been discussed.[375]

Schmid et al.[386] have evaluated the ionization energies of phosphorus donors in 6H-SiC, which was converted to n-type from as-grown p-type material by neutron transmission. The ionization energies of phosphorus in 4H-SiC have also been evaluated for CVD epitaxial growth[387] and for ion implantation with high dose of phosphorus.[388]

The doping behavior of phosphorus in CVD SiC layers was theoretically studied using ab initio supercell calculations. It was found that the most abundant defect is P at Si-site followed by P at C-site.[389] Electron spin resonance (EPR) spectra from phosphorus shallow donors occupying Si sites in 3C-, 4H- and 6H-SiC were identified using CVD grown films.[390]

PL lines due to shallow phosphorus donors have been evaluated.[391-393] Püche et al.[378] have found a series of peaks in the Raman spectra of phosphorus doped crystals and discussed the origin of these peaks.

(2) p-Type crystals

p-type 6H-SiC crystals were first grown using Al doped SiC source materials.[123, 127] The carrier concentration was varied from 1 × 10^{15} cm^{-3} to 3 × 10^{18} cm^{-3} and a maximum hole mobility of 23 cm^2/V·s could be obtained.

It is known that boron acts as a shallow acceptor impurity[394-396] and is incorporated from the source material during crystal growth by the PVT method. In fact, Hobgood et al.[152] obtained undoped p-type SiC crystals with a resistivity of 10^2-10^3 Ω.cm due to boron contamination during crystal growth.

p-type SiC is much in demand to realize devices using insulated gate bipolar transistors (IGBTs) and p-type doping has been performed by the modified Lely method.[397] In order to achieve low resistive p-type 4H- and 6H-SiC crystals, an additional gas pipe was installed to provide a continuous flux of aluminum atoms out of an external reservoir for the modified Lely system.[164,398-401] Carrier concentrations of the order of 10^{16}-10^{17} cm^{-3} have been achieved. In their first work, Al itself was used as the dopant material. Recently, they have used tri-methyl-alumium (TMA) for p-type doping and have been able to achieve Al incorporation greater than 10^{20} cm^{-3} and to reduce the unintentional compensation of the Al acceptors by nitrogen.

(3) Semi-Insulating (SI) crystals

For application to electronic devices operating at microwave frequencies, semi-insulating SiC substrates are required because they give low dielectric losses and reduced device parasitics.[402-404]

- V doping

Hobgood et al.[152] first obtained 50 mm diameter SI 6H-SiC by the PVT method with vanadium doping since vanadium was known to act as a deep donor.[405-408] Undoped SiC was p-type with a resistivity of 10^2-10^3 Ω.cm due to the residual boron impurity which acts as a shallow acceptor as mentioned above. By doping vanadium, SI crystals with resistivity in the region of 10^{15} Ω.cm as the extrapolated value at 300 K could be obtained. The semi-insulating behavior was attributed to the compensation of boron residual acceptors by the vanadium deep donor V^{4+} ($3d^1$) level located near the middle of the band gap at about 1.48 eV below the conduction band. Tsvenkov et al.[51] have also reported the commercial production of SI 4H-SiC by the PVT method. They have shown that the maximum oscillation frequency (f_{max}) of metal-semiconductor field effect transistors (MESFETs) using these substrates could reach up to 30 GHz. Augustine et al.[166, 167] have grown 50 mm diameter SI 4H-SiC single crystals, using the same mechanism as above with an activation energy of 1.18 eV.

The homogeneity of vanadium in the crystals has been improved. Bickermann et al.[409-411] have studied the factors which affect the yield of SI 6H-SiC crystal growth. Rasp et al.[412] have grown 6H-SiC crystals with co-doping of V and Al in order to improve the uniformity of resistivity. Racka-Dzietko et al.[413] have used an open seed backside arrangement to grow vanadium-doped SI 4H- and 6H- SiC single crystals.

- Undoped semi-insulating

It was also found that nominally undoped high purity semi-insulating (HP-SI) SiC of diameter between 50 mm and 100 mm can be grown by the PVT method[414-418] by improving the purity of grown crystals. Micthel et al.[414] first obtained 50 mm diameter HP-SI 4H-SiC single crystals without vanadium doping by the PVT method. This material was semi-insulating at room temperature and the resistivity was higher than 10^7

Ω.cm even at 450°C. The energy of the deep level responsible for the insulating behavior was 1.1 eV, which is different from the value 1.6 eV of vanadium. Jenny et al.[415-418] have shown that most impurities could be reduced to the limit of detection by SIMS and the residual impurities were only boron (typically 5×10^{15} cm^{-3}) and nitrogen (typically 3×10^{15} cm^{-3}). When the low temperature photoluminescence (LTPL) method was used to measure the nitrogen concentration, the lowest concentration obtained was 3×10^{14} cm^{-3}.[418] With this impurity reduction, SI crystals of resistivity higher than 10^9 Ω.cm could be obtained by the compensation of residual impurities with intrinsic defects. The origin of intrinsic defects proposed so far is described in Sec. 23.4.2 (6). MESFET devices fabricated using these crystals were free of backgating effects often seen in V-doped SI SiC crystals[417] and showed a power density of 5.2 W/mm, 63% power added efficiency (PAE) at 3.5 GHz.

The HP-SI SiC crystals have simultaneously been developed by the HT-CVD method with impurity reduction. Most impurities could be reduced below the detection limit of SIMS and the residual nitrogen and boron were about 4.5-7.5 $\times 10^{15}$ cm^{-3} and 0.7-1.2 $\times 10^{15}$ cm^{-3}. The HT-CVD methods themselves have been developed to grow high quality single crystals as explained in Sec. 23.3.1(4) [266-268] and were applied to grow HP-SI SiC single crystals.[269,270]

24.4 CHARACTERIZATION

24.4.1 Purity

The purity of bulk SiC crystals has been analyzed by various methods such as solid source mass spectroscopy (SSMS), SIMS, glow discharge mass spectroscopy (GDMS) and activation analysis.[148,150,419] It was reported that the source material commercially available by the modified Acheson method has a total background impurity reaching 100 ppma, with the main contaminants being V, Ti, Ni and Al.[150] During crystal growth, the reduction of impurities by 1/5-1/10 was observed as shown in Fig. 23.47.[150] When the source material is subjected to further purification, the purity of grown crystals was improved by up to two orders of magnitude. Photoluminescence showed only V and Ti to be the main impurities.

Buchan et al.[419] have studied the impurity incorporation in 15R- and 6H-SiC polytype crystals, growing both crystals at the same time by the PVT method. 64 elements were evaluated by GDMS and it was found that all impurities except B and Al were significantly reduced in the grown crystal relative to the SiC powder. The dominant impurity was N, measured at a concentration of 6×10^{18} cm^{-3}. Other impurities such as B, Al, Ti, V, Fe and Ni have been observed at levels above 1×10^{17} cm^{-3} which also contributed to the carrier concentration.

It is important to evaluate quantitatively the concentrations of residual impurities such as nitrogen at the lowest possible levels. This is especially so for the development of HPSI-SiC. The quantitative measurement method has been developed using low temperature photoluminescence (LTPL).[376,420,421] The calibration curve between the PL peak ratio (P_0/I_{77}) and the carrier concentration N_d-N_a has been determined.[421]

Zhou and Telle et al.[422] have studied the purifying mechanism in the Acheson pro-

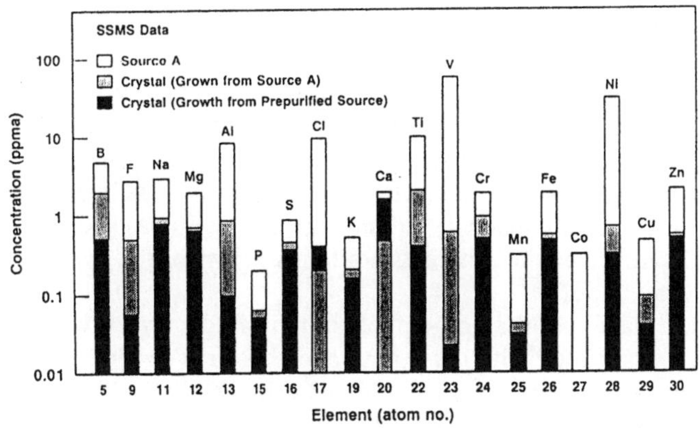

Fig. 23.47 Chemical purity of SiC starting source material and as-grown crystals as determined by spark source mass spectroscpy (SSMS) (reprinted from Ref. 150 with permission, copyright 1994 Elsevier).

cess with/without addition of NaCl, based on thermodynamical calculations compared with industrial observations. NaCl had been found to be the best additive for more than one century in commercial production, having the effect of evaporating Al_2O_3 and CaO by generating volatile metal chlorides.

High purity SiC has been achieved by HT-CVD, which allows nitrogen as donor impurity and boron and aluminium as acceptor impurities to be lowered to about 1×10^{15} cm^{-3} and other impurities to be made lower than the detection limit of SIMS.[270]

23.4.2 Defects

Various defects occur in SiC and since they affect the performance of devices as explained in Sec. 23.4.5, it is important to evaluate and reduce them.

These defects are classified as follows[423] and are also explained fundamentally in Ch. 4 of Vol. 1,

(i) point defects (vacancies, interstitials, complexes)
(ii) linear defects (edge dislocations, screw dislocations, extended dislocations)
(iii) planar defects (stacking faults, low angle grain boundaries, twins, inversion domain boundaries, antiphase boundaries)
(iv) volume defects (inclusions, voids)
(v) mixed defects (micropipes)

Most kinds of defects can be revealed by molten KOH etching and optical microscopy. Various characterization methods, such as transmission electron microscopy (TEM), atomic force microscopy (AFM), high resolution X-ray diffraction (HRXRD), scanning electron microscopy (SEM), cathodoluminescence (CL), energy dispersive X-ray analysis (EDXA), X-ray radiography, synchrotron white beam X-ray topography

(SWBXT) have also been applied.

(1) Macrodefects and structural defects

In the early stage of crystal growth by the modified Lely method, it has been found that various macrodefects occur. Some of these macrodefects are later found to be micropipes (MPs) (cored screw dislocations) and Si or C inclusions.[159,204,205]

Stein[424] has found that there are macrodefects in modified Lely grown crystals which are hexagonal tubelike cavities with a diameter of 10 to 50 μm and a length up to some millimeters. He attributed these defects not to instability of the growing interphase but to secondary evaporation.

Glass et al.[425] have found macrodefects in modified Lely grown crystals and analyzed the strain associated with these macrodefects such as pin-holes, pits and inclusions. Structural domains have been examined using X-ray rocking curve and reciprocal space mapping measurements. They have also examined Lely crystals compared to modified Lely crystals and found that the former are more perfect than the latter.[426]

Black et al.[427] have observed macrodefects such as hexagonal voids, MPs and inclusions using EDXA, X-ray radiography and found that these inclusions contain transition metal impurities such as Ti, Ni and Fe.

Tuominen et al.[428-430] have evaluated various defects such as dislocation networks, micropipes, basal plane defects, stacking faults, precipitates and cracks using synchrotron X-ray topography and HRXRD. They found that Lely grown crystals have no micropipes, few stacking faults and a dislocation density as low as 30 cm^{-2} compared with the modified Lely grown crystals. They have also classified the observed structural defects as grown-in defects and post-growth defects.

Nishio et al.[431] have evaluated the hexagonal planar and hollow defects in 6H-SiC grown by the modified Lely method and explained their formation based on the growth mechanism.

Takahashi et al.[432,433] have compared structural defects in 6H-SiC single crystals grown in the two orientations, [000-1] and [1-100] by the modified Lely method. It was found that in [000-1] grown crystals micropipes and screw dislocations were observed while stacking faults and edge dislocations were observed in the [1-100] grown crystals. [000-1] crystals contained misoriented subgrains with tilt and twist components while the [1-100] grown crystals contained only tilted subgrains.

In Fig. 23.48, a typical X-ray topograph is shown for 4H-SiC substrates, a conventional one with high micropipe and dislocation density and a repeated a-face (RAF) grown one with no micropipe and low dislocation density.[181]

(2) Micropipes (MPs)

Micropipes can be revealed by molten KOH etching as large hexagonal etchpits.[135,428] These large etchpits are believed to be due to the large Bergers vector as explained in Sec. 23.3.3 (1).

Since there is strong internal stress around micropipes, they can be observed by birefringent patterns.[134,434] The magnitude of the internal stress around the micropipe

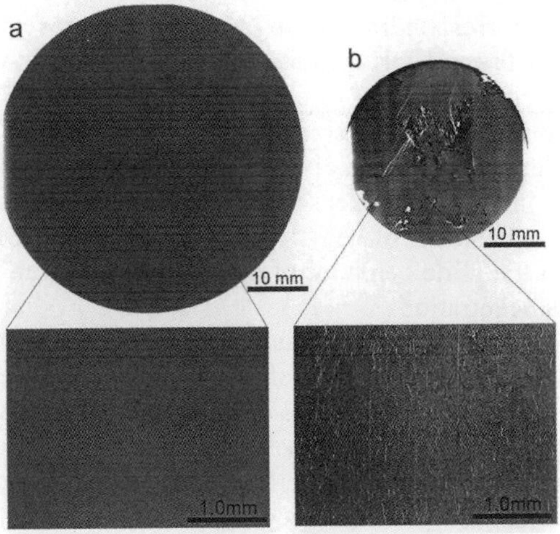

Fig. 23.48 Synchrotron monochromatic beam X-ray topographs. (a) 50 mm diameter (0001) 8° off-axis 4H-SiC grown by the repeat a-face (RAF) process (average EPD; 250 cm^{-2}, MD; 0 cm^{-2},) and (b) 30 mm diameter (0001) 4H-SiC grown by the conventional method (average EPD; 3 x 104 cm^{-2},, MD; 30 cm^{-2},) (reprinted from Ref. 181 with permission, copyright 2004 Nature).

was estimated to be around 113-166 MPa.

Destructive techniques based on high resolution digital imaging and laser scattering of KOH etched wafers have been developed.[435,436] Non-destructive techniques such as laser optical surface analyzer, ultraviolet system, TEM, SWBZT and double crystal X-ray diffraction imaging have been applied. The round robin test, using these techniques, has been performed at nine test sites[437] and a measurement standard deviation less than 33% was reported.

(3) Dislocations

Various mixed melts have been reported as dislocation etchants such as $NaNO_3$ + Na_2CO_3, K_2CO_2 + KNO_3, NaOH + Na_2O_2, NaOH, KOH.[30] Presently molten KOH is used very often. A eutectic mixture of 51 mol% NaOH-KOH etchant at 350 °C is used for selective etching of 1c screw dislocations on the C-face.[438]

Amelinckx et al.[439] examined the origin of various etch pits observed in 6H-SiC. By KOH etching, three types of hexagonal etchpits, one large, one medium and one small can be observed.[135] The large one accompanies hollow defects and is due to micropipes threading to the c-axis. The medium one is due to screw dislocations, the Burgers vector of which is along the c-axis. The small one is due to edge dislocations, the Burgers vector of which is along the c-axis. Shell like etchpits are located at the surface position of dislocation lines and are believed to be due to slide dislocations in the basal plane. Shell like etchpits therefore cannot be observed on the c face itself but on the off angled c face.[135] Katsuno et al.[440] have studied the etching kinetics of SiC using molten KOH.

Lineages of etchpits corresponding to small-angle boundaries[148] and mosaic structures with dislocation networks corresponding to subgrains could be revealed.[428] Wu et al.[441] have found a correlation between KOH etchpits and the synchrotron white beam X-ray topography image. The etching behavior of heavily nitrogen doped SiC was also studied. Kamata et al.[442] have shown that the size difference of the threading edge dislocation (TED) etch pits is dependent on the direction of the Burgers vectors, which was distinguished by grazing incidence high resolution topography. Katsuno et al.[443] have shown that the dislocation behavior of heavily nitrogen doped 4H-SiC does not show any evident influence on the glide behavior of basal plane dislocations (BPDs) up to mid 10^{19} cm^{-3} nitrogen concentrations.

Characterization by X-ray topography has been performed by various researchers[426,444-446] and various dislocations have been evaluated. Two types of dislocations have been observed, one is screw dislocations with a large Burgers vector which is at least 15 Å and which is observed as micropipes. The other is slide dislocation in the c plane, and is observed as a line structure from the center of the wafer to the periphery. Slide dislocations are observed by X-ray topography[135, 445] and synchrotron X-ray topography.[446] Glass et al.[426] have evaluated the misorientation domains in modified Lely grown crystals and concluded that they are due to the interaction of screw dislocations.

Yamaguchi et al.[447-450] have revealed inherent dislocation types and structures by synchrotron X-ray topography. These dislocations are summarized in Table 23.5 together with those identified by Ohtani.[451] Chen et al.[452,453] have used grazing incidence X-ray topographs together with ray-tracing simulation to analyze the distribution and the sense of threading screw dislocations. Dudley et al.[454] showed that back-reflection X-ray topography is a useful tool to evaluate the distribution and the magnitude of axial screw dislocations across the basal facets of SiC crystals where these defects appear as white circles surrounded by black rings.

The propagation of dislocations was analyzed three-dimensionally using a planar mapping of electron-beam-induced current (EBIC) method.[455] Screw dislocations, edge dislocations and basal plane dislocations were clearly observed and information on the decomposition and combination of dislocations was obtainable from the fluctuation of

Table 23.5 Classification of Various Dislocations

Dislocation type	Character	Line vector	Burgers vector
Micropipe	-	-	n<0001> (n≧2; 6H)
	-	-	n<0001> (n≧3; 4H)
Treading dislocation	Screw	//<0001>	n<0001> (n =1; 6H)
		//<0001>	n<0001> (n =2; 4H)
	Mixed	//<0001>	n<0001>+ 3/3<11-20> (n =1,2,...)
	Edge	//<0001>	1/3<11-20>
Basal plane dislocation	Screw	//<11-20>	1/3<11-20>
	Mixed	//<1-100>	1/3<11-20>
	Edge	⊥(0001}	n<0001> (n =1,2,...)

EBIC signals.

Raman scattering and polarizing optical microscopy have been applied to examine the local stress field around dislocations in 4H- and 6H- SiC.[456] The stress typically amounted to the order of 100 MPa

(4) Stacking Faults (SFs)

SFs can be revealed by molten KOH etching as etch pit lines[135,457,458] but a single line of etch pits arises from bunches of SF so that the number of SFs can not be estimated by optical microscopy alone. Raman scattering[459-461], photoluminescence (PL)[457,462-466], synchrotron white beam X-ray topography (SWBXT)[458] and transmission electron microscopy (TEM)[457] have also been applied to estimate the SF behaviors in epitaxial growth and in annealing.

Mitani et al.[459] have studied the structural properties of stacking disordered 3C-SiC crystals by visible and deep UV Raman measurements. The relationship between the SF density and Raman spectra was studied and the reduction in free-carrier density was strongly correlated with the SF structure and density. The density of SFs could be qualitatively estimated by comparison between experimental and simulated Raman profiles.[460]

Hoshino et al.[462] have applied room temperature PL to the nondestructive detection of in-grown SFs in off-axis 4H-SiC substrates and epitaxial layers. By using a deep UV light excitation, the stacking fault related intensity pattern can be observed because of its shallow penetration depth. A bar-shaped PL intensity pattern agreed well with the etch pit pattern due to the stacking fault.

(5) Inclusions

Carbon inclusions[159,467,468] and Si inclusions[159] were found to exist in PVT grown SiC single crystals and they were thought to be the origin of micropipes.[159] It is therefore important to evaluate these inclusions.[469]

Dudley et al.[470] have studied foreign material inclusions by synchrotron X-ray topography and TEM and found that screw dislocations of various Burgers vectors can be nucleated in SiC during the process of incorporation of inclusions which was found to be the origin of micropipe formation. A pair of opposite-sign screw dislocations is created as shown in Fig. 23.49 during the process of inclusion overgrowth.

Rost et al.[468] have evaluated the inclusions in modified Lely grown 6H-SiC single crystals by SEM and electron energy loss spectroscopy (EELS), and identified them as carbon particles. The dependency of the carbon particle concentration on growth parameters such as seed temperature, time, pressure and the distance between the source and the seed has been studied. A concentration model for the occurrence of carbon particles was discussed and showed that any change of composition of the vapor phase results in a change of the particle density.

Vetter and Dudley[471] have used white-beam back-reflection X-ray topography to characterize two types of SiC single crystals, a Baikov SiC platelet with inclusions but free of axial screw dislocations and a modified-Lely SiC wafer with low micropipe

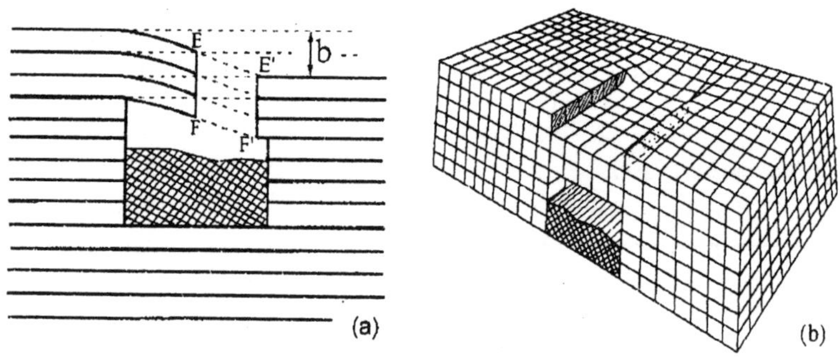

Fig. 23.49 Schematic diagram illustrating the creation of a pair of opposite-sign screw dislocations during the process of inclusion overgrowth. (a) Sectional view through the inclusion parallel to the dislocation line and (b) perspective view from above (reprinted from Ref. 470 with permission, copyright 1999 American Institute of Physics).

densities and free of inclusions. They have explained the contrast images of the X-ray topographs for these crystals.

Dong et al.[472] have shown that HRXRD is effective to observe various polytypes such as 6H- and 15R-SiC existing in 4H-SiC single crystals.

(6) Subgrain boundaries (low angle grain boundaries)

Katsuno et al.[473-475] have evaluated subgrain boundaries (low-angle grain boundaries) in PVT-grown 6H-SiC crystals by using high-resolution X-ray diffraction and molten KOH defect selective etching. It was found that subgrain boundaries were polygonized into <1-100> directions and most often occurred at the peripheral parts of crystals. It was also found that tilting of the (0001) lattice plane occurred at the subgrain boundaries and had a rotation axis parallel to both the boundary plane and the (0001) basal plane. The crystal mosaicity was found to originate in low angle grain boundaries.[474] They have also evaluated the mosaicity of 4H-SiC crystals grown in the [000-1] and [11-20] directions.[475] For the [000-1] grown crystals, the crystal mosaicity resulted from foreign polytype inclusions, while for the [11-20] grown crystals, there was a drastic change between the seed and the grown crystal.

(7) Other Defects

The "star" defect in 4H-SiC consists of a center region with high densities of threading dislocations (both edge and screw types) and six arms of dislocation arrays extending along <11-20> directions. Its structure and origin were discussed by Lee and Skowronski.[476]

Using Normarski optical microscopy and atomic force microscopy, Ohtani et al.[477,478] have studied the formation of macrosteps. Extremely regular equidistant step trains with the step height equal to the c-lattice parameter were observed on the 6H- and 4H-

SiC (0001) surface. Upon nitrogen doping, these regular step trains became unstable and they were transformed into meandering macrosteps of height greater than 10 nm. The mechanism of macrostep formation (step bunching) was discussed through the consideration of the interplay between repulsive step interaction and asymmetric step kinetics.

Hock et al.[479] have evaluated in-situ the thermally induced strain in the SiC crystal lattice during PVT bulk growth, using high energy x-ray diffraction as shown in Fig. 23.50. On increasing the growth rate, an increased thermoelastic effect was observed. Using numerical modelling, the effect was attributed to the heat of crystallization.

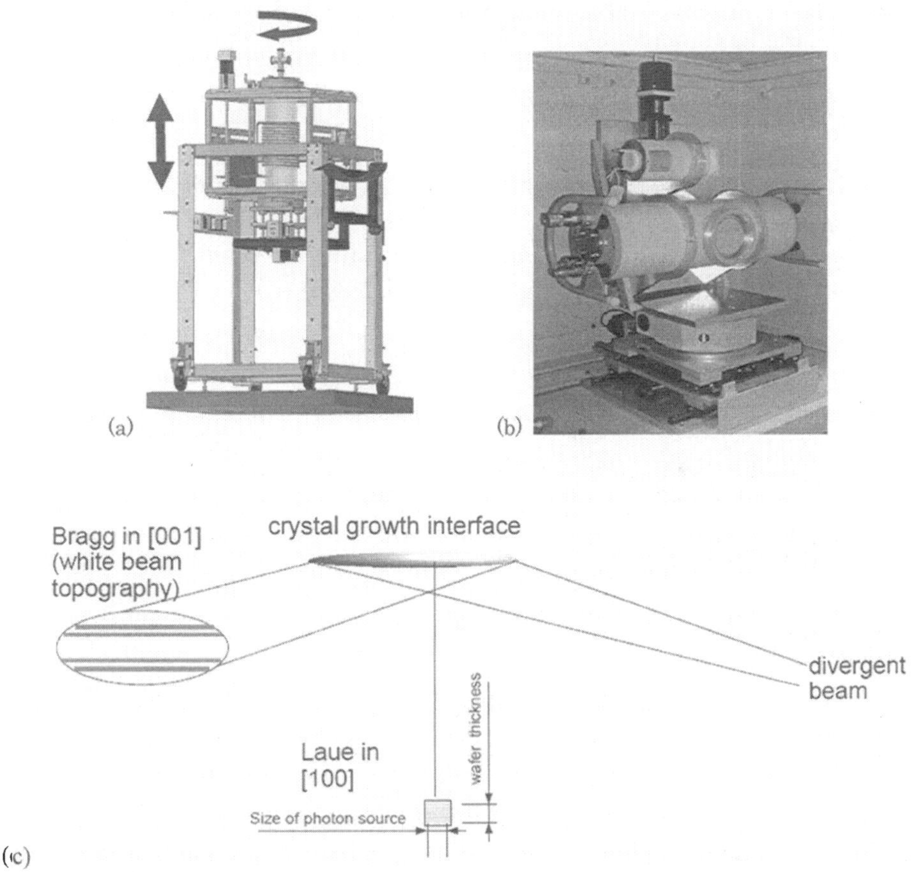

Fig. 23.50 (a) Schematic sketch of the PVT growth setup for in-situ x-ray diffraction with 3 degrees of freedom for crystal orientation, (b) image of the high energy x-ray sourc and (c) sketch of the measurement configuration (reprinted from Ref. 479 with permission, copyright 2010 Trans Tech Publlications).

(8) Point defects

Point defects in SiC include a variety of intrinsic defects, such as vacancies (monovacancies, V_{Si}, V_C, divacancies), vacancy pairs (V_{Si}-V_C), antisite defects (Si_C, C_{Si}), and antisite-vacancy pairs[407, 423, 480, 481] and they have been studied theoretically and experimentally not only in 6H- and 4H-SiC but also in 3C-SiC. The formation energies of these defects have been calculated by various methods[482-492] and the predominant point defects in various polytypes are predicted.

The existence of these defects in bulk 6H-, 4H- and 3H-SiC crystals has been studied experimentally, mainly by electron paramagnetic resonance (EPR)[480, 493-508], optically detected magnetic resonance (ODMR)[509, 510], electron energy loss spectroscopy (EELS)[511], positron annihilation spectroscopy (PAS)[502] and photoluminescence (PL).[494, 509, 511-515] Deep level transient spectroscopy (DLTS)[504, 516] has also been applied but it is limited because of the large band gap of SiC.

Various point defects have been identified as follows.

(i) Vacancies

Vacancy related defects such as monovacancy, divacancy and antisite-vacancy pairs studied by EPR are reviewed by Isoya et al.[493] and Bockstedte et al.[494]

(ii) Carbon vacancies

Positively charged carbon vacancies have been identified[495,496,507,508,510], which act as deep donors (+/0) E_V +1.47 ± 0.06 eV.[510] They are found as EI5 or EI6 centers in EPR.

(iii) Silicon vacancies

Silicon vacancy related defects have been identified in deep PL spectra.[509] Negatively charged V_{Si}[497], silicon vacancy-related V1 zero phonon line (ZPL) doublet at 860 nm[512] and negatively charged V_{Si} in HCVD-SiC[498] have been reported.

(iv) Divacancies

Divacancies such as $V_CV_{Si}^0$[499], V_{Si}-V_C^0[500] and vacancy aggregates[494] have been reported.

(v) Interstitials

Single carbon split-interstitials are suggested as EI_n centers.[501] Carbon interstitial-related defects have been reported from the study of local vibration mode (LVM) of optical centers.[517] Carbon interstitials are also detected by DLTS.[516]

(vi) antisite defects

Among antisite defects, an antisite pair such as Si_C-C_{Si} for D1 center[513], and di-carbon tri-carbon antisites known as P-T, HT3, HT4 centers[494] have been reported. Storasta et al.[514] have shown that the D1 center acts as a psudodonor correlating to a hole trap of E_V + 0.35 eV.

(vii) complexes

Among defect complexes, various defects such as V_C-C_{Si}[494,502-504,508,514], $V_CC_{Si}(Si_CC_{Si})$ as the DI defect[505], and V_{Si}-N_C[506], O-V_C oxygen-carbon vacancy complex[511] have been reported.

The importance of these point defects is increasing since they may affect the performance of electronic devices. In fact, it has been pointed out recently that point defect agglomeration and defect complex formation are important since they may greatly affect device performance.[494]

(9) Deep levels

There are various deep levels reported so far not only in bulk materials but also in epitaxial materials. These deep levels are summarized in Table 23.6.

Anikin et al.[518] have found deep levels at E_V+0.28 eV and E_V+0.61 eV in 6H-SiC epitaxial films prepared by the CVT and by ion implantation. Zhou et al.[519] found carrier traps at E_C-0.34 eV and E_C-0.68 eV in 3C-SiC grown on Si substrates by the CVD method using deep level transient spectroscopy (DLTS). Suttrop et al.[520] found the boron-related deep levels, that is the isolated boron acceptor at $E_B = E_V$+0.3 eV and the boron-related deep centers at $E_D = E_V$+0.58 eV. Uddin et al.[521,522] observed three electron traps at E_C-0.26 eV, E_C-0.35 eV, E_C-0.72 eV in 4H-SiC and one electron trap Ec-0.68 eV in 6H-SiC by DLTS and discussed their relationship with residual impurities such as V, Cr, Fe, and Ti.

Jang et al.[523] examined deep levels of 6H-SiC grown by a modified Lely method using isothermal current transient spectroscopy (ICTS) and DLTS and found electron traps located at 0.39-0.69 eV below the conduction band edge. Danno and Kimoto[524] have studied deep levels in p-type 4H-SiC epitaxial layers by DLTS.

Huh et al.[525] have studied hole traps in CVD grown 4H-SiC layers and found two boron-related traps at E_V+0.3 eV and E_V+0.6 eV by DLTS. They have also measured deep electron and hole traps in 6H-SiC grown by the halide chemical vapor deposition (HCVD) method using DLTS, PL and minority carrier spectroscopy (MCTS).[526] The dependence on the C/Si flow ratio and the growth temperature has been studied.

Kaminski et al.[527] have studied deep levels in PVT-grown N-doped 6H-SiC by DLTS and found six electron traps. The assignments of these levels are proposed to be complexes involving Si vacancies, those between carbon antisites and nitrogen atoms and vanadium contamination. Sasaki et al.[528] have studied deep levels in n-type 4H- and 6H-SiC epilayers and found four main centers which are related to carbon vacancies.

- V-doped Semi-Insulating (SI)

It was found that V doping is effective in preparing semi-insulating SiC and SI-SiC crystal growth has been performed.[152,166,167] This semi-insulating behavior is similar to that of Cr doping for GaAs (Sec. 8.3.1(4)) in Vol.1) and Fe doping for InP (Sec. 10.4.3(2) in Vol.1).

Vanadium itself was found to act as a deep donor and a deep acceptor and their energy levels were evaluated by various methods. They were measured by electromagnetic resonance (EPR) [405,529-531], electron spin resonance (ESR)[406,407] photo-ESR [406,408], magnetic circular dichromism of the absorption [532,533], optical absorption[405, 534-537], optical admittance spectroscopy (OAS) [537,538], high temperature Hall measurement [152,531,535,536,539], photoluminescence (PL) [405,529,540], high-resolution photoinduced transient spectroscopy (HRPITS) [539], thermally stimulated current (TSC)[535] and DLTS.[414,540]

Schnider et al.[404-408] found that V is a deep donor at the middle of the bandgap and then the electrical behavior has been investigated.[152,166,167] V was found to occupy silicon substitutional sites in one of the three possible charge states: positive ($3d^0$), neutral ($3d^1$) and negative ($3d^2$). It is an amphoteric impurity and produces two levels within

Table 23.6 Deep Levels in SiC

Material	Growth method	Deep level energy (eV)	Author	Year	Ref.
p-type, n-type					
6H-SiC:Al	CST	Ev+0.28, +0.61	Anikin et al.	1985	518
3C-SiC/Si	CVD	Ec-0.34, -0.68	Zhou et al.	1987	519
6H-SiC:B	LPE	Ev+0.3, +0.58	Suttrop et al.	1990	520
6H-SiC	LPE	Ec-0.71	Uddin et al.	1993	521
4H-SiC	PVT	Ec-0.26, -0.35, -0.72	Uddin et al .	1994	522
6H-SiC	PVT	Ec-0.68			
6H-SIC	PVT	Ec-0.39, -0.43, -0.69	Jang et al.	1994	523
4H-SiC:Al,B	CVD	Ev+0.84, +1.27, +1.44	Danno et al.	2006	524
4H-SiC	CVD	Ev+ 0.3, +0.6	Huh et al.	2006	525
6H-SiC	HCVD	Ec-0.36, -0.4, -0.5, -1.0 Ev+ 0.58, +0.65	Huh et al.	2006	526
6H-SiC:N	PVT	Ec -0.34, -0.40, -0.64, -0.67 eV Ec -0.69, -1.53	Kaminski et al.	2008	527
4H-SiC	CVD	Ec-0.36, -0.65, -0.79 Ec-1.0, -1.48	Sasaki et al.	2011	528
6H-SiC	CVD	Ec-0.45, -0.57, -0.80, -1.25	Sasaki et al.	2011	528
vanadium doped SI					
6H-SiC:V	PVT	Ev+1.31, 1.46 (donor)	Dörnen et al.	1992	534
6H-SiC:V	PVT	Ec-1.48 (donor)	Hobgood et al.	1995	152
6H-SiC:V	MPVT	Ec-1.35 (donor)	Jenny et al.	1995	535
4H-SiC:V	PVT	Ec-0.80 (acceptor)	Jenny et al.	1996	540
6H-SiC:V	PVT	Ec-0.66 (acceptor)	Jenny et al.	1996	540
6H-SiC:V	PVT	Ec-1.42 (donor) Ev+2.4 (acceptor)	Mitchel et al.	1998	536
4H-SiC:V	PVT	Ec-1.1	Zvaut et al.	2006	531
6H-SiC:V	PVT	Ec-0.85	Zvaut et al.	2006	531
undoped SI					
4H-SiC	PVT	Ec-1.1	Mithcel et al.	2000	414
4H-SiC	PVT	Ec-0.55	Mitchel et al.	2000	572
4H-SiC	PVT	Ev+1.1 (ID)	Zvanut et al.	2002	542
4H-SiC	PVT	Ev+0.9 (ID-1)	Konovalov	2002	560
4H-SiC	HTCVD	Ev+1.4 (donor)	Ellison et al.	2003	269
4H-SiC	HTCVD PVT	Ev+1.47	Son et al.	2003 2004	507 508
4H-SiC	HTCVD PVT	Ec-(1.24-1.51)	Son et al.	2004	508
4H-SiC	PVT	Ec-1.26 (X_h) E+1.36 (X_k)	Kalabukhoval	2006	544
6H-SiC	HTCVD	Ev+1.47	Carlsson et al.	2009	548

the bandgap of SiC. One is the donor level (0/+) ($V^{5+/4+}$) which resides approximately at the middle of the band gap.[406,534-536,538] The another is the acceptor level (-/0) (V^{3+}/V^{4+}) which is located at energy levels of about 0.8 eV[531,536,538,540] below the conduction band.

The energy levels of vanadium have been extensively studied and are summarized in Table 23.6.

Miao et al.[541] have calculated the energy levels of Ti, V and Cr by the density functional calculation using the full-potential linearized muffin-tin orbital method. The calculated energy levels were in good agreement with the PL experimental results.

- *Undoped Semi-Insulating (SI)*

It is also reported that without doping any vanadium, high purity (vanadium free) semi-insulating HP-SI SiC (also known as nominally undoped SI SiC) can be obtained when the concentration of residual impurities such as N and B is sufficiently reduced (below 10^{16} cm^{-3}) to the level at which intrinsic defects can compensate these residual impurities.[414-418]

The deep levels which may be the origin of the HP-SI SiC have been measured by various methods, such as EPR and photo-EPR [269,415 417,418,542-562], Fourier transform infrared (FTIR) spectroscopy [563], positron annihilation radiation Doppler broadening spectroscopy [554], positron annihilation spectroscopy (PAS) [565-567], photoluminescence (PL)[545,547,549,568-571], Zeeman spectroscopy[568], optical admittance spectroscopy (OAS) [414,416,418,543,572], Hall measurement [414-416,418,569], thermally stimulated current (TSC) technology [572,573], discharged current transient spectroscopy (DCTS) [574], persistent photoconductivity (PPC)[543], DLTS[416,417,575,576], time resolved four wave mixing (FWM) technique[577] and photoinduced current transient spectroscopy (PICTS).[578]

Various defects have been proposed and observed as the deep levels which compensate donor or acceptor impurities.

(i) Carbon vacancies

positively charged carbon vacancy V_C^+ [269,270,495,507,508, 79], neutral carbon vacancy V_C [446,546,548,558,561,562,565], negatively charged carbon vacancy V_C^-.[550,565]

(ii) Silicon vacancies

V_{Si}.[546, 550, 558, 565, 570]

(iii) Divacancies

V_CV_{Si} [508,546,548,558], SI-5 divacancy which was identified as the negative charged state of carbon vacancy-carbon antisite pair $V_CC_{Si}^-$ [551,557,559], neutral divacancy $V_CV_{Si}^0$ [552], vacancy cluster [564-566]

(iv) Antisite defects

silicon antisite Si_C [562], first-neighbor antisite pair Si_C-C_{Si}.[513]

(v) Complexes

V_C-C_{Si} [546,548,558], negative carbon vacancy - antisite pair $V_CC_{Si}^-$ [499,551,557,559], V_CV_{Si} [546,548,558], V_C-C_{Si}-V_C[545], the contamination of the carbon vacancies with hydrogen.[544]

(v) Unidentified defects

P center[552], UD-1-UD-3 [547,563,570,580], UD-4 [561,568], D1 center [569,571], ID-1 [560], ID center [542], ID-1, ID-2 [562] and SI1-SI5.[557]

The energy levels of some of these deep levels are summarized in Table 23.6 and various intrinsic defects responsible for HP-SI SiC are reviewed by Zvanut.[556] The annealing effect on the semi-insulating properties has been studied [547,569,571] in order to clarify the stability of the semi-insulation. To the extent that the semi-insulating property is the result of the compensation between the residual impurity and the intrinsic defect, it is vital to evaluate the instability because of the changes brought about in intrinsic defects by annealing.

The best semi-insulating material will be the compensation of a reduced residual impurity of the level of 10^{14} cm^{-3} and a small amount of vanadium of the around 10^{15} cm^{-3}. In this case, the concentration of vanadium is too low to affect the device parasitics. It must be noted that to activate vanadium as effective deep levels, it may be necessary to anneal at appropriate temperatures. This is actually realized in the case of InP as nominally undoped SI material, where in reality the compensation occurs of the residual shallow donor of the level of mid 10^{14} cm^{-3} with Fe of the level of 2×10^{15} cm^{-3} (Sec. 10.4.3(3) in Vol. 1).

23.4.3 Electrical Properties

(1) Transport properties

The mobilities of 6H-, 4H and 3C-SiC are as shown in Table 23.1. It is known that 4H- and 3C-SiC are preferable for use in electronic devices since their electron mobilities are higher than that of 6H-SiC. This is the reason why 4H-SiC is mainly used for electronic devices and 3C- SiC is being widely developed as explained in Sec. 23.3.1(4).

The mobilities of various polytypes were determined not only by conventional Hall measurement[246,403,520,581-590] but also by Raman scattering.[591-597] In Fig. 23.51 [589], the mobility of 4H-SiC at room temperature as a function of total impurity density is summarized and compared with the theoretical value. Fig. 23.52 shows the mobility of 3C-SiC as a function of temperature obtained from experimental and theoretical study.[586] In

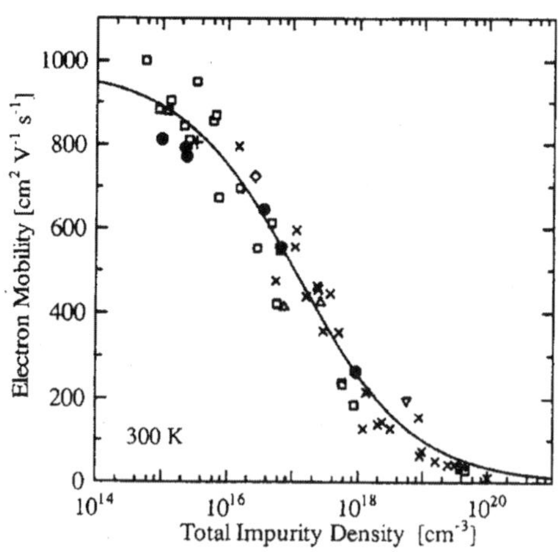

Fig. 23.51 Mobility as a function of the carrier density for 4H-SiC. The data obtained by Hall measurement. See references for each data set in the original article (reprinted from Ref. 590 with permission, copyright 2004 American Institute of Physics).

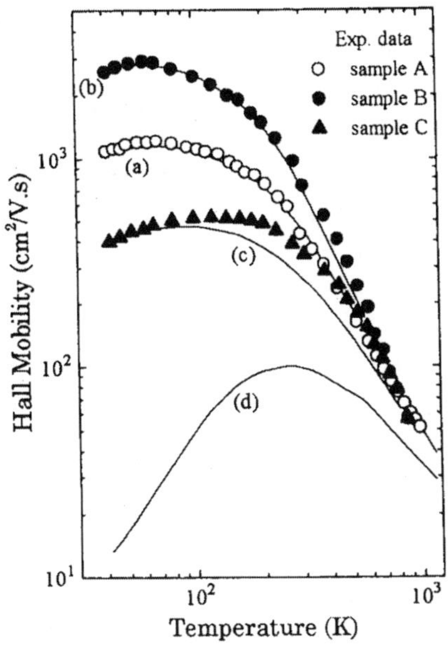

Fig. 23.52 Electron mobility as a function of temperature for β-SiC (3C-SiC). The data obtained by iterative calculation based on Hall measurement data (reprinted from Ref. 586 with permission, copyright 1998 Institute of Physics).

Fig. 23.53, the mobility of 4H-SiC obtained from Raman scattering is compared with that obtained from Hall measurement.

SiC also shows the high field drift velocity at high electric fields compared with other compound semiconductors (Fig. 1.18 in Vol. 1) and this is also the reason why SiC is preferable for electronic devices. The drift velocity has been theoretically calculated and compared with the experimental results[598, 599] as shown in Fig. 23.54.

(2) p, n control

It is known that there are residual impurities, nitrogen and boron, which act as a shallow donor and a shallow acceptor respectively. These residual impurities are reported to be reduced to the level of a few times 10^{14}-10^{15} cm^{-3} [58,196,418] by appropriate purification.

Carrier concentrations are normally evaluated by conventional Hall measurement and the dopant level can be deduced by analyzing the temperature dependence data of the carrier concentration by the free carrier concentration spectroscopy (FCCS) method.[31,600-602]

n-type SiC can be obtained when N_2 gas is added to Ar gas in the PVT crystal growth atmosphere. The carrier concentration can be controlled from 10^{16} cm^{-3} up to 10^{19} cm^{-3}. N acts as a shallow donor for various polytypes such as 3C, 4H, 6H and 15R. It is

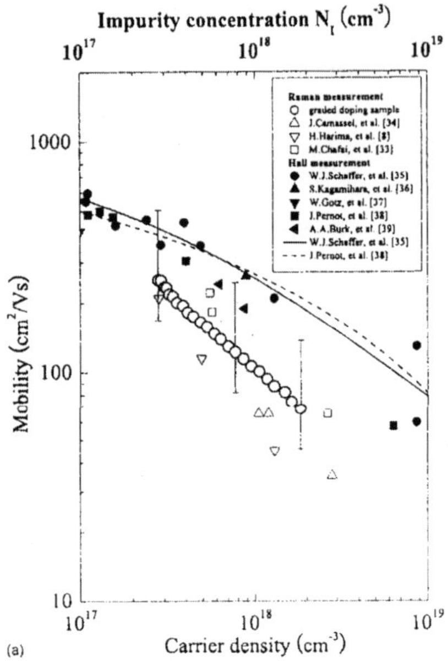

Fig. 23.53 (White symbols) Carrier mobility determined from line shape analysis of the longitudinal optical phonon-plasmon coupled (LOPC) mode of Raman scattering in 4H-SiC. (Black symbols) Hall measurement data. See references for each data set in the original article (reprinted from Ref. 597 with permission, copyright 2007 American Physical Society).

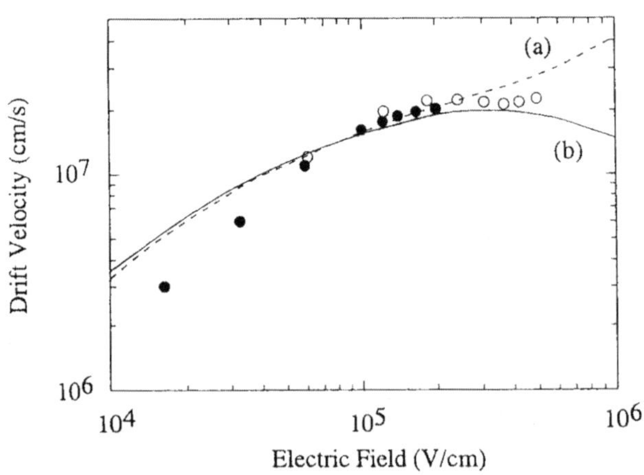

Fig. 23.54 High-field drift velocity at 300 K for (a) parabolic and (b) nonparabolic-band models. Solid and open circles are the experimental data (von Munch[43]) for 6H-SiC. (reprinted from Ref. 598 with permission, copyright 1993 Elsevier).

confirmed by ESR that N occupies the C site under normal doping conditions.[603, 604] For shallow donors, impurity levels of N [395,583,584,606,610-613], P [386-388,606] and As[607] have been reported as shown in Table. 23.7. In Fig. 23.55, the resistivity of 6H- and 4H-SiC is shown as a function of nitrogen doping.

p-type SiC can be obtained by doping Al [31,583,605,608,609] or B [395,605,608,609,614] and the ionization energies are shown in Table. 23.7. For doping Al, Al is mixed in raw materials or Al doped SiC powders are used as the material source for crystal growth. Carrier concentrations of the order of 1×10^{15}-3×10^{18} cm^{-3} are reported.[615]

Takahashi et al.[141] showed the influence of the seed plane on the electrical properties of the grown crystals. They showed that the undoped crystal grown on (000-1)C face showed n-type conduction while that grown on the (0001)Si face was highly resistive p-type. These differences were attributed to differences in impurity incorporation.

(3) Semi-Insulating (SI) properties

In the case of V-doped SI SiC, Hobgood et al.[152] first obtained SI SiC by doping vanadium as a deep donor to compensate a shallow acceptor, boron. The resistivity obtained is as shown in Fig. 23. 56 and it exceeds 10^{15} Ω.cm at room temperature.[152, 166, 167] Bickermann et al.[411] have calculated the Fermi level as a function of the compensation ratio (Fig. 23.57). It is shown that when the compensation ratio $\theta = (N_V/N_A\text{-}N_D)$ is larger than unity, SiC becomes semi-insulating with Fermi levels of about $E_V + 2.4$ eV where

Table 23.7 Ionization Energies of Dopants in SiC

(meV)

Polytype	Impurity	Ionization Energy	Ref.
4H-SiC	Donor		
	N	52.1(h), 91.8(k)	584
		42(h), 84(k)	606
	P	53(h), 93(k)	606
		52-53, 78-80	387
		53(h), 49(h), 26(h)	388
		109(k), 101(k), 74(k)	388
	As	66.8(h), 127(k)	607
	Acceptor		
	B	285	605, 608
	Al	183	608
		160-165	609
6H-SiC	Donor		
	N	150	395
		85.5(h), 125(k)	583, 610
	P	82-83, 105-114	387
		91.5, 81.8, 73.5	386
		70, 85	385
	Acceptor		
	B	390	395
		300	614
	Al	199	605
		200	583
3C-SiC	Donor		
	N	48(k)	612, 613

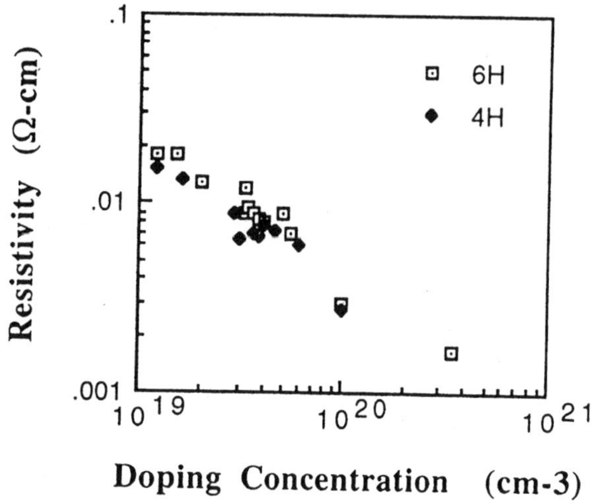

Fig. 23.55 Resistivity vs. nitrogen doping concentration for 6H and 4H-SiC boules (reprinted from Ref. 51 with permission, copyright 1995 Institute of Physics).

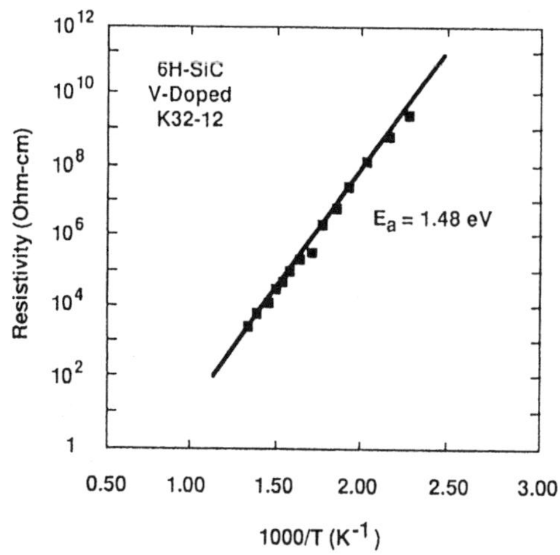

Fig. 23.56 Resistivity as a function of temperature for vanadium-doped 6H-SiC grown by the PVT method (reprinted from Ref. 152 with permission, copyright 1995 American Institute of Physics).

the residual donor dominates and of about E_V +1.6 eV where the residual acceptor dominates. Here, N_V is the concentration of V and N_A, that of the residual shallow acceptor and N_D, that of the residual shallow donor.

Semi-insulating properties have also been obtained without vanadium doping when

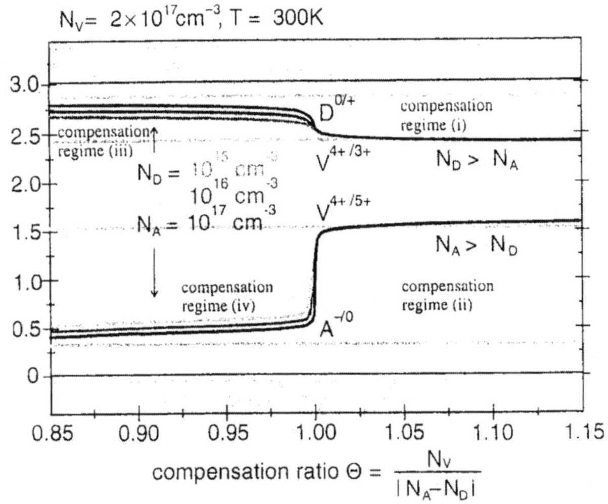

Fig. 23.57 Dependence of the Fermi level on the compensation ratio for different shallow impurity concentrations N_A and N_D in vanadium doped 6H-SiC (reprinted from Ref. 411 with permission, copyright 2003 Elsevier).

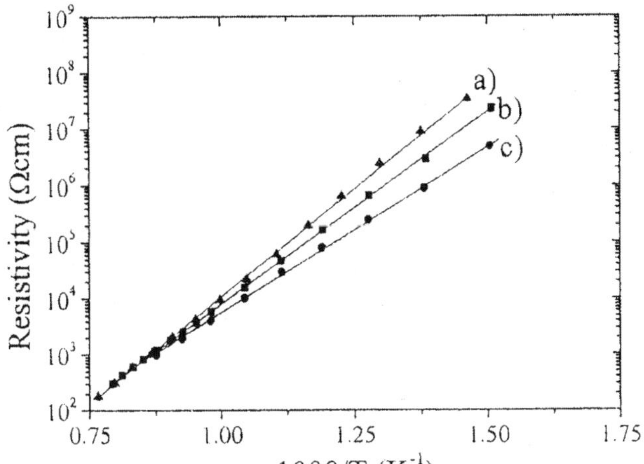

Fig. 23.58 Resistivity vs. temperature for three 4H-SiC HP-SI samples whose activation energies are (a) 1.53 eV, (b) 1.33 eV and (c) 1.19 eV (reprinted from Ref. 415 with permission, copyright 2002 Springer).

the residual impurity is reduced to the extent that intrinsic deep levels can compensate the residual impurities. This is called high purity semi-insulating (HP-SI) and was obtained not only by the PVT method but also by the HT-CVD method. In Fig. 23.58, a typical resistivity of HP-SI obtained by the PVT method is shown. The resistivity is known to be about 10^9 cm^{-3} Ω.cm at room temperature.[414]

Tamulatis et al.[616] have evaluated the carrier life time of V-doped SiC wafers and obtained the value of 130 ± 5 ps with a diffusion coefficient of 0.9 ± 0.5 cm^2s^{-1}. Muzykov et al.[617] have compared various methods to measure high resistivity such as the four point prove method, the van der Pauw method, the current-voltage technique and the contactless resistivity measurement.

It should be emphasized that 6H- and 4H-SiC becomes insulating without any electrons and holes, beyond semi-insulating when the Fermi level is in the middle of the bandgap since the bandgap is as large as 3.2 eV.

23.4.4 Optical Properties

(1) Raman Spectrum

Typical Raman spectra of various polytypes are shown in Fig. 23.59. The Raman spectrum is sensitive to the polytype stacking structure so that it is used to evaluate the mixing of polytypes.[618-622] It is also sensitive to the existence of stacking faults.[623, 624] When stacking faults exist, the form of the Raman band is distorted or a new Raman band appears. When the density of stacking faults is increased, the new spectrum due to the new polytype appears. The intensity of longitudinal optical phonon plasmon (LOPC) coupled mode spectrum is decreased as a function of the carrier concentration. The carrier concentration and the mobility can be estimated by fitting the LOPC mode spectrum to the theoretical band.[591-597] Raman spectroscopy is also effective in detecting defects in SiC.[625-627] It is also effective in evaluating the C concentration in PVT grown crystals.[197] The application of Raman spectroscopy and Raman microprobe is reviewed by Nakashima et al.[628,629]

(2) Photoluminescence (PL)

Photoluminescence is an effective characterization method, in which various optical transitions as shown in Fig. 5.8 of Sec. 5.4.1 in Vol. 1 can be observed and the purity and the existence of defects can be evaluated by comparing the intensity of these transitions.

- Exciton emission

Edge emissions where various exciton peaks can be detected are reported by various researchers.[391-393,583,630-634] In Fig. 23.60, an example of edge emission of 4H-SiC is shown. Various emissions related to excitons such as free excitons (FEs) and bound excitons (BEs) are seen. In this figure, the emission line indicated as I is due to free exciton emission and the intensity becomes higher with higher purity compared with bound excitons and D-A pair emission. The subscript shows the related phonon energy (meV) for each line. P_0 and Q_0 are zero phonon lines of neutral N donor bound exciton emissions and the former is an N atom at the hexagonal site and the latter is an N atom at the cubic site.

These bound exciton lines can be observed not only for N donor[58, 376, 634] but also for P donor[391-393], Al acceptor[31] and B acceptor.[631] When the impurity concentration is de-

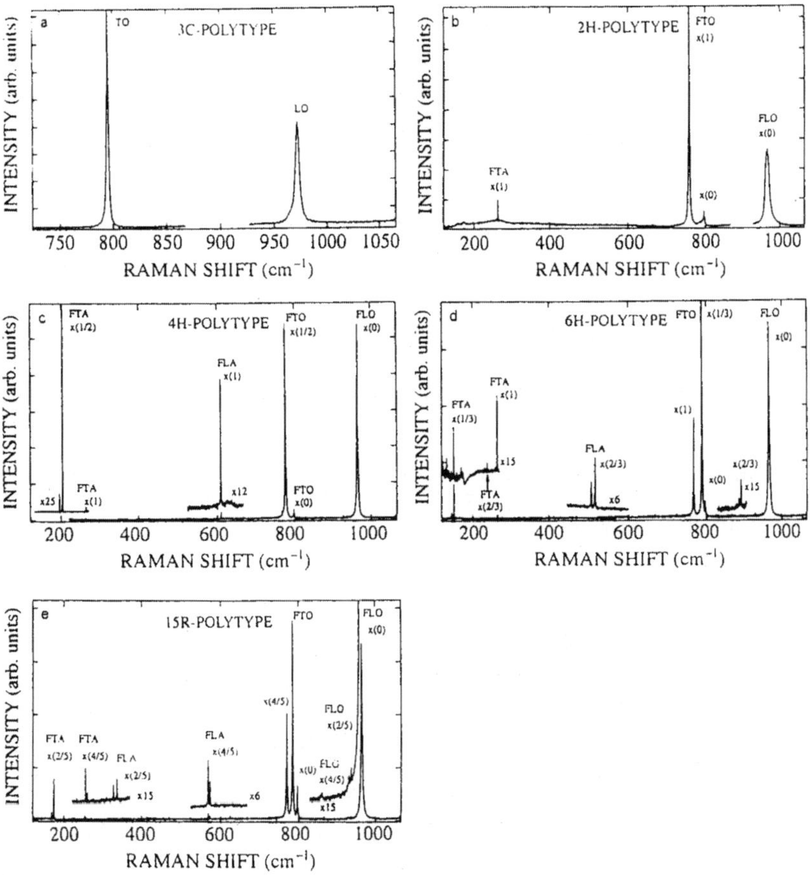

Fig. 23.59 Raman spectra measured in a quasi-scattering geometry for various SiC polytypes. (a) 3C, (b) 2H, (c) 4H, (d) 6H and (e) 15R (reprinted from Ref. 622 with permission, copyright 1997 John Wiley & Sons).

creased, the FE emission becomes strong while the impurity related BE emission is weakened. From the intensity ratio between the BE emission and FE emission, it becomes possible to quantify the impurity.[632] Edge emission spectra have been measured to evaluate material quality for Acheson crystals [633], Lely crystals [633], modified Lely crystals[632] and CVD epilayers.[632]

Actually, nitrogen donor concentrations can be precisely determined by photoluminescence.[376, 418, 420, 421, 634] In fact, in the case of nitrogen impurity, a good correlation line is obtained between the intensity ratio $I_{Q_0}/ I_{76.4}$ and the nitrogen concentration measured by the C-V method in the concentration range of 10^{14}-10^{16} cm^{-3}.[421] Low nitrogen concentrations as low as 3×10^{14} cm^{-3} were reported.[418]

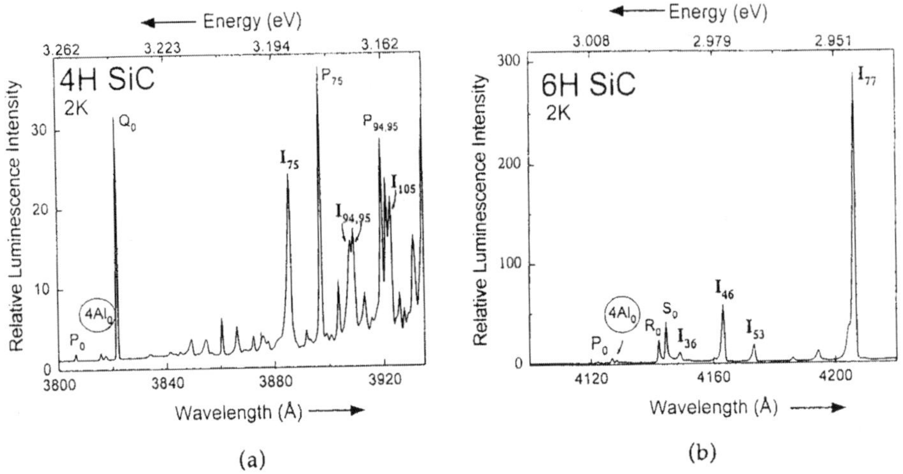

Fig. 23.60 Low temperature photoluminescence spectra of (a) a 4H-SiC epilayer grown on a 4H-SiC susbtrate and (b) a 6H-SiC epilayer on a 6H-SiC substrate (reprinted from Ref. 631 with permission, copyright 1997 John Wiley & Sons).

Chaussende et al.[108] have observed multiple bound exciton complex (MBEC) lines up to m = 5, accompanied by the main exciton line (m = 1) even in solution-growth 3C-SiC crystals, which proves the high purity of grown crystals.

- Donor-Acceptor (DA) pair emission

In Fig. 23.61, DA pair emission lines for Al doped 4H-SiC are shown. Many emission lines due to N donor-Al acceptor pair emissions depending on the pair distance are seen.[635] B acceptor and Ga acceptor also show D-A pair emissions with N donor.[635] Ivanov et al.[636] have observed the P donor-Al acceptor pair emission. Conteras et al.[400] used the Al-N pair for quantitative evaluation of the Al concentration. N-Al and N-B pairs in 3C-SiC have also been reported.[630]

- Defect emission

DI luminescence is well known as a defect emission.[630] It is seen in as-grown crystals, heat treated ion-implanted samples and radiation induced samples. In 4H-SiC, it is called the L1 line and is seen as a sharp emission line of 2.901 eV. The origin of this emission is attributed to the BE emission with Si vacancies. DII luminescence shows a sharp zero phonon line at 3.204 eV in 4H-SiC, and is attributed to C di-interstitials.

Stacking faults (SFs) are also detected by photoluminescence. In the region where SFs exist, the peaks at 1.7 eV and 0.95 eV are increased.[637] It is also reported that SFs show PL peaks at around 2.54 eV and 2.9 eV.[458,462]

- Deep level emission

Among deep levels, emissions due to transition metals such as V and Cr, and emissions

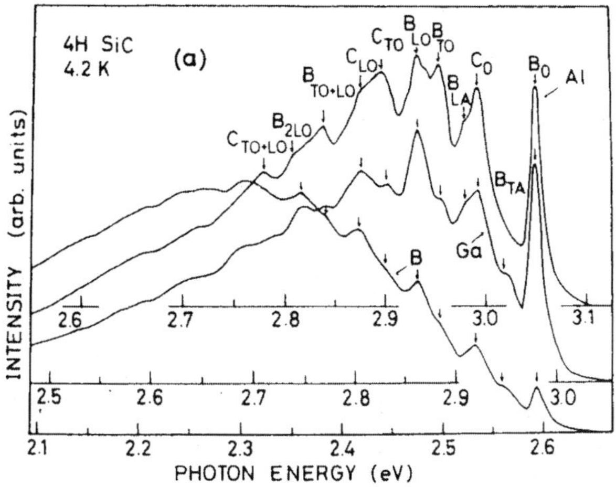

Fig. 23.61 Photoluminescence spectrum of Al-doped, Ga-doped and B-doped 4H-SiC at 4.2 K. The subscripts indicate the kinds of phonons and superscripts indicate the kinds of sites which acceptors replaced (reprinted from Ref. 635 with permission, copyright 1980 American Physical Society).

due to defects are observed.[150,2467,638-641] V has emissions in the range of 0.82-0.97 eV (Fig. 23.62). In Fig. 23.63, the emission of commercial 4H-SiC at 4.2 K is shown. Emissions indicated as α and β at around 0.82-0.97 eV are due to V at hexagonal Si site and cubic Si site. These V related PL peaks can be used to evaluate the V doping efficiency for SI crystals.[641] There are also undefined emissions such as UD-1, UD-2, UD-3 and UD-4 and the observation of these defects has been reported by various researcher.[553,561,642]

- *PL mapping*

PL mapping of SiC substrates and epitaxial layers has been performed to evaluate the uniformity of the material, using 1.80 eV and 1.06 eV peaks for substrates and 1.8 eV and 3.2 eV peaks for epitaxial layers.[639] It was shown that there is a big difference in the mapping features between substrates and epitaxial layers owing to the difference in the growth mechanism.

Microscopic mapping of deep level emission[640] has been studied for 4H-SiC using 0.94 eV, 1.15 eV and 1.34 eV peaks around the macrodefects called cracks.[428] The variation of the intensity around macrodefects is ascribed to the gettering effect of impurities due to the strain field.[640] Hirano et al.[637] have shown that SFs can be detected as bright lines for the 1.7 eV peak and dark lines for 0.95 eV by PL mapping.

23.4.5 Effect of defects on device performance

Since there is a variety of defects in SiC crystals and they strongly affect device performance, many studies have been performed to clarify the relationship between different

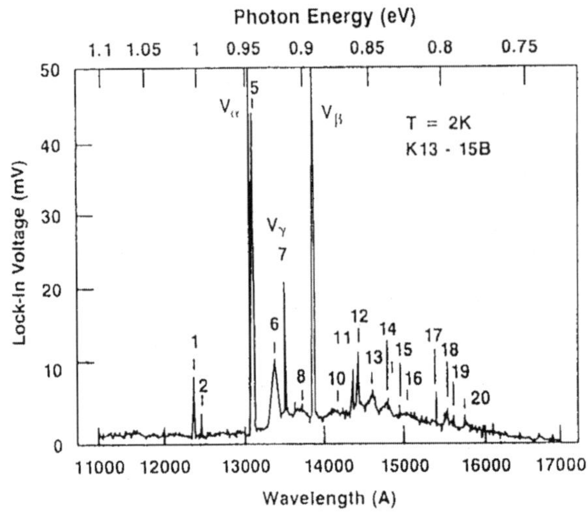

Fig. 23.62 Low temperature photoluminescence (LTPL) spectrum of undoped 6H-SiC exhibiting peaks due to vanadium impurity (reprinted from Ref. 150 with permission, copyright 1994 Elsevier).

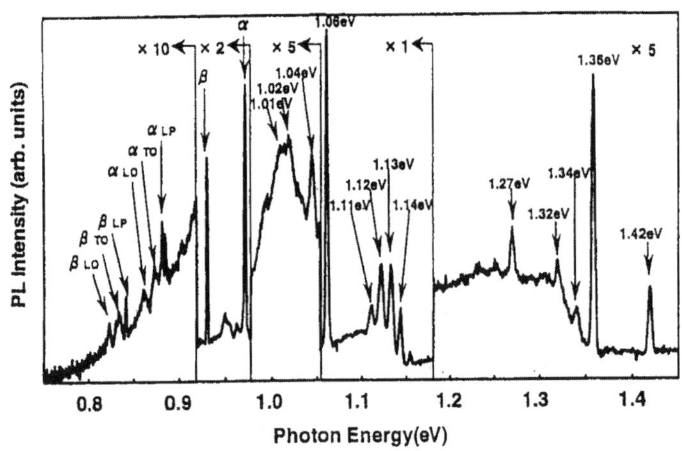

Fig. 23.63 High resolution spectrum of deep level PL from p-type 4H-SiC substrate at 4.3 K under 325 nm light excitation (reprinted from Ref. 639 with permission, copyright 2002 Trans Tech Publlications).

defects and device performance.

Typical effects of defects on device performance are:

(i) device failure of the reverse current - voltage characteristics due to micropipes,

(ii) forward bias drop in SiC pin diodes under forward bias operation,
(iii) gate oxide reliability of SiC MOSFETs.

These effects have been studied by various techniques such as transmission electron microscopy (TEM), synchrotron white beam X-ray topography (SWBXT), cathodoluminescence (CL), electroluminesence (EL), light emission imaging (LEI), optical emission microscopy (OEM) and photoluminescence (PL) compared with device performance data.

(1) Micropipes (MPs)

Neudeck and Powell[323] first reported the occurrence of device failure of the reverse curren-voltage characteristics at voltages below the avalanche breakdown field for epitaxially grown pn junction devices of 1 mm^2 or larger area in 6H- and 4H-SiC substrates. It was shown that this failure is due to micropipes in the substrates which propagated into the epitaxial layers and that reduction of the micropipe density would be required in order to realize the industrialization of SiC devices.[643-647]

Fazi et al.[444] have studied the defect structure of modified Lely crystals and Lely crystals and found that the latter have very low micropipe density. They have shown that a lower leakage current can be obtained in pn diodes fabricated on Lely crystals than in pn diodes fabricated on modified Lely crystals.

Kamata et al.[334,648] have studied the effect of micropipe closing (Sec. 23.3.3(1)) by the CVD method on the electrical properties of Schottky barrier diodes (SBDs) and found that closed screw dislocations which are formed by the dissociation of the micropipes did not affect significantly the breakdown voltage.

Tanaka et al.[649] have fabricated 1.4 kV mesa-type p^+-n diodes on CVD epilayers using a high quality in-house substrate cut from a high quality PVT crystal grown on the Lely seed crystal with no micropipes and with only minimal defects. They have obtained diodes with ideal avalanche breakdown without the leakage current increase and without forward bias degradation.

Kato et al.[650] have studied the crystal regions with low Schottky barrier height on 4H-SiC surfaces by the electrochemical deposition of ZnO. It was found these regions were formed not only by micropipes and various dislocations but also by other defects.

Kimoto et al.[651] have shown that device failure of the reverse current - voltage characteristics of pn diodes occurs even though micropipe free substrates have been used. Specific defects revealed as triangular-shaped defects and carrot-like grooves after reactive ion etching (RIE) greatly influence the device failure.

(2) Dislocations

Konstantinove and Bleichner[652] have observed the elelctroluminescence of a pn diode, which is due to defects related to dislocations. This is in contrast to he well-known dark-line defects due to the degradation of GaAs based LEDs.

It is also known that basal plane dislocations (BPDs) as well as stacking faults (SFs) generation are among the most severe lifetime limiting and performance killing defects in SiC bipolar electronic devices.[653] BPDs are known to dissociate into Schokley partial dislocations resulting in the formation of SFs.[654, 655]

The nucleation sites of SFs in SiC pin diodes during forward current operation have been intensively studied. The primary nucleation sites were found to be (0001) BPDs.[656-659] The other nucleation sites of SFs, threading dislocations[660, 661] and screw dislocations[662] have been reported. The effect of SFs generated from dislocations is described in the next section.

It is also found that even if there are no micropipes, dislocations such as screw dislocations affect the breakdown voltage.[644,663,664] Screw dislocations in SiC pn diodes lead to the onset of microplasma at a slightly lower reverse voltage than the theoretical breakdown voltage.[665] Wahab et al.[666] have found that screw dislocations under the active area directly affect the breakdown voltage. They found that only diodes with low screw dislocations and free from micropipes could block 2 kV or higher voltage. The relationship between the etchpits and leakage currents is also proved.[134]

In the case of devices with Shottky barriers, the effect of dislocations is not clear. Tsuji et al.[667] have analyzed the leakage currents in 4H-SiC shottkey barrier diodes using optical beam induced current (OBIC) measurement. They found that the bright spots on OBIC were located beneath the Schottky interfaces but they did not correlate with the closed-core screw dislocations.[668] Saitho et al.[669] have examined the relationship between dislocations and leakage current in SiC Shottky barrier diodes but did not find a clear correlation. Izumi et al.[670] have shown that threading closed core screw dislocations and edge dislocations are not as detrimental to device performance. Aigo et al.[671] have studied the effect of screw dislocations on the DC characteristics of metal-semiconductor field effect transistor (MESFETs) fabricated on epitaxial layers grown on 6H-SiC substrates. It was found that screw dislocations have a negligible effect on the performance of MESFETs while they caused some degradation of the Schottky gate characteristics.

Soloviev et al.[672] have fabricated avalanche photodiodes (APDs) and evaluated the effect of crystal defects on the reverse I-V characteristics. It was found that new defects are generated even in the dislocation free area by thermal breakdown.

Senzaki et al.[337, 673] have studied the effect of dislocations in the epitaxial layers grown on n-type 4H-SiC substrates on the reliability of thermal oxides used in metal-oxide-semiconductor field effect transistor (MOSFET) fabrication. It was found that the dielectric breakdown of thermal oxides occurs at the position of dislocations (Fig. 23.64) and that BPDs are the most common cause of this breakdown.

Threading screw dislocations in substrates are known to be the origin of the carrot defects in the epitaxial layers. These carrot defects are dislocation-stacking fault complex defects[674, 675] and are known to be fatal defects for SiC power devices. There was however a discrepancy between the density of carrot defects in the epitaxial layers which is much lower than that of screw dislocations. It is proposed that carrot defects can be generated when there are BPDs close to the screw dislocations.[676]

(3) Stacking faults (SFs)

The degradation of the forward voltage of p-i-n diodes is a serious issue for SiC device

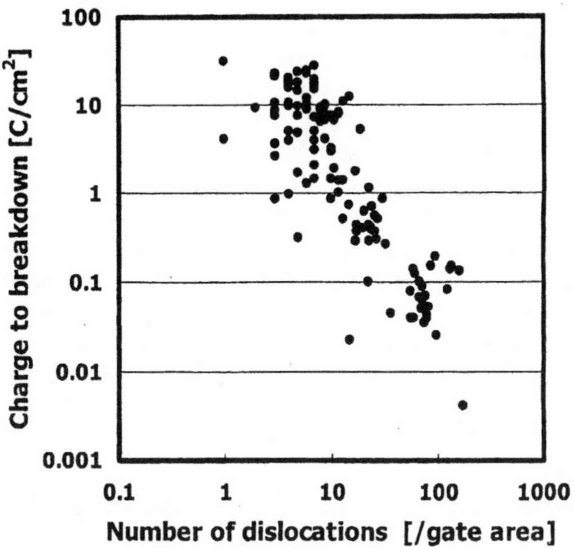

Fig. 23.64 Correlation between the number of dislocations under a gate oxide area and the values of the charge-to-breakdown (QBD) extracted from the constant voltage (9MV/cm) time-dependent dielectric breakdown (TDDB) measurement (reprinted from Ref. 337 with permission, copyright 2005 Trans Tech Publlications).

applications. This degradation phenomenon, first observed by Lendenmann et al.[677] is associated with the nucleation and development of SFs in the active region of SiC pin diodes under forward bias carrier injection.[336] These SFs are of the shear type and they extend rapidly under forward biasing. Most of them are single-layer faults with a triangular shape, although rhombic ones are also reported. The SFs extend by the motion of the bounding leading partial dislocation, the expansion of which is enhanced by a redirection of the e-h recombination energy into the motion of partial dislocations. This phenomenon is known as a recombination-enhanced dislocation glide (REDG) mechanism.[678] The formation of SFs has been studied theoretically.[354,679-685]

This degradation phenomenon has therefore been extensively studied by various researchers and it was revealed that the degradation happens both for 4H- and 6H-SiC.[686-706] Miao et al.[460] have studied the electronic driving force for the growth of SFs on n-type 4H-SiC in operating devices.

The nucleation sites of SFs in SiC pin diodes during forward current operation are not only BPDs[656-659] and other dislocations[660-662] as mentioned above in the former section but also basal segments of half-loop array defects [656], low-angle grain boundaries [707], agglomeration of point defects[708], surface damage[658] and clusters of impurities and precipitates.[700]

Malhan et al.[709] have compared (11-20) and (1-100) off-oriented in-house and commercially developed substrates with different structural defect densities to evaluate the forward current degradation phenomenon of bipolar pn diodes. It was found that substrates with lower EPDs for stacking faults showed less degradation.

Zhang et al.[710-713] have fabricated 10 kV 4H-SiC merged pin Schottky (MPS) diodes and first found that SFs generated from BPDs under forward biased stress increased the reverse leakage current.

Kojima et al.[714,715] have shown that SFs severely affect the reverse characteristics and the barrier height of 4H-SiC (11-20) Schottky barrier diodes (SBDs). Fujiwara et al.[716] have studied the effect of SFs on the 4H-SiC(0001) SBDs using epitaxial layers grown by the CVD method. It was found that in-grown SFs are of 8H structure and are generated mostly near the epilayer/substrate interface. It was also found that SFs lower the Schottky barrier height and decrease the breakdown voltage.

When 4H-SiC pn diodes are current flow deteriorated, SFs are formed and showed a emission group in the energy range of 2.89-3.01 eV.[687,701] They are attributed to excitons which are captured by local potential perturbation induced by SFs.

23.5 APPLICATIONS

Since SiC has a large bandgap, of 2.4-3.3 eV, it is promising for various applications such as blue LEDs, UV photodetectors, gas sensors, radiation detectors, electronic devices such as power devices and high frequency devices. SiC is also chemically inert and can be operated at high temperatures, so that it is expected to be effective in devices under harsh environments like chemical industry, nuclear industry and astronautics. Various applications of these devices are reviewed in Refs. 30, 31 and 717-724.

23.5.1 Substrates for GaN Based Light Emitting Diodes (LEDs), Lasers Diodes (LDs) and Electronic Devices

SiC was predicted as a promising substrate for GaN epitaxial growth because of its good lattice matching with GaN.[14,725] GaN-based LEDs were first developed on sapphire substrates (Sec. 6.2.2, Vol. 1). However, the lattice mismatching is as large as 13.8 % and sapphire is insulating so that LEDs with electrodes on both surfaces could not be fabricated. 6H-SiC has a lattice constant which is closer to that of GaN than that of sapphire and the lattice mismatching is only 3.4% and was expected to make good substrates for GaN-based LED fabrication. The quality of epitaxial growth on C-face and Si-face of SiC substrates[725] and the growth of (GaN/AlN) multibuffer layers on 6H-SiC are also studied.[726]

The application of SiC substrates to GaN based LEDs has been developed[727] and blue LEDs which compete with GaN based LEDs have been achieved[15] especially applying a structure for better light extraction.[728] These GaN/SiC LEDs have been industrially developed and commercially produced.[588,729-732] Fig. 23.65 shows a typical GaN/SiC LED structure. The advantage of GaN/SiC LEDs against GaN/sapphire LEDs is that better emission uniformity could be achieved. Since both electrodes can be formed on top and back surfaces and SiC has a high thermal conductivity, higher current densities can be achieved. High brightness LEDs have been developed and even short wavelength LDs are also developed.

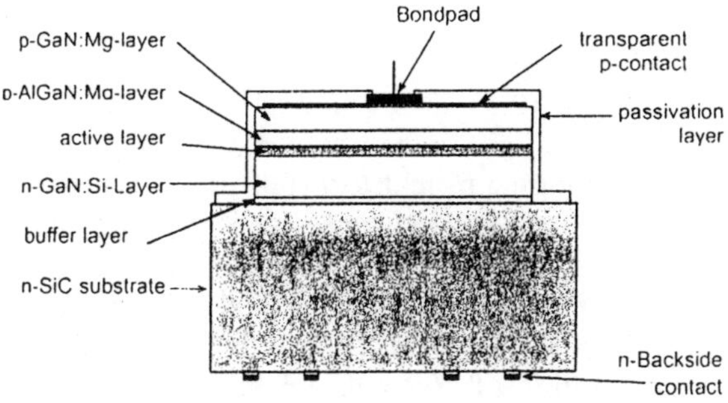

Fig. 23.65 Schematic view of an InGaN based light emitting diode (reprinted from Ref. 730 with permission, copyright 2001 Elsevier).

SiC substrates are also industrially applied as substrates for high frequency GaN electronic devices.[733,734] Since the thermal conductivity of SiC is large, it is effective as a substrate for power dissipation. Cordier et al.[735] have also studied the possibility of 3C-SiC/Si as the substrate for electronic devices.

23.5.2 Optical Devices and Sensors

(1) Blue and other color LEDs

SiC crystals were first developed in order to realize blue LEDs which were not commercialized in that era. Since 6H-SiC has a bandgap of 3.02 eV it is appropriate for blue LEDs. Even though SiC is an indirect material, the lack of blue LEDs pushed researchers to develop SiC materials and devices. The first blue LEDs were been prepared by growing n-type SiC with Al and N on p-type SiC with Al using LPE[736] and because of donor (Al) - accepter (N) pair recombination at about 470 nm, radiant values as high as 18.3 μW at 20 mA (V=3.0V) were achieved. Many efforts were then devoted to achieving blue LEDs mainly by LPE growth[2,69,70,72,124,127,128,736-742] but the luminosity was limited to about l m/W. After the realization of GaN based blue LEDs (Sec. 6.2.2, Vol. 1), the efforts for SiC LEDs were abandoned.

Other color LEDs have also been studied by Dmitrev et al. and others.[743-759] Even three color LEDs with blue (470 nm), green (510 nm) and red (650 nm) have been fabricated.[749] Since SiC bulk crystal growth technologies have advanced rapidly, SiC may still be a promising material for green and white LEDs if one can find some impurities or defects which permit direct transition as in the case of the Zn-O donor-and-acceptor (DA) emission in GaP (Sec. 7.5 in Vol.1).

Kamiyama et al.[760] have recently developed SiC LEDs based on N and B doping and realized DA emission. Based on this emission mechanism, they are developing new

LEDs based on two fluorescent SiC, N-Al doped SiC and N-B doped SiC to realize warm-white LEDs.[761]

(2) Ultraviolet (UV) photodetectors

Since SiC has a large bandgap (2.4-3.3) eV and is transparent to visible light, it is effective in detecting ultraviolet light in the wavelength range of 200-400 nm and has an absorption depth for light ranging from 0.1 to 1000 μm. Because of extremely small dark currents (less than 10^{-10} Acm^{-2}) and low noise, its detectability at room temperature reaches 10^{-14}-10^{-13} $WHz^{-1/2}$. Its high resistance to irradiation makes it possible for it to be operated at incident radiation power as high as 10^3-10^4 Wcm^{-2}.

The development of UV photodetectors is reviewed in Refs. 6 and 7. SiC UV photodiodes in the wavelength range of 200-400 nm have been developed, based on photoconductors[762], pn diodes and pin diodes[762-773], Schottky diodes[774-781] and avalanche photodiodes (APDs).[782-785] A typical UV photodiode is shown in Fig. 23.66 and various UV photodetectors are now commercially available.

UV photodetectors can be used as gas and flame sensors for oil combustion turbines, regulators for boilers, oil burners to observe combustion conditions, gas monitoring for car engines, urea injection control into diesel exhausts, because UV detectors are sensitive only to ultraviolet without detecting visible light.

(3) Sensors

In order to monitor the pollution emission of exhaust gases such as NO_x, carbon monoxide (CO), and hydrocarbons (C_xH_y), from high temperature combustion environ-

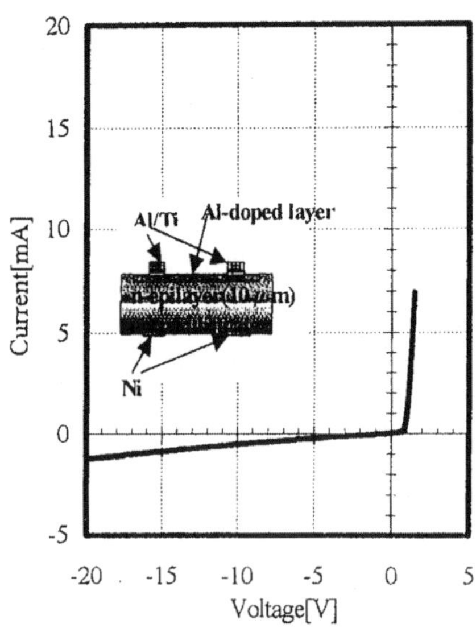

Fig. 23.66 Characteristics of a pn junction diode using Al doped epitaxial layer (reprinted from Ref. 770 with permission, copyright 2000 The Institute of Electronics, Information and Communication Engineers).

ments, such as power stations and automobiles, and to detect fuel leaks from hydrogen based launch vehicles and fuel cell and hydrogen automobiles, it is necessary to detect these gases sensitively and effectively. SiC is a stable material at high temperatures and is promising for gas detectors.

These gas sensors have been developed, using metal-oxide-semiconductor (MOS) capacitors, metal insulating (oxide)-silicon carbide (MISiC/MOSiC) capacitors[786-790], metal-insulating-silicon carbide FETs (MISiCFET)[791,792], Schottky barrier diodes (SBDs)[793-795] and pn diodes.[796] In these structures, gases adsorb on the catalytic surface metal electrodes and decomposed H (or H*) atoms diffuse into the device structures and change the diode forward current, the MOS capacitor or the FET drain current, thus it is possible to detect gas concentrations. Fig. 23.67(a) shows a typical gas sensor based on the catalytic gate field effect for hydrogen and oxygen and in Fig. 23.67(b), the sensing data is shown. It is also interesting to know that some of these sensors are at the stage of commercial production.

The development of these gas sensors is reviewed in Ref. 8 and 797. Sensors other than gas sensors such as micro electro-mechanical systems (MEMSs)[8] for sensing motion, acceleration, pressure and gas flow are also being developed.

(4) Radiation detectors

Because of the high bandgap energy and other physical properties (Table 23.1), SiC radiation detectors show extremely low dark currents and allow operation at high temperatures up to 700°C. SiC has a high-displacement threshold energy of 21.8 eV which gives it a high radiation hardness, two orders of magnitudes higher than that of Si. SiC is therefore a promising new material for radiation detectors compared with old materials as summarized in Table 6.5 in Vol. 1. Sellin and Vaitkus[9] have reviewed the device performance of SiC radiation detectors compared with other new materials such as GaN and amorphous silicon. Owens[798] also reviewed various wide-bandgap radiation detectors.

In fact, SiC detectors have negligible leakage current of the order of 1 nA/cm^2 three orders of magnitude lower than that of Si detectors. SiC has a high breakdown field of more than 2.7 MeV/cm, allowing devices to be operated under high applied electric fields, as seen in Fig. 23.68.[799]

SiC radiation detectors have been fabricated using bulk SiC crystals[800-806] and epitaxial SiC on SiC substrates.[807-822] Since the quality of bulk crystals was not satisfactory, detectors have been mainly fabricated using epitaxial SiC but the recent progress in the quality of bulk crystals is promising for the use of bulk crystals themselves. The crystals used were 6H-SiC in the past but 4H-SiC has mainly been used recently because of its better mobility.

The device structures studied have included Schottky type and pn type devices. Schifano et al.[817] have fabricated Schottky diode type detectors using 4H-SiC epitaxial materials and evaluated their characteristics. Grant et al.[806] have fabricated Schottky diode type detectors using bulk vanadium doped semi-insulating (SI) 4H-SiC crystals and evaluated their performance under high radiation conditions. Kalinina et al.[822] have developed ion-implanted pn junction radiation detectors.

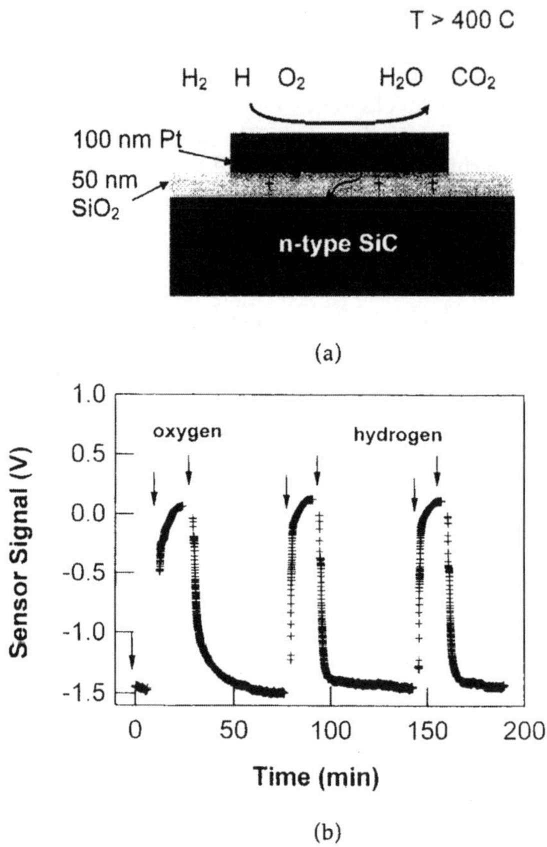

Fig. 23.67 (a) Cross-sectional view of a Pt/SiO_2/SiC catalytic gate field-effect gas sensor. (b) SiC gas sensor operating at 530 °C in a simulated energy plant environment. The sensor switches reversibly between alternating pulses of reducing (1% H_2 in N_2) and oxidizing (10% O_2 in N_2) gases (courtesy of R.N. Ghosh, Michigan State University). See Refs. 788 and 789 for details.

23.5.3 Electronic Devices

In Table 6.3 in Vol. 1, several figures of merit of devices are shown for various materials. As seen in the table, since SiC has a high electric breakdown field (E_C) 2.7 MV/cm, one order of magnitude higher than that of Si, a fast saturated velocity of electrons, two times faster that of Si, it is possible to make the drift layer thickness one order of magnitude thinner than Si and to increase the doping density two orders of magnitude higher than that of Si. The on-resistance (series resistance under operation) is therefore two-three orders of magnitude lower than that of Si, so that the conduction loss can be made 1/300 compared with Si devices. These calculations are shown in Fig. 23.69. SiC

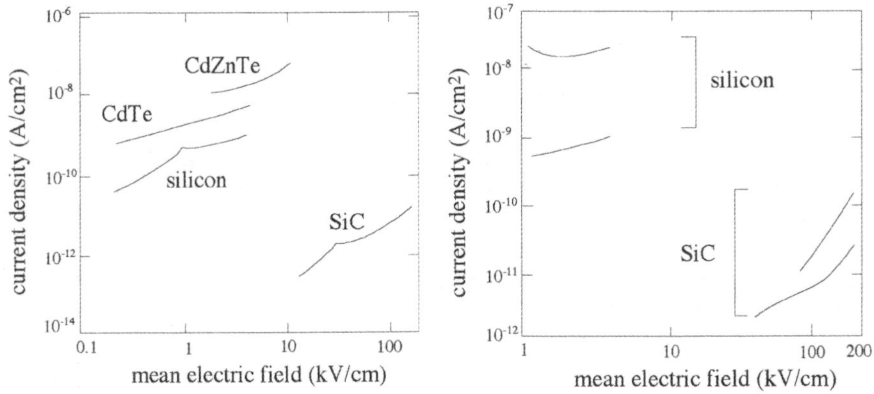

Fig. 23.68 (Left) Low leakage currents and the ability to withstand high electric-field strengths make SiC a strong candidate for next-generation detecotrs. (Right) The relatively small temperature dependence of leakage currnt for SiC detecors, when compared with those using silicon, makes them well-suited to hihg-tempearure applications (reprinted from Ref. 799 with permission, copyright 2005 Compound Semiconductors).

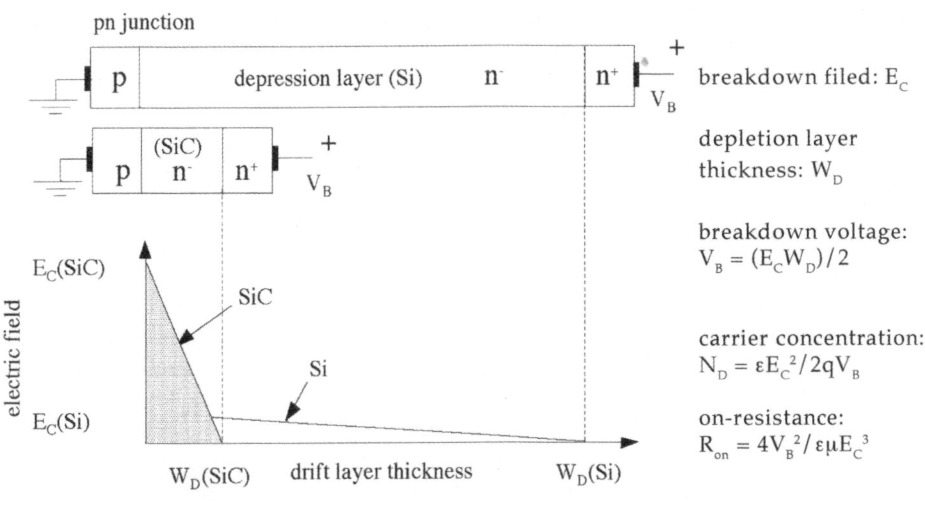

Fig. 23.69 Electric field distribution and the expansion of the depletion layer (reprinted from Ref. 31 with permission, copyright 2003 Ohmsha).

has a high Johnson index for high frequency devices. At high operating frequencies, SiC can tolerate an operating voltage ten times higher than Si and twice higher than GaAs. SiC also has a thermal conductivity of 4.9 W/cmK, three times greater than Si and ten times greater than GaAs. SiC is therefore a good material for high power devices operating at high frequencies and has the advantage that the cooling system can be made much smaller. Because of the wide band gap, SiC devices can be operated at higher temperatures of 600-800°C, with high radiation resistance, that is under harsh circumstances.

Among various polytypes, 4H-SiC whose bulk mobility is as high as 1000 cm^2/Vs and whose bandgap is as high as 3.26 eV is normally used in electronic devices. 3C-SiC is also attractive because of its isotropic electrical properties, high electron mobility, high saturation drift velocity and the possibility of high channel mobility due to localized interface traps in the conduction band. 15R is also a good candidate for electronic devices but the material itself is not yet available commercially.

Because of these superior SiC properties, SiC electronic devices are promising for high performance switching devices, inverter controllers, automobiles, power conditioners for solar cells and dispersed power sources, in order to meet the requirements for energy saving.

Electronic devices include unipolar devices (SBDs, MOSFETs, JFET/SITs, MESFETs) and bipolar devices (pn diodes, BJTs, IGBTs, GTOs). In the case of Si devices, since the breakdown electric field is low, the drift layer thickness has to be greater giving a high on-resistance. Therefore, for high breakdown voltage devices, bipolar devices are used in order to lower the on-resistance by conductivity modulation. In the case of SiC, since the breakdown electric field is one order of magnitude higher than that of Si, unipolar devices can be applied even for high breakdown voltage devices, even as high as several hundred V, reportedly up to 1200 V. The various SiC electronic devices are reviewed for power devices [1,10,11,13,32,33,823-826] and for high frequency devices.[12, 827]

(1) Diodes

- Schottky Barrier Diodes (SBDs)

In the case of Si, because of the high on-resistance, SBDs can be used only for the breakdown voltage region less than 200 V. For higher breakdown voltages, only pn diodes are used where the on-resistance can be reduced due to conductivity modulation by minority carrier injection. These pn diodes however have the drawback that the reverse recovery loss is large because of the capacitance due to the depleted layer in the pn junction. Since SiC has a low on-resistance and high breakdown electric field, high breakdown voltage SBDs can be fabricated. By using SiC-SBDs small reverse recovery losses and turn-on losses can be achieved and they can reduce the total energy loss in power electric circuits. SiC-SBDs have been therefore extensively studied.[50, 667, 828-836]

Kanaya et al.[50] fabricated Schottky barrier diodes on bulk SiC substrates and found a breakdown voltage of 1000 V for a carrier concentration of 2×10^{16} cm^{-3}.

Infinion has developed commercial SBDs as seen in Fig. 23.70.[832] These SBDs actually showed very low reverse recovery current characteristics, which are very favorable for low energy consumption for inverter applications. Powell et al.[833] have developed high current SBDs operating at 100 A with a breakdown voltage of 600 V with a

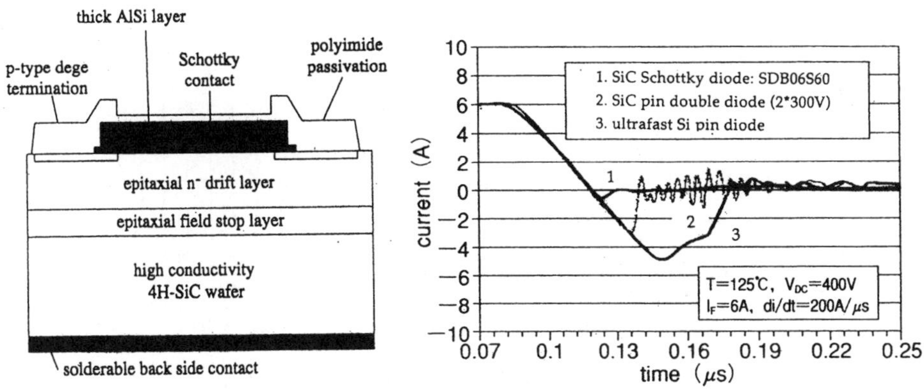

Fig. 23.70 (a) Basic structure of a SiC Schottkey barrier diode (SBD), (b) SiC switching in comparison to benchmark Si diodes. SiC pin double diode is also shown for comparison (from Ref. 832 with permission, copyright 2001 Compound Semiconductors).

production yield higher than 77%.

In fact, SBDs with breakdown voltages ranging from 300-1200 V are commercialized and are starting to be used for various switching circuits.

-pn diodes

Even for pn diodes, because of the high breakdown voltage and low on-resistance, SiC pn diodes show less reverse recovery current than for Si and are promising for diodes with breakdown voltage higher than 4 kV. Because the diffusion voltage is 2.5 V for a SiC pn junction, pn diodes can be used where the on voltage is greater than 2.5 V.

pn diodes were first developed for LEDs as explained in Sec. 23.5.2(1). After nitride based LEDs became industrially predominant replacing SiC LEDs, pn diodes for power device applications were developed[444,837-854] after the nineties. In Fig. 23.71, the characteristics of a 10 kV-200 A pn diode are shown. The pn junction was formed by epitaxial growth and after mesa etching, a p type junction termination extension (JTE) region was formed by ion-implantation. Sugawara et al.[850] have fabricated pn diodes with the highest breakdown voltage of 19.5 kV using an epitaxial layer thickness of 200 μm and an on voltage of 6.5 V.

(2) Power transistors

Power devices include unipolar devices such as MOSFETs, JFETs/SITs and bipolar devices such as gate-turn-off thyristors (GTOs) and insulated gate bipolar transistors (IGBTs). Since SiC has a high breakdown field and a low on-resistance as explained as above, even for high breakdown devices, unipolar devices can be utilized. This is not the case for Si which needs bipolar devices for high breakdown voltage devices in

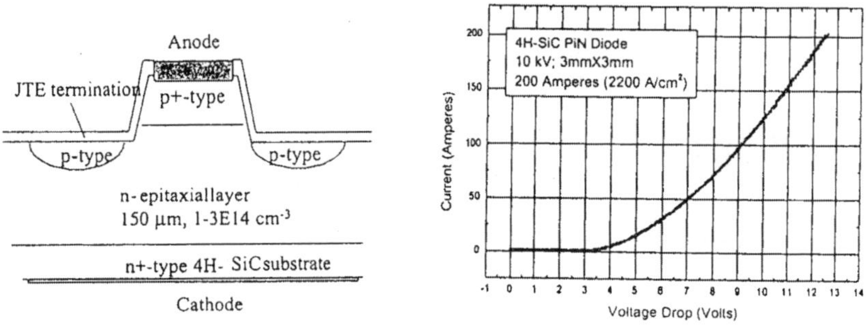

(a) (b)

Fig. 23.71 The characteristics of 10kV-200A 4H-SiC pin diode. (a) pin diode structure used in the fabrication of 10 kV diodes, (b) on-state characteristics of a 4H-SiC pin rectifier capable of blocking 10 kV (reprinted from Ref. 846 with permission, copyright 2002 Trans Tech Publlications).

order to lower the on-resistance. The recent development of SiC power transistors is reviewed in Refs. 855 and 856.

- *Metal-Oxide-Semiconductor Field Effect Transistors (MOSFETs)*

Inversion and depletion mode MOSFETs have been demonstrated by Palmour et al.[857] who successfully conducted the preliminary works on MOSFETs.[858-861] SiC MOSFETs are expected to be able to be used in high breakdown voltage (>1kV), low loss and high speed switching devices even though they are unipolar devices, and are expected to replace Si IGBTs in the future.[861]

Various types of MOSFETs as in the case of Si have been examined using SiC. Typical ones are double-implanted MOSFETs (DiMOSFETs) [862-865], trench type MOSFETs (UMOSFETs) [866, 867], accumulation channel MOSFETs (ACCUFETs) [868-871], buried channel MOSFETs (BCMOSFETs) [872-876], implantation and epitaxy MOSFETs (IEMOSFETs), static channel expansion MOSFETs (SEMOSFETs) [877], reduced surface field MOSFETs (RESURF MOSFETs) [878], metal-insulator-semiconductor FETs (MISFETs)[879] and superjunction MOSFET.[880] In Fig. 23.72, typical MOSFET structures are shown. In the case of MOSFETs, the normally-off mode works well because the p-type layer is inverted by the gate voltage. On the other hand, they acquire a high on-resistance because of carrier scattering by electrons trapped at the interface levels. [881-884]

The on-resistance is the sum of various resistances in the device structure. When other resistances become negligibly small compared with the drift resistance, the drift resistance becomes the theoretical limit of SiC. In reality, because the interface level density is one order of magnitude larger for SiC/oxide interfaces than that of Si/oxide, the channel mobility is small. The channel resistance is therefore larger than the drift resistance. In Fig. 23.73, the on-resistance of SiC MOSFETs is shown compared with competitive Si IGBTs. It is seen that the on-resistance is smaller than the Si unipolar theoretical limit but very large compared with that of the SiC limit.

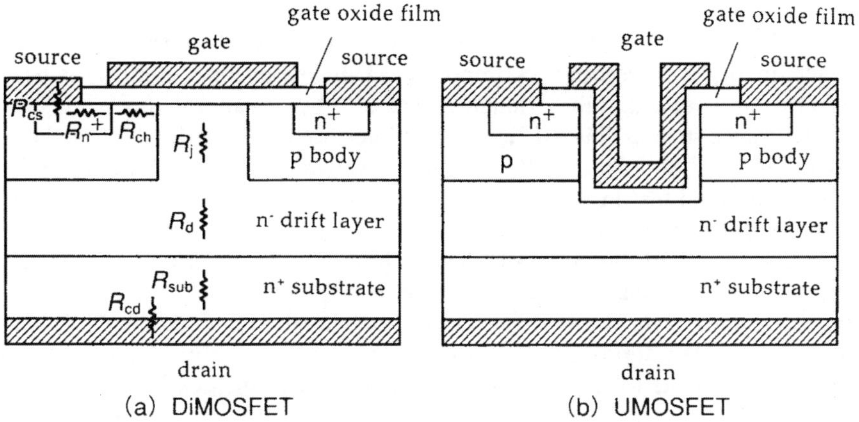

Fig. 23.72 Structures of MOSFETs (reprinted from Ref. 30 with permission, copyright 2003 Nikkan Kogyo Co.).

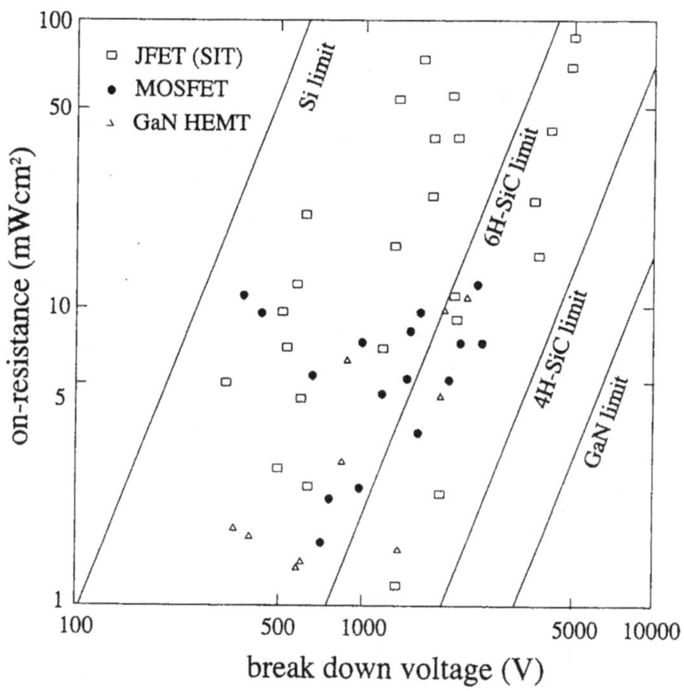

Fig. 23.73 Comparison of the on-resistance of SiC, Si and GaN devices.

In the development of SiC MOSFETs, technology to reduce the channel resistance is key. In fact, the channel mobility has been several tens of cm^2/Vs, very low compared with the ideal value of 200 cm^2/Vs due to the interface levels. Improvement of MOS interface quality is therefore very important. This low channel mobility is being increased by improvement of the oxidation process [885], heat treatment after oxidation [886-888], using SiC planes such as (11-20) faces where interface levels are less[889, 890] and nitrogen implantation.[891, 892] Shorner et al.[893] compared inversion channel mobility for 4H-, 6H-, 15R-SiC and found that in 15R-SiC, the inversion channel mobility is less affected by interface level traps.

The other issue with MOSFETs is the reliability of gate material[894] and the effect of substrate defects is being studied to improve their reliability.

Recently, high channel mobility of nearly one hundred cm^2/Vs[885,888,889-891] has been achieved and industrial production has also been announced. By using C face 4H-SiC, it was shown that the lowest on-resistance as low as 1.8 mΩ·cm^2 (at the breakdown voltage of 600 V) could be obtained.[31] There have been various announcements of industrial production of SiC MOSFETs such as 1200 V n-channel enhancement mode MOSFET with R_{on} = 80 mΩ (Cree), 1200 V, 10A, 10 mΩ·cm^2 (Mitsubishi Electric), IEMOSFETs with 1100 V, 4.3 mΩ·cm^2 (AIST), with on-resistance lower than that of Si IGBTs, 300 A trench MOSFETs (Rohm).

- Junction FETs (JFETs)/Static Induction Transistors (SITs)

JFETs or SITs[895, 896] are majority carrier devices in which the current is controlled by the depletion layer expanded between the pn junction as shown in Fig. 23.74. Since the bulk material itself is used as the channel, these devices can realize the high mobility of the material and have very low on resistance.

Various device structures are being extensively studied. Since there are no oxide interfaces or pn junctions between the source and the drain, JFETs and SITs are mainly normally-on (depletion) type devices so it is necessary to develop normally-off type devices.

Various vertical and lateral JFET/SIT, structures have been studied, such as vertical JFETs (VJFETs) [897-899], static expansion channel JFETs (SEJFETs) [900-902], buried gate SITs (BGSITs)[903-905] and reduce surface finish (RESURF-JFETs).[906]

A very low on-resistance down to 1.01 mΩ·cm^2 has been achieved in buried gate type SITs.[904] Normally-off devices such as 5. 3 kV 4H-SiC SEJFEs with a specific on resistance of 69 mΩcm^2 have also been reported and the commercialization of these devices has started.

For high frequency applications, SITs with average power of 200 W, and pulse peak of 1 kW have been developed.[907] The use of these SIT high frequency modules for high definition TV (HDTV) has been demonstrated.[908]

In Fig. 23.73, the characteristic on-resistance data for SiC unipolar switching devices are summarized. It is seen that on-resistance close to the SiC limit can be obtained for breakdown voltages higher than 2 kV, but is still large for the low breakdown voltage region.

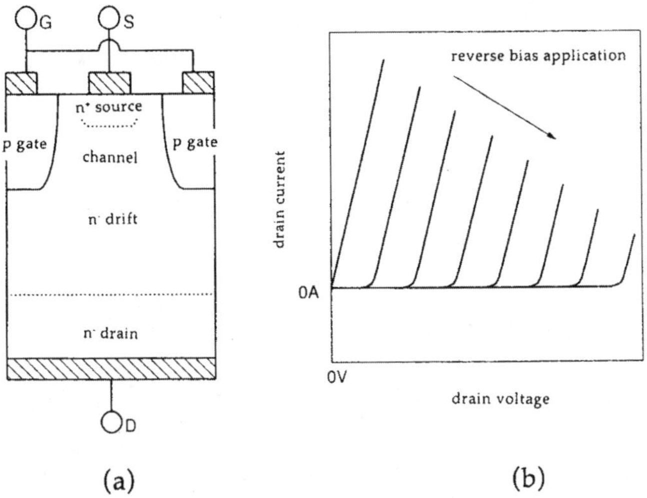

Fig. 23.74 (a) Strucrure of vertical JFET/SIT and (b) its static characteristics (reprinted from Ref. 30 with permission, copyright 2003 Nikkan Kogyo Co.).

-Bipolar transistors

Among bipolar devices, bipolar transistors are being studied for the low breakdown voltage region and thyristors have been studied for the high breakdown voltage region. Bipolar transistors such as bipolar junction transistors (BJTs), insulating gate transistors (IGBTs), static induction accumulation field-effect transistors (SIAFETs), bipolar mode JEETs and power thyristors such as gate turn-off thyristors (GTOs) have been studied as summarized by Chow.[909] For BJTs, a current gain β as high as 120 has been achieved.[910] Cao et al.[911] have fabricated 4H-SiC GTOs and evaluated the performance with a switching current density up to 10000 A/cm.

Bipolar transistors with a breakdown voltage of 1300 V and on-resistance of 8 mΩ·cm^2 and those with a breakdown voltage of 1750 V and on-resistance of 12 mΩ·cm^2 have been reported.[912]

Sugawara et al.[850] have reported thyristors with SiC commutated gate turn-off thyristor (SICGT) structure and shown 12.7 kV devices with on voltage of 6.6 V (at 100 A/cm^2).

Bipolar devices however have the on-state degradation effect due to basal plane dislocations (BPDs) as mentioned in detail in Sec. 23.4.5, so this material-origin problem must be solved in their future development.

(3) High frequency devices

As mentioned in the beginning of Sec. 23.5.3, SiC is a promising material for high power high frequency devices.[12,913,914] SiC metal-semiconductor FETs (MESFETs) are expected to have ten times higher performance than those of GaAs. SiC high frequency devices have therefore applications for mobile communication stations at several GHz,

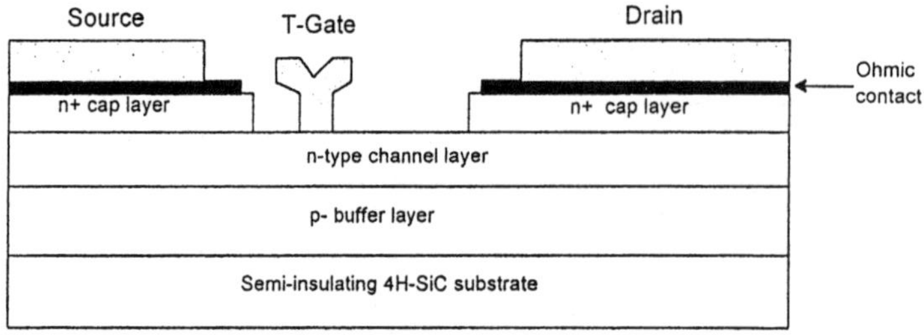

Fig. 23.75 Cross section of a SiC power MESFT (reprinted from Ref. 914 with permission, copyright 2002 Trans Tech Publlications).

transmitters for satellite communications, ground digital transmitters (several hundred MHz), radars, microwave ovens, and electronic tubes such as traveling wave tubes (TWTs) and magnetrons. As mobile communications develop, the energy consumption at communication stations is increasing exponentially and there is a great need for devices of low power consumption.

A typical MESFET structure is shown in Fig. 23.75.[915] For high frequency devices, semi-insulating substrates as explained in Sec. 23.3.4(3) are preferred in order to prevent noise due to parasitic capacitances in the substrates between the source and drain electrodes.

SiC MESFETs have been extensively developed[404,916-927] using 4H-SiC and 6H-SiC, with SI and conductive SiC substrates. MESFETs with performances such as high output power density of 4 W/mm and power added efficiencies of 50% have been achieved and industrial production has been announced mainly for WiMax base station applications covering 3.6-3.8 GHz. Even MMICs are also being developed. 3D tri-gate MESFETs have been developed recording a cut-off frequency (f_T) of 20.6 GHz and a maximum oscillation frequency (f_{max}) of 82.4 GHz.[927]

REFERENCES

1. *Silicon Carbide; Power Devices and Sensors*, ed. by P. Friedriches (Wiely, 2010).
2. H. Matsunami, M. Ikeda, A. Suzuki and T. Tanaka, *IEEE Trans. Electron Devices* **ED-24** (1977) 958.
3. S. Nishio, A. Ibaraki, H. Matsunami and T. Tanaka, *Jpn. J. Appl. Phys.* **19** (1980) L353.
4. K. Koga and T. Yamaguchi, *Jpn. Assoc. Crystal Growth((in japanese)* **17** (1990) 176.
5. S.I. Vlaskina, *Semicon. Phys.; Quntum Electron. Optoelectron.* **5** (2002) 71.
6. G.A. Shaw, A.M. Siegel, J. Model, A. Geboff, *Proc.SPIE* **7320** (2009) 73200J-1.
7. E. Monroy, F. Omnes and F. Calle, *Semicon. Sci. Technol.* **18** (2003) R33.
8. N.G. Wright and A.B. Horsfall, *J. Phys. D; Appl. Phys.* **40** (2007) 6345.
9. P.J. Sellin and J. Vaitkus, *Nucl. Instr. Meth. Phys. Appl. A* **557** (2006) 479.
10. P.L. Dreike, D.M. Fleetwood, D.B. King, D.C.Sprauer and T.E. Zipperian, *IEEE Trans. Comp. Pack. and Manuf. Tech.* **A-17** (1994) 594.
11. C.M. Johnson, N.G. Wright, M.J. Uren, K. P. Hilton, M. Rahimo, D.A. Rahimo, D.A. Hinchley, A.P. Knights, D.J. Morrison, A.B. Horsfall, S. Ortolland and A.G. O'Neill, *IEE Proc. Circuits Devices* **148** (2001) 101
12. R.W. Trew, J.-B, Yan and P.M. Mock, *Proc. IEEE* **79** (1991) 598.
13. J. W. Palmour, *Proc. 28th Inter. Energy Conver. Engineer. Conf.* (Atlanta, American Chemical Society, 1993) p. 1249.
14. T. Sasaki, T. Matsuoka and A. Katsui, *Appl. Surf. Sci.* **41** (1989) 504.
15. V. Härle, B. Hahn, H.-J. Lugauer, S. Bader, G. Bruderl, J. Bauer, D. Eidert, U. Strauss, U. Zehner, A. Lell and N. Hiller, *phys. stat. sol. (a)* **180** (2000) 5.

16. *Silicon Carbide-A High Temperature Semiconductors*, eds. J.R. O'Conner and J. Smiltens, (Pergamon Presss, Oxford, 1960)
17. W.F. Knippenberg, *Philips Res. Repts.* **18** (1963) 161.
18. E. Pettenpaul, W.V. Münch and G. Ziegler, *Inst. Res. Rept.* **21** (1966) 113.
19. *Silicon Carbide-1968*, ed. by H.K. Henisch and R. Roy (Pergamon Press, 1968)
20. *Silicon Carbide-1973*, ed. by R.C. Marshal, J.W. Faust and C.E. Ryan (University of South Carolina Press, 1974)
21. Y. M. Tairov and V. F. Tsvetkov, *J. Crystal Growth* **43** (1978) 209.
22. R.B. Campbell, *IEEE Trans. Ind. Electron.* **IE-29** (1982) 124.
23. H. Matsunami, *Optoelectronics-Dev. Technol.* **2** (1987) 29.
24. J.H. Edgar, *J. Mater. Res.* **7** (1992) 235.
25. *Proc. Int. Conf SiC and Related Materials*, (Howard University Materials Science Research Center, Washington DC, 1993).
26. *Silicon Carbide and Related Materials 2005 (ISCRM 2005), Mater. Sci. Forum*, **527-529** (2006).
27. *Silicon Carbide and Related Materils 2006 (ECSCRM 2006), Mater. Sci. Forum* **556-557** (2007).
28. *Silicon Carbide and Related Materials 2007 (ISCRM 2007), Mater. Sci. Forum* **600-603** (2009).
29. *Silicon Carbide and Related Materials 2009 (ISCRM 2009), Mater. Sci. Forum* **645-648** (2010).
30. H. Matsunami, *Technology of Semicondcutor SiC and its Applications (in japanese)* (Nikkan Kogyo Co., Japan, 2003).
31. K. Arai and S. Yoshida, *Fundamentals and Applications of SiC Devices (in japanese)*(Ohmsha, Japan, 2003)
32. B.J. Baliga, *Silicon Carbide Power Devices* (World Sci. Publ., 2006)
33. *Recent Development of Power Electronics*, eds. by H. Ohashi, T. Kimoto (CMC Publ., 2009).
34. R.I. Scace and G.A. Slack, *J. Chem. Phys.* **30** (1959) 1551.
35. R.I. Scace and G.A. Slack, in *Silicon Carbide-A High Temperatrue Semiconductor*, eds. by J.R. O'Connor ad J. Smiltens (Pergamon Press, 1960) p. 24.
36. O. Ruff, *Trans. Electrochem. Soc.* **68** (1935) 87.
37. J. Drowart, G. De Maria and M.G. Inghram, *J. Chem. Phys.* **29** (1958) 1015.
38. Y. M. Tairov and Y. A. Vodakov, in *Electroluminescence Topics in Applied Physics, Vol. 17*, ed. by J.L. Pankove (Spriger-Verlag, Berlin, 1977) p. 31.
39. *Landolt-Börnstein New Series. III-17h, Physics of Ternary Compounds,*ed. O. Madelung (Springer-Verlag, 1982).
40. *Properties of Silicon Carbide, EMS datareview series, No.13*, ed. by C.L. Harris (INSPEC, 1995).
41. M. Nakabayashi, T. Fujimoo, M. Katsuno and N. Ohtani, *Mater. Sci. Forum* **527-529** (2006) 699.
42. G.A. Slack, *J. Appl. Phys.* **35** (1964) 3460.
43. W. V. Munch and E. Pettenpaul, *J. Appl. Phys.* **48** (1977) 4823.
44. W. V. Munch and E. Pettenpaul, *J. Appl. Phys.* **48** (1977) 4831.
45. V.I. Gavrilenko, A.V. Postnikov, N.I. Klyui and V.G. Litovchenko, *phys. stat. sol. (b)* **162** (1990) 477.
46. W. J. Schaffer, G. H. Negley, K. G. Irvine and J. W. Palmour, *A1alerial lv'esearch Society Symp. Proc. Vol. 55* (Material Research Society, Pittsburgh, 1994).
47. R. F. Davis, *Int. J. Mater. Product Technol.* **4** (1989), 81.
48. H. Nagasawa, M. Abe, K. Yagi, T. Kawaara and N. Hatta, *phys. stat. sol. (b)* **245**(2008) 1272.
49. M. Kanaya and J. Takahashi, *FC Report* **11** (1993), 17.
50. M. Kanaya, N. Ohtani, J. Takahashi, T. Nishikawa and M. Katsuno, *Oyo Buturi (Monthly Publ. Jpn. Soc. Appl. Phys.) (in japanese)* **64** (1995) 642.
51. V.F. Tsvetkov, S.T. Allen, H.S. Kong and C.H. Carter, Jr., *Inst. Phys. Conf. Ser.* **142** (1995) 17.
52. R.C. Glass, D. Henshall, V.F. Tsvetkov and C.H. Carter, Jr., *MRS Bull.* **22** (1997) 30.
53. H. Matsunami and T. Kimoto, *Jpn. J. Assoc. Crystal Growth (in japanese)* **27** (2000) 186.
54. St.G. Müller, R.C. Glass, H.M. Hobgood, V.F. Tsvetkov, M. Brady, D. Henshall, J.R. Jenny, D. Malta and C.H. Carter Jr., *J. Crystal Growth* **211** (2000) 325.
55. N. Ohtani, T. Fujimoto, M. Kasuno. T. Aigo and H. Yashiro, *J. Crystal Growth* **237** (2002) 1180.
56. N. Ohtani, *Proc. 5th Int. Symp. Adv. Sci. Technol. Silicon Mater.* (Hawaii, 2008) p. 149.
57. N. Ohtani, *Oyo Buturi (Monthly Publ. Jpn. Soc. Appl. Phys.) (in japanese)* **74** (2005) 220
58 J.J. Sumaeris, J.R. Jenny and A.R. Powell, *MRS Bull.* **30** (2005) 280.
59. P. J. Wellmann, T.L. Straubinger, P. Desperrier, R. Müller, U. Kunecke, S.A. Sakwe, H. Schmitt, A. Winnacker, E. Blanquet, J.-M. Dedulle and M. Pons, *Mater. Sci. Forum* **483-485** (2005), 25.
60 P. Wellmann, R. Müller, S.A. Sakwe, U. Künecke, P. Hens, M. Stockmeier, K. Konias, R. Hock, A. Magerl and M. Pons, *Mat. Res. Soc. Symp. Proc.* **1069** (2008) 3.
61. A. G. Acheson, Engl. Pat., **17911**(1982)
62. F. Halden, *Proc. Conf. Silicon Carbide* (Pergamon Press, 1960) p. 115.
63. L.B. Griffiths and A.I. Mlavsky, *J. Electrochem. Soc.* **7** (1964) 805.
64. Y.M. Tairov, I.I. Khlebnikov and V.T. Tsvetkov, *phys. stat. sol. (a)* **25** (1974) 349.

65. R.T. Yakimova and A.A. Kalnin, *phys. stat. sol. (a)* **32** (1975) 297.
66. Y.M. Tairov, N.S. Peev, N.A. Smirnova and A.A. Kalnin, *Cryst. Res. Technol.* **21** (1986) 1503.
67. S.K. Lilov, Yu.M. Tairov and V.F. Tsvetkov, *J. Crystal Growth* **46** (1979) 269.
68. G. Ziegler, P. Lanig, D. Theis and C. Weyrich, *IEEE Trans. Electron. Devices* **ED30** (1983) 277.
69. L. Hofmann, D. Ziegler, D. Theis and C. Weyrich, *J. Appl. Phys.* **53** (1982) 6962.
70. M. Ikeda, T. Hayakawa, S. Yamagiwa, H. Matsunami and T. Tanaka, *J. Appl. Phys.* **50** (1979) 8215.
71. V.A. Dmitriev, P.A. Ivanov, V.I. Levin, I.V. Popov, A.M. Strel'chuk, Yu.M. Tairov, V.F. Tsvetkov and V.E. Chelnokov, *Sov. Tech. Phys. Lett.* **13** (1987) 489.
72. Y. Matsushita, T. Nakata, T. Ueani, T. Yamaguchi and T. Niina, *Jpn. J. Appl. Phys.* **29** (1990) L343.
73. V.A. Dmitriev, *Phys. B* **185** (1993) 440.
74. R. Yakimov, M. Tuominen, A.S. Bakin, J.-O. Fornell, A. Vehanen and E. Janzen, *Inst. Phys. Conf. Ser.* **142** (1996) 101.
75. R. Rendakova, V. Ivantsov and V. Dmitriev, *Mater. Sci. Forum* **264-268** (1998) 163.
76. N. Kusnetsov, A. Morozov, D. Bauman, V. Ivantsov, V. Sukhoveev, I. Nikitina, A. Zubrilov, S. Rendakova, V.A. Dimitriev, D. Hofman and P. Masri, *Mater. Sci. Forum* **338-342** (2000) 229.
77. M.G. Mauk, B.W. Feyock, A. Sharma, R.G. Hunsperger, *J. Crystal Growth* **225** (2001) 322.
78. A. Tanaka, T. Ataka, E. Ohkura and H. Katsuno, *J. Crystal Growth* **269** (2004) 413.
79. K. Kusunoki, K. Kamei, N. Yashiroand R. Hattori, *Mater. Sci. Forum* **615-617** (2009) 137.
80. M. Imade, S. Takeuchi, M. Uemura, M. Yoshimura, Y. Kitaoka, T. Sasaki, Y. Mori, S. Itoh, H. Okuda and M. Yamazaki, *Mater. Sci. Forum* **645-648** (2010) 45.
81. M. Imade, T. Ogura, M. Uemura, F. Kawamura, M. Yoshimura, Y. Kitaoka, Y. Mori, T. Sasaki, M. Yamazaki and S. Suwabe, *Mater. Sci. Forum* **615-617** (2009) 55.
82. V. Ivantsov and V. Dimitriev, *Mater. Sci. Forum* **264-268** (1998) 73.
83. M. Müller, M. Bickermann, D. Hofmann, A.D. Weber and A. Winnacker, *Mater. Sci. Forum* **264-268** (1998) 69.
84. D. Hofmann and M. H. Müller, *Mater. Sci. Engineer. B* **61-62** (1999) 29.
85. T. Ogura, M. Imade, F. Kawamura, M. Yoshimura,Y. Kitaoka, Y. Mori and T. Sasaki, *J. Jpn. Assoc. Crystal Growth (in japanese)* **33** (2006) 186.
86. R.N. Hall, *J. Appl. Phys.* **29** (1958) 914.
87. H. Kleykamp, G. Schimacher, *Ber. Bunsnges. Phys. Chem.* **9** (1993) 799.
88. K. Kusunoki, K. Kamei, Y. Ueda, S. Naga, Y. Ito, M. Hasebe, T. Ujihara and K. Nakajima, *Mater. Sci. Forum* **483-485** (2005) 13.
89. K. Kusunoki, K. Kamei, N. Okdada, N. Yshiro, T. Ujihara and K. Nakajima, *Mater. Sci. Forum* **527-529** (2006) 119.
90. N. Yashiro, K. Kusunoki, K. Kamei and A. Yauchi, *Mater. Sci. Forum* **556-557** (2007) 303.
91. N. Yashiro, K. Kusunoki, K. Kamei and A. Yauchi, *Mater. Sci. Forum* **645-648** (2010) 33.
92. N. Yashiro, K. Kusunoki, K. Kamei, M. Hasebe, T. Ujihara and K. Nakajima, *Mater. Sci. Forum* **527-529** (2006) 115.
93. R. Tanaka, K. Seki, S. Komiyama, T. Ujihara and Y. Takeda, *Mater. Sci. Forum* **600-603** (2009) 59.
94. K. Kusunoki, K. Kamei, N. Yshiro, T. Tanaka and A. Yauchi, *Mater. Sci. Forum* **600-603** (2009) 187.
95. T. Tanaka, N. Yashiro, K. Kusunoki, K. Kamei and A. Yauchi, *Mater. Sci. Forum* **600-603** (2009) 191.
96. T. Ujihara, R. Maekawa, R. Tanaka, K. Sasaki, K. Kuroda and Y. Takeda, *Mater. Sci. Forum* **600-603** (2009) 63.
97. K. Seki, R. Tanaka, T. Ujihara and Y. Takeda, *Mater. Sci. Forum* **615-617** (2009) 27.
98. T. Tanaka, K. Seki, T. Ujihara and Y. Takeda. *Mater. Sci. Forum* **615-617** (2009) 37.
99. F. Mercier, I.G. Galben-Sandulache, M. Marinova, G. Zoulis, T. Ouisse, E.K. Polychroniadis and D. Chaussend, *Mater. Sci. Forum* **645-648** (2010) 59.
100. F. Mercier, D. Chaussend, J-M. Dedulle, M. Pos and R. Madar, *Mater. Sci. Forum* **615-617** (2009) 41.
101. K. Danno, H. Saitoh, A. Seki, H. Daikoku, Y. Fujiwara, T. Ishii, H. Sakamoto and Y. Kawai, *Mater. Sci. Forum* **645-648** (2010) 13.
102. Alexander, K. Seki, S. Kazawa, Y. Yamamoto, T. Ujihara and Y. Takeda, *Mater. Sci. Forum* **679-680** (2011) 24.
103. D. Chaussende, L. Latu-Romain, L. Auvray, M. Ucar, M. Pons and M. Madar, *Mater. Sci. Forum* **225** (2005) 483.
104. M. Soueidan and G. Ferro, *Adv. Func. Mater.* **16** (2006) 975.
105. M. Soueidan, G. Ferro, B. Nsouli, M. Roumie, E. Polychroniadis, M. Kazan, S. Juillaguet, D. Chaussende, N. Habka, J. Stoemenos, J. Camassel and Y. Monteil, *Crystal Growth & Des.* **6** (2006) 2598.
106. M. Soueidan, G. Ferro, O. Kim-Hak, F. Robaut, O. Dezellus, J. Dazor, F. Cauwet. J. Vial and B. Nsouli, *Acta Mater.* **55** (2007) 6873.
107. M. Soudeidan, G. Ferro, O. Kim-Hak, F. Cauwet and B. Nsouli, *Crystal Growth & Des.* **8** (2008) 1044.

108. D. Chaussende, F. Mercier, A. Bouelle, F. Conchon, M. Soueidan, G. Ferro, A. Mantzari, A. Andreadou, E.K. Polychroniadis, C. Balloud, S. Juillaguer, J. Camassel and M. Pons, *J. Crystal Growth* **310** (2008) 976.
109. F. Mercier, O. Kim-Hak, J. Lorenzzi, J-M. Dedulle, G. Ferro and D. Chaussend, *Mater. Sci. Forum* **645-648** (2010) 67.
110. D. Chaussend, M. Pons and R. Madar, *Mater. Sci. Forum* **527-529** (2006) 111.
111. J.Eid, J.L. Santailler, B. Ferrand, P. Ferret, J. Pesenti, A. Basset, A. Passero, A. Mantzari, E.K. Polychronaidis, C. Balloud, P. Soares and J Camassel, *Mater. Sci. Forum* **527-529** (2006) 123.
112. J.A. Lely, *Ber. Deut. Keram. Ges.* **32** (1955)229.
113. D.R. Hamilton, in *Silicon Carbide- A High Temperature Semiconductors*, eds. by J.R. O'Conner and J. Smiltens (Pergamon Presss, Oxford, 1960) p. 453.
114. Y. Inomata, Z. Inoue, M. Mitomo and H. Suzuki, *J. Ceram. Assoc. Jpn. (in japanese)* **76** (1968) 29.
115. K. Inomata, *Oyo Buturi (Monthly Publ. Jpn. Soc. Appl. Phys.) (in japanese)* **40** (1971) 880.
116. S.K. Lilov and I.Y. Yanchev, *Adv. Mater. Opt. Electr.* **1** (1992) 203.
117. S.K. Lilov, *Mater. Sci. Engineer. B* **27** (1994) 69.
118. S.K. Lilov, *J. Mater. Sci. Lett.* **14** (1995) 1546.
119. Y. M. Tairov and V.F. Tsvetkof, *Silicon Carbide - 1973*, eds. R.C. Marshal, J.W. Faust and C.E. Ryan (Columbia S.C., 1974).
120. V.I. Levin, Y.M. Tairov, M.G. Travadzhyan and V.F. Tsvetkov, *Inorganic Mater.* **14** (1978) 830.
121. Y. M. Tairov and V. F. Tsvetkov, *J. Crystal Growth* **52** (1981) 146.
122. N.Y. Shushlebina, Y.M. Shasakov, E.K. Bogonosova and E.E. Golobokina, *Inorganic Mater.* **13** (1977) 682.
123. K. Koga, Y. Ueda, T. Nakata, T Yamaguchi and T. Niina, *J. Vacuum Sci. Japan (in japanese)* **30** (1987) 886.
124. Y. Matsushita, T. Nakata, Y. Ueda, T. Uetani, Y. Fujikawa, K. Koga and T. Niina, *Proc. Japan Display '89* (Kyoto, 1989) p. 696.
125. K. Koga, T. Nakata, Y. Ueda and T. Niina, *Ext. Abstr. 17th Conf. Solid State Dev. Mater.* (Jpn. Soc. Appl. Phys., Tokyo, 1985) p. 249.
126. T. Nakata, K. Koga, Y. Matsushita, Y. Ueda and T.Niino, *Amorphous and Crystalline Silicon Carbide and Related Materials II, Springer Proc. in Physics* **43** (Springer, Berlin, 1989) p. 26.
127. K. Koga, T. Nakata, Y. Ueda, Y. Matsushita, Y. Fujikawa, T. Uetani and T. Niina, *Ext. Abstr. Fall Meeting Electrochem. Soc.* **89** (1989) 689.
128. K. Koga and T. Yamaguchi, *Prog. Crystal Growth Charact.* **23** (1991) 127.
129. Y. Matsushita, T. Nakata, T. Uetani, T. Yamaguchi and T. Niina, *Jpn. J. Appl. Phys.* **29** (1990) L343.
130. W. S. Yoo, S. Nishino and H. Matsunami, *J. Crystal Growth* **99** (1990) 278.
131. W. S. Yoo and H. Matsunami, *J. Appl. Phys.* **70** (1991) 7124.
132. N. Kuroda, K. Shibahara, W.S. Yoo, S. Nishino and H. Matsunami, *Ext. Abstr. 19th Conf. Solid State Dev. Mater.* (Japan Society of Applied Phys., Tokyo, 1987) p. 227.
133. M. Kanaya, J. Takahashi, Y. Fujiwara and A. Moritani, *Appl. Phys. Lett.* **58** (1991) 56.
134. J. Takahashi, M. Kanaya and T. Hoshino, *Inst. Phys. Conf. Ser.* **137** (1994) 13.
135. J. Takahashi, M. Kanaya and Y. Fujiwara, *J. Crystal Growth* **135** (1994) 61.
136. J. Takahashi, N. Ohtani, M. Katsuno amd S. Shinoyama, *J. Crystal Growth* **181** (1997) 229.
137. J. Takahashi and N. Ohtani, *phys. stat. sol. (b)* **202** (1997) 163.
138. N. Ohtani, M. Kasuno and T. Fujimoto, *Jpn. J. Appl. Phys.* **42** (2003) L277.
139. M. Katsuno, N. Ohtani, T. Fujimoto and H. Yashiro, *Mater. Sci. Forum* **457-460** (2004) 83.
140 M. Katsuno, M. Nakabayashi, T. Fujimoto and H. Yashiro, *J. Electron. Mater.* **34** (2006) 91.
141. J. Takahashi, N. Ohtani and M. Kanaya, *Jpn. J. Appl. Phys., Part1* **34** (1995) 4694.
142. N. Ohtani, M. Kasuno, M. Nakabayashi, T. Fujimoto, H. Tsuge, H. Yashiro, T. Aigo, H. Hirano, T. Hoshino and K. Tatsumim, *J. Crytal Growth* **311** (2009) 1475.
143. N. Ohtani, T. Fujimoto, M. Kasuno, T. Aigo and H. Yashiro, *Mater. Sci. Forum* **389-393** (2002) 29.
144. N. Ohtani, T. Fujimoto, M. Kasuno, T. Aigo and H. Yashiro, *J. Crytal Growth* **237-239** (2002) 1180.
145. M. Nakabayashi, T. Fujimoto, M. Kasuno, N. Ohtani, H. Tsuge, H. Yashiro, T. Aigo, T. Hoshino, H. Hirano and K. Tatsumim, *Mater. Sci. Forum* **600-603** (2009) 3.
146. M. Nakabayashi, T. Fujimoto, M. Kasuno, H. Tsuge, T. Aigo, S. Satoh, H. Yashiro, T. Hoshino, H. Hirano and W. Ohashi, *Mater. Sci. Forum* **645-648** (2010) 9.
147. D.L. Barrett, R.G. Seidensticker, W. Gaiga, R.H. Hopkins and W.J. Choke, *J. Crystal Growth* **109** (1991) 17.
148. D. L. Barrett, J. P. MeHugh, H. M. Hobgood, R. H. Hopkins, P. G. McMullin, R. C. Clarke and W. J. Choyke, *J. Crystal Growth* **128** (1993) 358.
149. H.M. Hobgood, J.P. McHugh, J. Greggi, R.H. Hopkins and M. Skowronski, *Inst. Phys. Conf. Ser.* **137** (1994) 7.

150. H. M. Hobgood, D. L. Barrett, J. P. MeHugh, R. C. Clarke, S. Sriram A. A. Burk, J. Greggi, C. D. Brandt, R. H. Hopkins and W. J. Choyke, *J. Crystal Growth* **137** (1994) 181.
151. C. D. Brandt, A. K. Agarwal, G. Augustine, A. A. Burk, R. C. Clarke, R. C. Glass, H. M. Hobgood, J. P. MeHugh, P. G. McMullin, R. R. Siergiej, T. J. Smith, S. Sriram, M. C. Driver and R. H. Hopkins, *Inst. Phys. Conf. Ser.* **141** (1995) 373.
152. H. McD. Hobgood, R.C. Glass, G. Augustine, R.H. Hopkins, J. Jenny, M. Skowronski, W.C. Mitchel and M. Roth, *Appl. Phys. Lett.* **66** (1995) 1364.
153. D. Hobgood, M. Brady, W. Brixius, G. Fechko, R. Glass, D. Henshall, J. Jenny, R. Leonard, D. Malta, St.G. Müller, V. Tsvetkov, C. H. Carter Jr., *Mater. Sci. Forum* **338-342**(2000) 3.
154. R.T. Leonard, Y. Khlebnikov, A.R. Powell, C. Basceru, M.F. Brady, I. Khlbnikov, J.R. Jenny, D.P. Malta, M.J. Paisley, V.F. Tsvetkov, R. Zilli, E. Deyneka, H. McD. Hobgood, V. Balakrishna and C.H. Carter, Jr., *Mater. Sci. Forum* **600-603** (2009) 7.
155. Cree Research Inc., *Industrial Session at the 8th Euro. Conf. SiC Relat. Mater.* (ECSCRM2010, Norway, Oslo).
156. http//www.cree.com/press/press_detail.asp?i=1283181742965.
157. A.A. Burk, D. Tsvetkov, D. Barnhardt, M.J. O'Loughlin, L. Garrett, P. Towner, J. Eaman, E. Deyneka, Y. Khlebnikov and J. Palmour, *Abst. Silicon Carbide and Related Materials 2011 (ISCRM 2011)* p. 286.
158. D. Hofmann, R. Eckstein, L. Kadinski, M. Kölbl, M. Müller, St. G. Müller, E. Schmitt, A. Weber and A. Winnacker, *Mat. Res. Soc. Symp. Proc.* **483** (1998) 301.
159. D. Hofmann, M. Bickermann, R. Eckstein, M. Kölbl, St. G. Müller, E. Schmitt, A. Weber and A. Winnacker, *J. Crystal Growth* **198/199** (1999) 1005.
160. N. Sculze, J. Gajowski and G. Pensl, *Appl. Phys. Lett.* **72** (1998) 1632.
161. P. J. Wellmann, M. Bickermann, D. Hofmann, L. Kadinski, M. Selder, T.L. Straubinger and A. Winnacker, *J. Crystal Growth* **216** (2000), 263.
162. D. Schulz, J. Doerschel, M. Lechner, H. J. Rost, D. Siche and J. Wollweber, *J. Crystal Growth* **246** (2002) 31.
163. Z.G. Herro, B.M. Epelbaum, M. Bickermann, P. Masri and A. Winnacker, *J. Crystal Growth* **262** (2004) 105.
164. P. Wellmann, P. Desperrier, R. Müller, T. Straubinger, A. Winnacker, F. Baillet, E. Blanquet, J.M. Dedulle and M. Pons, *J. Crystal Growth* **275** (2005) e555.
165. P. Wellmann, R. Müller and M. Pons, *Mater Sci. Forum* **527-529** (2006) 75, T. L. Straubinger, M. Bickermann, P.J. Wellmann and A. Winnacker, *J. Crystal Growth* **240** (2002) 117.
166. G. Augustine, H. M. Hobgood, V. Balakrishna, G. Dunne, R.H. Hopkins, *phys. stat. sol. (b)* **202** (1997) 137.
167. G. Augustine, V. Balakrishna and C. D. Brandt, *J. Crystal Growth* **211** (2000) 339.
168. H. Yamaguchi, S. Nishizawa, T. Kato, N. Oyanagi, W. Bahng, S. Yoshida, K. Arai, Y. Machitani and T. Kikuchi, *Rev. Sci. Instr.* **71** (2000) 2829.
169. T. Kato, N. Oyanagi, H. Yamaguchi, Y. Tanaka, S. Nishizawa and K. Arai., *Mater. Sci. Forum* **338-342** (2002) 457.
170. S. Nishizawa, Y. Kitou, W. Bahng, N. Oyanagi, M. N. Khan and K. Arai, *Mater. Sci. Forum* **338-342** (2000) 99.
171. W. Bahng, Y. Kitou, S. Nishizawa, H. Yamaguchi, M. N. Khan, N. Oyanagi, S. Nishio and K. Arai, *J. Crystal Growth* **209** (2000) 767.
172. T. Kato, N. Oyanagi, H. Yamaguchi, S. Nishizawa, M.N. Khan, Y. Kitou and K. Arai., *J. Crystal Growth* **222** (2001) 579.
173. T. Kato, S. Nishizawa and K. Arai, *J. Crystal Growth* **233** (2001) 219.
174. S. Nishizawa, T. Kato, Y. Kitou, N. Oyanagi and K. Arai, *Mater. Sci. Forum* **389-393** (2002) 43.
175. Y. Kitou, W. Bahng, T. Kato. S. Nishizawa and K. Arai, *Mater. Sci. Forum* **389-393** (2002), 83.
176. N. Oyanagi, H. Yamaguchi, T. Kato, S. Nishizawa and K. Arai, *Mater. Sci. Forum* **389-393** (2002) 87.
177. S. Nishizawa, H. Yamaguchi, T. Kato, N. Oyanaga, Y. Kitou and K. Arai, *J. Jpn. Assoc. Crystal Growth (in japanese)* **29** (2002) 431.
178. S. Nishizawa, T. Kato, Y. Kitou, F. Hirose, N. Oyanagi, H. Yamaguchi and K. Arai, *J. Elec. Inf. Comm. Acc. (in japanese)* **J86-C** (2003) 334.
179. K. Nakayama, Y. Miyanagi, H. Shiomi, S. Nishio, T. Kimoto and H. Matsunami, *Mater Sci. Forum* **389-393** (2002) 123.
180. T. Furusho, R. Kobayashi, T. Nishiguchi, M. Sasaki, K. Hirari, T. Hayashi, H. Kinoshita and H. Shiomi, *Mater Sci. Forum* **527-529** (2006) 35.
181. D. Nakamura, H. Gunjishima, S. Yamaguchi, T. Ito, A. Okamoto, H. Kondo, S. Onda and K. Takatori, *Nature* **430** (2004) 1009.

182. D. Nakamura, *Mater. Sci. Forum* **527-529** (2006) 3.
183. O. Filip, B. Epelbaum, J. Li, M. Bickermann, X. Xu and A. Winnacker, *Mater. Sci. Forum* **600-603** (2009) 23.
184. I. Yeo, W. Lee, W. Lee, B. Shin, I. Kim, J. Choi, K. Ku, Y. Kim and S. Nishio, *Mater Sci. Forum* **645-648** (2010) 37.
185. E. Tymicki, K. Grasza, K. Racka-Dzietko, M. Raczkiewicz, T. Lukasiewicz, M. Gala, K. Kosciewicz, R. Diduszko and R. Bozek, *Mater. Sci. Forum* **645-648** (2010) 17.
186. K.A. Jones, *Marter. Sci. Forum* **600-603** (2009) 369.
187. K. Grasza, E. Tymicki and J. Kisielewski, *Mater Sci. Forum* **527-529** (2006) 87.
188. M.A. Fanton, Q. Li, A.Y. Polyakov, R.L. Cavalero, R.G. Ray, B.E. Weiland and M. Skowronski, *Mater Sci. Forum* **527-529** (2006) 103.
189. J. Kim, S. Seo, K. Kim, J. Song, T. Kim and M. Oh, *Mater Sci. Forum* **527-529** (2006) 91.
190. E. Tupitsyn, A. Galyukov, M. Bogdanov, A. Kulik, M. Ramm, Y. Makarov, T. Sudarshan, *Mater. Sci. Forum* **600-603** (2009) 27.
191. S.F. Avramenko, M.Y. Valak, V.S. Kiselev, O.T. Segeev and M.Y. Skorokhod, *Met. Phys. Adv. Tech.* **15** (1996) 729.
192. S.F. Avramenko, V.S. Kiselev, M.Ya Valakh, V.G. Visotski, *Semicon. Phys. Quant. Electron. Optron.* **2** (1999) 76.
193. K. Ku, J. Kim, J. Seo, J. Lee, M. Kyun, W. Lee, G. Lee, I. Kim and B. Shin, *Mater Sci. Forum* **527-529** (2006) 83.
194. M. Ankin, K. Chourou, M. Pons, J.M. Bluet, R. Madar, P. Grosse, C. Fuare, G Basset and Y. Grange, *Mater. Sci. Engineer. B* **61-62** (1999) 73.
195. E. Tymicki, K. Grasza, W. Hofman, R. Diduszko and R. Bozek, *Mater. Sci. Forum* **600-603** (2009) 15.
196. M.A. Fanton, R.L Cavalero, R.G. Ray, B.E. Weiland, W. Everson, D. Snyder, R. Gamble and E. Oslosky, *Mater Sci. Forum* **527-529** (2006) 47.
197. S. Rohmfeld, M. Hundhausen, L. Lely, N. Schulze and G. Pensl, *Mater. Sci. Forum* **338-342** (2000) 579.
198. M. Anikin, M. Pons, K. Chourou, O. Chaix, J.M. Bluet, V. Lauer and R. Madar, *Mater. Sci. Forum* **264-268** (1998) 45.
199. T. Fujimoto, H. Tsuge, M. Kasuno, N. Ohtani, H. Yashiro and M. Nakabayashi, *Mater. Sci. Forum* **457-460** (2004), 79.
200. A.R. Powell, R.T. Leonard, M.F. Brady, St. G. Muller, V.F. Tsvetkov, R.Trussel, J.J. Sumakeris, H. D. Hobgood, A.A. Burk, R.C. Glass and C.H. Carter, Jr., *Mater Sci. Forum* **457-460** (2004) 41.
201. C. Basceri, I. Khlebnikov, Y. Khlebnikov, P. Muzykov, M. Sharma, G. Startifym N, Silan and C. Balkas, *Mater Sci. Forum* **527-529** (2006) 39.
202. A. Gupta, E. Semenas, E. Emorhokpor, J. Chen, I. Zwieback, A. Souzis and T. Anderson, *Mater Sci. Forum* **527-529** (2006) 43.
203. A. Gupta, I. Zwieback, A. Souzis, M. Yoganathan and T. Anderson, *Mater Sci. Forum* **600-603** (2009) 35.
204. Yu.N. Makarov, D.P. Litvin, A.V. Vasiliev, A.S. Segal, S.S. Nagalyuk, H. Helave, M.I. Voronova and K.D. Scherbachov, *Mater Sci. Forum* **645-648** (2010) 25.
205. T. Straubinger, E. Schmitt, s. Storm, M. Vogel, A-D. Weber and A. Wohlfart, *Mater Sci. Forum* **645-648** (2010) 3.
206. K. Furukawa, Y. Tajima, H. Saitoh, Y. Fujii, A. Suzuki and S. Nakajima, *Spring Meeting of Jpn. Appl. Phys.* **39** (1992) 288.
207. K. Furukawa, Y. Tajima, H. Saito, Y. Fujii, A. Suzuki and S. Nakajima, *Jpn. J. Appl. Phys.* **32** (1993) L645.
208. J. Yang, S. Nishino, J. A. Powell and P. Pirouz, *Inst. Phys. Conf. Ser.* **137** (1994) 25.
209. K. Semmerlrot, M. Krieger, G. Pensl, H. Nagasawa, R. Püsche, M. Hundhausen, L. Leym M. Nerding and H.P. Strunk, *Mater. Sci. Forum* **457-460** (2004) 151.
210. K. Semmelroth, M. Krieger, G. Pensl, H. Nagasawa, R. Pusche, M. Hundhausen, L. Ley, M. Nerdinga nd H.P. Strunk, *J. Crystal Growth* **308** (2007) 241.
211. R. Püsche, M. Hundhausen, L. Ley, K. Sommelroth, F. Schmid, G. Pensl and H. Nagasawa, *J. Appl. Phys.* **96** (2004) 5569.
212. A. Freudenberg, J. Wollweber, R. Nitschke, V. Alex and J. Doerschel, *J. Crystal Growth* **275** (2005) e467.
213. D. Chaussede, F. Baillet, L. Charpentier, E. Prenot, M. Pons and R. Madar, *J. Electrochem. Soc.* **150** (2003) G653.
214. D. Chaussende, M. Ucar, L. Auvray, F. Baillet, M. Pons and R. Madar, *Crystal Growth & Des.* **5** (2005) 1539.

215. D. Chaussende, L. Latu-romain, L. Auvray, M. Ucar, M. Pons and R. Madar, *Mater Sci. Forum* **483-485** (2005) 225.
216. D. Chaussende, P. Wellmann, M. Ucar, M. Pons and R. Madar, *Mater Sci. Forum* **527-529** (2006) 63.
217. A. Mantzari, F. Mercier, M. Soueidan, D. Chaussende, G. Ferro and E.K. Polychroniadis, *Mater. Sci. Forum* **600-603** (2009) 67.
218. L. Latu-Romain, D. Chaussemde, P. Chaudoueet, F. Robaut, G. Berthome, M. Pons and R. Madar, *J. Crystal Growth* **275** (2005) e609.
219. L. Latu-Romain, D. Chaussende, C. Balloud, S. Juillaguet, L. Rapenn, E. Pernt, J. Camassel, M. Pons and R. Madar, *Mater Sci. Forum* **527-529** (2006) 99.
220. D. Chaussende, J. Eid, F. Mercier, R. Madar and M. Pons, *Mater Sci. Forum* **615** (2009) 31.
221. A. Boulle, D. Chaussende, L. Latu-Romain, F. Conchon, O. Masson and G. Guinebretiere, *Appl. Phys. Lett.* **89** (2006) 091902.
222. D. Chaussende, F. Mercier, R. Madar and M. Pons, *Mater Sci. Forum* **600-603** (2009) 71.
223. I.-G. Galben-Sandulache, G.L. Sun, J.-M. Dedulle, T. Ouisse, R. Madar, M. Pons and D. Caussende, *Mater Sci. Forum* **645-648** (2010) 55.
224. G.L. Sun, I.-G. Galben-Sandulache, T. Ouisse, J.-M. Dedulle, M. Pons, R. Madar and D. Chaussende, *Mater Sci. Forum* **645-648** (2010) 63.
225. G. Ferro, *Mater Sci. Forum* **645-648** (2010) 49.
226. E. Pearson, T. Takai, T. Halicioglu and A. Tiller, *J. Crystal Growth* **70** (1984) 33.
227. D. Hofmann, R. Eckstein, M Kolblm Y. Makarov, St.G. Müller, E. Schmitt, A. Winnacker, R. Rupp, R. Stein and J. Volkl, *J. Crystal Growth* (1997) 669.
228. P. Raback, R. Nieminen, R. Yakimova, M. Tuominen and E. Janzen, *Mater. Sci. Forum* **264-268** (1998) 65.
229. Yu. Egorov, A.O. Galyukov, S.G. Gurevich, Yu.N. Makarov, E.N. Mokhov, M.G. Ramm, M.S. Ramm, A.D. Roenkov, A.S. Segal, Yu.A. Vodakov, A.N. Yorob'ev and A.I. Zhmakin, *Mater. Sci. Forum* **264-268** (1998) 61.
230. M. Pons, M. Ankin, K. Chourou, J.M. Dedulle R. Madar, E. Blanquet, A. Pisch, C. Bernard, P. Grosse, C. Fuare, G. Basser and Y. Grange, *Mater. Sci. Engineer. B* **61-62** (1999) 18.
231. S.Yu. Karpov, A.V. Kulik, I.A. Zhmakin, Yu.N. Makarov, E.N. Mokhov, M.G. Ramm, M.S. Ramm, A.D. Roenkov and Yu.A. Vodakov, *J. Crystal Growth* **211** (2000) 347.
232. M. Selder, L. Kadinski, Yu. Makarov, F. Durst, P. Wellmann, T. Straubinger, D. Hofmann, S. Karpov and M. Ramm, *J. Crystal Growth* **211** (2000) 333.
233. O. Klein, P. Philip, J. Sprekels and K. Wilmanski, *J. Crystal Growth* **222** (2001) 823.
234. Q.-S. Chen, H. Zhang, V. Prasad, *J. Crystal Growth* **224** (2001) 101.
235. Q.-S. Chen, H. Zhang, V. Prasad, *J. Crystal Growth* **230** (2001) 239.
236. R.-H. Ma, Q.-S. Chen, H. Zhang, V. Prasad, C.M. Balkas and N.K. Yushin, *J. Crystal Growth* **211** (2000) 352.
237. R.-H. Ma, H. Zhang, S. Ha and M. Skowronski, *J. Crystal Growth* **252** (2003) 523.
238. D.I. Cherednichenko, R.V. Drachev and T.S. Dudarchan, *J. Crystal Growth* **262** (2004) 175.
239. Q.-S. Chen, P. Gao and W.R. Hu, *J. Crystal Growth* **266** (2004) 320.
240. R.V. Drachev, D.I. Cherednichenko and T.S. Sudarshan, *J. Crystal Growth* **265** (2004) 179.
241. I.D. Matukov, D.S. Kalinin, M.V. Bogdanov, S.Yu. Karpov, D.Kh. Ofengeim, M.S. Ramm, J.S. Barash, E.N. Mokhov, A.D. Roenkov, Yu.A. Vodakov, M.G. Ramm, H Helava and Yu.N. Makarov, *J. Crystal Growth* **266** (2004) 313.
242. R. Drachev, E. Deyneka, C. Rodes, J. Schupp and T. Sudarshan, *Mater Sci. Forum* **527-529** (2006) 15.
243. H. Matsunami, S. Nishio and H. Ono, *IEEE Trans. Electron Dvices* **ED-28** (1981) 1235.
244. S. Nishio, J.A. Powell and H.A. Will, *Appl. Phys. Lett.* **42** (1983) 460.
245. S. Nishio, J. Suhara, H. Ono and H. Matsunami, *J. Appl. Phys.* **61** (1987) 4889.
246. H. Matsunami and T. Kimoto, *Mater. Sci, Enginner. B* **20** (1997) 125.
247. H. Matsunami, *J. Jpn. Assoc. Crystal Growth (in japanese)* **33** (2006) 12.
248. K. Shibahara, S. Nishio and H. Matsunami, *J. Crystal Growth* **78** (1986) 538.
249. K. Shibahara, S. Nishio and H. Matsunami, *J. Jpn. Assoc. Crystal Growth (in japanese)* **13** (1986) 226.
250. K. Shibahara, S. Nishio and H. Matsunami, *Appl. Phys. Lett.* **50** (1987) 1888.
251. H. Nagasawa and K. Yagi, *J. Jpn. Assoc. Crystal Growth (in japanese)* **24** (1997) 270.
252. H. Nagasawa, K. Yagi, T. Kawahara, N. Hatta and M. Abe, *Microelectron. Eng. B* **83** (2006) 185.
253. H. Nagasawa, K. Yagi, T. Kawahara, *J. Crystal Growth* **237-239** (2002) 1244.
254. H. Nagasawa, T. Kawahara and K. Yagi, *Mater. Sci. Forum* **389-393** (2002) 319.
255. H. Nagasawa, K. Yagi, T. Kawahara and N. Hatta, *Mat. Res. Soc. Symp. Proc.* **742** (2003) 47.
256. H. Nagasawa, *Ceramics (in japanese)* **40** (2005) 19.
257. H. Nagasawa, K. Yagi, T. Kawahara and N. Hatta, *Chem. Vap. Deposition* **12** (2006) 502.

258. T. Chassagne, G. Ferro, C. Gourberyre, M. Le Berre, D. Barbier and T. Monteil, *Mater. Sci. Forum* **353-356** (2001) 155.
259. J. Camassel, *J. Vac. Sci. Technol.* **B16** (1998) 1648.
260. C. Zgheib, L.E. McNeil, M. Kazan, P. Masri, F.M. Morales, O. Ambacher and J. Perzoldt, *Appl. Phys. Lett.* **87** (2005) 041905.
261. T. Chassagne, G. Ferro, H. haas, H. Mank, A. Leycuras, Y. Moneil, F. Socares, C. Balloud, Ph. Arcade, C. Blanc, H. Peyre, S. Juillaguer and J. Camassel, *phys. stat. sol. (a)* **202** (2005) 524.
262. P.G. Neudeck, A.J. Trunek, D.J. Spry, J.A. Powell, H. Du, M. Skowronski, X.R. Huang and M. Dudley, *Chem. Vap. Deposition* **12** (2006) 531.
263. G. Ferro , T. Chassagne, A. Leycuras, F. Cauwet and Y. Monteil, *Chem. Vap. Depositon* **12** (2006) 483.
264. M. Zielinski, A. Leycuras, S. Ndiaye and T. Chassagne, *Appl. Phys. Lett.* **89** (2006) 131906.
265. M. Zielinski, S. Ndiaye, T. Chassagne, S. Julliaguet, R. Lewandowska, M. Portail, A. Leycuras and J. Camassel, *phys. stat. sol. (a)* **204** (2007) 981.
266. O. Kordina, C. Hallin, A. Elliso, A.S. Bakin, L.G. Ivanov, A. Henry, R. Yakimova, M. Touminen, A. Vehanen and E. Janzen, *Appl. Phys. Lett.* **69** (1996) 1456.
267. A. Ellison, T. Kimoto, I.G. Ivanov, Q. Wahab, A. Henry, O. Kordina, J. Zhang, C.G. Hemmingsson, C.-Yu. Gu, M.R. Leys and E. Janzen, *Mater. Sci. Forum.* **264-268** (1998) 103.
268. A. Ellison, J. Zhang, J. Peterson, A. Henry, Q. Wahab, J.P. Bergman, Y.N. Makarov, A. Vorob'ev, A. Vehanen and E. Janzen, *Mater. Sci. Engineer. B* **61-62** (1999) 113.
269. A. Ellison, B. Magnusson, N.T. Son, L. Storasta and E. Janzen, *Mater. Sci. Forum.* **433-436** (2003) 33.
270. A. Ellison B. Magnusson, B. Sundqvist, G. Pozina, J.P. Bergman, E. Janzen and A. Vehanen., *Mater. Sci. Forum* **457-460** (2004) 9.
271. S. Nigam, H.J. Chung, S.W. Huh, J. Grim, A.Y. Polyakov, M.A. Fantan, B. Weiland, D.W. Snyder and M. Skowronski, *Mater. Sci. Forum* **527-529** (2006) 27.
272. A.Y. Polyakov, M.A. Fantan, M. Skowronski, H.J. Chung, S. Nigam and S.W. Huh, *Mater. Sci. Forum* **527-529** (2006) 21.
273. Y. Kitou, E, Makino, K. Inaba, M. Nagakubo and S. Onda, *Mater. Sci. Forum* **527-529** (2009) 107.
274. Y. Kitou, E, Makino, K. Inaba, N. Hosokawa, H. Hiramatsu, J. Hasegawa, S. Onda, H. Tsuboi, H. Takaba and A. Miyamoto, *Mater. Sci. Forum* **600-603** (2009) 47.
275. Yu.N. Marakov, R.A. Talalaev, A.N. Vorob'ev, M.S. Ramm and M.V. Bogadov, *Mater. Sci. Fo rum* **600-603** (2009) 51.
276. Y.M. Tairov and V.F. Tsvetkov, *J. Crystal Growth* **36** (1976) 147.
277. Y.M. Tairov and V.F. Tsvetkov, *J. Crystal Growth* **46** (1979) 403.
278. Y.A. Vodakov, E.M. Morkhov, M.G. Ramm and A.D. Rcnknow, *Krist. Tech.* **14** (1979) 729.
279. S.K. Lilov, *J. Crystal Growth* **46** (1979) 269.
280. V.B. Sgields, K. Fekade and G. Spencer, *Appl Phys. Lett.* **62** (1993) 1919.
281. H. Matsunami, in *Technology of Semicondcutor SiC and its Applications* (Nikkan Kogyo Co., Japan, 2003) p. 53.
282. A.E. van Arkel and J.H. de Boer, *Z. anog. Chem.* **148** (1925) 345.
283. J.N. Pring and W.J. Fielding, *J. Chem. Soc.* **95** (1909) 1497.
284. S.N. Gorin and L.M. Ivanova, *phys. stat. sol. (b)* **202** (1997) 221.
285. W.M. Vetter and M. Dudley, *J. Crystal Growth* **260** (2004) 2011
286. M. Bruel, *Electron. Lett.* **31** (1995) 1201.
287. M. Bruel, *Nucl. Instr. Met. Phys. Res. B* **108** (1996) 313.
288. Q.Y. Tong, U. Gosele, C. Yuan, A.J. Steckel and M. Reich, *J. Electrochem. Soc.* **142** (1995) 232.
289. L. D. Cioccio, Y.L. Tiec, F. Letertre, C. Jaussaud and M. Bruel, *Electron. Lett.* **32** (1996) 1144.
290. E. Huonnard-Bruyere, F. Letertre, L.Di Cioccio, H.J. Von Bardeleben, J.L. Cantin, T. Ouiss, T Billon and G. Guillot, *Mater. Sci. Forum* **338-342** (2000) 715.
291. F. Leteertre, B. Ghyselen and B. Aspar, *III-Vs Review* **15** (2002) 46.
292. G. Chichignoud, L. Auvray, E. Blanquet, M. Ankin, E. Pernot, J-M. Bluet, P. Chaydouet, M. Mermoux, C. Moisoon, F. Letertre, M. Pons and R. Madar, *Mater Sci. Forum* **527-529** (2006) 71.
293. R. Yakimova, M. Syvajarvi, T. Iakimov, H. Jacobson, R. Raback, A. Vehanen and E. Janzen, *J. Crystal Growth* **217** (2000) 255.
294. R. A. Stein, P. Lanig and S. Leibenzeder, *Mater. Sci. and Engineer. B* **11** (1992) 69.
295. R. A. Stein and P. Lanig, *J. Crystal Growth* **131** (1993) 71.
296. Yu.M. Tairov, V.F. Tsvetkov, *Prog. Crystal Growth Character.* **7** (1983) 111.
297. E.Y. Tupitsyn, A. Arulchakkaravarthi, R.V. Drachev and T.S. Sudarshan, *J. Crystal Growth* **299** (2007) 70.
298. E. Schmitt, T. Straubinger, M. Rasp, M. Vogel and A. Wohlfart, *J. Crystal Growth* **310** (2008) 966.
299. E. Schmitt, T. Straubinger, M. Rasp, M. Vogel and A. Wohlfart, *Mater Sci. Forum* **600-603** (2009) 11.

300. A. Itoh, H. Akita, T. Kimoto and H. Matsunami, *Appl. Phys. Lett.* **65** (1994) 1400.
301. H.-J. Rost, J. Doerschel, K. Irmscher, M. Robbergm D. Schulz and D. Sicher, *J. Crystal Growth* **275** (2005) e451.
302. J. Chen, S.C. Lien, Y.C. Shin, Z.C. Feng, C.H. Kuan, J.H. Zhao and W.J. Lu, *Mater Sci. Forum* **600-603** (2009) 39.
303. H.-J. Rost, M., Schmidbauer, D. Siche and R. Fornari, *J. Crystal Growth* **290** (2006) 137.
304. D. Siche, M. Albrecht, J. Dershel, K. Irmsher, H.-J. Rost, M. Rossberg and D. Schulz, *Mater. Sci. Forum* **483-485** (2005) 39.
305. T. Fujimoto, M. Katsuno, N. Ohtani, T. Aigo and H. Yashiro, *Mater. Sci. Forum* **389-393** (2002) 47.
306. X. Li, S. Jiang, X. Hu, J. Dong, J. Li, X. Chen, L. Wang, X. Xu and M. Jiang, *Mater. Sci. Forum* **527-529** (2006) 95.
307. E. Sanchez, A. Kopec, S. Poplawski, R. Ware, S. Holmes, S. Wang and A. Timmerman, *Mater. Sci. Forum* **389-393** (2002) 71.
308. M. Aoki, M. Miyazaki, T. Nishiguchi, H. Kinoshita and M. Yoshimoto, *Mater. Sci. Forum* **600-603** (2009) 365.
309. F.C. Frank, *Acta Crystal* **4** (1951) 497.
310. A.R. Verma, *Phil. Mag.* **43** (1952) 441.
311. W. Si, M. Dudley, R. Glass, V. Tsvetkov and C.H. Carter, *Mater. Sci. Forum* **264-268** (1998) 429.
312. I. Sunagawa and P. Bennema, *J. Crystal Growth* **53** (1981) 490.
313. J. Heindl, W. Dorsch, R. Eckstein, D. Hofmann, T. Mareck, St.G. Müller, H.P. Strunk and A. Winnacker, *J. Crystal Growth* **179** (1997) 510.
314. N. Cabrera and M.M. Levine, *Phil. Mag.* **1** (1956) 450.
315. J. Heindel, W. Dorsch, H.P. Strunk, St. G. Müller, R. Eckstein, D. Hofmann and A. Winnacker, *Phys. Rev. Lett.* **80** (1998) 740.
316. E. Pernot, J. Hartwig, M. Pons and R. Madar, *Mater. Sci. Forum* **527-529** (2006) 435.
317. S.I. Maximenko, P. Pirouz and T.S. sudarshan, *Mater. Sci. Forum* **527-529** (2006) 439.
318. J. Giocondi, G.S. Rohrer, M. Skowrnski, V. Alakrishna, G. Augustine, H.M. Hobgood and R.H. Hopkins, *J. Crystal Growth* **181** (1997) 351.
319. P. Pirouz, *Phil. Mag. A***78** (1998) 727.
320. J. Heindl, H.P. Strunk, V.D. Tsvenkov and C.H. Carter Jr., *phys. stat. sol. (a)* **162** (1997) 251.
321. N. Ohtani, M. Katsuno, T. Fujimoto, T. Aigo and H. Yashiro, *J. Cystal Growth* **226** (2001) 254.
322. N. Ohtani, M. Katsuno, T. Fujimoto, T. Aigo and H. Yashiro, *Mater. Sci. Forum* **389-393** (2002) 99.
323. P. J. Neudeck and J. A. Powell, *Electron Devices Lett.* **15** (1994) 63.
324. K.Y. Fujikawa, Y. Ueda and T. Yamaguchi, in *Amorphous and Crystalline Silicon Carbide IV, Springer Proceedings in Physics, Vol. 71* (Springer-Verlag, Berlin, 1992) p. 96.
325. N. Ohtani, J. Takahashi, M. Katsuno, H. Yashiro and M. Kanaya, *Mater. Res. Soc. Symp.* **510** (1998) 37.
326. H.-J. Rost, M. Schmidbauer and D. Sicher, *Mater. Sci. Forum* **527-529** (2006) 9.
327. K. Grasza and E. Tymicki, *Mater. Sci. Forum* **600-603** (2009) 31.
328. I. Klebnikov, V.P. Madangarli, M.A. Khan, T.S. Sundarshan, *Mater. Sci. Forum* **264-268** (1998) 167.
329. R. Yakimova, M. Syvajarvi, and E. Janzen, *Mater. Sci. Forum* **264-268** (1998) 159.
330. R. Yakimova, M. Syvajarvi, S. Redankova, V.A. Dimitriev, A. Henry and E. Janzen, *Mater. Sci. Forum* **338-342** (2000) 237.
331. O, Filp, B. Epelbaum, M. Blickermann and A. Winnacker, *J. Crystal Growth* **271** (2004) 142.
332. B. M. Epelbaum and D. Hofmann, *J. Crystal Growth* **225** (2001) 1.
333. I. Kamata, H. Tsuchida, T. Jikimoto and K. Izumi, *Jpn. J. Appl. Phys.* **39** (2000) 6496.
334. I. Kamata, H. Tsuchida, T. Jikimoto and K. Izumi, *Mater. Sci. Forum* **389-393** (2002) 1137.
335. I. Kamata and H. Tsuchida, *Oyo Buturi (Monthly Publ. Jpn. Soc. Appl. Phys.) (in japanese)* **71** (2002) 576.
336. J.P. Bergman, H. Lendenmann, P.A. Nilsson, U. Lindefelt and P. Skytt, *Mater. Sci. Forum* **353-356** (2000) 299.
337. J. Senzaki, K. Kojima, T. Kato, A. Shimozatoa and K. Fukuda, *Mater. Sci. Forum* **483-485** (2005) 661.
338. G. Savini, M.I. Heggie and S. Oberg, *Mater. Sci. Forum* **527-529** (2006) 359.
339. T.A.G. Eberlin, R. Jones, A.T. Blumenau, S. Oberg and P.R. Briddon, *phys. stat. sol. (c)* **4** (2007) 2923.
340. A.T. Blumenau, C.J. Fall, R. Jones, M.I. Heggie, P.R. Briddon, T. Rrauenheim and S. Oberg, *J. Phys.:Condend. Matter.* **14** (2002) 12741.
341. A.T. Blumenau, C.J. Fall, R. Jones, M.I. Heggie, S. Oberg, T. Rrauenheim and P.R. Briddon, *Phys. Rev. B* **68** (2003) 174108.
342. N. Ohtani, M. Katsuno, M. Nakabayashi, H. Tsuge, T. Fujimoto, H. Yashiro, M. Sawamura, T. Aigo and T. Hoshino, *Mater. Res. Symp. Proc.* **911** (2006) 3.
343. N. Ohtani, M. Katsuno, H. Tsuge, T. Fujimoto, M. Nakabayashi, H. Yashiro, M. Sawamura, T. Aigo and T. Hoshino, *J. Crystal Growth* **286** (2006) 55.

344. N. Ohtani, M. Katsuno, H. Tsuge, T. Fujimoto, M. Nakabayashi, H. Yashiro, M. Sawamura, T. Aigo and T. Hoshino, *Microelectron. Engineer.* **83** (2006) 142.
345. N. Ohtani, M. Katsuno, H. Tsuge, T. Fujimoto, M. Nakabayashi, H. Yashiro, M. Sawamura, T. Aigo and T. Hoshino, *Jpn. J. Appl. Phys.* **45** (2006) 1738.
346. N. Ohtani, M. Katsuno, T. Fujimoto, M. Nakabayashi, H. Tsuge, H. Yashiro, T. Aigo, H. Hirano, T. Hoshino and W. Ohashi, *Jpn. J. Appl. Phys.* **48** (2009) 065503.
347. T. Nishiguchi, T. Furusho, T. Isshiki, K, Nishio, H. Shiomi and S. Nishio, *Mater. Sci. Forum* **600-603** (2009) 329.
348. J.L. Demenet, M. Amer, A. Mussi and J. Rabier, *Mater. Sci. Forum* **645-648** (2010) 335.
349. T.A.G. Eberlin, R. Jones, A.T. Blumenau, S. Oberg and P.R. Briddon, *phys. stat. sol. (c)* **4** (2007) 2923.
350. M. Katsuno, N. Ohtani, M. Nakabayashi, T. Fujimoto, H. Yashiro, H. Tsugem, T. Aigo, T. Hoshino, H. Hirano and W. Ohashi, *Mater. Sci. Forum* **645-648** (2010) 311.
351. P. Wellmann, D. Quellen, R. Muller, S.A. Sakwe and U. Kunecke, *Mater. Sci. Forum* **527-529** (2006) 79.
352 P. Wu, X. Xu, V. Rengarajan amd I. Zwieback, *Mater. Sci. Soc. Symp. Proc.* **1069** (2008) 83.
353. C. Cheng, R.J. Needs and V. Heine, *J. Phys. C* **21** (1988) 1049.
354. M.H. Hong, A.V. Samant and P. Pirouz, *Phil. Mag. A* **80** (2000) 919.
355. U. Lindefelt, H. Iwata, S. Oberg and P.R. Briddon, *Phys. Rev. B* **67** (2003) 155204.
356. M. Camarda, P. Delugas, A. Canino, A. Severino, N. Piluso, A. La Magna and F. La Via, *Mater. Sci. Forum* **645-648** (2010) 283.
357. H. Iwata, U. Lindefelt, S. Oberg and P.R. Briddon, *Mater. Sci. Forum* **389-393** (2002) 529.
358. J. Takahashi, N. Ohtani, M. Katsuno and S. Shinoyama, *Mater. Sci. Forum* **264-268** (1998) 25.
359. M. Katsuno, M. Nakabayashi, T. Fujimoto, N. Ohtani, H. Yashiro, H. Tsuge, T. Aigo, N. Hoshino and K. Tatsumi, *Mater. Sci. Forum* **600-603** (2009) 341.
360. M. Nakabayashi, T. Fujimoto, M. Katsuno, H. Tsuge, T. Aigo, S. Satoh, H. Yashiro, T. Hshino, H. Hirano and W. Ohashi, *Mater. Sci. Forum* **645-648** (2010) 9.
361. K. Seki, K. Morimoto, T. Ujihara, T. Tokunaga, K. Sasaki, K. Kuroda and Y. Takeda, *Mater. Sci. Forum* **645-648** (2010) 363.
362. D. Siche, M. Albrecht, H.-J. Rost and A. Sendzik, *Mater. Sci. Forum* **556-557** (2006) 21.
363. V. Jakubavicius, J. Palisaitis, R. Vasiliauskas, R. Yakimova and M. Syväjärvi, *Mater. Sci. Forum* **645-648** (2010) 375.
364. V.D. Heydemann, N. Schulze, D.L. Barrett and G. Pensl, *Appl. Phys. Lett.* **69** (1995) 3728.
365. K. Onoue, T. Nishikawa, M. Katsuno, N. Ohtani, H. Yashiro and M. Kanaya, *Inst. Phys. Conf. Ser.* **142** (1996) 65.
366. K. Onoue, T. Nishikawa, M. Katsuno, N. Ohtani, H. Yashiro and M. Kanaya, *Jpn. J. Appl. Phys.* **35** (1996) 2240.
367. N. Ohtani, M. Katsuno, J. Takahashi, H. Yashiro and M. Kanaya, *J. Appl. Phys.* **83** (1996) 4487.
368. N. Ohtani, M. Katsuno, J. Takahashi, M. Kanaya and S. Shinoyama, *Mater. Sci. Forum* **264-268** (1998) 49.
369. D. Schulz, K. Irmscher, J. Dolle, W. Eiserbeck, T. Müller, H.-J. Rost, D. Siche, G. Wagenr and J. Wollweber, *Mater. Sci. Forum* **338-342** (2000) 87.
370. D.W. Hansen, G. Chung and M.J. Loboda, *Mater. Sci. Forum* **527-529** (2006) 59.
371. K. Kim, S. Seo, J. Kim, J. Song, M. Oh, W. Bahng and E. Kim, *Mater. Sci. Forum* **527-529** (2006) 55.
372. N. Ohtani, M. Katuno, M. Nakabayashi, T. Fujimoto, H. Tsuge, H. Yashiro, T. Aigo, H. Hirano, T. Hoshino and K. Tatsumi, *J. Crystal Growth* **311** (2009) 1475.
373. M. Bockstete, A. Mattausch and O. Pankratov, *Mater. Sci. Forum* **527-529** (2006) 621.
374. R. Rurali, P. Godignon, J. Rebollo, E. Hemandas and P. Ordejon, *Appl. Phys. Lett.* **82** (2003) 4298.
375. M. Bockstete, A. Mattausch and O. Pankratov, *Appl. Phys. Lett.* **83** (2004) 58.
376. A. Henry, O. Kordina, C. Hallin, C. Hemmingsson and E. Janzen, *Appl. Phys. Lett.* **65** (1994) 2457.
377. H.E. Smith, K.G. Eyink, W.C. Mitchel, M.C. Wood and M.A. Fanton, *Mater. Sci. Forum* **527-529** (2006) 617.
378. R. Püshce, M. Hundhausen, L. Ley, K. Semmelrote, G. Pensl, P. Desperrier, P.J. Wellmann, E.E. Haller, J.W. Ager and U. Starke, *Mater. Sci. Forum* **527-529** (2006) 579.
379. I.G. Ivanov, A. Henry and E. Janzen, *Phys. Rev. B* **71** (2005) 241201.
380. I.G. Ivanov, C. Persson, A. Henry and E. Janzen, *Mater. Sci. Forum* **600-603** (2009) 445.
381. E.N. Kalabukhova, S.N. Lukin, E.N. Mokhov, *Sov. Phys. Solid State* **35** (1993) 361.
382. N.T. Son, A. Henry, J. Isoya and E. Janzen, *Mater. Sci. Forum* **483-485** (2005) 515.
383. E. Rauls, U. Gertmann and Th. Frauenheim, *Appl. Surf. Phys.* **243** (2004) 345.
384. E. Rauls, M.V.B. Pinheiro, S. Greulich-Weber and U. Gerstmann, *Phys. Rev. B* **70** (2004) 085202.
385. E. Rauls, U. Gerstmann, S. Greulich-Weber, K. Semmelroth, G. Pensl and E.E. Haller, *Mater. Sci. Forum* **527-529** (2006) 609.

386. F. Schmid, K. Semmelroth, M. Krieger, H.B. Weber, G. Pensl and E.E. Haller, *Mater. Sci. Forum* **600-603** (2009) 441.
387. R. Wang, I. Bhat and T.P. Chow, *Mater. Sci. Forum* **433-436** (2003) 145.
388. S. Rao, T.P. Chow and I. Bhat, *Mater. Sci. Forum* **527-529** (2006) 597.
389. T. Hornors, A. Gali, R.P. Devaty and W.J. Choyke, *Mater. Sci. Forum* **527-529** (2006) 605.
390. J. Isoya, M. Katagiri, T. Umeda, N.T. Son, A. Henry, A. Gali, N. Morishita, T. Ohshima, H. Itoh and E. Janzen, *Mater. Sci. Forum* **527-529** (2006) 593.
391. F. Yan, R.P. Devaty, W.J. Choyke, A. Gali, I.B. Bhat and D.J. Larkin, *Mater. Sci. Forum* **527-529** (2006) 585.
392. A. Henry and E. Janzen, *Mater. Sci. Forum* **527-529** (2006) 589.
393. S.G Sridhara, L.I. Clemen, D.G. Nizhner, R.P. Devaty, W.J. Choyke and D.J. Larkin, *Mater. Sci. Forum* **264-268** (1998) 465.
394. W. Suttrop, G. Pensl, and P. Lanig, *Appl. Phys. A* **51** (1990) 231.
395. G.A. Lomakina, *Sov. Phys. Solid State* 7 (1965) 475.
396. M.M. Lanakin, A.A. Lebedev, A.L. Syrkin and A.V. Suvorov, *Sov. Phys. Semicond.* **19** (1985) 69.
397. N. Sculze, J. Gajowski and G. Pensl, *Mater. Sci. Forum* **353-356** (2001) 45.
398. T. L. Straubinger, M. Bickermann, P.J. Wellmann and A. Winnacker, *J. Crystal Growth* **240** (2002) 117.
399. T. L. Straubinger, M. Bickermann, M. Rasp, R. Weingartner, P.J. Wellmann and A. Winnacker, *Mater. Sci. Forum* **389-393** (2002), 131.
400. S. Conteras, M. Zielinski, L. Konczewics, C. Blanc, S. Juellaquet, R. Müller, U. Künecke, P. Wellmann and J. Camassel, *Mater. Sci. Forum* **527-529** (2006) 633.
401. P. Hens, U. Künecke and P. Wellmann, *Mater. Sci. Forum* **600-603** (2009) 19.
402. C. Brylinski, *Diamond Relat. Mater.* **6** (1997) 1405.
403. A. Schöner, S. Karlsson, T. Schmitt, N. Nodell, M. Linnarsson and K. Rottner, *Mater. Sci. Engi neer. B* **61-62** (1999) 389.
404. O. Noblanc, C. Arnondo, C. Dua, E. Chartier and C. Brylinski, *Mater. Sci. Forum* **338-342** (2000) 1247.
405. J. Schnider, H.D. Muller, K. Maier and F. Fuchs, *Appl. Phys. Lett.* **56** (1990) 1184.
406. K. Maier, J. Schneider, W. Wilkening, S. Leibenzeder and R. Stein, *Mater. Sci. Engineer. B* **11** (1992) 27.
407. J. Schnider and K. Maier, *Physica B* **185** (1993) 199.
408. S. Sriram, R. C. Clarke, A. A. Burk, H. M. Hobgood, P. G. McMullin, P. A. Orphanos, R. R. Siergiej, T. J. Smith, C. D. Brandt, M. C. Driver and R. H. Hopkins, *IEEE Electron Device Lett.* **15** (1994) 458.
409. M. Bickermann, B.M. Epelbaum, D. Hofmann, T.L. Sraubinger, R. Weingartner, and A. Winnacker, *J. Crystal Growth* **233** (2001) 211.
410. M. Bickermann, D. Hofmann, T.L. Sraubinger, R. Weingartner amd P.J. Winnacker, *Appl. Sur. Sci.* **184** (2001) 84.
411. M. Bickermann, R. Weingartner amd A. Winnacker, *J. Crystal Growth* **254** (2003) 390.
412. M. Rasp, Th.L. Straubinger, E. Schmitt, M. Bickermann, S. Reshanov, H. Sadowski, *Mater. Sci. Forum* **433-436** (2003) 55.
413. K. Racka-Dzietko, E. Tymicki, K. Grasza, M. Raczkewicz, R. Jakiela, M. Kozubal, E. Jurkewicz-Wegner, A. Brzozowski, R. Diduszko, M. Piersa, K. Kosciewicz, M. Pawlowski and J. Krupka, *Mater. Sci. Forum* **645-648** (2010) 21.
414. W.C. Mitchel, A. Saxler, R. Perrin, J. Goldstein, S.R. Smith, A.O. Evwraye, J.S. Solomon, M. Braddy, V. Tsvetkov and C.H. Carter, Jr., *Mater. Sci. Forum* **338-342** (2000) 21.
415. J.R. Jenny, St. G. Müller, A. Powel, V.F. Tsvetkov, H.M. Hobgood, R.C. Glass and C.H. Carter, Jr., *J. Electron. Mater.* **31** (2002) 366.
416. J.R. Jenny, D.P. Malta, St. G. Müller, A. Powel, V.F. Tsvetkov, H.M. Hobgood, R.C. Glass and C.H. Carter, Jr., *J. Electon. Mater.* **32** (2003) 432.
417. St. G. Müller, M.F. Brady, W.H. Brixius, R.C. Glass, H.D. McHobgood, J.R. Jenny, R.T. Leonard, D.P. Maltam A.R. Powell, V.F. Tsvetkov, S.T. Allen, J.W. Palmour and C.H. Carter, Jr., *Mater. Sci. Forum* **433-436** (2003) 39.
418. J.R. Jenny, D.P. Malta, M.R. Calus, St.G. Müller, A.R. Powell, V.F. Tsvetkov, H. McD. Hobgood, R.C. Glass and C.H. Carter, Jr., *Mater. Sci. Forum* **457-460** (2004) 35.
419. N.I. Buchan, D.N. Henshall, W.S. Yoo, P.A. Mailloux and M.A. Tishler, *Inst. Phys. Conf. Ser.* **137** (1994) 113.
420. L.L. Clemen, M. Yoganathan, W.J. Choyke, R.P. Devaty, H.S. Kong, J.A. Elmond, D.J. Larkin, J.A. Powell and A.A. Burk, Jr., *Inst. Phys. Conf. Ser.* **137** (1994) 251.
421. E.R. Glaser, B.V. Shanabrook, W.E. Carlos, H.J. Chung, S. Nigam, A.Y. Polyakov and M. Skowronski, *Mater. Sci. Forum* **527-529** (2006) 613.

422. L.-Y. Zhou and R. Telle, *Mater. Sci. Forum* **645-648** (2010) 41.
423. G. Manolis, J.C.C. Lorenzzi, *Phys. Adv. Mater. Winter School* (2008)
424. R. A. Stein, *Physica B* **185** (1993) 211.
425. R. C. Glass, L. O. Kjellberg, V. F. Tsvetkov, J. E. Sundgren and I. Janzell, *J Crystal Growth* **132** (1993) 504.
426. R. C. Glass, C.I. Harris, V. F. Tsvetkov, P. F. Fewster, J. E. Sundgren and E. Janzen, *Inst. Phys. Conf. Ser.* **137** (1994), 165.
427. D. Black, L. Robins, N. Buchan and D. Henshall, *Inst. Phys. Conf. Ser.* **137** (1994) 337.
428. M. Tuominen, R. Yakimova, R. C. Glass, T. Turome and E. Janzen, *J. Crystal Growth* **144** (1994) 267.
429. M. Tuominen, E. Prieur, R. Yakimova, R.C. Glass, T. Tuomi and E. Janzen, *J. Crystal Growth* **165** (1996) 233.
430. M. Tuominen, A. Ellison, T. Tuomi, R. Yakimova, S. Milita and E. Janzen, *J. Crystal Growth* **225** (2001) 23.
431. S. Nishio, T. Higashino, T. Tanaka and J. Saraie, *J. Crystal Growth* **147** (1995) 339.
432. J. Takahashi, N. Ohtani and M. Kanaya, *J. Crystal Growth* **167** (1996) 596.
433. J. Takahashi, N. Ohtani and M. Kanaya, *Inst. Phys. Conf. Ser.* **142** (1996) 443.
434. T. Kato, H. Ohsato, A. Okamoto, N. Sugiyama and T. Okuda, *Mater. Sci. Engineer. B* **57** (1999) 147.
435. E. Emorhokpor, T. Kerr, I. Zwieback W. Elkington, M. Dudley, T. Anderson and J. Chen, *Mat. Res. Symp. Proc.* **815** (2004) 127.
436. J. Wan, S.-H. Park, G. Chung, E. Carlson and M.J. Loboda, *Mater. Sci. Forum* **527-529** (2006) 447.
437. E. Emorhokpor, E. Carlson, J. Wan, A. Weber, C. Basceri, J. Jenny R. Sandhu, J. Oliver, F. Burkeen, A. Somanchi, V. Velidandla, F. Orazio, A. Blew, M. Goorsku, M. Dudley and W.M. Vetter, *Mater. Sci. Forum* **527-529** (2006) 443.
438. D. Zhuang and J.H. Edger, *Mater. Sci. Engineer. R* **48** (2005) 1.
439. S. Amelinckx, G. Strumane and W.W. Webb, *J. Appl. Phys.* **31** (1960) 1359.
440. M. Katsuno, N. Ohtani, J. Takahashi, H. Yahsiro, M. Kanaya and S. Shinoyama, *Mater. Sci. Forum* **264-268** (1998) 837.
441. P. Wu, M. Yoganathan I. Zwieback, Y. Chen and M. Dudley, *Mater. Sci. Forum* **600-603** (2009) 333.
442. I. Kamata, M. Nagano and H. Tsuchida, *Mater. Sci. Forum* **645-648** (2010) 303.
443. M. Katsuno, N. Ohtani, M. Nakabayashi, T. Fujimoto, H. Yashiro, H. Tsuge, T. Aigo, T. Hoshino, H. Hirano and W. Ohasi, *Mater. Sci. Forum* **645-648** (2010) 311.
444. C. Fazi, M. Dudley, S. Wang and M. Ghezzo, *Inst. Phys. Conf. Ser.* **137** (1994) 487.
445. H. Posen and A. Bruce, in *Silicon Carbide-1973*, (Univ. of South Carolina Press, Columbia, SC, 1974) p. 238.
446. S. Wang, M. Dudley, C. Carter, Jr., D. Asbury and C. Fazi, *Mat. Res. Soc. Symp. Proc.* **307** (1993) 249.
447. S. Yamaguchi, D. Nakamura, I. Gunjishima and Y. Hirose, *Mater. Sci. Forum* **527-529** (2006) 407.
448. D. Nakamura, S. Yamaguchi, I. Gunjishima, Y. Hirose and T. Kimoto, *J. Crystal Growth* **304** (2007) 57.
449. H. Yamaguchi, H. Matsuhata and I. nagai, *Mater. Sci. Forum* **600-603** (2009) 313.
450. H. Matsuhata, H. Yamaguchi, I. Nagai, t. Ohno, R. Kosugi and A. Kinoshita, *Mater. Sci. Forum* **600-603** (2009) 321.
451. N. Ohtani, in *Recent Development of Power Electronics*, eds. by H. Ohashi, T. Kimoto (CMC Publ., 2009) p. 14.
452. Y. Chen, X. Huang, G. Dhanaraj, M. Dudley, E.K. Sanchez and M.F. MacMillan, *Mater. Sci. Forum* **600-603** (2009) 297.
453. Y. Chen, N. Zhang, X. Huang, D.R. Black and M. Dudley, *Mater. Sci. Forum* **600-603** (2009) 301
454. M. Dudley, S. Wang, W. Huang, C.H. Jr. Carter, V.F. Tsuvetkov and C. Fazi, *J. Phys. D. Appl. Phys.* **28** (1995) A63.
455. Y. Yanagisawa, T. Hatayama, H. Yano, Y. Uraoka and T. Fuyuki, *Mater. Sci. Forum* **527-529** (2006) 423.
456. D. Matsuoka, H. Yamamoto, S. Nishino, N. Hasuike, K. Kisoda and H. Harima, *Mater. Sci. Forum* **600-603** (2009) 337.
457. M. Lancin, G. Regula, J. Douin, H. Idrissi, L. Ottaviani and B. Pichaud, *Mater. Sci. Forum* **527-529** (2006) 379.
458. J. Hassan, A. Hery and J.P. Bergman, *Mater. Sci. Forum* **645-648** (2010) 307.
459. T. Mitani, S. Nakashima, H. Okumura and H. Nagasawa, *Mater. Sci. Forum* **527-529** (2006) 343.
460. M.S. Miao and W.R. Lambrecht, *Mater. Sci. Forum* **527-529** (2006) 351.
461. O.J. Glemnocki, M. Skowronsku, S.M. Prokes, D.K Gaskill and J.D. Caldwell, *Mater. Sci. Forum* **527-529** (2006) 343.
462. N. Hoshino, M. Tajima, T. Nishiguchi, K. Ikeda, T. Hayashi, H. Kinoshita and H. Shiomi, *Jpn. J. Appl. Phys.* **46** (2007) L973.

463. T. Miyanagi, H. Tsuchida, I. Kamata, T. Nakamura, R. Ishii, K. Nakayama and Y. Sugawara, *Mater. Sci. Forum* **527-529** (2006) 373.
464. G. Feng, J. Suda and T. Kimoto, *Appl. Phys. Lett.* **92** (2008) 221906.
465. G. Feng, J. Suda and T. Kimoto, *Appl. Phys. Lett.* **94** (2009) 091910.
466. G. Feng, J. Suda and T. Kimoto, *Mater. Sci. Forum* **645-648** (2010) 287.
467. E.K. Sanchez, V.D. Heydemann, G.S. Rohrer, M. Skowronski, J. Solomon, M.A. Capano and W.C. Mitchel, *Mater. Sci. Forum* **264-268** (1998) 433.
468. H.-J. Rost, J. Dolle, J. Doerschel, D. Siche, D. Schultz and J. Wollweber, *J. Crystal Growth* **225** (2001) 317.
469. M. Tuominen, R. Yakimova, A. Vehanen and E. Janzen, *Mater. Sci. Engineer. B* **67** (1999) 228.
470. M. Dudley, X.R. Huang, W. Huang, A. Powell, S. Wang, P. Neudeck and M. Skowronski, *Appl. Phys. Lett.* **75** (1999) 784.
471. W.M. Vetter and M. Dudley, *J. Appl. Crystal* **37** (2004) 200.
472. J. Dong, L. Wang. X. Hu, X. Li, J. Li, S. Jiang, X. Chen, X. Xu and M. Jiang, *Mater. Sci. Forum* **527-529** (2006) 451.
473. M. Katsuno, N. Ohtani, T. Aigo, T. Fujimoto, H. Tsuge, H. Yahsiro and M. Kanaya, *J. Crystal Growth* **216** (2000) 256.
474. M. Katsuno, N. Ohtani, T. Aigo, H. Yashiro and M. Kanaya, *Mater. Sci. Forum* **338-342** (2000) 493.
475. M. Katsuno, N. Ohtani, T. Fujimoto, T. Aigo and H. Yahsiro, *Mater. Sci. Forum* **389-393** (2002) 55.
476. J.W. Lee and M. Skowronski, *Mater. Sci. Forum* **527-529** (2006) 403.
477. N. Ohtani, M. Katsuno, J. Takahashi, H. Yashiro and M. Kanaya, *Phys. Rev. B* **59** (1999) 4592.
478. N. Ohtani, M. katsuno, T. Aigo, T. Fujimoto, H. Tsuge, H. Yashiro and M. Kanaya, *J. Crystal Growth* **210** (2006) 613.
479. R. Hock, K. Konias, L. Perdicaro, A. Magerl, P. Hens and P.J. Wellmann, *Mater. Sci. Forum* **645-648** (2010) 29.
480. M.E. Zvanut, *J. Phys.: Condens. Matter.* **16** (2004) R1341.
481. A. Gali, M. Bockstedte, N.T. Son and E. Janzen, *Mater. Res. Soc. Stymp. Proc.* **1069** (2008) 65.
482. J. Bernholc, S.A. Kajihara, C. Wang and A. Abtonelli, *Mater. Sci. Engineer B* **11** (1992) 265.
483. H.W.L. Alves, Braz. *J. Phys.* **27** (1997) 121.
484. J. Bernholc, A. Fissel, J. Furthmuller, U. Grosser and A. Zywietz, *J. phys: Condens. Matter.* **13** (2001) 9027.
485. A.P. Garshin, E.N. Mokhov and V.E. Shvaiko-Shvaikovskii, *Refrac. Ind. Ceram.* **44** (2003) 199.
486. M. Bockstdete, M. Heid and O. Pankratov, *Phys. Rev. B* **67** (2003) 193102.
487. M. Salvador, J.M. Perlado, A. Mattoni, F. Bernardini and L. Colombo, *J. Nucl. Mater.* **329-333** (2004) 1219.
488. E. Bernardii, A. Mattoni and L. Colombo, *Eur. Phys. J. B* **38** (2004) 437.
489. A. Gali, M. Bockstedte, N.T. Son, T. Umeda, J. Isoya and E. Janzen, *Mater. Sci. Forum* **523-526** (2006) 523.
490. T.A.G. Eberlein, R. Jones, S. Oberg and P.R. Briddon, *Phys. Rev. B* **74** (2006) 144106.
491. M. Bockstedte, A. Gali, T. Umeda, N.T. Son, J. Isoya and E. Janzen, *Mater. Sci. Forum* **527-529** (2006) 539.
492. T. Hornors, A. Gali, N.T. Son and E. Janzen, *Mater. Sci. Forum* **556-557** (2007) 445.
493. J. Isoya, T. Umeda, N. Mizuochi, N.T. Son, E. Janzen and T. Ohshima, *phys. stat. sol. (b)* **245** (2008) 1298.
494. M. Bockstedte, A. Gali, A. Mattasch, O. Pankratov and J.W. Steeds, *phys. stat. sol. (b)* 245(2008) 1281.
495. N.T. Son, P.N. Hai and E. Janzen, *Phys. Rev. B* **63** (2001) 201201.
496. T. Umeda, J. Isoya, N. Morisita, T. Ohshima, T. Kamiya, A. Gali, P. Deak, N.T. Son and E. Janzen, *Phys. Rev. B* **70** (2004) 235212.
497. N. Mizuochi, S. Yamasaki, N. Morishita, T. Ohshima, H. Itoh, T. Umeda and J. Isoya, *Phys. Rev. B*, **72** (2005) 235208.
498. M.E. Zvanut, H.J. Chung, A.Y. Polyakov and M. Skowronski, *Mater. Sci. Forum* **556-557** (2007) 473.
499. N.T. Son, T. Umeda, J. Isoya, A. Gali, M. Bockstedte, B. Magnusson, A. Ellison, N. Morishita, T. Ohshima, H. Itoh and E. Janzen, *Mater. Sci. Forum* **527-529** (2006) 527.
500. I.V. llyin, M.V. Muzafarova, E.N. Mokhov, V.I. Sankin, P.G. Baranov, S.B. Orlinskii and J. Schmidt, *Mater. Sci. Forum* **527-529** (2006) 535.
501. M.V.B. Pinheiro, E. Rauls, U. Gerstmann, S. Greulich-Weber, J.-M. Spaeth and H. Overhof, *Mater. Sci. Forum* **527-529** (2006) 551.
502. X. Kerbiriou, M.F. Barthe, S. Esnouf, P. Desgardin, G. Blondiaux and G. Petiere, *Mater. Sci. Forum* **527-529** (2006) 571.

503. T. Umeda, N.T. Son, J. Isoya, N. Morishita, T. Ohsima, H. Itoh and E. Janzen, *Mater. Sci. Forum* **527-529** (2006) 543.
504. N.Y. Garces, W.E. Carlosm, E.R. Glaser, S.W. Huh, H.J. Chung, S. Nigam, A.Y. Polyakov and M. Skowronski, *Mater. Sci. Forum* **527-529** (2006) 547.
505. M.V.B. Pinheiro, E. Rauls, U. Gerstmann, S. Greulich-Weber, H. Overhof and J.-M. Spaeth, *Phys. Rev. B* **70** (2004) 245204.
506. M.V. Muzafarova, I.V. llyin. E.N. Mokhov, V.I. Sankin and P.G. Baronov, *Mater. Sci. Forum* **527-529** (2006) 555.
507. N.T. Son, B. Magnusson, Z. Zolnai, A. Ellison and E. Janzen, *Mater. Sci. Forum* **433-436** (2003) 45.
508. N.T. Son, B. Magnusson, Z. Zolnai, A. Ellison and E. Janzen, *Mater. Sci. Forum* **457-460** (2004) 437.
509. E. Sörman, N.T. Son, W.M. Chen, O. Kordina, C. Hallin and E. Janzen, *Phys. Rev. B* **61** (2000) 2613.
510. N.T. Son, B. Magnusson and E. Janzen, *Appl. Phys. Lett.* **81** (2002) 3945.
511. S. Muto, H. Sugiyama, T. Kimura and T. Tanabe, *Jpn. J. Appl. Phys.* **43** (2004) 1076.
512. W. Sullivan and J.W. Steeds, *Mater. Sci. Forum* **527-529** (2006) 481.
513. A. Gali, P. Deak, E. Rauls, N.T. Son, I.G. Ivanov, F.H.C. Carlsson, E. Janzen and W.J. Choyke, *Phys. Rev. B* **67** (2003) 155203.
514. L. Storasta, F.H.C. Carlsoon, S.G. Sridhara, J.P. Bergman, A. Henry, T. Eglison, A. Hallen and E. Jansen, *Appl. Phys. Lett.* **78** (2001) 46.
515. J.W. Steeds, *Mater. Sci. Forum* **600-603** (2009) 437.
516. G. Alfieri, U. Grossner, E.V. Monakhov, B.G. Svensson, J.W. Steeds and W. Sullivan, *Mater. Sci. Forum* **527-529** (2006) 485.
517. J.W. Steeds, *Mater. Sci. Forum* **600-603** (2009) 433.
518. M. M. Anikin, A. A. Levedev, A. L. Syrkin, and A. V. Suvorov, *Sov. Phys. Scmicond.* **19** (1985) 69.
519. P. Zhou, M.G. Spencer, G.L. Harris and K. Fekade, *Appl. Phys. Lett.* **50** (1987) 1384.
520. W. Suttrop, G. Pensl, and P. Lanig, *Appl. Phys. A* **51** (1990) 231.
521. A. Uddin and T. Uemoto, *Jpn. J. Appl. Phys.* **32** (1993) L1670.
522. A. Uddin, H. Mitsuhashi and T. Uemoto, *Jpn. J. Appl. Phys.* **33** (1994) L908.
523. S. Jang, T. Kimoto and H. Matsunami, *Appl. Phys. Lett.* **65** (1994), 581.
524. K. Danno and T. Kimoto, *Mater. Sci. Forum* **527-529** (2006) 501.
525. S.W. Huh, J.J. Sumakeris, A.Y. Polyakov, M. Skowronski, P.B. Klein, B.V. Shanabrook and M.J. O'Loughlin, *Mater. Sci. Forum* **527-529** (2006) 493.
526. S.W. Huh, A.Y. Polyakov. H.J. Chung, S. Nigam, M. Skowronski, E.R. Glaser, W.E. Carlos, M.A. Fanton and N.B. Smirov, *Mater. Sci. Forum* **527-529** (2006) 489.
527. P. Kaminski, M. Kozubal, K. Grasza and E. Tymiscki, *Mat. Res. Soc. Symp. Proc.* **1069** (2008) 89.
528. S. Sasaki, K. Kawahara, G. Feng, G. Alfieri and T. Kimoto, *J. Appl. Phys.* **109** (2011) 013705.
529. E.N. Kalabukhova, D.V. Savchenko, G. Greulich-Weber, M.F. Bulanyi, S.A. Omelchenko, O.V. Khmelenko. A.A. Gorban and E.N. Mokhov, *Mater. Sci. Forum* **527-529** (2006) 651.
530. N.T. Son, P. Carlsoon, A. Gallstrom, B. Magnusson and E. Janzen, *Mater. Sci. Forum* **600-603** (2009) 401.
531. M.E. Zvanut, W. Lee, H. Wang, W.C. Mitchel and W.D. Mitchell, *Mater. Sci. Forum* **527-529** (2006) 517.
532. M. Kunzer, H.D. Müller and U. Kaufmann, *Phys. Rev. B* **48** (1993) 10846.
533. J. Reinke, H. Weihrich, S. Gruelich-Weber and J.-M. Spaeth, *Semicond. Sci. Technol.* **8** (1993) 1862.
534. A. Dörnen, Y. Latushko, W. Suttrop, G. Pensl, S. Leibenzeder and R. Stein, *Mater. Sci. Forum* **83-87** (1992) 1213.
535. J.R. Jenny, M. Skowronski, W.C. Mitchel, H.M. Hobgood, R.C. Glass, G. Augustine and R.H. Hopkins, *J. Appl. Phys.* **58** (1995) 3839.
536. W.C. Mitchel, R. Perrin, J. Goldstein, M. Roth, M. Ahoujja, S.R. Smith, A.O. Evwaraye, J.S. Solomon, G. Landis, J. Jenny, H.M. Hobgood, G. Augustine and V. Balakrishna, *Mater. Sci. Forum* **264-268** (1998), 545.
537. W.C. Mitchel, M.D. Roth, A.O. Evwaraye, P.W. Yu, S.R. Smith, J. Jenny, M. Skowronski, H.M. Hobgood, R.C. Glass, G. Augustine and R.H. Hopkins, *Inst. Phys. Conf. Ser.* **142** (1996) 313.
538. W. Lee and M.E. Zvant, *Mater. Sci. Forum* **527-529** (2006) 647.
539. P. Kaminski, R. Kozowski, M. Miczuga, M. Pawlowski, M. Kozubal and J. Zelazko, *Joint Fifth Int. Conf. Solid State Crystal and Eighth Pol. Conf. Crystal Growth* (2009)
540. J.R. Jenny, M. Skowronski, W.C. Mitchel, H.M. Hobgood, R.C. Glass, G. Augustine and R.H. Hopkins, *Appl. Phys. Lett.* **68** (1996) 1963.
541. M.S. Miao and W. Lambrecht, *Mater. Sci. Forum* **527-529** (2006) 641.

542. M.E. Zvanut and V.V. Konovalov, *Appl. Phys. Lett.* **80** (2002) 410.
543. E.N. Kalabukhova, S.N. Lukin,D.V. Savchenko, A.A. Sitnikov, W.C. Mitchel, S.R. Smith and S. Greulich-Weber, *Mater. Sci. Forum* **527-529** (2006) 563.
544. E.N. Kalabukhova, S.N. Lukin, D.V. Savchenko, W.C. Mitchel, S. Greulich-Weber, E. Rauls and U. Gerstmann, *Mater. Sci. Forum* **527-529** (2006) 559.
545. W.E. Carlos, E.R. Glaser, N.Y. Garces, B.V. Shanabrook and M.A. Fanton, *Mater. Sci. Forum* **527-529** (2006) 531.
546. N.T. Son, P. Carlsson, B. Magnusson and E. Janzen, *Mater. Sci. Forum* **556-557** (2007) 465.
547. N.Y. Garces, E.R. Glaser, W.E. Carlos and M.A. Fanton, *Mater. Sci. Forum* **600-603** (2009) 389.
548. P. Carlsson, N.T. Son, B. Magnusson, A. Henry and E. Janzen, *Mater. Sci. Forum* **600-603** (2009) 381.
549. E.N. Kalabukhova, S.N. Lukin, W.C. Mitchel, A. Saxler and R.L. Jones, *Physica B* **308-310** (2001) 698.
550. E.N. Kalabukhova, S.N. Lukin, D.V. Savchenko, W.C. Mitcheland W.D. Michel, *Mater. Sci. Forum* **457-460** (2006) 501.
551. M.E. Zvanut, G. Ngetich, H.J. Chung, A.Y. Polyakov, M Skowronski, N.Y. Garces and E.R. Glaser, *Mater. Sci. Forum* **600-603** (2009) 385.
552. N.T. Son, P. Carlssson, J.U. Hassan, E. Janzen, T. Umeda, J. Isoya, A. Gali, M. Bockstedte, N. Morishita, T. Ohshima and H. Itoh, *Phys. Rev. Lett.* **96** (2006) 055501.
553. V.S. Vanier and V.A. Ilin, *Sov. Phys. Solid State* **22** (1980) 2011.
554. V.S. Vanier, V.I. Veinger, V.A. Ilin and V.F. Tsvetkov, *Sov. Phys. Solid State* **23** (1981) 2126.
555. N.T. Son, P.N. Hai, A. Shuja, W.M. Chen, J.L. Lindstrom, B. Monemar and E. Janzen, *Mater. Sci. Forum* **338-342** (2000) 821.
556. M.E. Zvanut, *J. Phys.: Condens. Matter.* **16** (2004) R1341.
557. T. Umeda, N.T. Son, J. Isoya, E. Janzen, T. Ohshima, N. Morishita, H. Itoh, A. Gali and M. Bockstedte, *Phys. Rev. Lett.* **96** (2006) 145501.
558. N.T. Son, P. Carlsoon, J. ul Hassan, B. Magnusson and E. Janzen, *Phys. Rev. B* **75** (2007) 155204.
559. P. Carlsson, N.T. Son, t. Umedam J. Isoya and E. Janzen, *Mater. Sci. Forum* **556-557** (2007) 449.
560. V.V. Konovalov and M.E. Zvanut, *J. Electron. Mater.* **31** (2002) 351.
561. D. Alvarcz, V.V. Konovalov and M.E. Zvanut, *J. Elcctron. Mater.* **32** (2003) 444.
562. V.V. Konovalov and M.E. Zvanut, *Phys. Rev.* **68** (2003) 012102.
563. B. Magnusson and E. Janzen, *Mater. Sci. Forum* **483-485** (2005) 341.
564. R. Aavikko, K. Saarinen, B. Magnusson and E. Janzen, *Mater. Sci. Forum* **527-529** (2006) 575.
565. R. Aavikko, K. Saarinen, B. Magnusson and E. Janzen, *Mater. Sci. Forum* **483-485** (2005) 469.
566. R. Aavikko, K. Saarinen, B. Magnusson and E. Janzen, *Phys. Rev. B* **75** (2007) 085208.
567. A. Thuaire, A. Henry, B. Magnusson, J.P. Bergman, W.M. Chen, E. Janzen, M. Mermoux and E. Bano, *Mater. Sci. Forum* **527-529** (2006) 461.
568. S.I. Maximenko, J.A. Freitas, Jr. N.Y. Garces, E.R. Glaser and M.A. Fanton, *Mater. Sci. Forum* **600-603** (2009) 429.
569. A. Gällstrom, B. Magnusson, P. Carlsson, N.T. Son, A. Henry, F. Beyer, M. Syväjärvi, R. Yakimova and E. Janzen, *Mater. Sci. Forum* **556-557** (2007) 371.
570. S.I. Maximenko, J.A. Freitas, Jr., N.Y. Garces, E.R. Glaser and M,A, Fanton, *J. Electron. Mater.* **38** (2009) 551.
571 W.C. Mitchel, W.D. Michell, S.R. Smith, G. Landis, A.O. Evwaraye, Z.-Q. Fang, D.C. Look and J.R. Sozelove, *Mater. Sci. Forum* **527-529** (2006) 505.
572. Z-Q. Fang, B. Claflin, D.C. Look, L. Polenta, J. Chen, T. Anderson and W.C. Mitchel, *Mater. Sci. Forum* **527-529** (2006) 509.
573. M. Takahashi and H. Matssuura, *Mater. Sci. Forum* **600-603** (2009) 393.
574. A.P. Zhang, L.B. Rowland, E.B. Kaminsky, J.B. Tucker, R.A. Beaupre, J.W. Kretchmer, J.L. Garrett, A. Vertiatchikh, G. Koley, H.Y. Cha, A.F. Allen, J. Cook, J. Foppes and B.J. Edward, *J. Electron. Mater.* **32** (2003) 437.
575. R.R. Ciechonski, M. Syväjärvi, A. Kakanakova-Georgieva and R. Yakomova, *J. Electron. Mater.* **32** (2003) 452.
576. K. Neimontas, A. Kadys, R. Aleksiejunas, K. Jarasiunas, G. Chung, E. Sanchez and M. Loboda, *Mater. Sci. Forum* **527-529** (2006) 469.
577. P. Kaminski, R. Kozlowski, M. Miczuga, M. Pawowski, M. Kozubal and J. Zelazko, *Mater. Sci. Soc. Symp. Proc.* **1069** (2008) 33.
578. T. Umeda, J. Isoya, M. Morishita, T. Ohshima and T. Kamiya, *Phys. Rev. B* **69** (2004) 121201(R).
579. M.E. Zvanut, W. Lee, H. Wang, W.C. Michel and W.D. Michell, *Mater. Sci. Forum* **527-529** (2006) 527.

580. B. Magnusson, R. Aavikko, K. Saarinen, N.T. Son and E. Janzen, *Mater. Sci. Forum* **527-529** (2006) 455.
581. D.L. Barrett and R.B. Campbell, *J. Appl. Phys.* **388** (1967) 53.
582. A. Rosengreen, *Mater. Res. Bull.* **4** (1969) S335.
583. G. Pensl and W. J. Choyke, *Physica B* **185** (1993) 264.
584. W. Götz, A. Schollel, G. I'ensl, W. Sutirop, W. J. Choyke, R. Stein and S. Leibenzeder, *J. Appl. Phys.* **73** (1993) 3332.
585. W.J. Schaffer, G.H. Negley, K.G. Irvine and J.W. Palmour, *Mat. Res. Soc. Symp. Proc.* **339** (1994) 595.
586. K. Tsukioka, *Inst. Phys. Conf. Ser.* **142** (1996) 397.
587. A.A. Burk. Jr., M.J. O'Loughlin, R.R. Sieriej, A.K. Agarwal, S. Sriram, R.C. Clarke, M.F. MacMilan, V. Balakrishna and C.D. Brandt, *Solid-State Electron.* **43** (1999) 1459.
588. C.H. Carter, Jr., V.F. Tsvetkov, R.C. Glass, D. Henshall, M. Brady, S.G. Müller, O. Kordina, K. Irvine, J.A. Edmond, H.-S. Kong, R. Singh, S.T. Allen and J.W. Palmour, *Mater. Sci. Engineer. B* **61-62** (1999) p. 1.
589. J. Permot, S. Contretas, J. Camassel, J.L. Robert, W. Zawadzki, E. Neyret and L. Di Cioccio, *Appl. Phys. Lett.* 77 (2000) 4359.
590. S. Kagamihara, H. Matsuura, T. Hatakeyama, T. Watanabe, M. Kushibe, T. Shinohe and K. Arai, *J. Appl. Phys.* **96** (2004) 5601.
591. H. Harima, S. Nakashima and T. Uemura, *J. Appl. Phys.* **78** (1995) 1996.
592. S. Nakashima and H. Harima, *Inst. Phys. Conf. Ser.* **142** (1996) 269.
593. M. Chafai, A. Jaouhari, A. Torres, R. Anton, E. Martin, J. Jimenez and W. C. Mitchel, *J. Appl. Phys.* **90** (2001) 5211.
594. E. Kurimoto, M. Hangyo, H. Harima, K. Kisoda, T. Nishiguchi, S. Nishino, S. Nakashima, M. Katsuno and N. Ohtani, *Mater. Sci. Forum* **457-460** (2004) 621.
595. S. Nakashima, Y. Nakatake, Y. Yano, H. Harima, N. Ohtani and M. Katsuno, *Mater. Sci. Forum* **389-393** (2002) 633.
596. S. Nakashima, H. Harima, N. Ohtani and M. Katsuno, *J. Appl. Phys.* **95** (2004) 3547.
597. S. Nakashima, T. Kitamura, T. Mitani, H. Okumura, M. Katsuno and N. Ohtani, *Phys. Rev. B* 76 (2007) 245208.
598. K. Tsukioka, D. Vasileska and D.K. Ferry, *Phys. B* **185** (1993) 466.
599. K. Tsukioka, K. Haruna, M. Goto, K. Kobayashi and H. Obara, *Bulletin of Tamagawa Univer sity Academic Institute* **5** (1999) 95.
600. H. Mattuura, Y. Masuda, Y. Chen and S. Nishino, *Jpn. J. Appl. Phys.* **39** (2000) 5069.
601. H. Matsuura, *Mater. Sci. Forum* **389-393** (2002) 679.
602. H. Matsuura, *New J. Phys.* **4** (2002) 12.1.
603. H. H. Woodbury and G. W. Ludwig, *Phys. Rev.* **124** (1961) 1083.
604. S.G. Weber, *phys. stat. sol. (a)* **162** (1997) 95.
605. G. Pensl, V.V. Afanas'ev, M. Bassler, M. Schadt. T. Troffer. J. Heindl, H.P. Strunk, M. Maier and W.J. Choyke, *Inst. Phys. Conf. Ser.* **142** (1996) 275.
606. M.A. Capano, J.A. Cooper, Jr., M.R. Melloch, A. Sakler and W.C. Mitchel, *J. Appl. Phys.* **87** (2000) 8773.
607. J. Senzaki, K. fukuda, Y. Ishida, Y. Tanaka, H. Tanoue, N. Kobayashi, T. Tanaka and K. Arai, *Mater. Res. Soc. Symp.* **622** (2000) T.6.7.1.
608. T. Troffer, M. Schadt, T. Frank, H. Itoh, G. Pensl, J. Heindl, H.P. Strunk and M. Maier, *phys. stat. sol.* **162** (1997) 277.
609. Y. Tanaka, N. Kobayashi, H. Okumura, R. Suzuki, T. Ohdaira, M. Hasegawa, M. Ogura, S. Yoshida and H. Tanoue, *Mater. Sci. Forum* **338-342** (2000) 909.
610. W. Suttrop, G. Pensl, W. J. Choyke, R. Stein and S. Leibenzeder, *J. Appl. Phys.* **72** (1992) 3708.
611. M.V. Alekseenko, A.G. Zabrodskii and M.P. Timofeev, *Sov. Phys. Semicond.* **21** (1987) 494.
612. B. Segall, S.A. Alterovitz, E.J. Haugland and L.G. Matus, *Appl. Phys. Lett.* **49** (1986) 584.
613. B. Segall, S.A. Alterovitz, E.J. Haugland and L.G. Matus, *Appl. Phys. Lett.* **50** (1987) 1533.
614. J. Reinke, S. Greulich-Weber, J. -M. Spaeth, E. N. Kalabukhova, S. N. Lukin and E. N. Mokhov, *Inst. Phys. Conf. Ser.* **137** (1994), 211.
615. C. Goldberg and J.W. Ostroski, in *Silicon Carbide- A High Temperature Semiconductors*, eds. by J.R. O'Conner and J. Smiltens (Pergamon Presss, Oxford, 1960) p. 453.
616. G. Tamulatis, I. Yimaz, M.S. Shur, T. Anderson and R. Gaska, *Appl. Phys. Lett.* **84** (2004) 335.
617. P.G. Muzykov, Y.I. Khlebnikov, S.G. Regula, Y. Gao and T.S. Sudarshan, *J. Electron. Mater.* **32** (2003) 505.
618. S. Nakashima, H. Katahama, Y. Nakakura and A. Mitsuishi, *Phys. Rev. B* **33** (1986), 5721.
619. S. Nakashima and K. Tahara, *Phys. Rev. B* **40** (1989) 6339.
620. S. Nakashima, H. Ohta and M. Hangyo, *Phil. Mag. B* **70** (1994) 971.
621. S. Nakashima, K. Kisoda and J.-P. Gauthier, *J. Appl. Phys.* **75** (1994) 5354.

622. S. Nakashima and H. Harima, *phys. stat. sol. (a)* **162** (1997) 39.
623. S. Rohmfeld, M. Hundhausen and L. Lely, *Phys. Rev. B* **58** (1998) 9858.
624. S. Nakashima, Y. Nakatake, H. Harima, M. Katsuno and N. Ohtani, *Appl. Phys. Lett.* **77** (2000) 3612.
625. S. Nakashima, Y. Nakatake, Y. Ishida, T. Takahashi and H. Okumura, *Phys. B* **308-310** (2001), 684.
626. S. Nakashima, Y. Nakatake, Y. Ishida, T. Takahashi and H. Okumura, *Mater. Sci. Forum* **389-393** (2002) 629.
627. S. Nakashima and T. Mitani, *Mater. Sci. Forum* **527-529** (2006) 333.
628. S. Nakashima and H. Harima, in *Handbook of Vibrational Spectroscopy, Vol.4*, eds. J.M. Chalmers and P.R. Griffiths (John Willey & Sons, Chichester, 2002) p. 2637.
629. S S. Nakashima and M. Hangyo, *IEEE J. Quant. Electron.* **25** (1989) 965.
630. W.J. Choke and L. Patrick, in *Silicon Carbide-1973*, ed. by R.C. Marshal, J.W. Faust and C.E. Ryan (University of South Carolina Press, 1974) p. 261.
631. R.P. Devaty and W.J. Choke, *phys. stat. sol. (a)* **162** (1997) 5.
632. J. Camassel, S. Juillaguet, N. Planes, A. Raymond, P. Grosse, G. Basset, C. Faure, M. Couchaud, J.M. Bluet, K. Chourou, M. Anikin and R. Madar, *Mater. Sci. Engineer. B* **61-62** (1999) 258.
633. J.A. Powell, D.J. Larkin, L.G. Matus, W.J. Choyke, J.L. Bradshaw, L. Henderson, M. Yoganathan, J. Yang and P. Pirouz, *Appl. Phys. Lett.* **56** (1990) 1442.
634. I.G. Ivanov, C. Hallin, A. Henry, O. Kordina and E. Janzen, *J. Appl. Phys.* **80** (1996) 3504.
635. M. Ikeda, H. Matsunami and T. Tanaka, *Phys. Rev. B* **22** (1980) 2842.
636. I.G. Ivanov, A. Henry and E. Janzen, *Mater. Sci. Forum* **527-529** (2006) 601.
637. R. Hirano, M. Rjim and K.M. Itoh, *Mater. Sci. Forum* **645-648** (2010) 355.
638. A. Ellison, T. Kimoto, I.V. Ivanov, Q. Wahab, A. Henry, O. Kordina, J. Zhang, C.G. Hemmingsson, C.-Yu, Gu, M.R. Leys and E. Janzen, *Mater. Sci. Forum* **264-268** (1998) 108.
639. M. Tajima, M. Tanaka and N. Hoshino, *Mater. Sci. Forum* **389-393** (2002) 597.
640. M. Tajima, Y. Kumagaya, T. Nakata, M. Inoue and A. Nakamura, *Mater. Sci. Forum* **264-268** (1998) 481.
641. Y. Koshka, M. Mazzola, S. Yingquan and C.U. Pttman, Jr., *J. Electron. Mater.* **30** (2001) 220.
642. A. Gallstrom, B. Magnusson, A. Thuaire, P. Paskov, A. Henrry and E. Janzen, *Mater. Sci. Forum* **600-603** (2009) 397.
643. R.K. Malhan, H. Nakamura, S. Onda, D. Nakamura and K. Hara, *Denso Tech. Rev.* **8** (2003) 77.
644. P.G. Neudeck, *Mater. Sci. Forum* **338-342** (2000) 1161.
645. J.W. Palmour, C.H. Carter, Jr. and L.A. Lipkin, in *Symp. Handbook of the 21st Int. Symp. Compound Semiconductors* (San Diego, 1994) p. 15.
646. J.W. Palmour, L.A. Lipkin, R. Singh, D.B. Slater, Jr., A.V.Suvorov and C.H. Carter, Jr., *Diamond Relat. Mater.* **6** (1997) 1400.
647. A.R. Powell, J.J. Sumakeris, R.T. Leonard, M.F. Brady, St. G. Müller, V.F. Tsvutkov, H.D. Hobgood, A.A. Burk, M.J. Paisley, R.C. Glass and C.H. Carter, Jr., *Mat. Res. Soc. Symp. Proc.*, **815** (2004) 3.
648. I. Kamata, H. Tsuchida, T. Jikimoto and K. Izumi, *Jpn. J. Appl. Phys.* **40** (2001) L1012.
649. Y. Tanaka, S. Nishizawa, K. Fukuda, K. Arai, T. Ohno, N. Oyanagi, T. Suzuki and T. Yatsuo, *Appl. Phys. Lett.* **83** (2003) 377.
650. M. Kato, K. Ogawa and M. Ichimura, *Mater. Sci. Forum* **600-603** (2009) 373.
651. T. Kimoto, N. Miyamoto and H. Matsunami, *IEEE Trans. Electron Devices* **46** (1999) 471.
652. A.O. Konstantinov and H. Bleichner, *Appl. Phys. Lett.* **71** (1997) 3700.
653. H. Lendenmann, F. Dahlquist, J.P. Bergman, H. Bleichner and C. Hallin, *Mater. Sci. Forum* **389-393** (2002) 1259.
654. P. Pirouz and J.W. Yang, *Ultramicroscopy* **51** (1993) 189.
655. H. Jacobson, J. Birche, R. Roweleit, L.R. Syvajavi, J.P. Bergamn, A. Ellison, T. Tuomi and E. Janzen, *J. Appl. Phys.* **91** (2002) 6354.
656. S. Ha, M. Skowronski and H. Lendenmann, *J. Appl. Phys.* **96** (2004) 393.
657. S.I. Maximenko and T. S. Sudarshan, *J. Appl. Phys.* **97** (2005) 074501.
658. Z. Zhang. A.E. Gekov, P. Sadagopan, S.I. Maximenko and T.S. Sudarshan,*Mater. Sci. Forum* **527-529** (2006) 371.
659. J.J. Sumakeris, M. Das, H. McD. Hobgood, S. Müller, M.J. Paisley, S. Ha, M. Skowronski, J.W. Palmour and C.H. Carter, Jr., *Mater. Sci. Forum* **457-460** (2004) 1113.
660. S. Ha, M. Benamara, M. Skowronski and H. Lendenmann, *Appl. Phys. Lett.* **83** (2003) 4957.
661. S.I. Maximenko, P. Pirouz and T. S. Sudarshan, *Appl. Phys. Lett.* 87 (2005) 033503.
662. H. Jacobson, J.P. Bergman, C. Hallin, E. Janzen, T. Tuomi and H. Lendenmann, *J. Appl. Phys.* **95** (2004) 1485.
663. P.G. Neudeck, W. Huang and M. Dudley, *Solid-State Electron.* **42** (1998) 2157.

664. Q. Wahab, A. Ellison, C. Hallin, A. Henry J. Di Persio, R. Martinez and E. Janzen, *Mater. Sci. Forum* **338-342** (2000) 1175.
665. P.G. Neudeck, W. Huang and M. Dudley, *IEEE Electron Dev.* **46** (1999) 478.
666. Q. Wahab, A. Ellison, A. Henry and E. Janzen, *Appl. Phys. Lett.* **76** (2000) 2725.
667. T. Tsuji, S. Izumi, A. Ueda, H. Fujisawa, K. Ueno, H. Tsuchida, I. Kamata, T. Jikimoto and K. Izumi, *Mater. Sci. Forum* **389-393** (2002) 1141.
668. D. Defives. O. Noblanc, C. Dua, C. Brylinski, M. Barthula, V. Aubry-Fortuna and F. Meyer, *IEEE Electron Dev.* **46** (1999) 449.
669. H. Saitoh, T. Kimoto and H. Matusnami, *Mater. Sci. Forum* **457-460** (2004) 997.
670. S. Izumi, I. Kamata, T. Tawara, H. Fujisawa, and T. Tsuchida, *Mater. Sci. Forum* **457-460** (2004) 1085.
671. T. Aigo, M. Katsuno, T. Fujimoto, H. Yashiro and N. Ohtani, *Jpn. J. Appl. Phys.* **41** (2002) 2855.
672. S. Soloviev, P. Sandvik, A. Arthur, K. Matocha, S. Maximenko and T.S. Sidarshan, *Mater. Sci. Forum* **527-529** (2006) 427.
673. J. Senzaki, A. Shimozato, M. Okamoto, K. Kojima, K. Fukuda, H. Okumura and K. Arai, *Mater. Sci. Forum* **600-603** (2009) 787.
674. H. Tsuchida, I. Kamata and M. Ngano, *J. Crystal Growth* **306** (2007) 254.
675. M. Benamara, Z. Zhang, M. Skowronski, P. Ruterana, G. Nouet, J.J. Sumakeris, M.J. Paisley and M.J. O'Loughlin, *Appl. Phys. Lett.* **86** (2005) 021905.
676. J. Hassen, J.P. Bergman, P.J. McNally and E. Janzen, *Absrt. ICSCRM 2007*, Th-IP-1 p. Th-32.
677. H. Lendenmann, F. Dahlquist, N. Johansson, R. Soderholm, P.A. Nilsson, J.P. Bergman and P. Skytt, *Mater. Sci. Forum* **353** (2001) 727.
678. K. Maeda and S. Takeuchi, in *Dislocations in Solids Vol. 10* (North-Holland, Amsterdam, 1996) p. 444.
679. J.D. Weeks, J.C. Tully and L.C. Kimerling, *Phys. Rev. B* **12** (1975) 3286.
680. H. Sumi, *Phys. Rev. B* **29** (1984) 4616.
681. M.S. Miao, S. Limpijumnong and W.R. Lambrecht, *Appl. Phys. Lett.* **79** (2001) 4360.
682. T.A. Kuhr, J.Q. Liu, H.J. Chung, M. Skowronski and F. Szmulowicz, *J. Appl. Phys.* **92** (2002) 5863.
683. H.P. Iwata, U. Lindefelt, S. Oberg and P.R. Brindon, *J. Appl. Phys.* **94** (2003) 4972.
684. W.R.L. Lambrecht and M.S. Miao, *Phys. Rev. B* **73**(2006) 155312.
685. A. Galeckas and J. Linnros, *Phys. Rev. Lett.* **96** (2006) 025502.
686. A. Galeckas, J. Linnrostar, S.B. Breitholtz and H. Bleichner, *Mater. Sci. Forum* **353-357** (2001) 389.
687. S.G. Sridhara, F.H.C. Carlsson, J.P. Bergman and E. Janzen, *Appl. Phys. Lett.* **79** (2001) 3944.
688. A. Galeckas, J. Linnros, B. Breitholtz and H. Bleichner, *J. Appl. Phys.* **90** (2001) 980.
689. R.E. Stahlbush, M. Fatemi, J.B. Fedison, S.D. Arthur, L.B. Rowland and S. Wang, *J. Electron. Mater.* **31** (2002) 370.
690. A. Galeckas, J. Linnros and P. Pirouz, *Appl. Phys. Lett.* **81** (2002) 883.
691. J.Q. Liu, M. Skowronski, C. Hallin, R. Soderholm and H. Lendenmann, *Appl. Phys. Lett.* **80** (2002) 749.
692. P.O. Persson, L. Hultman, H. Jacobson, J.P. Bergman, E. Janzen, J.M. Molina-Aldareguia, W.J. Clegg and T. Tuomi, *Appl. Phys. Lett.* **80** (2002) 4852.
693. M. Vetter, J.Q. Liu, M. Dudley, M. Skowronski, H. Lendenmann and C. Hallin, *Mater. Sci. Engineer. B* **98** (2003) 220.
694. M. Zhang, P. Pirouz and H. Lendenmann, *Appl. Phys. Lett.* **83** (2003) 3320.
695. M.E. Twigg, R.E. Stahlbuch, M. Fatemi, S.D. Arthur, J.B. Fedison, J.B. Tucker and S. Wang, *Appl. Phys. Lett.* **82** (2003) 2410.
696. H. Jacobson, J.P. Bergman, C. Hallin, E. Janzen, T. Tuomi and H. Lendenmann, *J. Appl. Phys.* **95** (2004) 1485.
697. S Ha, M. Sowronski, J.J. Sumakeris, M.J. Paisley and M.K. Das, *Phys. Rev. Lett.* **92** (2004) 175504.
698. P. Pirouz, M. Zhang, A. Galeckas and J. Linnros, *Mat. Res. Soc. Symp. Proc.* **815** (2004) 91.
699. R.E. Stahlbush, M.E. Twigg, J.J. Sumakeris, K.G. Irvine and P.A. Losse, *Mat. Res. Soc. Symp. Proc.* **815** (2004) 103.
700. S.I. Maximenko and T. S. Sudarshan, *J. Appl. Phys.* **97** (2005) 074501.
701. S.I. Maximenko, P. Pirouz and T. S. Sudarshan, *Appl. Phys. Lett.* **87** (2005) 033503.
702. Y. Wang, L. Chen, M.K. Mikhov, G. Samson and B.J. Skromme, *Mater. Sci. Forum* **527-529** (2006) 363.
703. S.I. Maximenko, P. Pirouz and T.S. Sudarshan, *Mater. Sci. Forum* **527-529** (2006) 367.
704. M.T. Twigg, R.E. Stahlbush, P.A. Losee, C. Li, I.B. Bhat and T.P. Chow, *Mater. Sci. Forum* **527-529** (2006) 383.
705. T. Kawahara, N. Hatta, K. Yagi, H. Uchida, M. Kobayashi, M. Abe, H. Nagasawa, B. Zippelius and G. Pensl, *Mater. Sci. Forum* **645-648** (2010) 339.

706. J.D. Caldwell, A.J. Giles, R.E. Stahbush, M.G. Ancona, O.J. Glembocki, K.D Hobart, B.A. Hull and K.X. Liu, *Mater. Sci. Forum* **645-648** (2010) 277.
707. M. Skowronski, J.Q. Liu, W.M. Vetter, M. Dudley, C. Hallin and H. Lendenmann, *J. Appl. Phys.* **92** (2002) 4699.
708. B.J. Skromme, K. Palle, C.D. Poweleit, L.R. Bryant, W.M. Vetter, M. Dudley, K. Moore and T. Gehoski, *Mater. Sci. Forum* **389-393** (2002) 455.
709. R.K. Malhan, H. Nakamuram S. Onda, D. Nakamura and K. Hara, *Mater. Sci. Forum* **433-436** (2003) 917.
710. Z. Zhang. A.E. Gekov, P. Sadagopan, S.I. Maximenko and T.S. Sudarshan, *Mater. Sci. Forum* **527-529** (2006) 371.
711. J.D. Caldwell, R.E. Stahlbuch, E.A. Imhoff, O.J. Glembocki, K.D. Hobart, M.J. Tadjer, Q. Zhang, M. Das and . Agarwal, *Mater. Rec. Soc. Symp. Proc.* **1069** (2008) 195.
712. S.-H. Ryu, Q. Zhang, H. Fatima, S. Heney, R. Stahlbush and A. Agarwal, *Mat. Res. Soc. Symp. Proc.* **1069** (2008) 251.
713. Q. Zhang, A. Agarwal, A. Burk, M. O'Loughlin, J. Palmour, R. Stahlbush and C. Scozzie, *Mater. Sci. Forum* **645-648** (2010) 331.
714. K. Kojima, T. Ohno, T. Fujimoto, M. Katsuno, N. Ohtani, J. Nishio, Y. Ishida, T. Takahashi, T. Suzuki, T. Tanaka and K. Arai, *Mater. Sci. Forum* **433-436** (2003) 925.
715. K. Kojima, T. Ohno, T. Fujimoto, M. Katsuno, N. Ohtani, J. Nishio, T. Suzuki and T. Tanaka, *Appl. Phys. Lett.* **81** (2002) 2974.
716. H. Fujiwara, T, Kimoto, T. Tojo and H. Matsunami, *Appl. Phys. Lett.* **87** (2005) 051912.
717. E. Pettenpaul, W. von Münch and G. Ziegler, *Inst. Phys. Conf. Ser.* **53** (1980) 21.
718. R. F. Davis, G. Kelner, M. Shur, J.W. Palmour and J.A. Edmond, *Proc. IEEE* **79** (1991) 677.
719. J.A. Edmond, H.S. Kong, and C.H. Carter, Jr., in *Springer Proc. Phys. Vol. 71, Amorphous and Crystalline Silicon Carbide IV*, eds. C.Y. Yang, M.M. Rahman and G.L. Harris (Springer, Berlin 1992) p. 344.
720. P.A. Ivanov and V.E. Chelnokov, *Semicond. Sci. Technol.* **7** (1992) 863.
721. J. W. Palmour, J. A. Edmond, H. S. Kong, and C. H. Carter, Jr., in *Springer Proc. Phys. Vol. 71, Amorphous and Crystalline Silicon Carbide IV*, eds. C.Y. Yang, M.M. Rahman and G.L. Harris (Springer, Berlin, 1992) p. 289.
722. J.W. Palmour, J.A. Edmond, H.S. Kong and C.H. Carter, Jr., *Physica B* **185** (1993) 461.
723. P.A. Ivanov and V.E. Chelnokov, *Semiconductors* **29** (1995) 1003.
724. *SiC Materials and Devices (Semiconductors and Semimetals) Vol. 52*, ed. P. Yoon-Soo, (Academic Press, 1998).
725. S.Y. Ren and J.D. Dow, *Appl. Phys. Lett.* **69** (1996) 251.
726. M. Horie, Y. Ishihara, J. Yamamoto, M. Kurimoto, T. Takano and H. Kawanishi, *phys. stat. sol. (a)* **192** (2002) 151.
727. *Compound Semiconductor* **6** (2000) 11.
728. *Compound Semiconductor* **7** (2001) 7.
729. D. Eisert, U. Strauss, S. Bader, H.-J. Lugauer, M. Fehrer, B.H. Harle, J. Nayer, U. Zehnder, N. Stath and V. Harle. *Proc. Int. Workshop on Nitride Semiconductors (IWN 2000), IPAP Conf. Ser.* **1** (2000) 841.
730. U. Zehnder, A. Weimar, U. Strauss, M. Fehrer, B. Hahn, H.-J. Lugauer and V. Harle, J. Crystal Growth **230** (2001) 497.
731. J. Baur, B. Hahn, M. Fehrer, D. Eisert, W. Stein, A. Pllössl, F. Kühn, H. Zull, M. Winter and V. Härle, *phys. stat. sol.(a)* **194** (2002) 399.
732. U. Strauss, H.-J. Lugauer, A. Weimar, J. Baur, G. Brüderl, D. Eisert, F. Kühn, U. Zehnder and V. Härle, *phys. stat. sol. (c)* **0** (2002) 276.
733. M. Nagahara, T. Kikkawa, N. Adachi, Y. tateno, S. Kto, M. Yokoyama, S. Yokoawa, T. Kimura, Y. Yamaguchi, N. Hara and K. Hoshin, *IEDM Tech. Dig.* (2002) p. 693.
734. K. Joshin, T. Kikkawa, H. Hayaashi, T. Maniwa, S. Yokokawa, M. Yokoyama, N. Adachi and M. Takikawa, *IEDM Tech. Dig.* (2003) p. 983.
735. Y. Cordier, M. Portail, S. Chernot, O. Tottereau, M. Zielinsku and T. Chassagne, *J. Crystal Gorwth* **310** (2008) 4417.
736. R.W. Brander and R.P. Sutton, Brit. *J. Appl. Phys.* **2** (1969) 309.
737. A. Suzuki, M. Matsunami and T. Tanaka, *J. Electrochem. Soc.* **122** (1975) 1741.
738. W. V. Münch, W. Kurzinger and I. Pfaffenender, *Solid State Eelctron.* **19** (1976) 871.
739. A. Suzuki, L. Hoffman, G. Ziegler, D. Thesis and C.Weyrich, *J. Appl. Phys.* **47** (1976) 4546.
740. A Suzuki, Y. Fujii, H. Saito, Y. Tajima, K. Furukawa, and S. Nakajima, *J. Crystal Growth* **115**(1991) 623.
741. J.A. Edmond, H.S. Kong and C.H. Carter, Jr., *Phys. B* **185** (1993) 453.

742.Y. Matsushita, T. Uetani, T. Kunisato, J. Suzuki, Y. Ueda, K. Yagi, T. Yamaguchi and T. Niina, *Jpn. J. Appl. Phys.* **34** (1995) 1833.
743. G.F. Kholuyanov and V.A. Vodakov, in *Silicon Carbide*, eds. R.C. Marshall, J.W Faust, Jr. and C.E. Ryan, (University of Sourth Carolina Press, 1973) p. 574.
744. V.M. Gusev and K.D. Demakov, M.G. Kasagonova, M.B. REifman and V.G. Stolyarova, *Sov. Phys. Semicond.* **9** (1975) 820.
745. V.M. Gusev and Demakov, *Sov. Phys. Semicond.* **15** (1981) 2430.
746. V.A. Dmitrev, L.M. Kogan, Ya.V. Morozenko, I.V. Popov and V.E. Chelnokov, *Sov. Tech. Phys. Lett.* **11** (1985) 101.
747. V.A. Dmitrev, P.A. Ivanov, Ya.V. Morozenko, I.V. Popov and V.E. Chelnokov, *Sov. Tech. Phys. Lett.* **11** (1985) 246.
748. V.A. Dmitrev, L.M. Kogan, Ya.V. Morozenko, V.S. Rodkin and V.E. Chelnokov, *Sov. Tech. Phys. Lett.* **12** (1986) 157.
749. V.A. Dmitrev, Ya.V. Morozenko, I.V. Popov, A.V. Suvorov, A.L. Syrkin and V.E. Chelnokov, *Sov. Tech. Phys. Lett.* **12** (1986) 221.
750. V.A. Dmitrev, L.M. Kogan, Ya.V. Morozenko, B.V. Tasrenkov, V.E. Chelnokov and A.E. Cherenkov, *Sov. Phys. Semicond.* **23** (1988) 23.
751. B.I. Vishnevskaya, V.A. Dmitrev, I.D. Kovalenko, L.M. Kogan, Ya.V. Morozenko, V.S. Rodkin, A.L. Syrkin, B.V. Tasrenkov, V.E. Chelnokov, *Sov. Phys. Semicond.* **22** (1988) 414.
752. A.S. Barash, Yu. A. Vodakov, E.N. Koltsova, A.A. Maltsev, *Sov. Tech. Phys. Lett.* **14** (1988) 964.
753. A.S. Barash, Yu. A. Vodakov, E.N. Koltsova, A.A. Maltsev, E.N.and A.D. Roenko, *Sov. Tech. Phys. Lett.* **14** (1988) 2222.
754. B.I. Vishnevskaya, V.A. Dmitrev, I.D. Kovalenko, L.M. Kogan, Ya.V. Morozenko, V.S. Rodkin, B.V. Tasrenko and V.E. Chelnokov, *Sov. Phys. Semicond.* **22** (1988) 664.
755. L.A. Tang, J.A. Edmond, J.W. Palmour and C.H. Carter, *Extend. Abstr. 176th Meet. Electrochem. Soc.* (Holywood, Fl, 1989) p.705.
756. B.I. Vishnevskaya, V.A. Dmitrev, L.M. Kogan, Ya.V. Morozenko, V.E. Chelnokov and A.E. Cherenkov, *Sov. Tech. Phys. Lett.* **16** (1990) 908.
757. Yu. A. Vodakov, E.N. Mokhov, A.D. Roenko, V.I. Semenov, R.G. Verenchikova, A.O. Konstantinov, V.I. Sokolov and V.G. Oding, *Proc. ICACSC90*, ed. G.L. Harris, M.G. Spencer and C.Y-W Yang (Washington DC, 1990) p. 320.
758. V.A. Dmitriev, Ya.V. Morozenko, B.V. Tzarenkov and V.E. Chelnokov, *Displays* **13** (1992) 97.
759. N. Zheludev, *Nature Photonics* **1** (2007) 189.
760. S. Kamiyama, T. Maeda, Y. Nakamura, *J. Appl. Phys.* **99** (2006) 093108.
761. S. Kamiyama, M. Iwaya, T. Takeuchi, I. skasaki, M. Syvajarvi and R. Yakimova, *J. Semicon.* **32** (2011) 013004.
762. R.B. Campbell and H.C. Chang, *Tech. Rep.* AFAPL-TR-67-23 (1967).
763. J.A. Edmond, H.S. Kong and C.H. Carter, Jr., *Phys. B* **185** (1993) 453.
764. P. Glasow, G. Ziegler, W. Suttrop, G. Pensl and R. Helbig, *SPIE* **868** (1988) 40.
765. P.A. Glasov, Amorphous and Crystalline Silicon Carbide ed. G.L. Harris and C. Y-W Yang, in *Springer Proc. In Physics, Vol. 34* (Springer, 1989) p. 13.
766. D. M. Brown, E. T. Downey, M. Ghezzo, J. W. Kretchmer, R. J. Sania, Y. S. Liu, J. A. Edmond, G. Grati, J. M. Pimbley and W. E. Schneider, *IEEE Trans. Electron Dev.* **ED-40** (1993) 325.
767. K. Park, T. Kimoto and H. Matsunami, *J. Korean Phys. Soc.* **30** (1997) 123.
768. D. M. Brown, E. T. Downey, J. W. Kretchmer, G. Michon, E. Shu and W. E. Schneider, *Solid Stat. Electon.* **42** (1998) 755.
769. J.T. Torvik, J.I. Pankove and B.J. Zeghbroeck, *IEEE Trans. Electron Dev.* **46** (1999) 1326.
770. K. Nakashima, *Tech. Rep. IEICE* **ED2000-27** (2000) 67.
771. T. Toda, M. Hata, Y. Nomura, Y. Ueda, M, Sawada and M. Shono, *Jpn. J. appl. Phys.* **43** (2004) L27.
772. A.L. Beck, G. Karve, S. Wang, J. Ming, X. Guo and J.C. Campbell, *IEEE Photo. Technol. Kett.* **17** (2005) 1507.
773. J. Wei, *Cryst. Res. Technol.* **41** (2006) 186.
774. R.G. Verenchikova and V.I. Sankin, *Sov. Tech. Phys. Lett.* **14** (1988) 756.
775. F. Yan, X. Xin, S. Aslam, Y. Zhao, D. Franz, J.H. Zhao and M. Weiner, *IEEE J. Quant. Electron.* **40** (2004) 1315.
776. T.V. Blank, Y.A. Goldberg, E.V. Kalinina, O.V. Konstantinov, A.O. Konstantinov and A. Hallen, *Semicon. Sci. Technol.* **20** (2005) 710.
777. X. Xin, F. Yan, T.W. Koeth, C. Joseph, J. Hu and J.H. Zhao, *Electron. Lett.* **41** (2005) 1192.
778. A. Sciuto, F. Roccaforte, S.D. Franco and A.Raineri, *Appl. Phys. Lett.* **89** (2006) 081111.

779. J. Hu, X. Xin, J.H. Zhao, F. Yan, B. Guan, J. Seely and B. Kjornrattanawanich, *Optical Lett.* **31** (2006) 1591.
780. A. Sciuto, F. Roccaforte, S.D. Franco and A.Raineri, *Mater. Sci. Forum* **600-603** (2009) 1215.
781. A. Sciuto, F. Roccaforte, S.D. Franco and A. Raineri, *Appl. Phys. Lett.* **90** (2007) 223507.
782. B.K. Ng, J.P.R. David, R.C. Tozer, G.J. Rees, Y. Feng, J.H. Zhao and M. Weiner, *IEEE Trans. Electron. Dev.* **50** (2003) 1724.
783. P.M. Sandvik, S.I. Solviev, A.V. Vert, A. Bolotnikov, J. Frechette, S. Verghese, P. Grssmann and G.A. Shaw, *LEOS Ann. Meet. Conf. Proc.* (IEEE, 2009) p. 291.
784. H.-Y Cha, H.-K. Sung, H. Kim, C.-H. Cho and P.M. Sandvik, *IEICE Trans. Electron.* **E93.C** (2010) 648.
785. H.-Y. Cha, *J. Korean Phys. Soc.* **56** (2010) 672.
786. A. Arbab, A. Spetz and I. Lundstrom, *Sens. Achtuat. B* **18-19** (1994) 562.
787. W. Moriz, V. Fillipov, A. Vasiliev, L. bartholomaus and A. Terentjev, *J. Fluorine Chem.* **93** (1999) 61.
788. P. Tobias, S. Ejakov, B. Golding and R.N. Ghosh, *IEEE Sensors J.* **2** (2003) 543.
789. R. Loloee, B. Chorpening, S. Beer, R.N. Ghosh, *Sensors and Actuators* B **129** (2008) 200.
790. V. Tilak, K. Matocha and P. Sandvik, *phys. stat. sol.* **3** (2006) 548.
791. A.L. Spetz, L. Uneusm H. Svenningstrorp, P. Tobuasm L.-G. Ejedahl, O. Larsoon, A. Göras, S. Savage, C. Harris, P. Martensson, R. Wigren, P. Salomonsson, B. Häggendahl, P. Ljung, M. Mattsson and I. Lundström, *phys. stat. sol.* **185** (2001) 15.
792. A.L. Spetz, L. Uneus, H. Suenningstorp, H. Wingbrant, C.I. Harris, P. Salomosson, P. Tengstrom, P. Martensson, P. Ljung, M. Mattsson, J.H. Vissr, S.G. Ejalov, D. Kubinski, L.-G. Ekedahl, I. Lundstrom and S.M. Savage, *Mater. Sci. Forum* **389-393** (2002) 1415.
793. P. Tobias, A. Baranzahi, A.L. Spetz, O. Kordina, E. Janzen and I. Lundstrom, *IEEE Electron Dev. Lett.* **18** (1997) 287.
794. G.W. Hunter, P.G. Neudeck, M. Gray, D. Andorojna, L.-Y. Chen, R.W. Hoffman, Jr., C.C. Liu and Q.H. Wu, NASA/TM-2000-209653 (2000).
795. A. Trinchi, K. Galatsis, W. Wlodarski and Y.X. Li, *IEEE Sens. J.* **3** (2003) 548.
796. K. Nakashima, Y. Okuyama, S. Ando, O. Eryu, K. Abe, H. Yokoi and T. Oshima, *Mater. Sci. Forum* **389-393** (2002) 1427.
797. N.G. Wright, A.B. Hrosfall and K. Vassilevski, *Materialstoday* **11** (2008) 16, R.N. Ghosh and r. Loloee, *IEEE Sensors 2007 Conference* (2007) p. 612.
798. A. Owens, *J. Synchrotron Rad.* **13** (2006) 143.
799. R. Stevenson, *Compound Semiconductors* **Jan./Feb.** (2005) 29.
800. C.P. Capera, P. Malinaric, P.B. Campbell and J. Ostroski, *IEEE Trans. Nucl. Sci.* **11** (1964) 262.
801. G.F. Kholuyanov and B.V. Gavrilovii, *Sov. Phys. Semicond.* **2** (1968) 472.
802. V.A. Tikhomirova, O.P. Fedoseeva and G.F. Kholuyanov, *Sov. Phys. Semicond.* **6** (1972) 831.
803. V.A. Tikhomirova, O.P. Fedoseeva and G.F. Kholuyanov, *Sov. Phys. Semicond.* **34** (1973) 122.
804. M. Rogalla, K. Runga, A. Scholdner-Rembold, *Nucl. Phys. B (Proc. Suppl.)* **78** (1999) 516.
805. W. Cunnigham, J. Melone, M. Horn, V. Kazukauskas, P. Roy, F. Doherty, M. Glaser, J. Vaitkus and M. Rahman, *Nucl. Instr. Meth. Phys. Res. A* **509** (2003) 127.
806. J. Grant, W. Cunningham, A. Blue, V. O'Shea, J. Vaitkus, E. Gaubas and M. Rahman, *Nucl. Instr. Meth. Phys. Res. A* **546** (2005) 213.
807. F.H. Ruddy, A. R. Dullo, J.G. Seidal, S. Seshadri and L.B. Rowland, *IEEE Trans. Nucl. Sci.* **45** (1998) 536.
808. F. Nava, P. Vanni, C. Lanzieri and C. Canali, *Nucl. Instr. Meth. Phys. Res. A* **437** (1999) 354.
809. M. Bruzzi, F. Nava, S. Pini and S. Russo, *Appl. Sur. Sci.* **184** (2001) 425.
810. B. Bruzzi, F. Nava, S. Russo, S. Sciortino and P. Vanni, *Diamond Relat. Mater.* **10** (2001) 657.
811. G. Verzellesi, P. Vanni, F. Nava and C. Canali, *Nucl. Instr. Meth. Phys. Res. A* **476** (2002) 717.
812. M. Bruzzi, S. Lagaomarsino, F. Nava and S. Sciortino, *Diamond. Relat. Mater.* **12** (2003) 1205.
813. F. Nava, G. Wagner, C. Lanzieri, P. Vanni and E. Vittone, *Nucl. Instr. Meth. Phys. Res. A* **510** (2003) 273.
814. F. Nava, P. Vanni, M. Bruzzi, S. Lagomarsino, S. Sciortino, G. Wagner and C. Lanzieri, *IEEE Trans. Nucl. Sci.* **51** (2004) 238.
815. A.M. Ivanov, E.V. Kalinina, A.O. Konstantinov, G.A. Onushkin, N.B. Strokan, G.F. Kholuyanov and A. Hallen, *Tech. Phys. Lett.* **30** (2004) 575.
816. G. Bertuccio, R. Casiraghi, A. Certronio, C. Lanzieri and F. Nava, *Nucl. Instr. Meth. Phys. Res. A* **522** (2004) 413.
817. R. Schifano, A. Vinattieri, M. Bruzzi, S. Miglio, S. Lagomarsino and S. Sciortino, *J. Appl. Phys.* **97** (2005) 103539.
818. N.B. Strokan, A.M. Ivanov, E.V. Kalinina, G.F. Kholuhanov, G.A. Onushkin, D.V. Davydov and G.N. Violina, *Semiconductors* **39** (2005) 364.

819. G.N. Violina, E.V. Kalinina, G.F. Kholuyanov, V.G. Kossov, R.R. Yafaev, A. Hallen and A.O. Konstantinov, *Semiconductors* **36** (2002) 710.
820. F. Moscattelli, A. Scorzoni, A. Poggi, M. Bruzzi, S. Lagomarsino, S. Mersi, S. Sciortino, M. Lazar, A.D. Piacido and r. Nipoti, *Mater. Sci. Forum* **483-485** (2005) 1021.
821. E.V. Kalinina, V.G. Kossv, N.B. Strokan, A.M. Ivanov, R.R. Yafaev and G.F. Kholuyanov, *Semiconductors* **40** (2006) 1096.
822. E.V. Kalinina, N.B. Strokan, A.M. Ivanov, A.A. Sitnikova, A.V. Sadokhin, A.Yu. Azarov, V.G. Kossov and R.R. Yafaev, *Semiconductors* **42** (2008) 86.
823. K. Hara, *Mater. Sci. Forum* **264-268** (1998) 901.
824. K. Rottner, M. Frischholz, T. Myrtveit, D. Mou, K. Nordgren, A. Henry, C. Hallin, U. Gustafsson and A. Schoner, *Mater. Sci. Engineer. B* **61-62** (1999) 330.
825. C.M. Johnson, *35th Annual IEEE Electronics Specialists Conf.* (Aachen, Germany, 2004) p. 572.
826. Z.C. Feng, *SiC Power Materials: Devices and Applications* (Springer, 2004).
827. A.W. Morse, P.M. Esker, S. Sriram, J.J. Hawkins, L.S. Chen, J.A. Ostop, T.J. Smith, C.D. Davis, R.R. Barron, R.C. Clarke, R.R. Siergiej and C.D. Brandt, *IEEE MTT-S Digest* (1997) 53.
828. J A. Edmond, J. Ryu, J. T. Glass, and R. F. Davis, *J. Electrochem. Soc.* **135** (1988) 359.
829. T. Urushidani, S. Kobayashi, T. Kimoto and H. Matsunami, *Ext. Abstr. 25th Int. Conf. Solid State Dev. Mater.* D-5-4 (Japan Society of Applied Physics, Makuhari, 1993) p. 814.
830. T. Kimoto, T. Urushidani, S. Kobayashi and H. Matsunami, *IEEE Elec. Dev. Lett.* **14** (1993) 548.
831. Q. Wahab, T. Kimoto, A. Ellison, C. Hllin, M. Tuominen, R. Yakimova, A. Henry, J.P. Bergman and E. Janzen, *Appl. Phys. Lett.* **72** (1998) 445.
832. R. Rupp, *Compound Semicond.* **April** (2001) p. 45.
833. A.R. Powell, R.T. Leonard, M.F. Brady, St.G. Muller, V.F. Tsvetkov, R. Trussell, J.J. Sumakeris, H.McD. Hobgood, A.A. Burk, R.C. Glass and C.H. Carter, Jr. *Mater. Sci. Forum* **41** (2004) 457.
834. T. Nakamura, T. Miyanagi, I. Kamata, T. Jikimoto and H. Tsuchida, *IEEE Electron Dev. Lett.* **26** (2005) 99.
835. A. Agarwal, R. Singh, S. Ryu, J. Richmond, C. Capell, S. Schwab, B. Moore and J. Palmour, CPWR-TECH1.
836. Z. Lin, Z. Yi-Men, Z. Yu-Ming, H. Chao and M. Yong-Ji, *Chines Phys. B* **18** (2009) 3490.
837. L.G. Matus, J.A. Powell and C.S. Salupo, *Appl. Phys. Lett.* **59** (1991) 1770.
838. M. Karlsteen, Q. Wahab, M. Willander and J.-E. Sundgren, *Diamnd. Relat. Mater.* **1** (1992) 486.
839. O. Kordina, J.P. Bergman, A. henry, E. Janzen, S. Savage, J. Andre, L.P. Ramberg, U.Lindefelt, W. Hermansson and K. Bergman, *Appl. Phys. Lett.* **67** (1995) 1561.
840. H. Mitlehner, P. Friedrichs, d. Peters, R. Schorner, U. Weinert, B. Weis and D. Stephani, *Proc. of 1998 Int. Symp. Power Semicon. Dev. & ICs* (Kyoto, 1998) p. 12.
841. K. Rottner, M. Frischholz, t. Myrtveit, D. Mou, K. Nordgren, A. Henry, C. Hallin, U. Gustafsson and A. Schoner, *Mater. Sci, & Engineer. B* **61-62** (1999) 330.
842. H. Lendenmann, F. Dahlquist, N. Johansson, J.P. Bergman, H. Bleichner, C. Ovren, *Ext. Abstr. 1st Int. Workshop Ultra-Low-Loss Power Dev. Technol.* (Nara, 2000) p. 125.
843. H. Lendenmann, A. Mukhitdinov, F. Dahlquist, H. Bleichner, M. Irwin, R. Soderholm and P. Skytt, *Proc. 2001 Int. Symp. Power Semicond.Dev. & ICs* (Osaka, 2001) p. 31.
844. H. Lendenmann, F. Dahlquist, N. Johansson, R. Soderholm, P.A. Nilsson, J.P. Bergman and P. Skytt, *Mater. Sci. Forum* **353** (2001) 727.
845. K. Fujihira, S. Tamura, T. Kimoto and H. Matsunami, *IEEE Trans. Electron Dev.* **49** (2002), 150.
846. R. Singh, K.G. Irvine, J.T. Richmond and J.W. Palmour, *Mater. Sci. Forum* **389-393** (2002) 1265.
847. Y. Sugawara, D. Takayama, K. Asano, R. Singh, J. Palmour and T. Hayashi, *Proc. 2001 Int. Symp. Power Semicond.Dev. & ICs* (Osaka, 2001) p. 27.
848. M. Braun, B. Weis, W. Bartche and H. Mitlehner, *Int. Power Electron. Motion Control Conf. (EPE-PEMC)* (2000) p. 1.
849. Y. Sugawara, D. Takayama, K. Asano, R. Singh, H. Kodama, S. Ogata and T. Hayashi, *Proc. 2002 Int. Symp. Power Semicond. Dev. & ICs* (Santafe, 2002) p. 245.
850. Y. Sugawara, *Mater. Sci. Forum* **457-460** (2004) 963.
851. Y. Tanaka, K. Fukuda, K. Arai, K. Kojima, T. Suzuki and T. Yatsuo, *Appl. Phys. Lett.* **84** (2004) 1774.
852. S. Ha and J.P. Bergman, *MRS Bulletin* **30** (2005), 305.
853. C. Li, P. Losee, J. Seiler, I. Bhat and T.P. Chow, *J. Electron. Mater.* **34** (2005) 450.
854. S. Ogata, D. Takayama, K. Asano and Y. Sugawara, *Electron. Engineer. Jpn.* **171** (2010) 1.
855. P. Fridrichs, H. Mitlehner, W. Bartsh, K.O. Dohnke, R. Kaltsshmidt, U. Weinert, B. Weis and D. Stephani, *Proc. ISPSD '2000* (2000) p. 213.
856. J.H. Zhao, *MRS Bulletin* **30** (2005) 293.
857. J.W. Palmour, H.S. Kong and R.F. Davis, *J. Appl. Phys.* **64** (1988) 2168.

858. R.E. Avila, J.J. Kopansku and C.D. Fung, *Appl. Phys. Lett.* **49** (1987) 442.
859. K. Shibarara, S, Nishio and H. Matsunami, *Jpn. J. Appl. Phys.* **23** (1984) L862.
860. Y. Kondo, T. Takahashi, K. Ishii, Y. hayashi, E. Sakuma, S. Misawa, H. Daimon, M. Yamanaka and s. Yoshida, *IEEE Electron Device Lett.* **EDL-7** (1986) 404.
861. B.J. Baliga, *Proc. IEEE* **82** (1994) 1112.
862. V.R. Vathulya and M.H. White, *Solid-State Electron.* **44** (2000) 309.
863. O. Kusumoto, T. Yokogawa, K. Yamashita, K. Takahashi, M. Kitabatake, M. Uchida and R. Miyabaga, *Mater. Sci. Forum* **389-393** (2002) 1211.
864. S. Ryu, A. Agarwal, J. Richmond, J. Palmour and N, Saks, *IEEE Electron Dev. Lett.* **23** (2002) 321.
865. S. Ryu, B. Hull, s. Dhar, L. Cheng, Q. Zhan, J. Richmond, M. Das, A. Agarwal, J. Palmour, A. Lelis, B. Geil and C. Scozzie, *Mater. Sci. Forum* **645-648** (2010) 969.
866. S. Onda, R. Kumar and K. Hara, *phys. stat. sol. (a)* **162** (1997) 369.
867. J. Tan, J.A. Cooper and M.R. Melloch, *IEEE Electron Dev. Lett.* **19** (1998) 487.
868. P.M. Shenoy and B.J. Baliga, *IEEE Electron Dev. Lett.* **18** (1997) 589.
869. R. Chilukuri, P.M. Shenoy and B. Baliga, *Proc. 1998 Int. Symp. Power Semicond. Dev. & ICs* (Kyoto, 1998) 115.
870. R. Kumar, J. Kojima and T. Yamamoto, *Ext. Abstr. 1999 Int. Conf. Solid State Dev. Maters.* (Tokyo) 1999) p. 146.
871. M. Kitabatake, *FED Review (in japanese)* **1** (2002) 1.
872. S.T. Sheppard, M.R. Melloch and J.A. Cooper, Jr., *IEEE Electron Dev. Lett.* **41** (1994) 1257.
873. S. Harada, S. Suzuki, J. Senzaki, R. Kosugi, K. Adachi, K. Fukuda and K. Arai, *IEEE Electron Dev. Lett.* **22** (2001) 272.
874. S. Harada, M. Okamoto, T. Yatsuo, K. Adachi, K. Fukuda and K. Arai, *IEEE Electron Dev. Lett.* **25** (2004) 292.
875. S. Harada, M. Kato, K. Suzuki, M. Okamoto, T. Ytsuo, K. Fukuda and K. Arai, *IEDM 2006* (2006) p. 650.
876. S. Harada, Y. Hayashi, K. Takao, A. Kinoshita, M. Kao, M. Okamoro, T. Kato, S. Nishizaqa, T. Yatsuo, K. Fukuda, H. Ohashi and K. Arai, *Proc. 19th Int. Symp. Power Semicon. Dev. & ICs* (Korea, 2007) 289.
877. Y. Sugawara, K. Asano, T. Takayama, S. Ryu, R. singh, J. Palmour and T. Hayashi, *Mater. Sci. Forum* **289-393** (2002) 1199.
878. S. Suzuki, S. Harada, T. Yatsuo, R. Kosugi, J. Senzaki and K. Fukuda, *Mater. Sci. Forum* **433-436** (2003) 753.
879. L.A. Lipkin and J.W. Palmour, *Mater. Sci. Forum* **338-342** (2000) 1093.
880. Y. Oonishi, A. Ooi and t. Shimatou, *Fuji Electrics Bulletin (in japanese)* **82** (2009) 389.
881. V.V. Afanasev, M. Bassler, G. Pensl and M. Schulz, *phys. stat. sol.* **162** (1997) 321.
882. S. Harada, R. Kosugi, J. Senzaki, W. Cho and K. Fukuda, *J. Appl. Phys.* **91** (2002) 1568.
883. S. Suzuki, S. Harada, R. Kosugi, J. Senzaki, W. Cho and K. Fukuda, *J. Appl. Phys.* **92** (2002) 6230.
884. S. Dhar, S. Wang, J.R. Williams, S.T. Pantelides and L.C. Feldman, *MRS Bull.* **30** (2005) 288.
885. K. Fukuda, M. Kato, K. Kojima and J. Senzaki, *Appl. Phys. Lett.* **84** (2004) 2088.
886. F. Fukuda, S. Suzuki and T. Tanaka, *Appl. Phys. Lett.* **76** (2000) 1585.
887. G.Y. Chung, C.C. Tin, J.R. Williams, K. McDonald, .K. Chanana, R.A. Weller, S.T. Pantelides, L.C. Feldman, O.W. Holland, M.K. Das and J.W. Palmour, *IEEE Elecron Dev. Lett.* **22** (2001) 176.
888. R. Kosugi, S. Suzuki, M. Okamoto, S Harada, J. Senzaki and K. Fukuda, *IEEE Elecron Dev. Lett.* **23** (2002) 136.
889. H. Yano, T. Hirano, T. Kimoto, H. Matsunami, K. Asano and Y. Sugawara, *IEEE Electron Dev. Lett.* **20** (2001) 611.
890. H. Yano, T. Hiro, T. Kimoto, H. Matsunami, K. Asano and Y. Sugawara, *Mater. Sci. Forum.* **338-342** (2000) 1105.
891. S. Ogino, T. Oikawa and K. Ueno, *Mater. Sci. Forum.* **338-342** (2000) 1101.
892. K. Ueno and T. Okikawa, *IEEE Electron Dev. Lett.* **20** (1999) 624.
893. R. Schorner, P. Friedrichs, D. Peters and d. Stephani, *IEEE Electron Dev. Lett.* **20** (1999) 241.
894. T.P. Ma, *1st Int. Workshop Ultr-Low-Lwss Power Dev. Tech.* (Nara, Japan, 2000) p. 61.
895. B.J. Baliga, *Proc. 10th Symp. Static Induc. Dev.* (1997) p. 697.
896. J. Nishizawa, T. Terasaki and J. Shibata, *IEEE Trans. Electron. Dev.* **ED-22** (1975) 185.
897. P. Fridrichs, H. Mitlehner, W. Bartsh, K.O. Dohnke, R. Kaltsshmidt, U. Weinert, B. Weis and D. Stephani, *Proc. ISPSD '2000* (2000) p. 213.
898. H. Onose, A. Watanabe, T. Someya and Y. Kobayashi, *Proc. ICSCRM 2001* (2001) p. 1227.
899. H. Onose, A. Watanabe, T. Someya and Y. Kobayashi, *Proc. ISPSD'02 (2002)* p. 189.
900. K. Asano, Y. Sugawara, S. Ryu, E. Singh, J. Palmour, T. Hayashi and D. Takayama, *Proc. ISPSD'02(2002)* p. 23.

901. K. Asano, Y. Sugawara, T. Hayashi, S. Ryu, R. Singh, J. Palmour and D. Takayama, *Proc. ISPSD'02(2002)* p. 61.
902. A. Katsumori, T. Hayashi, D. Takayama, Y. Sugawara, S. Ryu and J.W. Palmour, *IEEE Trans. Indus. Appl.* **125** (2005) 26.
903. G. Kelner, S. Binari, M. Shur, K. Sleger, J. Palmour and H. Kong, *Mater. Sci. Engineer. B* **11** (1992) 121.
904. Y. Tanaka, K. Yano, M. Okamoto, A. Takatuki, K. Fukuda, M. Kasaga, K. Arai and T. Yatsuo, *Mater. Sci. Forum.* **527-529** (2006) 1219.
905. Y. Tanaka, M. Okamoro, A. Takatsuki, K. Arai, T. Yatsuo and K. Yano, *IEEE Electron Lett.* **27** (2006) 908.
906. K. Fujikawa, T. Masuda, H. Tamaso, K. Shibata, S. Harada, S. Hatsukawa, H. Tokuda, A. Saegusa and Y. Namikawa, *SEI Tech. Rev.* **167** (2005) 109.
907. T. Iwasaki, T. Ohno, K. Asano, Y. Sugawara and T. Yatsuo, *Proc. ICSCRM '95* (1995) p. 513.
908. D.J. Spry, P.G. Neudeck, L.-Y. Chen, G.M. Beheim, R.S. Okojie, C.W. Chang, R.D. Meredith, T.L. Ferrier and L.J. Evance, *Mater. Sci. Forum.* **600-603** (2009) 1079.
909. T.P. Chow, *MRS Bulletin* **30** (2005) 299.
910. S. Krishnaswami, A.K. Agarwal, C. Capell, J. Richmond, S.-H. Ryu, J.W. Palmour, S. Balachandan, T.P. Chow, S. Bayne, B. Geil, K.A. Jones and C.S. Scozzie, *Mater. Sci. Forum.* **483-485** (2004) 901.
911. L. Cao, B. Li and J.H. Zhao, *Solid State Electron.* **33** (2000) 347.
912. J.H. Zhao, J. Zhang, P. Alexandrov, X. Li and T. Burke, *Mater. Sci. Forum* **457-460** (2004) 1173.
913. A.W. Morse, P.M. Esker, S. Sriram, J.J. Hawkins, L.S. Chen, J.A. Ostop, T.J. Smith, C.D. Davis, R.R. Barron, R.C. Clarke, R.R. Siergirj and C.D. Brandt, *IEEE MTT-S Digest* (1997) p. 53.
914. D. Miller and M. Drinkwine, *2003 Int. Conf. Compound Semicond. Mfg. (GaAs MANTECH)* (2003) #7-1.
915. S. Allen, R. Sadler, T. Alcorn, J. Sumakeris, R. Glass, C.H. Carter, Jr. and J. Palmour, *Mater. Sci. Forum* **264-268** (1998) 953.
916. S. Sriram, R.C. Clarke, A.A. Burk, Jr., H.M. Hobgood, P.G. McMullin, P.A. Orphanos, R.R. Siergiej, T.J. Smith, C.D. Brandt, M.C. Driverand R.H. Hopkins, *IEEE Electron Dev. Lett.* **15** (1994) 458.
917. C. Weitzel, *IEEE Electron Dev. Lett.* **16** (1995) 451.
918. D. Alok and B.J. Baliga, *Electron. Lett.* **32** (1996) 1929.
919. S. Sriram, G. Augustine, A. A. Burk, Jr., R.C. Glass, H. M. Hobgood, P. A. Orphanos, L.B. Rowland, T. J. Smith, C. D. Brandt, M. C. Driver and R. H. Hopkins, *IEEE Electron Dev. Lett.* **17** (1996) 369.
920. J. Palmour, S. Allen, S. Sheppard, W. Probble, R. Sadler, T. Alcorn, W. Ring and C. Carter, Jr., *57th An. Dev. Res. Conf.* (1999) p. 38.
921. N. Rorsman, J. Eriksson and H. Zirath, *Mater. Sci. Forum 338-342* (2000) 1259.
922. N. Sghaier, A.K. Souifi, J.-M. Bluet, M. Berenguer, G. Guillot, O. Noblanc and C. Brylinski, *Mat. Res. Symp. Proc.* **640** (2001) H4.4.1.
923. S. Sriam, A. Ward, J. Henning and S.T. Allen, *MRS Bull.* **30** (2005) 308.
924. H.J. Na, S.Y. Jung, J.H. Moon, J. H. Yim, H.K. Song, J.B. Lee and H.J. Kim, J. *Semicon. Technol. Sci.* **5** (2005) 113.
925. H.J. Na, J.H. Moon, J.H. Yim, J.B. Lee and H.J. Kim, *Microelectron. Engineer.* **83** (2006) 160.
926. M. Südow, K. Andersson, N. Billström, J. Grahn, H. Hjelmgren, J. Nilsson, P. Nilssonm J. Ståhl, H. Zirath and N. Rorsman, *IEEE Trans. Microwave. Theo. Tech.* **54** (2006) 4072.
927. J. Zhang, B. Zhang and Z. Li, *Semicond. Sci. Technol.* **23** (2009) 1.

24. LEAD COMPOUNDS

24.1 INTRODUCTION

The lead salt compounds such as $Pb_{1-x}Sn_xTe$, $Pb_{1-x}Sn_xSe$ and $Pb_{1-x}S_xSe$ are narrow-gap semiconductors. These semiconductors have been investigated since 1965.[1] By varying the composition, the bandgap can be varied in the range of almost zero to 0.3 eV. They have important potential applications in the wavelength range from 3 to 30 μm, such as IR lasers and detectors. It is also interesting that these lasers can be tuned easily by varying temperature and current. The fact that they show a narrow line width in the 10-4 cm^{-1} range is promising in such applications as high-resolution spectroscopy and gas-monitoring. The pseudo-binary compound $Pb_{1-x}Sn_xTe$ is of special interest because of its fast response in detecting infrared radiation in the range of 8-14 μm. The applications of these lead compounds are reviewed in Refs. 2-8.

24.2 PHYSICAL PROPERTIES

All lead compounds have rock salt structures as shown in Fig. 24.1. They are different from most of other compound semiconductors whose crystal structures are zincblende or wurtzite. Physical properties of various lead compounds are reviewed in Refs. 9-10 and some of them are summarized in Table 24.1.

The energy bandgap of PbTe is 0.31 eV and can be controlled by alloying it with SnTe as shown in Fig. 24.2.[11] Using this alloy, the laser emission wavelength can be controlled. PbSe has a similar band structure to PbTe and the bandgap is slightly less than that of PbTe. The bandgap of PbSe can be narrowed with SnSe alloying as shown in Fig. 24.3.[12] PbS has a similar band structure to PbTe and the bandgap is wider than that of PbTe. In $PbS_{1-x}Se_x$ alloys, the bandgap changes linearly according to $E_g(eV) = 0.289-0.144x$.[3]

Energy bandgaps and lattice constants for various lead salt compounds are shown in Fig. 24.4.[6] It can be seen that the wavelength range of 3-30 μm can be covered by various compounds and that the lattice constants can be matched by heterostructures.

The temperature-composition phase diagrams of PbTe and SnTe are shown in Fig. 24.5.[13] The phase diagram of PbTe-SnTe alloy is shown in Fig. 24.6.[14] It exhibits complete solid solubility over the entire composition range (isomorphous system). The separation of liquidus and solidus lines is small so that the ratio of tin content in the grown solid and liquid melt is small. The p-X-T diagrams have been studied in the past by several researchers.[15-17] Recently, the precise phase diagram has been summarized by Kusnetsov and Zlomanov.[18]

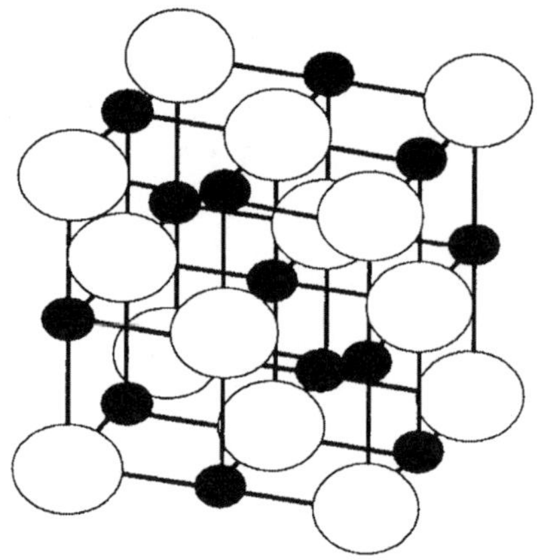

Fig. 24.1 Crystal structure of lead compounds, rock salt (NaCl) structure. White circle is Na and black circle is Cl.

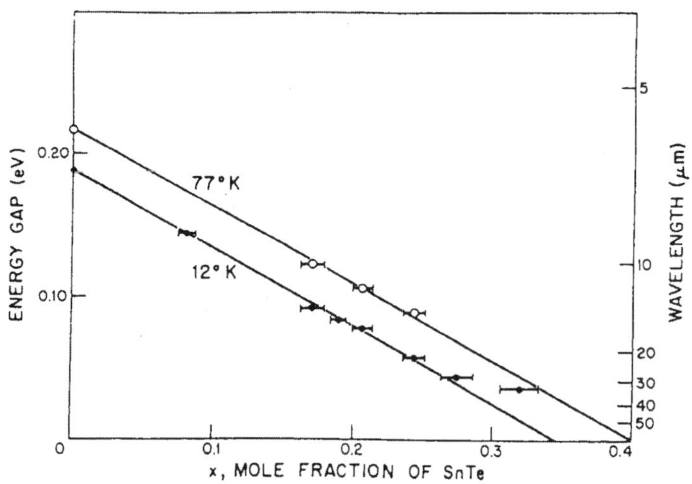

Fig. 24.2 Energy gap and wavelength of laser emission versus mole fraction of SnTe for $Pb_{1-x}Sn_xTe$ diodes (reprinted from Ref. 11 with permission, copyright 1969 IEEE).

24.3 CRYSTAL GROWTH

Lead single crystals have been grown by various methods as summarized in Table 24.2.

24.3.1 Melt Growth

(1) Liquid Encapsulated Czochralski (LEC) method

The LEC method (Vol. 1, Sec. 2.2.3(2)) developed for the crystal growth of III-V materials has been used to grow PbTe [19,20], SnTe [20], PbSe [19] and $Pb_{1-x}Sn_xTe$ [14] single crystals

Table 24.1 Physical Properties of Lead Compounds

	PbS	PbSe	PbTe
Crystal Structure	rock salt	rock salt	rock salt
Lattice Constant (Å)	5.936	6.124	6.462
Density (g/cm^3)	7.597	8.274	8.242
Melting Point (°C)	1111	1082	924
Linear Expansion Coefficient ($\times 10^{-6}$cm/deg)	20.27	19.40	19.80
Thermal Conductivity (W/cm·K)(at 76K)	0.04-0.08	0.055	0.07-0.08
Dielectric Constant	17.2	22.9	32.8
Refractive Index	4.3	5.0	6.7
Band gap at Room Temperature (eV)	0.41	0.27	0.31
Optical Transition Type	direct	direct	direct
Intrinsic Carrier Concentration at R.T. (cm^{-3})	$\sim 2 \times 10^{15}$	$\sim 3 \times 10^{16}$	$\sim 1.5 \times 10^{16}$
Electron Mobility at R.T. (cm^2/V·sec)	600	1200	1800
Hole Mobility at R.T. (cm^2/V·sec)	700	1000	900

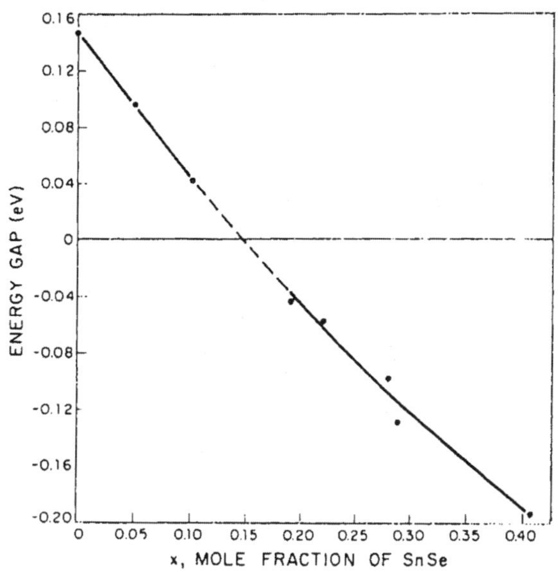

Fig. 24.3 Energy gap versus mole fraction of SnSe for $Pb_{1-x}Sn_xSe$ diode lasers (reprinted from Ref. 12 with permission, copyright 1969 American Institute of Physics).

using B_2O_3 as the encapsulant. The growth arrangement is shown in Fig. 24.7. Since the liquidus and solidus are very close in the PbTe-SnTe phase diagram as shown in Fig. 24.6, it has been found that single crystals can be grown with only a small change in composition along the length of the crystals.

(2) Bridgman method and other melt growth methods

PbTe, PbSe and $Pb_{1-x}Sn_xTe$ crystals have been grown by the Bridgman method.[21-38] Lawson[21] was the first to grow PbTe and PbSe single crystals by the Bridgman method.

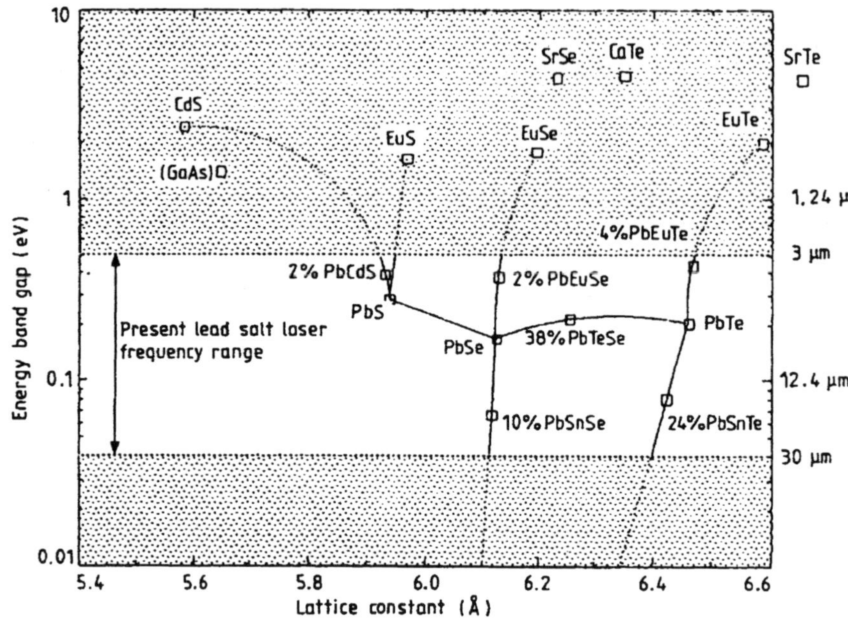

Fig. 24.4 Energy gap against lattice constant at 77 K for various lead salt compounds (reprinted from Ref. 3 with permission, copyright 1990 Institute of Physics).

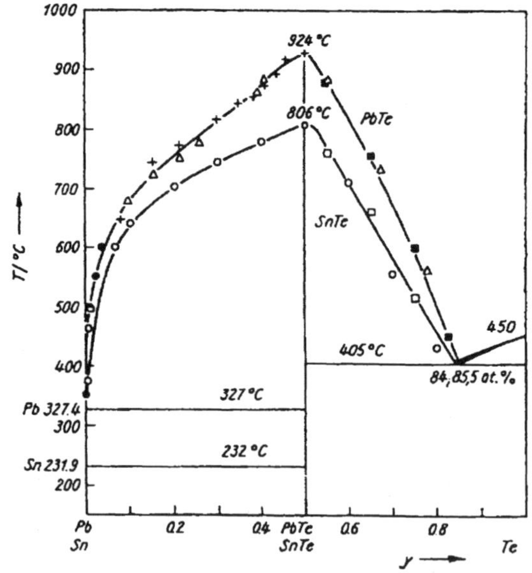

Fig. 24.5 Phase diagram of the systems Pb-Te and Sn-Te (reprinted from Ref. 99 with permission, copyright 1980 John Wiley & Sons).

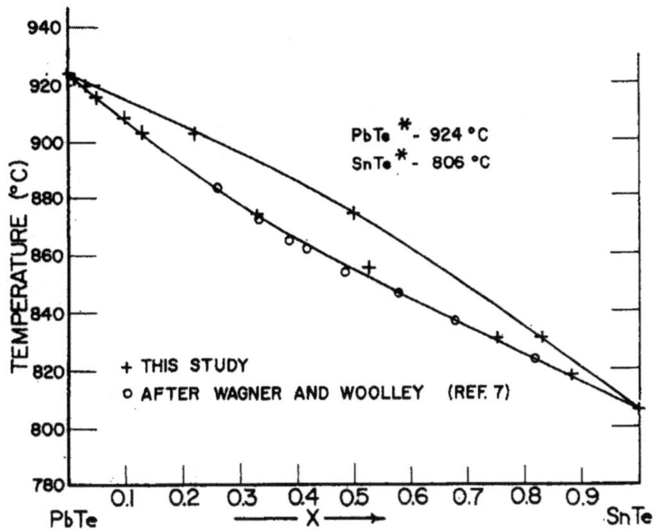

Fig. 24.6 PbTe-SnTe phase diagram. PbTe and SnTe melting points are calibration points from the literature (reprinted from Ref. 14 with permission, copyright 1968 The American Institute of Mining, Metallurgical and Petroleum Engineers).

The crystals grown were 3 cm in diameter and 6 cm in length.

Machonis et al.[22] have grown $Pb_{1-x}Sn_xTe$ single crystals and measured their electrical properties. When $x > 0.2$, the grown crystals showed p-type conductivity. The effect of annealing on the carrier concentrations has been examined by Calawa et al.[23] The effect of growth conditions on segregation in $Pb_{1-x}Sn_xTe$ crystal growth has been studied by Fano et al.[26] It was found that the Pb/Sn ratio affects the segregation while Te was found constant in all grown ingots.

The effect of the crystal growth conditions on small grain boundaries has been studied by Kinoshita et al.[28] and it was found that thermal stress is the main factor in the occurrence of small grain boundaries. It was found that the axial temperature gradient during the crystal growth and cooling process must be lower than 5°C/cm.

Nishizawa's group[29-32] have grown PbTe single crystals by the Bridgman method under controlled Te or Pb pressures (Fig. 24.8). They found that they could obtain p-type crystals when $P_{Te} \geq 3.03$ Torr but n-type crystals when $P_{Te} \leq 0.26$ Torr. They also have grown Tl-doped PbTe single crystals under controlled Te or Pb pressures. They found that high hole concentrations of $1.5\text{-}2 \times 10^{19}$ cm^{-3} can be obtained when the crystal is grown with Tl in the melt (0.65 at%) under Pb vapor pressures. These crystals showed small full width at half maximum (FWHM) of X-ray rocking curve (XRC) (20-50 arcsec) and an etch pit density (EPD) of $8\text{-}10 \times 10^5$ cm^{-2}, much lower than undoped PbTe crystals.[30]

Shtanov[33] has studied precisely the effect of the shape of the conical bottom of the

Table 24.2 Crystal Growth of Lead Compounds

Method	Contents	Authors	Year	Ref.
VB	First growth of PbTe and PbSe by the Bridgman method	Lawson	1951	21
PVT	First growth of PbSe controlling the Se vapor pressure	Prior	1961	46
LEC	First growth of PbTe and PbSe using B_2O_3 encapsulant	Metz et al.	1962	19
VB	Growth of $Pb_{1-x}Sn_xTe$ and the characterization of electrical properties	Machonis et al.	1964	22
ZM	Growth of PbTe doped with halogens	Kaidanov et al.	1966	39
VB	Growth of PbSe, $Pb_{1-x}Sn_xSe$ and $Pb_{1-x}Sn_xTe$ crystals	Calawa et al.	1968	23
LEC	Growth of $Pb_{1-x}Sn_xTe$ using B_2O_3 encapsulant	Wagner et al.	1968	14
PVT	Growth of $Pb_{1-x}Sn_xTe$ crystals with x up to 0.27	Butler et al.	1968	47
VB	Mechanism of inclusion formation of $Pb_{1-x}Sn_xTe$ crystals	Butler et al.	1969	25
LEC	Growth of PbTe and SnTe using B_2O_3 encapsulant	Baughman et al.	1969	20
PVT	Growth of $Pb_{1-x}Sn_xTe$ single crystals and first evaluation of dislocation density	Ohtsuki et al.	1971	48
VLS	First growth of $Pb_{1-x}Sn_xTe$ single crystals by the VLS method, 9 mm in diameter and 60 mm in length	Mateika	1971	76
CVT	Growth of PbTe and $Pb_{1-x}Sn_xTe$ single crystals using iodine as the transport agent	Bezrodnaya et al.	1972	87
RC	Growth of Bi-doped $Pb_{1-x}Sn_xTe$ single crystals by long term annealing procedure	Kimura	1972	89
PVT	Growth of PbTe wth a capillary tube	Knittel	1974	49
PVT	Growth of PbTe ,$Pb_{1-x}Sn_xTe$, $Pb_{1-x}S_xSe$, $Pb_{1-x}Sn_xSe$ by horizontal unseeded PVT	Harman et al.	1974	50
PVT	Growth of $Pb_{1-x}Sn_xTe$ crystals with x up to 0.26 weighing 125 g	Parker	1974	51
PVT	Growth of $Pb_{1-x}Sn_xTe$ on BaF_2 substrates	Pandey	1974	52
PVT	Growth of PbSnSe and $Pb_{1-x}Sn_xTe$	Kasai et al.	1974	65
VLS	Growth of PbSe single crystals	Zlomanov et al.	1974	77
CVT	Growth of PbTe and SnTe single crystals using S_2, Se_2 and Te_2 as the transport agent	Wäsch	1974	88
PVT	Growth of $Pb_{0.8}Sn_{0.2}Te$ measuring 19 mm in diameter and 25 mm long	Bradford et al.	1975	54
PVT	Growth of $Pb_{1-x}Sn_xTe$ by nucleation on the source ingot	Lo et al.	1976	55
PVT	Quality improvement by preheting of the charge	Parker et al.	1976	53
PVT	Control of carrier concentrations by Pb and Se composition for PbSe crystal growth	Maier et al.	1976	56
PVT	Growth of $Pb_{1-x}Sn_xTe$ crystals with x up to 0.25, weighing up to 100 g and measuring 40 mm in diameter	Tamari et al.	1978	57, 58
VB	Segregation in $Pb_{1-x}Sn_xTe$ crystal growth	Fano et al.	1979	26
VLS	Growth of $Pb_{1-x}Sn_xTe$ crystals	Andryushenko	1979	78-80
VLS	The effect of Ga on the electrical properties of PbTe crystals	Lazarenko et al.	1980	81
PVT	Growth of PbTe and thermodynamics	Zoutendyk	1982	59
PVT	Growth of PbTe and $Pb_{1-x}Sn_xTe$ on BaF_2, 20 mm in diameter weighing up to 100 g	Golacki et al.	1982- 1986	60 61
THM	Growth of PbTe using Te and Pb as solvants	Triboulet et al.	1982	43
VLS	Growth of $Pb_{1-x}Sn_xTe$ by the horizontal VLS method	Kinoshita et al.	1982- 1983	82, 83
THM	Growth of Tl-doped PbTe with Te solvant	Gill et al.	1983	44
PVT	Thermodynamical analysis of the growth of $Pb_{1-x}Sn_xTe$	Saunina et al.	1985	62

Table 24.2 Crystal Growth of Lead Compounds (contiued)

Method	Contents	Authors	Year	Ref.
PVT	Growth of $Pb_{1-x}S_xTe$ and the determination of the homogeneity region	Demin et al.	1986	66
VB	Analysis of small grain boundaries in $Pb_{1-x}Sn_xTe$ crystals	Kinoshita et al.	1987	28
PVT	Self-selecting vapor growth (SSVG) method for lead compound single crytals	Szczerbakow	1987-2005	70, 72-
VLS	Model of the growth of $Pb_{1-x}Sn_xTe$	Fabbri et al.	1990	85
VLS	Growth of $Pb_{1-x}S_xTe$ single crystals	Yaneko et al.	1990	84
ZM	Growth of $Pb_{1-x}Sn_xTe$ single crystals under microgravity	Iwai et al.	1994	40
DS	Growth of $Pb_{1-x}Sn_xTe$ single crystals under microgravity	Yamada et al.	1994	41
PVT	Growth of $Pb_{1-x}Ge_xTe$ single crystals over the whole composition range	Leszczynski et al.	1994	67
PVT	Growth of $Pb_{1-x}Sn_xSe$ ($x<0.4$)	Szczerbakow et al.	1994	68
VB	Growth of PbTe single crystals under controlled Te and Pb vapor pressure	Nugraha et al.	1996	29
VB	Effect of magnetic field on convection and segregation of $Pb_{1-x}Sn_xTe$ single crystal growth	Yao et al.	1997	42
VB	Lattice constants of $Pb_{1-x-y}Sn_xMn_yTe$ alloys	Miotkowska et al.	1999	35
THM	Growth of PbTe single crystals by the cold THM (CTHM)	Munoz et al.	1999	45
PVT	Thermodynamical analysis of PbTe growth	Palosz	2000	63, 64
VB	Growth of Tl doped PbTe single crystals under controlled Te and Pb vapor pressure	Nugraha et al.	2001	30
PVT	Growth kinetics of PbTe and GeTe single crystals	Yashina et al.	2002	69
VB	Growth of PbTe and $Pb_{1-x}Sn_xTe$ single crystals under controlled Te and Pb vapor pressure	Yasuda et al.	2003	31
VLS	Growth of various lead compounds	Yashina et al.	2003	86
VB	Effect of various conditions on $Pb_{1-x}Sn_xTe$ single crystal growth	Shtanov	2004	33
VB	Growth of Gd-doped PbTe single crystals	Nicolic et al.	2008	36
VB	Growth of $Pb_{1-x}Sn_xSe$ with uniform tin distribution	Shtanov et al.	2009	34
VB	Growth of Cr-doped PbTe and $Pb_{1-x}Sn_xTe$ single crystals	Nicolic et al.	2009	37
VB	Growth of Pd-doped PbTe and $Pb_{1-x}Sn_xTe$ single crystals	Nicolic et al.	2009	38

VB: Vertical Bridgman, ZM; Zone Melting, DS: Directional Solidification, LEC: Liuqid Encapsulated Czochralski, THM: Traveling Heater Method, PVT: Physical Vapor Transport, VLS: Vapor-Liquid-Solid, CVT: Chemical Vapor Transport, RC: Recrystallization

ampoule, the melt superheating, the axial temperature gradient, and the pulling rate of the ampoule, on the microstructure of IV-VI crystals. They determined the growth conditions required to prevent polycrystallization. Shtanov and Yashina[34] have applied the T-x-y phase diagram to control the alloy composition during Bridgman growth, taking $Pb_{1-x}Sn_xSe$ crystal growth as an example. They obtained uniform crystals with $x = (0.23\text{-}0.33) \pm 0.01$ from an Se rich melt. Mitokowska et al.[35] have grown ternary and quaternary alloys based on PbTe-SnTe-MnTe by the Bridgman method and measured the lattice constants of these crystals.

Doping with rare earth elements and with transition elements is attractive since they give interesting physical properties such as giant magnetoresistance, persistent photoconductivity and Fermi level pinning. Nikolic et al.[36-38] have grown PbTe and $Pb_{1-x}Sn_xTe$ crystals doped with transition metals such as Gd, Cr and Pd and evaluated their electri-

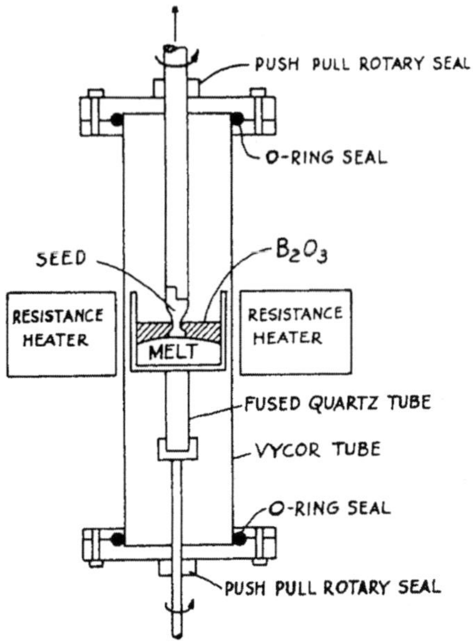

Fig. 24.7 Czochralski growth apparatus with B_2O_3 encapsulation (reprinted from Ref. 14 with permission, copyright 1968 The American Institute of Mining, Metallurgical and Petroleum Engineers).

cal and optical properties.

Kaidanov et al.[39] have applied the zone melting method to grow PbTe crystals doped with halogens. $PbCl_2$, $PbBr_2$ and PbI_2 with excess lead have been doped in PbTe crystals and the effective distribution coefficients have been evaluated. Crystal growth of $Pb_{1-x}Sn_xTe$ under microgravity has been studied by the zone melting (ZM) method[40] and the unidirectional freezing method.[41] Yao et al.[42] have studied the effect of magnetic field on convection and solute segregation during Bridgman growth by numerical simulation and compared the result with experimental observation under microgravity in space.

24.3.2 Solution Growth

The traveling heater method (THM) has been applied to grow PbTe crystals. Triboulet et al.[43] have grown PbTe single crystals by the THM (Vol. 1, Sec. 2.3.2) using the growth arrangement as shown in Fig. 24.9. Gille and Rudolph[44] have grown Tl-doped PbTe single crystals with diameters of 12 mm as substrates for laser diodes. Munoz et al.[45] have grown PbTe crystals by the cold traveling heater method (CTHM) with diameters of 15 mm. It has been shown that crystals with high electron mobilities could be grown due to the high purity obtained. This high purity is due to growth at low temperatures of ~ 1000°C, with consequent reduction in contamination from the ampoule.

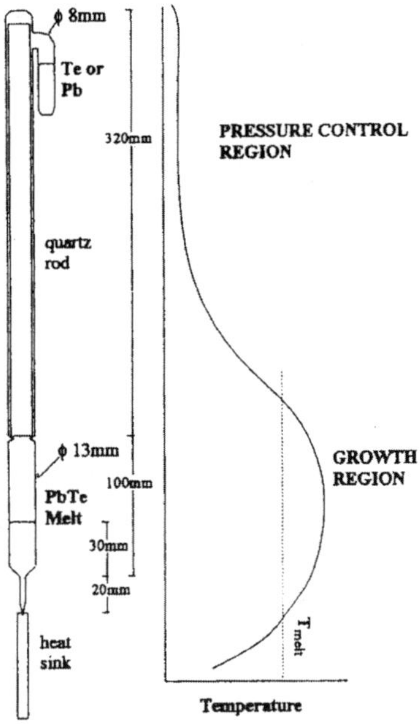

Fig. 24.8 Illustration of ampoule and temperature profile for Bridgman growth under controlled vapor pressure (reprinted from Ref. 29 with permission, copyright 1996 Elsevier).

24.3.3 Vapor Phase Growth

(1) Physical Vapor Transport (PVT) method

Prior[46] has grown PbSe single crystals by a horizontal physical vapor transport method. Butler et al.[47] have grown single crystals of $Pb_{1-x}Sn_xTe$ with x up to 0.27 by a vertical configuration and fabricated laser diodes with emission wavelength as long as 28 μm. Ohtsuki et al.[48] have grown single crystals of $Pb_{1-x}Sn_xTe$ by a horizontal configuration and obtained crystals with low dislocation densities of 2×10^4 cm^{-2}. Knittel[49] has grown PbTe crystals of constant composition using a capillary tube interconnecting the growth region and the source reservoir region in order to maintain the source material pressure above the liquidus temperature. Harman and McVittie[50] have grown PbTe crystals by the horizontal unseeded PVT method as shown in Fig. 24.10 and controlled the carrier concentration by excess of Pb and Te, varying the source composition. They reported that the dislocation density is less than 10^3 cm^{-2} over the entire area of about 2 cm^2 slices cut from the grown crystals.

Parker et al.[51, 52] have grown $Pb_{1-x}Sn_xTe$ with x up to 0.26 using a seed crystal with the same composition as the crystal to be grown by the crystal growth arrangement as

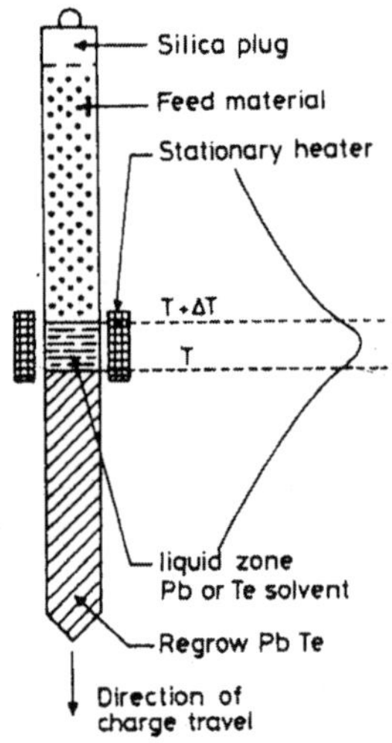

Fig. 24.9 Arrangement of the THM method (reprinted from Ref. 42 with permission, copyright 1982 Springer).

Fig. 24.10 Photographs of PbTe coarse powder and sealed quartz ampoule before insertion in the growth furnace (upper tube) and that of PbTe crystals after removal from the growth furnace (lower tube) (reprinted from Ref. 50 with permission, copyright 1968 Springer).

shown in Fig. 24.11. In Fig. 24.12, a crystal grown by this method is shown. Pendey[53] has grown $Pb_{1-x}Sn_xTe$ single crystals using BaF_2 seed substrates with varying Pb, Sn and Te compositions using the crystal growth assembly shown in Fig. 24.13. The crystal growth rate was 2 g/day and all grown crystals showed p-type conductivity.

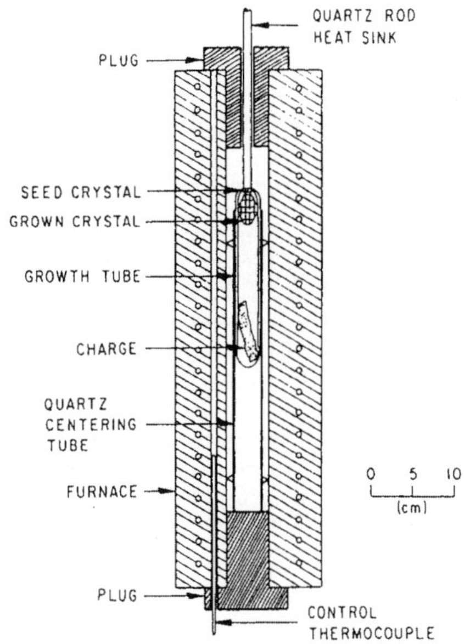

Fig. 24.11 Crystal growth furnace and growth tube (reprinted from Ref. 51 with permission, copyright 1974 Springer).

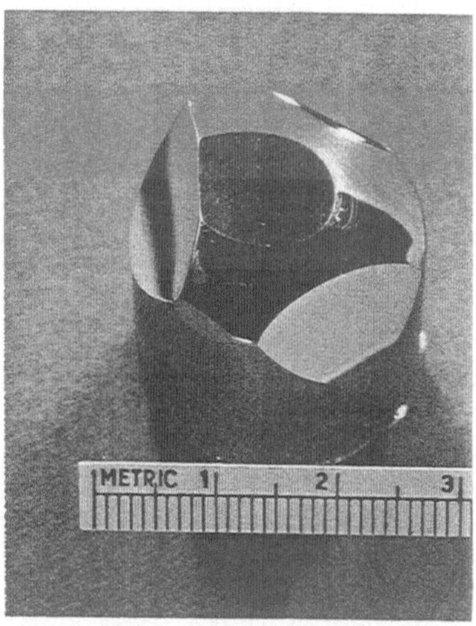

Fig. 24.12 Photomicrograph of $Pb_{0.79}Sn_{0.21}Te$ crystal grown on (111) seed with a constant opening to the seed (reprinted from Ref. 52 with permission, copyright 1976 Springer).

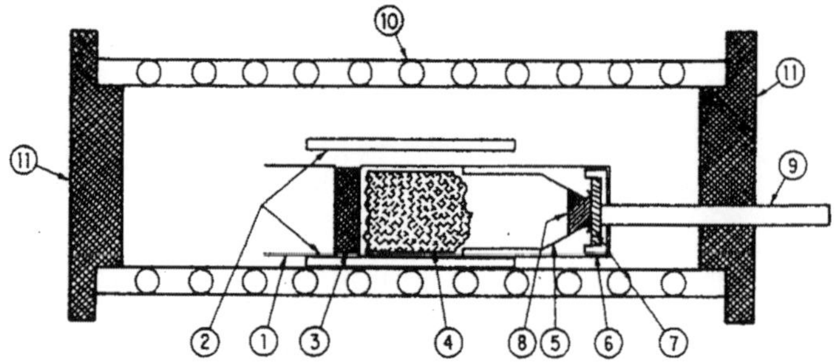

Fig. 24.13 Crystal growth assembly; (1) outer chamber, (2) thermal liner, (3) plug (for sealing the evacuated ampoule), (4) source alloy, (5) inner growth chamber, (6) substrate holder, (7) BaF_2 substrate, (8) crystal (growth), (9) heat sink rod, (10) furnace and (11) ceramic plugs (reprinted from Ref. 53 with permission, copyright 1974 Elsevier).

Bradford and Wentworth[54] have grown $Pb_{1-x}Sn_xTe$ single crystals using heat pipes in order to obtain homogeneous temperature distribution. They have grown single crystals measuring 19 mm in diameter and 25 mm long, weighing up to 60 g. Lo et al.[55] have grown $Pb_{1-x}Sn_xTe$ single crystals (x = 0.13) by a horizontal vapor growth method in which the horizontal temperature gradient is adjusted to produce nucleation on the source ingot. The crystals have dislocation densities as low as 2×10^4 cm^{-2}. Laser diodes of wavelength 10.6 μm could be fabricated using these crystals.

Maier et al.[56] have grown PbSe crystals in a closed tube configuration. They studied the effect of Pb or Se addition on the carrier type and concentration and compared the experimental results with the p-X-T diagram previously obtained.[16]

Tamari and Shitrikman[57,58] have grown single crystals of $Pb_{1-x}Sn_xTe$ ($0<x<0.25$) weighing up to 100 g and measuring up to 40 mm in diameter by sublimation onto the quartz base of a closed ampoule. In Fig. 24.14, the growth conditions are schematically explained and in Fig. 24.15, the grown crystals are shown.

Zoutendyk and Akutagawa[59] have grown PbTe single crystals from a Te-rich solid source and examined the transport rate from the viewpoint of a diffusion-limited vapor transport mechanism.

Golaski et al.[60,61] have grown PbTe and $Pb_{1-x}Sn_xTe$ single crystals on oriented BaF_2 sticks in the growth arrangement shown in Fig. 24.16. The growth was performed without contacting the ampoule wall and dislocation density as low as 10^2-2×10^3 cm^{-2} could be achieved for crystals measuring 20 mm in diameter and weighing 100 g.

Saunia et al.[62] have studied the effect of thermodynamic conditions on the growth rate, structural quality and electrical properties of $Pb_{1-x}Sn_xTe$ single crystals. The growth rate was found to be maximum when the vapor pressure of the components was approximately equal to the stoichiometric ratio. Palosz et. al.[63,64] have calculated the partial vapor pressure of constituents for PbTe and PbSe. The growth mechanism was discussed based on these theoretical results.

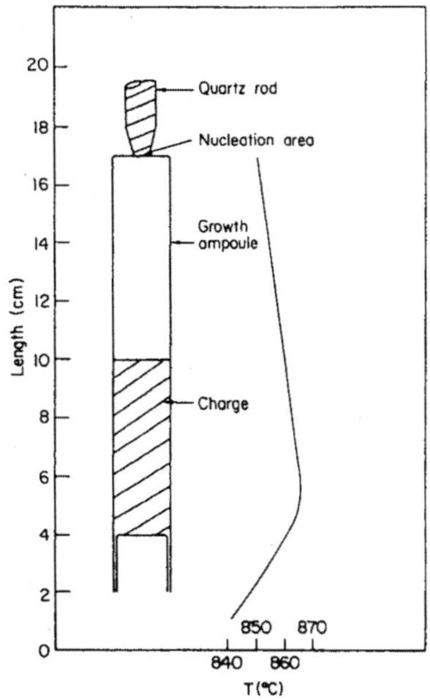

Fig. 24.14 Schematic representation of the growth ampoule. The position of the ampoule is shown relative to the temperature profile of the furnace (reprinted from Ref. 57 with permission, copyright 1978 Elsevier).

Fig. 24.15 Single crystals of $Pb_{1-x}Sn_xTe$ grown by nucleation control on quartz (reprinted from Ref. 57 with permission, copyright 1978 Elsevier).

Growth of various lead compounds other than $Pb_{1-x}Sn_xTe$ has also been studied. Kasai et al.[65] have grown $Pb_{1-x}Sn_xSe$ and $PbSe_{1-x-y}Sn_xTe_y$ single crystals by the horizontal sublimation method. Demin et al.[66] have grown $PbS_{1-x}Te_x$ single crystals and exam-

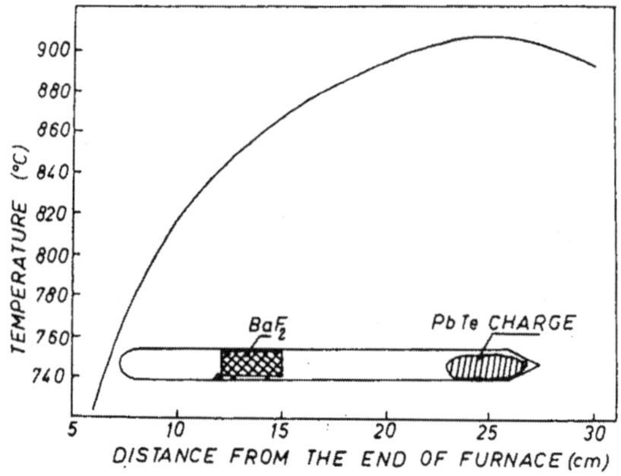

Fig. 24.16 Schematic diagram of the experimantal set-up and temperature profile of the furnace for PbTe (reprinted from Ref. 60 with permission, copyright 1982 Elsevier).

ined homogeneity region of the solid solutions. Leszczysnski et al.[67] have gown $Pb_{1-x}Ge_xTe$ single crystals and evaluated their structural and electrical properties. Szczerbakow[68] has grown $Pb_{1-x}Sn_xTe$ single crystals with $0 < x < 0.4$ and evaluated their lattice parameters as a function of composition.

The kinetics of crystal growth for PdTe and GeTe crystals has been studied by Yashina et al.[69] The growth rate was analyzed as a function of the F value, which was expressed as follows from process parameters.

$$F = \frac{\Delta PS\sqrt{T_s}}{P_{Ar}Ln} \tag{24.1}$$

where $\Delta P = P_e(T_S) - P_e(T_g)$, $P_e(T_S)$ is the partial pressure of the crystallizing substance before crystals appear in the growth chamber, and $P_e(T_g)$ is the equilibrium pressure corresponding to the minimum temperature in the growth chamber, P_{Ar} is the argon pressure, S and L are the ampoule cross section and length, and n is the number of crystals. The growth mechanism was discussed depending on the region of the F value.

Szczerbakow and Durose[70] have reviewed the self-selecting vapor growth (SSVG) method [71,72], which has been applied to various lead compounds. In this method, the temperature gradient is set as low as possible and a single crystal is grown from a grain which spontaneously starts to grow to a large one. Fig 24.17 shows how this method differs from the Piper-Polish method[73] and the Markov-Devydov method.[74] Fig. 24. 18 shows an example of the growth of a $Pb_{0.94}Sn_{0.06}Se$ single crystal. This method has now been scaled up using a vertical furnace as shown in Fig. 24.19.

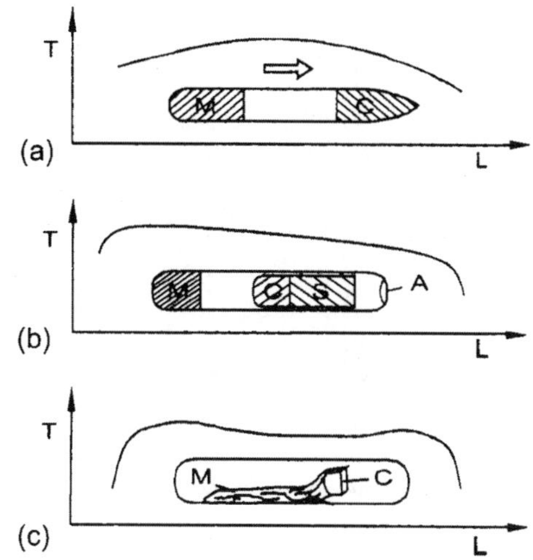

Fig. 24.17 Three kinds of schemes for crystal growth from the vapor. (a) Direct vapor transport of the source material (M) in the so-cold Piper-Polish method[73], (b) growth on to a substrate or seed using the Markov-Davydov method[74],(c) self-selecting vapor growth on a bed comprising a polycrystalline granular or powdered source material (M)[70] (reprinted from Ref. 70 with permission, copyright 2005 Elsevier).

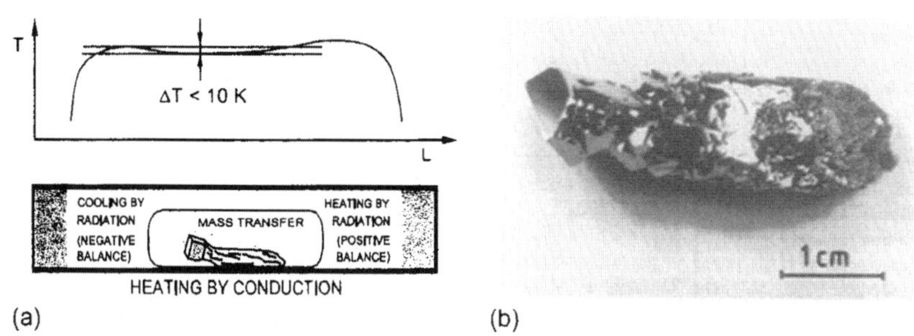

Fig. 24.18 (a) Schematic diagram of a self-selecting vapor growth (SSVG).[70] (b) An example of $Pb_{0.04}Sn_{0.96}Te$ crystal growth.[71] (reprinted from Ref. 70 with permission, copyright 2005 Elsevier).

(2) Vapor-Liquid-Solid (VLS) /Vapor-Melt-Solid (VMS) method

The VLS method also known as the VMS method was first invented by Wager and Ellis[75] for the crystal growth of Si whiskers and has been applied to the crystal growth of $Pb_{1-x}Sn_xTe$[76] and PbSe.[77] A typical growth arrangement is shown in Fig. 24.20.

This method has then been developed by Andryushenko et al.[78-81] in a vertical ar-

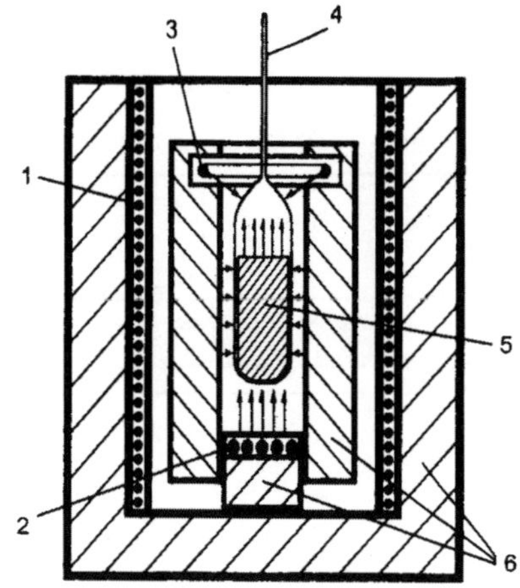

Fig. 24.19 Schematic diagram of a vertical SSVG system. 1: main heater, 2: heater generating the process-driving temperature difference, 3: heater protecting against condensation in the ampoule top, 4: capillary for condensing traces of an excess element, 5: crystal, 6: thermal insulation (reprinted from Ref. 70 with permission, copyright 2005 Elsevier).

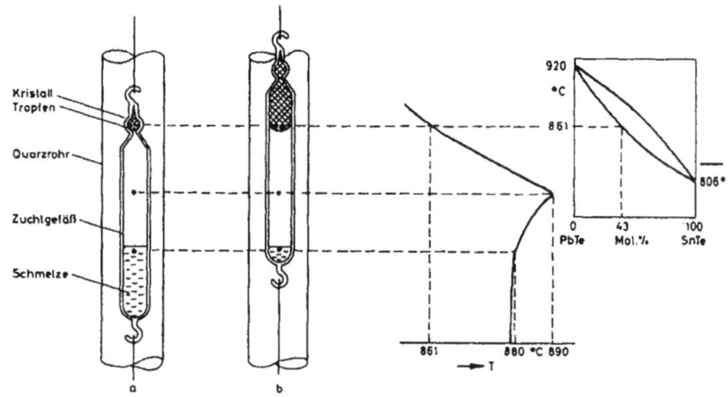

Fig. 24.20 Vapor-liquid-solid (VLS) growth of lead-zinc-telluride alloy with the temperature distribution in the furnace and the phase diagram. (a) growth beginning (b) growth end (reprinted from Ref. 76 with permission, copyright 1971 Elsevier).

rangement. Kinoshita et al.[82,83] further developed this method in a horizontal arrangement (Fig. 24.21) and have applied it to the crystal growth of $Pb_{1-x}Sn_xTe$. Yaneko[84] has studied the growth mechanism of $Pb_{1-x}Sn_xTe$ crystals with various growth temperature

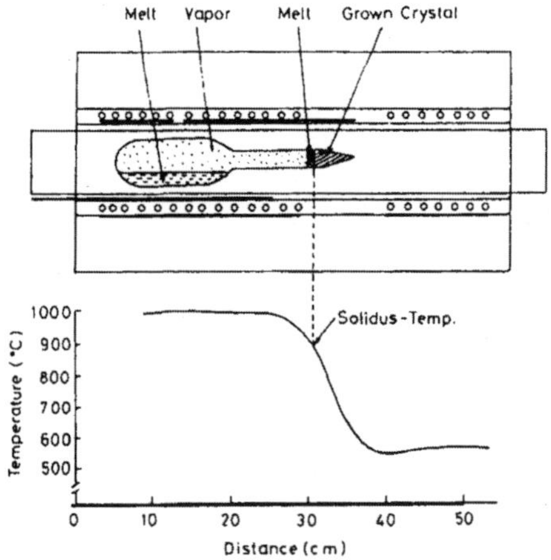

Fig. 24.21 Horizontal VMS assembly (reprinted from Ref. 82 with permission, copyright 1982 Elsevier).

gradients. Fabrri et al.[85] have made a theoretical study of the growth mechanism and calculated the composition uniformity for mixed crystals as a function of growth conditions.

The unique point of this method is in the fact that vapor phase growth is performed under the condition that the liquid phase exists in equilibrium with the vapor phase and the solid, that is the surface of grown crystals. When the growth is performed under this equilibrium condition, the constituent material dissolves in the liquid and the crystal is grown from the liquid on the grown crystal surface. Since the crystal is grown from the liquid, it does not need screw dislocations as the sites for crystal growth (the Frank screw dislocation mechanism) as is the case of vapor phase growth.[75] This equilibrium condition can be achieved when the temperature of the melt surface is held as the vapor pressure of the melt equilibrates with the liquid phase at the grown crystal surface as shown in Fig. 24.20. This method is interesting not only in decreasing the dislocation density but also in obtaining a uniform composition for mixed crystals. In fact, a uniform composition can be obtained for $Pb_{1-x}Sn_xTe$ crystal growth[76,78,82-84], one of which is shown in Fig. 24.22.

Yashina et al.[86] have studied this method systematically for various IV-VI materials such as PbTe, PbSe, SnTe, GeTe, $Pb_{1-x}Sn_xTe$ and $Pb_{1-x}Ga_xTe$ and have revealed the growth mechanism by examining the effect of growth parameters on the growth rate. They have also shown that uniform distribution of a dopant such as Ga can be achieved by their method, though the dopant is nonuniform in the conventional VLS method.[81]

(3) Chemical Vapor Transport (CVT) method

Growth of PbTe and $Pb_{1-x}Sn_xTe$ crystals using iodine as the transport agent has also been reported.[87] Octahedron and cube-octahedron single crystals have been grown with evaporation zone temperatures of 780-890°C and crystallization zone temperatures of

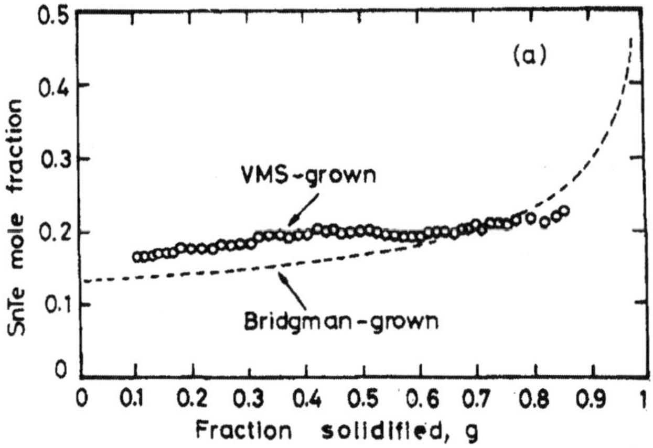

Fig. 24.22 Sn distribution of vapor-melt-solid (VMS)-grown crystals along growth axis compared with that of Bridgman-grown crystals (reprinted from Ref. 83 with permission, copyright 1983 Elsevier).

450-610°C. PbTe and SnTe single crystals have also been grown by using the chalcogenides S_2, Se_2 and Te_2 as transport agents.[88]

20.3.4 Recrystallization

Growth of Bi doped $Pb_{1-x}Sn_xTe$ crystals by recrystallization has been performed by Kimura.[89] The source material was prepared by direct reaction between the source metals and was annealed in equilibrium with the melt at temperatures of 853-863°C for five to fourteen days. It was found that large single crystals of size one centimeter could be grown and the lead-tin ratio could be controlled to be homogeneous in the crystal. The carrier concentration of bismuth-doped crystals (order of $10^{19}cm^{-3}$) could be reduced to be as low as $10^{17}cm^{-3}$ by isothermal annealing and the hole mobility was increased up to 3.5×10^4 cm^2/V.

24.4 CHARACTERIZATION

24.4.1 Purity

The purity of $Pb_{1-x}Sn_xTe$ crystals grown by the LEC method has been analyzed by Wagner and Willardson.[14] The major p-type impurity was sodium and the major n-type impurities were chlorine and aluminum. The distribution coefficients of various impurities have been determined as shown in Table 24.3.

The purity of PbTe grown by the CTHM has been measured by Munoz et al.[44] and it was found that excellent purity comparable to commercially available $Cd_{1-x}Zn_xTe$ crystals could be achieved by this low temperature solution growth method. Li and C however remained as major impurities, Li presumably from 5N Pb source material and C from the graphite layer covering the internal walls of the growth ampoule. Shtanov[33]

Table. 24.3 Distribution Coefficients of Major Impurities (reprinted from Ref. 14 with permission, copyright 1968 The American Institute of Mining, Metallurgical and Petroleum Engineers).

	PbTe	$Pb_{0.97}Sn_{0.03}Te$	$Pb_{0.87}Sn_{0.13}Te$
Na	-5	-5	-5
Mg	3	3	3
Al	10	7	5
Si	-4	-4	-4
S	0.01	0.01	0.01
Cl	0.2	0.2	0.2
Zn	0.5	0.5	0.5
Ga	0.3	0.3	0.3

studied the purity of Bridgman grown $Pb_{1-x}Sn_xTe$ by neutron activation analysis and solid source mass spectroscopy (SSMS). Most impurities were less than atomic ppm level except Na, Cu, Zn and Mn .

It has recently become known that doping by In, Ga and Cr results in anomalous behavior[90-92] such as Fermi level pinning[93] and persistent conductivity.[94] Doping by transition metals is also interesting because the Fermi level position can be tuned by the magnetic field. The doping of rare earth metals is also attractive because of the discovery of the giant magneto resistance effect.[95] Much research has therefore recently been carried out on these phenomena and crystal growth with various impurities is being actively pursued.[36-38]

24.4.1 Defects

(1) Dislocations

Various etchants have been used for dislocation measurement. Houston and Norr[96] and Norr[97] have examined two types of etchants. One is NaOH (5 g) + I_2 (0.2 g) + H_2O (10 ml) at 94-98°C for 5 min and another is 0.5 % $NaIO_3$ solution (10 ml) + NaOH (5 g). Ohtsuki et al.[48] have used a mixture of H_2O (10ml) + KOH (10 gm) + $C_3H_8O_3$ (1 ml) + H_2O_2 (0.5 ml) as the dislocation etchant and found that the dislocation of vapor phase grown crystals was one order of magnitude lower than that of Bridgman grown crystals.

Formation of dislocations during PVT is discussed by Parker[52] and Tamari and Shtrikman.[57,58] Parker[52] revealed that the cooling rate after crystal growth has a large effect on the dislocation density. He also showed that less wall contact led to very low dislocation density crystals being achieved even though the cooling rate was very high. Tamari and Shtrikman[57,58] also showed that slow furnace cooling reduces the dislocation density.

Rudolph[98] has discussed the formation of dislocation cells. It is known that PbTe grown by the Bridgman method shows a similar dislocation cell structure as in the case of CdTe but there are more dislocation pits in the cell interior than with CdTe.

Mülberg[99] and Kinoshita and Sugii[28] have examined the relationship between cell grain size and growth conditions and reached the same equation as follows.

$$d = a\,(n\,G)^{-1/2} \sim t^{-1/2} \qquad (24.\ 2)$$

where n is the crystallization velocity, G the axial temperature gradient and a is the proportionality factor of about 10^2. Since there is no dependence on the G/n ratio, it was concluded that the subgrain structures are due to dynamic polygonization and not to morphological instability of the growing interface caused by constitutional supercooling.

Nugraha et al.[29] have studied the effect of the tellurium vapor pressure on the etch pit density (EPD) as shown in Fig. 24.23. It was found that the EPD reaches a minimum when the tellurium vapor pressure approaches the point where the transition in conductivity type occurs.

(2) Small angle grain boundaries (SAGBs)

Kinoshita and Sugii[28] have studied precisely the origin of SAGBs and classified it as follows.

i) composition deviation or fluctuation

- nonstoichiometry
- x fluctuation of $Pb_{1-x}Sn_xTe$

ii) thermal or mechanical stress

- axial temperature gradient at the solid/liquid interface
- radial temperature gradient, that is the flatness of the interface
- ampoule shape
- cooling rate after crystal growth

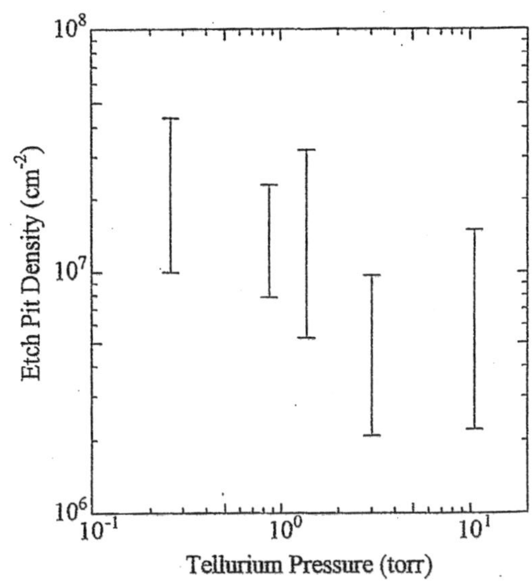

Fig. 24.23 Etch pit densities (EPDs) of PbTe crystals grown by the Bridgman method under controlled tellurium vapor pressure as a function of applied tellurium pressure (reprinted from Ref. 29 with permission, copyright 1996 Elsevier).

- mechanical stress for handling crystals
- structural perfection

From the precise study, it was concluded that SAGBs are due to dislocations and the main origin is the thermal stress during crystal growth and cooling. It is desirable to keep axial temperature gradient less than 5°C/cm during crystal growth and cooling rate less than 5°C/h.

Yashina[86] has applied the VLS method to grow various lead compounds. It was shown that the crystal quality is very dependent on the growth rate, that is the temperature difference ΔT between the source temperature and the growth temperature. When ΔT <10-15 K, there are few defects, while when ΔT is higher than this value, many SAGBs are clearly observed as shown in Fig. 24.24.

(3) Striations

Burke et al.[100] have studied compositional striations in $Pb_{1-x}Sn_xTe$ crystals grown by the liquid encapsulated Czochralski (LEC) method. The striations have been revealed by Norr etching. It was found that the spacing of the striations was much too large and not sufficiently periodic to be explained by rotational effect. The intensity and the spacing of the striations were not changed by stopping the rotation during growth. The nonrotational origin of the striation was discussed in relation to compositional variation. The alloy compositional variation was as much as 0.01 over distances of the order of 100 μm.

(4) Precipitation and inclusions

Scanlon[101] has studied the precipitation kinetics of Te and Pb in PbTe crystals in the temperature range from 180-400°C. The dislocation and its interacting field was considered to be a simple cylinder sink for precipitation. The activation energy associated with precipitation was found to be 0.88 eV.

Butler and Harman[25] have studied metallic inclusions and cellular structures in Bridgman grown $Pb_{1-x}Sn_xTe$ crystals and explained them by the mechanism of constitutional supercooling. Based on this mechanism, they suggested the way to avoid inclusions, is to grow from the excess Te side with a steeper temperature gradient and a low growth rate. Mühlberg and Hesse[102] made a transmission electron microscope (TEM) observation on PbTe single crystals grown under Te-rich conditions and found Te precipitates. Their occurrence has been explained as a diffusion-determined process due to the retrograde course of the solidus line.

Parker[51] studied the small hole and void formation during PVT crystal growth. It was found that nonstoichiometry is the main reason for void formation due to excess Pb or Sn in charges. Void formation could be reduced by using an equilibrated stoichiometric or slightly Te rich charge and by using a constant diameter growth tube instead of a tapered diameter tube together with (111) seed crystals. Tamari and Shtrikman[57, 58] have found that presublimation and preannealing of the charge material was very effective in reducing these small holes and voids. They also suggested that these defects are due to excess Pb or Sn metals in charges and they can be eliminated by annealing with Te excess charges.

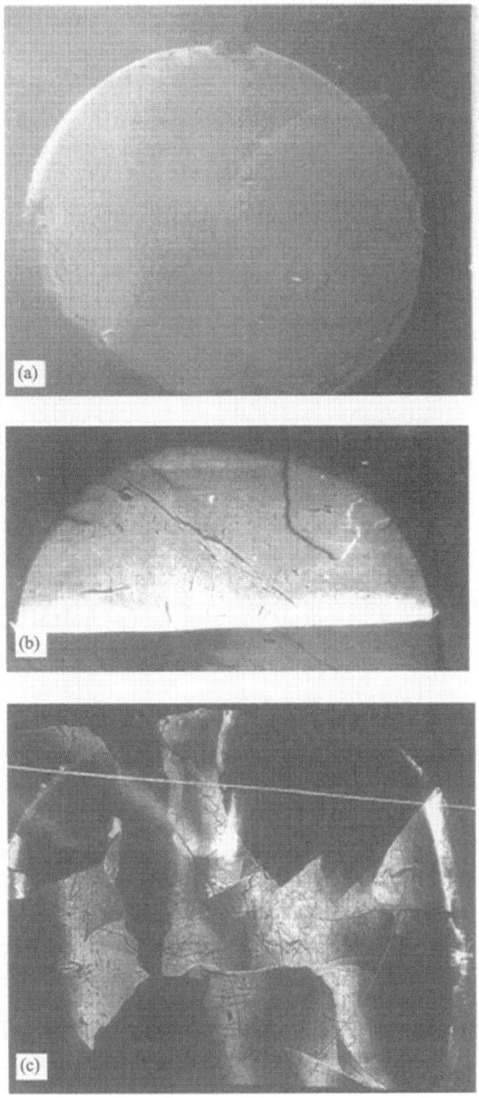

Fig. 24.24 Typical X-ray topographs (600) of $Pb_{1-x}Sn_xTe$ crystals obtained by the VLS method at different values of ΔT and, hence, growth rate: (a) ΔT = 2-6 K, (b) ΔT = 13-16 K, (c) ΔT = 30-40 K (slice diameter 20 ± 1 mm) (reprinted from Ref. 86 with permission, copyright 2003 Elsevier).

(5) Stoichiometry

Strauss and Brebrick[103] have discussed nonstoichiometry in relation to the phase diagram and point defects for various IV-VI compounds. Since the bandgap of lead compounds is small, point defects have a profound influence on the electrical properties.

Rudolph[104] has overviewed the effect of nonstoichiometry on point defects, dislocations, precipitates and inclusions, summarizing various compounds such as IV-VI (PbTe), II-VI (CdTe) and III-V (GaAs, InP) in order to discuss our general understanding of the phenomena related to nonstoichiometry.

Parker et al.[51] have pointed out that stoichiometry has a large effect on crystal growth rate and crystal perfection in PVT growth. Under Te excess there was a low growth rate with Te inclusions while stoichiometric or slightly metal rich conditions gave a successful growth rate. Metal excess however caused metal hillocks in the grown crystals.

Nishizawa's group[29-32] have studied stoichiometry control by the Bridgman method under controlled vapor pressure. Based on these studies, Suto et al.[105] have calculated the optimum vapor pressure for stoichiometry in PbTe and $Pb_{1-x}Sn_xTe$ as follows.

For PbTe,

$$P_{Te}{}^2 = 3.81 \times 10^9 \exp[-2.26 \text{ eV/kT}] \text{ Torr} \quad (24.3).$$

For $Pb_{0.88}Sn_{0.12}Te$,

$$P_{Te}{}^2 = 8.251 \times 10^9 \exp[-2.52 \text{ eV/kT}] \text{ Torr} \quad (24.4).$$

They found that these calculations were in good agreement with the experimental results.

(6) Native defects

The effect of native defects on electrical properties was discussed by Saunina et al.[62] The conductivity of $Pb_{1-x}Sn_xTe$ is due to carriers resulting from the ionization of native donor and acceptor defects associated with deviation from stoichiometry. The equilibrium concentration of native defects at high temperatures was achieved by quenching crystals to room temperature. p-type conduction in $Pb_{1-x}Sn_xTe$ is due to z-fold ionized metal vacancies (N_a^-) and n-type is associated with metal interstitials or with tellurium vacancies (N_d^+). The concentrations of ionized native defects of both types are related by the equation as,

$$K_d = [N_a^-][N_d^+] \quad (24.5).$$

Parada et al.[106] have calculated the electronic energy levels associated with vacancies in PbTe using a Green's function method. The unperturbed Bloch functions were obtained from a relativistic augmented-plan-wave (APW) energy-band calculation. It was concluded that Pb vacancies produce p-type PbTe and Te vacancies produce n-type PbTe.

Recently, it was found that doping of IV-VI compounds with mixed valence impurities results in a large modification of electrical properties due to the formation of impurity induced deep defect levels.[90] Skiperov et al.[107] have studied the deep defect states in diluted magnetic semiconductors $Pb_{1-x}Sn_xTe$;Yb and concluded that a deep impurity level due to Yb is formed whose position depends strongly on the alloy's composition. A charge carrier energy spectrum model was proposed to explain the impurity level shift as a function of the alloy composition.

As mentioned already, the doping of impurities such as In, Ga, and Cr exhibits anomalous behavior such as the Fermi level pinning effect.[90,93] Ahmad et al.[108] have studied the effect of III-impurity doping on deep defect states associated with III-impurities. The Fermi surface pinning and the excellent high temperature thermoelectric behavior were explained by an ab initio electronic structure calculation.

24.4.2 Electrical Properties

It is well known that deviation from stoichiometry has a strong effect on the electrical properties of IV-VI crystals. The vacancies in the tellurium sublattice are donors and those in the lead sublattice are acceptors. This means that an excess of lead causes n-type conductivity and an excess of tellurium causes p-type conductivity.[3,13,62] Based on these vacancies, PbTe and $Pb_{1-x}Sn_xTe$ show high carrier concentrations above 10^{18} cm^{-3}. In fact, this deviation from stoichiometry can be explained by the x-T diagram (Fig. 24.25)[13] and the p-X-T diagram.[16,17,109] The P-T-x phase diagrams have been studied by many researchers by measuring the electrical properties. One of them for PbSe is shown in Fig 24.26. The relationship between nonstoichiometry and the electrical properties is clearly shown as a function of the vapor pressure and the vapor pressure for stoichiometry is shown as a function of temperature in the figure.

Typical electrical properties of lead compounds are summarized in Ref. 10. In lead compounds, it is known that the carrier concentration is constant from room tempera-

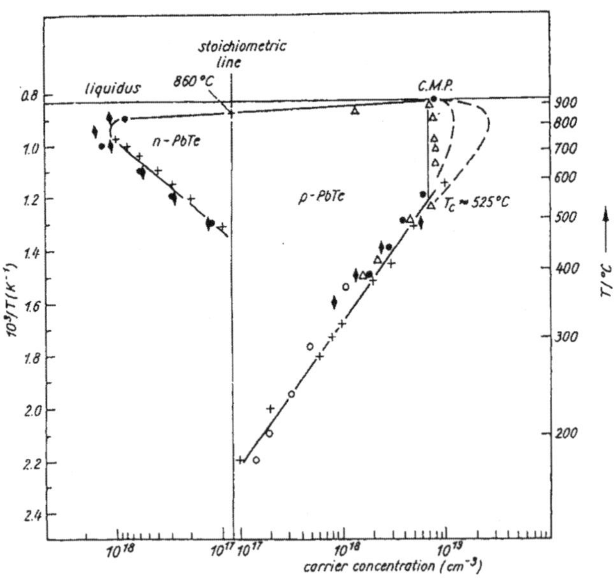

Fig. 24.25 Stability region of PbTe (reprinted from Ref. 13 with permission, copyright 1980 John Wiley and Sons).

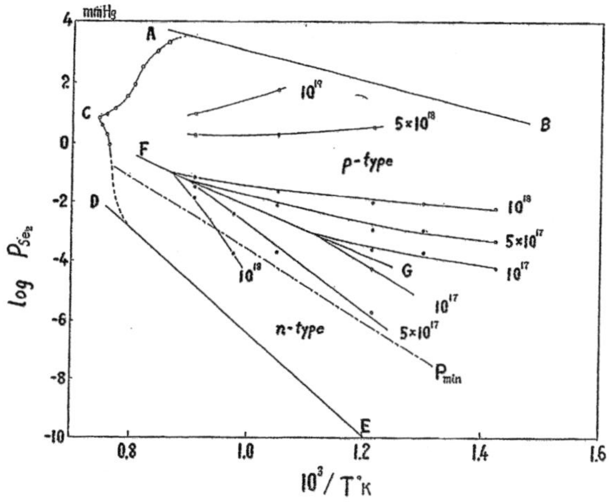

Fig. 24.26 p-T projection of the stability range with equi-carrier concentration curves and congruently subliming composition curve P_{min} (reprinted from Ref. 16 with permission, copyright 1963 Physical Society of Japan).

ture down to low temperatures so that the extrinsic center ionization energy E_i is very small. The mobility of both electrons and holes is known to exhibit unusual features. The mobility as a function of temperature has the same general shape. As the temperature decreases, the mobility increases as T^{-a}, where a is a number dependent on the carrier concentration. This behavior is explained by the combination of one-phonon acoustic scattering and a temperature dependent effective mass.

When PbTe is doped with various impurities such as III group elements, rare earth elements and transition metal elements, abnormal electrical properties are obtained and these phenomena are currently being studied.[91-95] Akimov[90] has reviewed these electrical properties for various lead compounds from theoretical and characterization viewpoints.

p-type PbTe crystals with Tl doping have been grown by THM and carrier concentrations of 10^{18} to 2×10^{19} cm^{-3} were obtained.[43] Nugraha et al.[30] have grown Tl doped PbTe crystals by the Bridgman method and measured the Hole mobility as a function of the hole carrier concentration as shown in Fig. 24.27. It was found that the Hole mobility is largely dependent on the applied Te and Pb vapor pressures. The result obtained was explained by the formation of the complex defect, Tl_{Pb}-Pb_{Te}.

24.4.3 Optical Properties

The absorption edge of lead compounds is reported by Prakash[110] as shown in Fig. 24.28. α-α_{min} where α_{min} represents free carrier and impurity absorption. The absorption coefficient is an exponential function of the photon energy over several decades of absorption. The refractive index of lead compounds as a function of photon energy is shown in Fig. 24. 29 for various temperatures.[111]

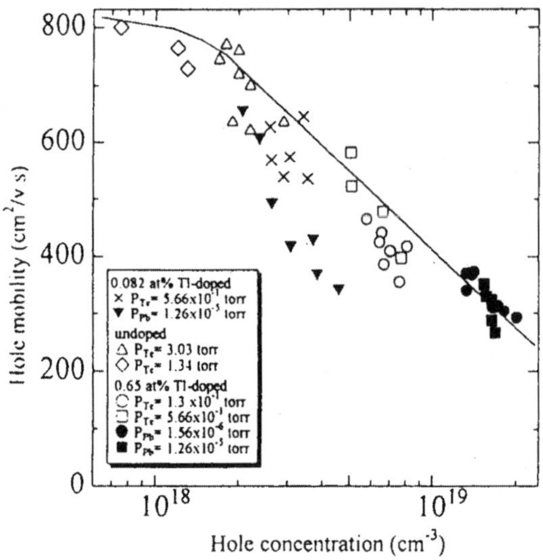

Fig. 24.27 Hall concentration dependence of Hall mobility at 300 K for p-type undoped and Tl-doped PbTe crystals grown under applied Te and Pb vapor pressures (reprinted from Ref. 30 with permission, copyright 2001 Elsevier).

Since doping with certain impurities gives abnormal effects, various studies have been performed to measure the optical properties. Nikolic et al. have studied the far infrared reflectivity spectra of PbTe and $Pb_{1-x}Sn_xTe$ doped with Cr, Gd and Pd[36-38] and found that these impurities increase free carrier mobilities. Optical mobilities as high as 3600-4760 cm^2/Vs have been obtained.

24.5 APPLICATIONS

Infrared lasers[112,113] and detectors[114] based on lead compounds can cover long wavelengths ranging between 3 and 30 μm and are promising for various applications.[3] These applications are as follows.

- atmospheric pollution measurement, detection of sub-ppb concentrations of various gases in airborne systems
- noninvasive medical diagnostics, clinical breath analysis[115]- determination of metabolic malfunctions measuring the isotope ratio $^{13}CO_2/^{12}CO_2$, in a non-invasive, non-radioactive and non poisonous method.
- automobile engine exhaust diagnostics[115]
- industrial process monitoring and control, monitoring of trace amounts of CO_2 or H_2O in process gases like Ar and N_2
- evaluation of high-quality surfaces by scattered light detection
- environmental control stations, smoke stack monitoring, detection of trace amounts of NH_3 in the exhaust stream of catalytic converters

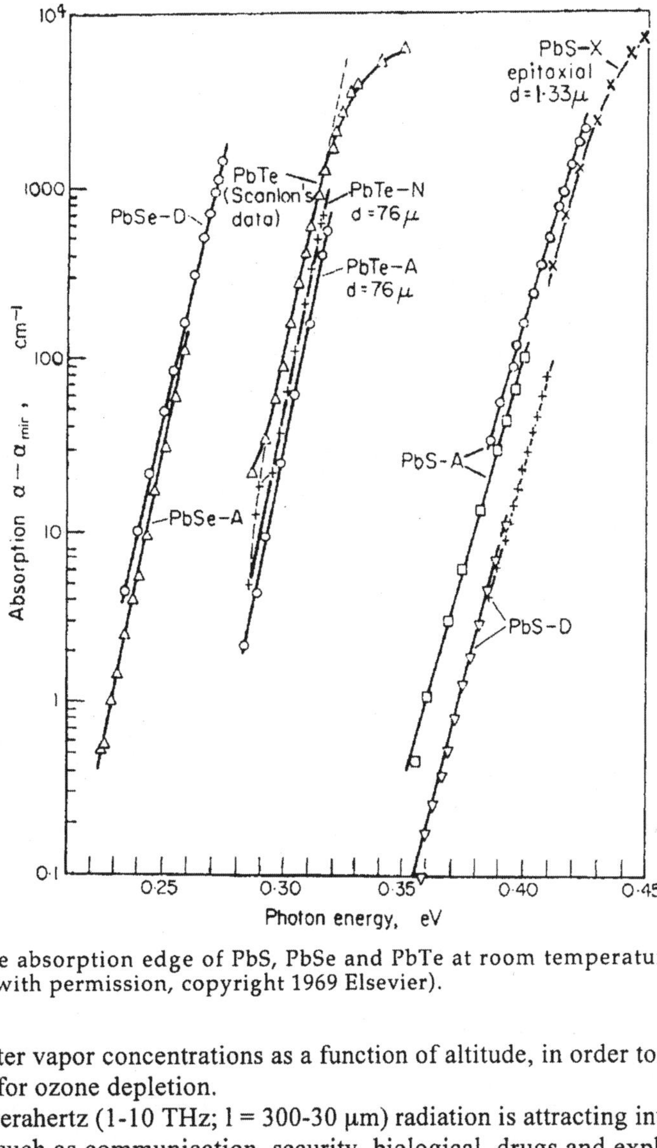

Fig. 24.28 The absorption edge of PbS, PbSe and PbTe at room temperature (reprinted from Ref. 10 with permission, copyright 1969 Elsevier).

- trace of water vapor concentrations as a function of altitude, in order to estimate the OH radicals for ozone depletion.

Recently terahertz (1-10 THz; l = 300-30 μm) radiation is attracting interest[116-119] in many fields such as communication, security, biological, drugs and explosion detection, imaging and astronomy. Terahertz optical communication is attractive since by using 10 μm far infrared lasers, optical fiber communications can use fibers with much less loss than the present quartz fibers.

24.5.1 IR Laser Diodes (IR-LDs)

As reviewed by Horikoshi [112], $Pb_{1-x}Sn_xTe$ lasers have been studied for lasers in the wavelength region longer than 3 μm. The first long wavelength lasing was reported by

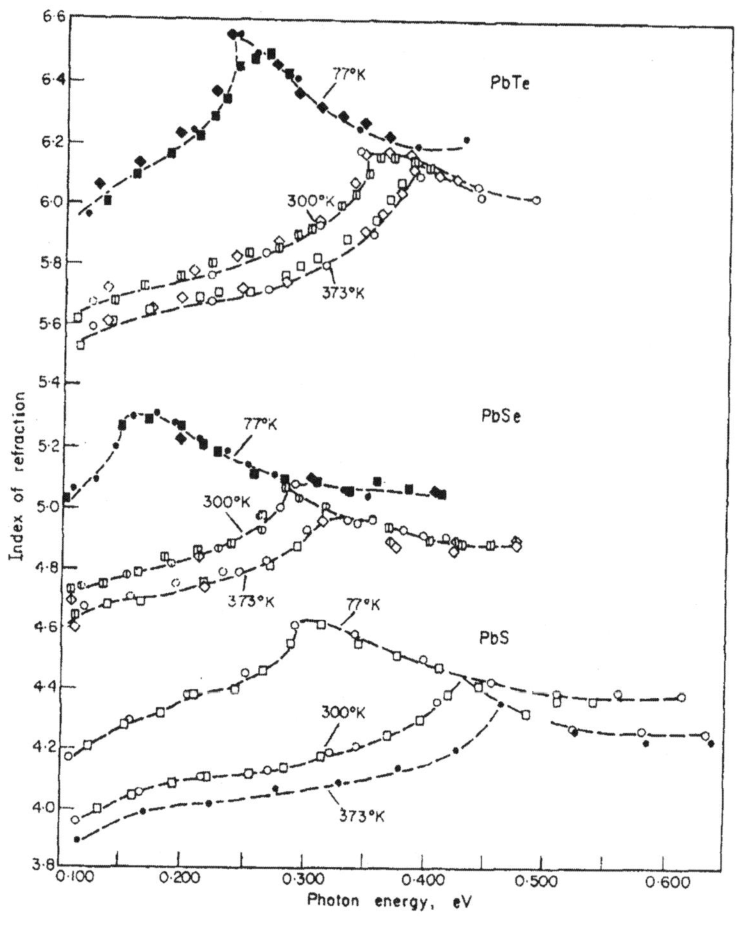

Fig. 24.29 Variation of the refrective index of PbS, PbSe and PbTe with photon energy at 77, 300 and 373 K (reprinted from Ref. 10 with permission, copyright 1969 Elsevier).

Butler et al.[120] with PbTe and PbSe pn junction laser diodes. Hereafter, pn junction lasers, single hetero (SH) lasers and double hetero (DH) lasers have been studied by many researchers using various lead compound based materials.[121-129]

Yasuda et al.[130] have developed middle far-infrared $Pb_{1-x}Sn_xTe$ lasers using PbTe substrates. The temperature difference method under controlled vapor pressure (TDM-CVP) has been applied for the growth of PbTe-$Pb_{1-x}Sn_xTe$-PbTe double heterostructures on PbTe substrates.

Vertical cavity surface emitting lasers (VSELs) have been fabricated by various researchers.[131-136] Rahim et al.[136] have recently developed a surface emitting laser with 5 μm wavelength, quantum wells composed of different lead chalcogenides.

24.5.2 Far Infrared Photodetectors

$Pb_{1-x}Sn_xTe$ photodetectors in the 8-12 μm spectral range have been developed in the past by various researchers as reviewed in Refs. 137-140 comparing with $Hg_{1-x}Cd_xTe$ detectors.

DeVaux et al.[141] have studied the incremental resistance-area product of $Pb_{0.79}Sn_{0.21}Te$ detectors sensitive in the 8-12 μm spectral range. The identification of the resistance-area product is to improve the thermally limited detectivity of $Pb_{1-x}Sn_xTe$ detectors. Kenndy et al.[142] have fabricated $Pb_{1-x}Sn_xTe$ detectors having detectivity of $1.5x10^{11}$ cm^2/W with 45 % quantum efficiencies. Wang et al. have fabricated $Pb_{1-x}Sn_xTe$ diodes and detector arrays covering the 8-14 μm spectral band at 85 K using liquid phase epitaxy (LPE) layers on PbTe and $Pb_{1-x}Sn_xTe$ substrates.[143,144]

$Pb_{1-x}Sn_xTe$ photodetectors however have drawbacks such as high dielectric constant resulting in high diode capacitance which limits the frequency response. Secondly, they have high thermal coefficients of expansion. This has limited their applicability with silicon multiplexers. Since $Pb_{1-x}Sn_xTe$ photodetectors however have an advantage in that the maximum available doping level is high, their study has continued.[145-147]

Recent developments of epitaxial growth techniques such as liquid phase epitaxy (LPE), metal organic chemical vapor deposition (MOCVD) and molecular beam epitaxy (MBE) promise to realize high performance $Pb_{1-x}Sn_xTe$ detectors and theoretical performances have been studied.[148-150] Arnold et al.[151] have fabricated resonent-cavity-enhanced photodetectors in the mid-infrared region (4.13 μm) and (8.41 μm) using epitaxialy grown IV-VI materials on Si substrates.

Recently, terahertz detectors are attracting much interest. Terahertz detectors are reviewed by Sizov and Rogalski.[116] There are many reports on terahertz range detectors using $Pb_{1-x}Sn_xTe$:In (x ~0.25, In ~2 atm.%) persistent response with current responsivity about 10^3 A/W at the wavelengths of 90 and 116 μm.[152] The persistent photoconductivity is based on the effect of III-group doping such as Ga and In in narrow gap lead telluride compounds. The physical reason for the persistent photoconductivity is ascribed to the existence of a barrier between the localized deep and free electron states. This phenomenon is presently under investigation. Klikov et al.[153] have studied the increase in the electric current through $Pb_{1-x}Sn_xTe$:In exposed to laser radiation with a wavelength of $\lambda = 336.8$ μm.

24.5.3 Thermoelectric Generators

Recently, lead compound materials have become attractive as thermoelectric materials because of their high thermopower (Seebeck coefficient) at elevated temperatures (~700 K).[154,155] Especially, $Pb_{1-x}Sn_xTe$ is attractive since alloy scattering lowers its thermal conductivity resulting in a higher figure of merit ZT, where $Z=S^2d/k$, S is the thermopower (Seebeck coefficient), d the electrical conductivity, k the thermal conductivity and T is the absolute temperature. This alloying effect is known as the Joeff concept.

PbTe and $Pb_{1-x}Sn_xTe$ materials are promising as thermoelectric generators in the mid-temperature range (550K<T<850K)[156] and are considered suitable for applications converting waste heat to electricity.[157] The efficiency of $Pb_{1-x}Sn_xTe$ based thermopower modules has been calculated as a function of temperature.[158] Dong et al.[159,160] have studied In-doped $Pb_{1-x}Sn_xTe$ to find ways of increasing the electrical conductivity thus improving the ZT. McGuire et al.[161] have studied the effect of high pressure and high tempearture treatment on the thermoelectric properties of PbTe and explained the data from the viewpoint of the removal of lattice defects and possibly Pb vacancies. Dughaish[162] has made a theoretical study of the effectiveness of PbTe in thermoelectric power generators.

REFERENCES

1. P.M. Nikolic, *Br J. Appl. Phys.* **16** (1965) 1075.
2. *Semiconducotrs and Semimetals, Vol. 5*, eds. R.K. Willardson and A.C. Beer (Academic Press, New York, 1970).
3. T.C. Harman and I. Melngailis, in *Applied Solid State Science, Vol. 4*, ed. Raymond Wolfe (Academic Press, New York, 1974) p. 1.
4. A. Rogalski and J. Piotrowski, *Prog. Quant. Electron.* **12** (1988) 87.
5. M. Tacke, B. Spanger, A. Lanbrecht, P.P. Norton and H. Bottner, *Appl. Phys. Lett.* **53** (1988) 2260.
6. H. Preier, *Semicond. Sci. Technol.* **5** (1990) S12.
7. M. Tacke, *Infrared Phys. Tech.* **36** (1995) 447.
8. *Lead Chalcogenides: Physics and Applications*, ed. D. Khokhlov (Gordon & Breach, London, 2002).
9. N.Kh. Abrikosov, V.F. Bankina, L.V. Poretskaya, L.E. Shelimova and E.V. Skudnova, *Semiconduting II-VI, IV-V and IV-VI Compounds* (Plenum Press, New York, 1969).
10. R. Dalven, *Infrared Phys.* **9** (1969) 141.
11. J.F. Butler and T.C. Harman, *IEEE J. Quant. Electron.* **5** (1989) 50.
12. T.C. Harman, A.R. Calawa, I. Melngailis and J.O. Dimmock, *Appl. Phys. Lett.* **14** (1969) 333.
13. T.C. Harman, *J. Nonmetals* **1** (1973) 183.
14. J.W. Wagner and R. Willardson, *Trans. AIME* **242** (1968) 366.
15. W.W. Scanlon, in *Solid State Physics-Advances in Research and applications Vol. 9*, eds, F. Seitz and D. Turnbull (Academic Press, 1959) p. 83.
16. K. Igaki and N. Ohashi, *J. Phys. Soc. Japan* **18(Suppl. II)** (1963) 143.
17. M. Fujimoto and Y. Sato, *Jpn. J. Appl. Phys.* **5** (1966) 128.
18. V.L. Kouznetsov and V.P. Zlomanov, *Izv. Akad. Nauk SSSR, Neorg, Mater.* **35**(1999) 263.
19. E.P.A. Metz, R.C. Miller and R. Mazelsky, *J. Appl. Phys.* **33** (1962) 2016.
20. R.J. Baughman and R.A. Lefever, *Mater. Res. Bull.* **4** (1969) 721.
21. W.D. Lawson, *J. Appl. Phys.* **22** (1951) 1444.
22. A.A. Machonis, T.B. Cadoff, *Trans AIME* **230** (1964) 333.
23. A.R. Calawa, T.C. Harman, M. Finn and P. Youtz, *Trans. AIME* **242** (1968) 374.
24. G.A. Nanney, P.H. Schmidt, J.H. Wernick and J.P. Garno, *J. Phys.* **29 Coll. C4** (1968) 34.
25. J.F. Butler and T.C. Harman, *Electrochem. Soc.* **116** (1969) 260.
26. V. Fano, R. Pergolari and L. Zanotti, *J. Mater. Sci.* **14** (1979) 535.
27. M.I. Karklina, M.S. Ablova, V.M. Muzhdaba, *Izv. Nauk. SSSR, Neorg. Mater.* **6** (1980) 985.
28. K. Kinoshita and K. Sugii, *J. Crystal Growth* **71** (1985) 283.
29. Nugraha, K. Suto, O. Itoh, J. Nishizawa, and Y. Yokota, *J. Crystal Growth* **165** (1996) 402.
30. Nugraha, W. Tamura, O. Itoh, T. Amemiya, K. Suto and J. Nishizawa, *J. Crystal Growth* **222** (2001) 38.
31. A. Yasuda, K. Suto, Y. Ohyama and J. Nishizawa, *Meter. Mater. Proc.* **15** (2003) 225.
32. J. Nishizawa and Y. Ohyama, *Encycl. Mater. Sci. Technol.* (2008) p. 845.
33. V.I. Shtanov, *Crystal. Rep.* **49** (2004) 343.
34. V.I. Shtanov and L.V. Yashina, *J. Crystal Growth* **311** (2009) 3257.
35. S. Mitkowska, d. Dynowska, I. Miotkowski, A. Szczerbakow, B. Wotlpwsla, J. Kachniarz and W. Paszkowicz, *J. Crystal Growth* **200** (1999) 483.
36. P.M. Nikolic, K.M. Paraskevopoulos, S.S. Vujatovic, M.V. Nikolic, A. Bojicic, T.T. Zorba, B. Stamenovic, V. Blagojevic, M. Jovic, M. Dasic and W. Konig, *Mater. Chem. Phys.* **112** (2008) 496.
37. P.M. Nikolic, K.M. Paraskevopoulos, S.S. Vujatovic, M.V. Nikolic, X. Chatzistavrou, E. Pavlidou, T. Ivetic, V. Blagojevic, A. Bojicic and d. Dinulovic, *J. Alloys and Compounds* **480** (2009) 893.
38. P.M. Nicolic, K.M. Parskevopoulos, S.R. Djukic, S.S. Vujatovic, N.J. Labus, T.T. Zorba, M. Jovic, M.V. Nikolic, A. Bojcic, V. Blagojevic, B. Stamenovic and W. Konig, *J. Alloys and Compounds* **475** (2009) 930.
39. V.I. Kaidonov, R.B. Mel'Nik and E. Sh. Fedorenko, *Izv. Akad. Nauk SSSR, Neorg. Mater.* **2** (1966) 2246.
40. I. Iwai and S. Yusaburo, *NASDA Tehnical Memorandum* (NASDA, 1994) p. 468.

41. T. Yamada and K. Kinoshita, *NASDA Tehnical Memorandum* (NASDA, 1994) p. 431.
42. M. Yao, A. Chait, A.L. Fripp and W.J. Debnam, *J. Crystal Growth* **173** (1997) 467.
43. R. Triboulet, G. Didier, A. Lasbley, F. Morales, B. Toulouse, C.M. Pelletier, R. Granger and S. Rolland, *Phys. Narrow Gap Semicond., Lecture Notes in Physics* **Vol. 152** (Springer, Berlin, 1982) p. 54.
44. P. Gille and P. Rudolph, *J. Crystal Growth* **64** (1983) 613.
45. V. Munoz, A, Lasvley, S. Klotz, R. Triboulet, *J. Crystal Growth* **196** (1999) 71.
46. A.C. Prior, *J. Electrochem. Soc.* **108** (1961) 82.
47. J.F. Butler and T.C. Harman, *Appl. Phys. Lett.* **12** (1968) 347.
48. O. Ohtsuki, K. Shinohara and O. Ryuzan, *Jpn. J. Phys.* **10** (1971) 515.
49. A.Z. Knittel, *J. Crystal Growth* **26** (1974) 33.
50. T.C. Harman and J.P. McVittie, *J. Electron. Mater.* **3** (1974) 843.
51. S.G. Parker, J.E. Pinell and R.E. Johnson, *J. Electron. Mater.* **3** (1974) 731.
52. S.G. Parker, *J. Electron. Mater.* **5** (1976) 497.
53. R.K. Pendey, *Solid State Commun.* **15** (1974) 449.1
54. A. Bradford and E. Wentworth, *Infrared Phys.* **15** (1975) 303.
55. W. Lo, G.P. Montgomery, Jr. and D.E. Swets, *J. Appl. Phys.* **47** (1976) 267.
56. H. Maier, D.R. Daniel and H. Preier, *J. Crystal Growth* **35** (1976) 121.
57. N. Tamari and H. Shtrikman, *J. Crystal Growth* **43** (1978) 378.
58. N. Tamari and H. Shtrikman, *J. Electron. Mater.* **8** (1979) 269.
59. J. Zoutendyk and W. Akutagawa, *J. Crystal Growth* **56** (1982) 245.
60. Z. Golacki, Z. Furmanik, M. Gorska, A. Szczerbakow and W. Zahorowski, *J. Crystal Growth* **60** (1982) 150.
61. Z. Golacki, M. Gorska, T. Warminski and A. Szczerbakow, *J. Crystal Growth* **74** (1986) 129.
62. T.V. Saunina, D.B. Chesnokova and D.A. Jaskov, *J. Crystal Growth* **71** (1985) 75.
63. W. Palosz, *J. Crystal Growth* **216** (2000) 273.
64. W. Palosz, H.A. Alexander and K. Grasza, *J. Crystal Growth* **216** (2000) 283.
65. I. Kasai, D.R. Daniel, H. Maier and H.D. Wurzinger, *J. Crystal Growth* **23** (1974) 201.
66. V.N. Demin, I.G. Burkreeva, A.M. Gas'kov, V.P. Zlomanov and A.V. Novoselova, *Izv. Akad. Nauk SSSR, Neorg, Mater.* **22** (1986) 33.
67. Leszczynski, A. Szczerbakow and G. Karczewski, *J. Crystal Growth* **135** (1994) 565.
68. A. Szczerbakow, *J. Crystal Growth* **139** (1994) 172.
69. L.V. Yashina, V.I. Dernovskii, V.P. Zlomanov and V.I. Shatanov, in *Growth of Crystals*, eds. E.I. Givargizow, A.M. Mel'nikova, (Klumar Academic/Consultants Bureaux, New York) **21** (2002) 11.
70. A. Szczerbakow and K. Durose, *Prog. Crystal Growth Charact. Mater.* **51** (2005) 81.
71. A. Szczerbakow, *J. Crystal Growth* **82** (1987) 709.
72. A. Szczerbakow, *Crystal Growth Des.* **1** (2001) 183.
73. W.W. Piper and S.J. Polich, *J. Appl. Phys.* **32** (1961) 1278.
74. E.V. Markov and A.A. Davydov, *Izv. Akad, Nauk SSSR Neorg. Mater.* 7 (1971) 575.
75. R.S. Wagner and W.C. Ellis, *Appl. Phys. Lett.* **4** (1964) 89.
76. D. Mateika, *J. Crystal Growth* **9** (1971) 249.
77. V.P. Zlomanov, E.V. Masyakin and A.V. Novoselova, *J. Crystal Growth* **26** (1974) 261.
78. E.D. Andryushenko, I.M. Varlamova, V.P. Zlomanov, A.V. Novoselova and O.I. Tananaeva, *Izv. Akad. Nauk SSSR, Neorg, Mater.* **15** (1979) 1569.
79. E.D. Andryushenko, Y.N. Dani'chuk, V.P. Zlomanov, A.V. Novoselova and I.O. Tananaeva, *Izv. Akad. Nauk SSSR, Neorg, Mater.* **15** (1979) 1573.
80. E.D. Andryushenko, Y.N. Dani'chuk, V.P. Zlomanov and M.E. Tamm, *Izv. Akad. Nauk SSSR, Neorg, Mater.* **15** (1979) 1682.
81. M.A. Lazarenko, A.M. Gas'kov, V.P. Zlomanov and A.V. Novoselova, *Vestnic MGU Khimia* **21** (1980) 518.
82. K. Kinoshita and S. Miyazawa, *J. Crystal Growth* **57** (1982) 141.
83. K. Kinoshita and K. Sugii, *J. Crystal Growth* **65** (1983) 379.
84. Z.G. Yanenko, V.I. Shtaov and V.P. Zlomanov, *Izv. Akad. Nauk SSSR, Neorg, Mater.* **26** (1990) 437.
85. M. Fabbri, I.N. Bandeira and L.C.M. Miranda, *J. Crystal Growth* **104** (1990) 435.
86. L.V. Yashina, V.I. Shtanov and Z.G.Yenenko, *J. Crystal Growth* **252** (2003) 68.
87. L.I. Bezrodnaya, N.I. Makarova, E.P. Strukova, Yu.S. Jharionovskii and S.G. Yudin, *Rost. Krist.* **9** (1972) 231.
88. F. Wäesch, *Krist. Tech.* **9** (1974) K107.
89. H. Kimura, *J. Electron. Mater.* **166** (1972) 1.
90. B.A. Akimov. A.V. Dmitriev, D.R. Khokhlov, L.I. Ryabova, *phys. stat.sol. (a)* **137** (1993) 9.
91. L.I Ryabova and D.R. Khoklov, *JETP Lett.* **80** (2004) 133.
92. Y. Gelbstein, Z. Dashvsky and M.P. Dariel, *Physica (Amsterdam)* **363B** (2005) 196.

93. N. Romcevic, M. Romcevic, D.R. Khokhlov, A.I. Belogorukov, L.I. Ivanchuk and W. König, *Infrared Phys. Technol.* **40** (1999) 453.
94. B.A. Volkov, L.I. Ryabova and D.R. Khoklov, *Physics-Uspekhi* **45** (2002) 819.
95. A.V. Morozov, A.E. Kozhanov, A.I. Artamkin, E.I. Sly'nko, V.E. Sly'nko, W.D. Dobrovolski, T. Story and D.R. Khoklov, *Semiconductors* **38** (2004) 27.
96. B.B. Houston and M.K. Norr, *J. Appl. Phys.* **12** (1961) 615.
97. M.K. Norr, *J. Electrohchem.* **109** (1962) 433.
98. P. Rudolph, *Cryst. Res. Technol.* **40** (2005) 7.
99. M. Mühlberg, *Krist. Tech.* **15** (1980) 565.
100. J.R. Burke, J.D. Jensen and B. Houston, *J. Electrochem. Soc.* **120** (1973) 431.
101. W.W. Scanlon, *Phys. Rev.* **126** (1962) 509.
102. M. Mühlberg and D. Hesse, *phys. stat. sol. (a)* **76** (1983) 513.
103. A.J. Strauss and R.F. Brebrick, *J. Phys.* **29 Coll. C4** (1968) 21.
104. P. Rudolph, *Cryst. Res. Technol.* **38** (2003) 542.
105. K. Suto, J. Nishizawa and A. Yasuda, *Mater. Sci. Semicond. Proces.* **6** (2003) 289.
106. N.J. Parada, *Phys. Rev.* **3** (1971) 2042.
107. E.P. Skipetrov, E.A. Zvereva, B.B. Kovaev, O.S. Volkova, A.V. Olubev and E.I. Slyn'ko, *Eur. Phys. J. Appl. Phys.* **29** (2005) 23.
108. S. Ahmad, K. Hoang and S.D. Mahanti, *Phys. Rev. Lett.* **96** (2006) 056403.
109. Y. Sato, M. Fujimoto and A. Kobayashi, *Jpn. J. Appl. Phys.* **2** (1963) 688.
110. V. Prakash, *Technical Report* **HP-13** (Harvard University, Cambridge, MA, 1967).
111. J.M. Zemel, J.D. Jensen and R.B. Schoolar, *Phys. Rev.* **140** (1965) A330.
112. Y. Horikoshi, *Semicon. Semimet.* **22** (1985) 93.
113. U.P. Schessl, J. John and J. McCann, *Long Wavelength Infrared Semiconductor Laser* (John Wiley and Sons, Inc., 2004)
114. G. Nimtz and B. Schlight, in *Narrow Gap Semiconductors* (Springer, Berlin, 1983) p. 43.
115. H. Wolf, R. Griser, U. Klocke, W. Riedel and R. Wissler, *Proc. 2nd Int. Symp. Monitor. Gaseous Pollutants Tunable Diode Lasers* (Freiburg, Reidel, 1988)
116. F. Sizov and A. Rogalski, *Prog. Quant. Electon.* **34** (2010) 278.
117. B. Ferguson and X.-C. Zhang, *Nat. Matr.* **1** (2002) 26.
118. M. Tonoichi, *Nat. Photon* **1** (2007) 97.
119. D. Mittleman, *Sensing with Teraheltz Radiation* (Springer, Berlin, 2003)
120. J.F. Butler, A.R. Calawa, R.J. Phelan, T.C. Harman, A.J. Strauss and R.H. Rediker, *Appl. Phys. Lett.* **5** (1964) 75.
121. S.H. Groves, K.W. Nill and A.J. Strauss, *Appl. Phys. Lett.* **25** (1974) 331.
122. L.R. Tomasetta and C.G. Fonstad, *Appl. Phys. Lett.* **25** (1974) 440.
123. W. Lo, G.P. Montgomery, Jr., D.E. Swets and J.C. Hill, *IEEE Trans. Electron Dev.* **ED-22** (1975) 1055.
124. K.J. Linden, K.W. Nill and J.F. Butler, *IEEE J. Quant. Electron.* **QE-13** (1977) 720.
125. K.L. Lindenm *Proc. SPIE* **438** (1983) 2.
126. D.L. Partin, J.P. Heremans and C.M. Thrush, *Superlatt. Microstr.* **2** (1986) 459.
127. M. Tacke, B. Spanger, A. Lanbrecht, P.P. Norton and H. Bottner, *Appl. Phys. Lett.* **53** (1988) 2260.
128. Z. Felt, M. McDonald, R.J. Woods, V. Archambault and P. Mak, *Appl. Phys. Lett.* **68** (1996) 738.
129. B. Spanger, U. Schiessl, A. Lambrecht, H. Boettner and M. Tacke, *Appl. Phys. Lett.* **53** (1988) 2582.
130. A. Yasuda, K. Suto and J. Nishizawa, *Extended Abstr.* (50th Spring Meeting, Jpn. Soc. Appl. Phys. and Related Socs., 2003) p. 1237.
131. S. Khosravani and Z. Shi, *Appl. Phys. Lett.* **78** (2001) 139.
132. F. Zhao, H. Wu, L. Jayasinghe and Z. Shi, *Appl. Phys. Lett.* **80** (2002) 1129.
133. J. Furst, H. Pashcer, T. Schwarzl, M. Boberl, W. Heissm G. Springholz and G. Bauer, *Appl. Phys. Lett.* **81** (2002) 208.
134. T. Schwarzl, G. Springholz, M. Boerl, E. Kaufmann, J. Roither and W. Heiss, *Appl. Phys. Lett.* **86** (2005) 031102.
135. F.Felder, M. Amold, M. Rahim, c. Ebneter and H. Zogg, *Appl. Phys. Lett.* **91** (2007) 101102.
136. M. Rahim, M. Arnold, F. Felder, K. Nehfar and H. Zogg, *Appl. Phys. Lett.* **91** (2007) 1511002.
137. I. Melngailis and T.C. Herman, in *Semiconductros and Semimetals Vol. 5*, ed. by R.K. Willadson and A.C. Berr (Academic Press, New York, 1970) p. 111.
138. T.C. Herman and I. Melngailis, in *Applied Solid Science*, ed. by R. Wolfe (Academic Press, 1974) p. 1.
139. M.V. Fazio, Doctor Thesis (Rice University, 1978).
140. A. Rogalski, in *Infrared Photon Detectors* (SPIE Optical Enginnering Press, Bellingham, 1995) p. 513.
141. L.H. DeVaux, H. Kimura, M.J. Sheets, F.J. Renda, J.R. Balon, P.S. Chia and A.H. Lockwood, *Infrared Phys.* **15** (1975) 271.
142. C.A. Kennedy, K.J. Linden and D.A. Soderman, *Proc. IEEE* **63** (1975) 27.

143. C.C. Wang and J.S. Lorenzo, *Infrared Phys.* **17** (1977) 83.
144. C.JC. Wang and M.E. Kim, *J. Appl. Phys.* **50** (1979) 3733.
145. H. Zogg, C. Maissen, J. Masek, T. Hoshino, S. Blunier and A.N. Tiwaru, *Semicon. Sci. Technol.* **6** (1991) C35.
146. T. Hoshino, C. Maissen, H. Zogg, J. Masek, S. Blunier, A.N. Tiwari, S. Teodoropol and W.J. Bober, *Infrared Phys.* **32** (1991) 169.
147. V.F. Chishko, v.T. Hyrapov, I.L. Kasatkin, V.V. Osipov and O.V. Smolin, *Infrared Phys.* **33** (1992) 275.
148. A. Ragalski and J. Rutkowski, *Infrared Phys.* **21** (1981) 191.
149. A. Rogalski and W. Larkowski, *Electron Technology* **18** (1985) 55.
150. A. Rogalski and R. Ciupa, *Opto-Electr. Rev.* **5** (1997) 21.
151. M. Arnold, D. Zimin and H. Zogg, *Appl. Phys. Lett.* **87** (2005) 141103.
152. D.R. Khokholov, I.I. Ivanchik, S.N. Raines, D.M. Watson and J.L. Pipher, *Appl. Phys. Lett.* 76 (2000) 2835.
153. A. Klimov, V. Shumsky and V. Kubarev, *Ferroelectrics* **347** (2007) 111.
154. J. König, J. Nurnus, R. Glahaar, H, Bottener and A. Lambrecht, M.S. Dressehous, M.Y. Lin, T. Koga, S.B. Cronin, Q. Rabin, M.R. Black and G. Dresselhous, *Semiconductors and Semimetals: Material Research III* (Academic Press, 2001) p. 1.
155. M.S. Dresselhous, M.Y. Lin, T. Koga, S.B. Cronin, O. Rabin, M.R. Black and G. Dresselhous, *Semiconducotrs and Semimetals: Materials Research III* (Academic Press, 2001) p. 1.
156. K.Qiu and A.C.S. Hayden, *J. Power Sources* **180** (2008) 884.
157. J. Yang and T. Cailla, *MRS Bull.* **31** (2006) 224.
158. Y. Golbstein Z. Dashevsky and M.P. Dariel, *Physica (Amsterdam)* **363B** (2005) 196.
159. Y. Dong, A. Malik and F.J. DiSalvo, *J. Solid. States Chem.* **183** (2010) 1817.
160. Y. Dong, M.A. McGuine, A. Malik, F.J. DiSalvo, *J. Solid. States Chem.* **182** (2009) 2602.
161. M.A. McGuine, A. Malik, F.J. SiSalvo, *J. Alloys Comp.* **460** (2008) 8.
162. J.H. Duhaish, *Physica (Amsterdam)* **322B** (2002) 20

25. CHALCOPYRITE COMPOUNDS

25.1 INTRODUCTION

Chalcopyrite crystals, I-III-VI_2 and II-IV-V_2 compounds, have semiconducting properties and are promising materials for solar cells[1,2], emitting devices[3], nonlinear optical devices[3-5] and optical filters.[5] This is because their bandgaps range from ultraviolet to the far infrared region, and they have low absorption coefficients and high electro-optical conversion coefficients. Their crystal structure is a chalcopyrite lattice structure as shown in Fig. 1.4 (Vol. 1). The main difficulty in their practical application stems from the difficulty in growing single crystals. The review of crystal growth, physical properties and applications of chalcopyrite crystals is found in Refs. 4 and 6-8.

25.2 PHYSICAL PROPERTIES

The physical properties of various chalcopyrite crystals are reviewed in Refs. 1-3, 6-11 and some of them are shown in Table 25.1. The lattice constants of various chalcopyrite crystals have been theoretically calculated and compared with experimental results.[12,13] In order to grow single crystals of chalcopyrite compounds, it is necessary to know their phase diagrams which have been studied by several authors.[5,14-20] In Figs. 25.1-5, the phase diagrams of $AgGaS_2$[5], $CuInSe_2$[14], $AgInTe_2$[14], $CuGaSe_2$[15] and $AgGaS_2$[16] are shown.

Table 25.1 Lattice Constants, Bandgap and Conductivity of Direct Transition Chalcopyrite Compounds

Compound	Lattice Constant (Å)		Bandgap (eV)	Melting Point (°C)
	a	c		
I-III-VI_2				
$CuInSe_2$	5.789	11.62	1.02	986
$CuAlS_2$	5.334	10.44	3.49	1302
$CuAlSe_2$	5.602	10.95	2.7	1202
$CuGaS_2$	5.356	10.43	2.49	1280
$CuGaSe_2$	5.614	11.03	1.68	1070
$AgGaS_2$	5.754	10.29	2.69	1040
II-IV-V_2				
$ZnSiP_2$	5.40	10.44	2.96	1370
$ZnGeP_2$	5.47	10.73	2.34	1025
$CdSiP_2$	5.68	10.88	2.45	1120
$CdGeP_2$	5.74	11.22	2.66	790

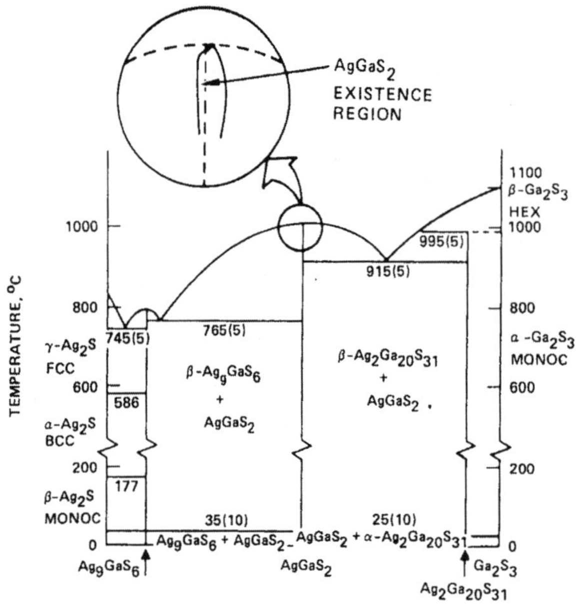

Fig. 25.1 Pseudobinary phase diagram of Ag_2S-Ga_2S_3 (reprinted from Ref. 5 with permission, copyright 1984 Elsevier).

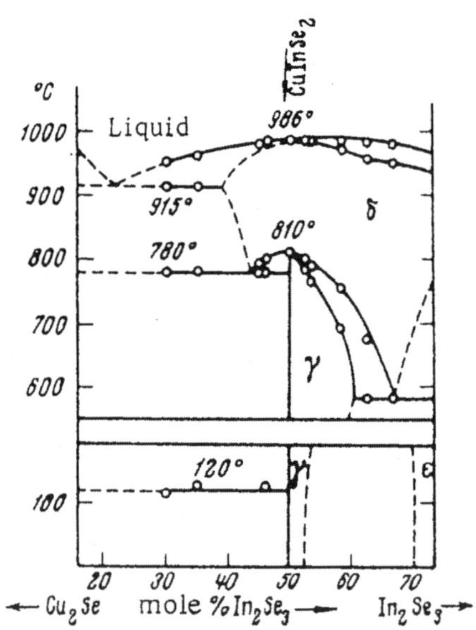

Fig. 25.2 Pseudobinary phase diagram of Cu_2Se-In_2Se_3 (reprinted from Ref. 14 with permission, copyright 1967 American Institute of Physics).

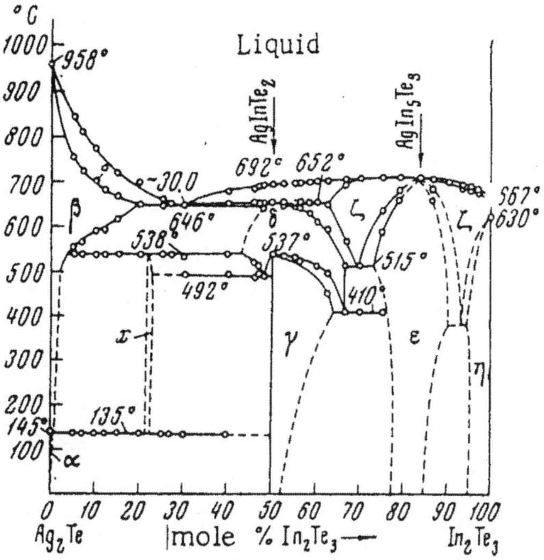

Fig. 25.3 Pseudobinary phase diagram of Ag_2Te-In_2Te_3 (reprinted from Ref. 14 with permission, copyright 1967 American Institute of Physics).

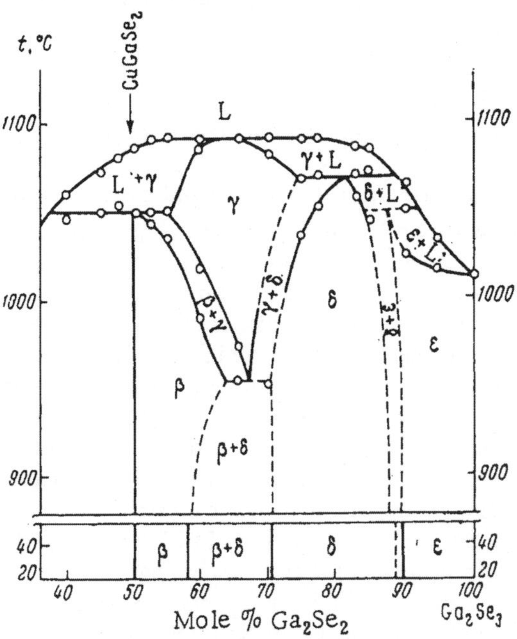

Fig. 25.4 Pseudobinary phase diagram of Cu_2Se-Ga_2Se_3 (reprinted from Ref. 15 with permission, copyright 1967 Nauka Publishing).

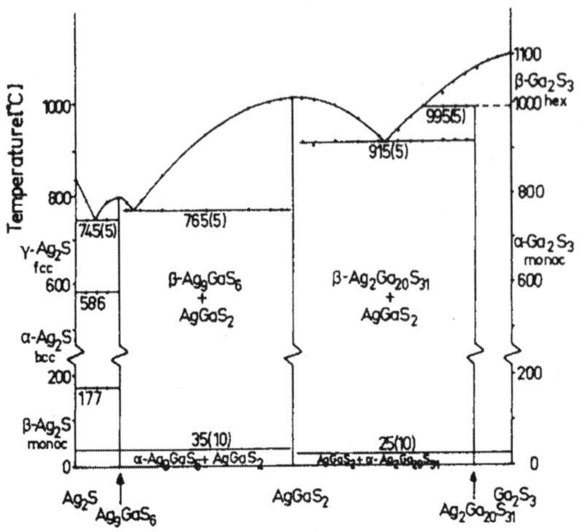

Fig. 25.5 Pseudobinary phase diagram of Ag_2Te-Ga_2S_3 (reprinted from Ref. 16 with permission, copyright 1976 Elsevier).

25.3 CRYSTAL GROWTH

Crystal growth of chalcopyrite compounds has been performed by various methods as summarized in Table 25.2.

25.3.1 Melt Growth

Various melt growth methods have been reviewed in Refs. 4, 7, 21 and 22.

(1) Gradient freezing, directional freezing and zone melting methods

Various authors have grown $CuInSe_2$ crystals by the directional freezing method[23-33] and by the zone melting method.[31, 32] Arsene et al.[33] applied a horizontal varying gradient zone freezing method as shown in Fig. 25.6. In the A zone, Se is set and in B zone, Cu and In are set, and the temperature gradient was varied as shown in the figure.

$AgInTe_2$ crystals have been grown by solidification from the melt. $AgInTe_2$ however has a zincblende phase (δ) similar to $CuInSe_2$ but does not have a eutectic melting point[14] because of the existence of a solid-liquid mixture phase as shown in Fig. 25.3. This phase diagram would suggest that it would be difficult to obtain large crystals.

Shunemann et al.[34] have grown quaternary crystals, $AgGaGeS_4$ and $AgGaGe_5Se_{12}$ by the horizontal gradient freezing method. These crystals can overcome the disadvantage of $AgGaS_2$ and $AgGaSe_2$, the former having a low laser damage threshold and the latter having insufficient birefringence for phase matching.

Bodbar et al.[35] have grown $CuIn_5Se_8$ single crystals, an example of I-III_5-VI_8 materials, which should be considered a new material. Starting components from high-purity elements were synthesized directly in an ampoule and crystals were grown by direct crystallization from the melt by cooling the furnace.

Table 25.2 Crystal Growth of Chalchopyrite Crystals

Method	Contents	Authors	Year	Ref.
GF	Growth of various I-III-VI_2 crystals	Austin et al.	1956	23
CVT	Growth of various I-III-VI_2 crystals using iodine reagent vapor transport	Honeyman et al.	1969 -1971	75 -76
ZL	Growth of $CuInSe_2$ by the zone leveling method	Parkes et al.	1973	31
SG	Growth of $CuGaS_2$ from In solution	Yamamoto et al.	1972	54
VB	Growth of $AgGaS_2$ of 13 mmϕ × 60 mm	Route et al.	1974	36
LEC	Growth of $AgGaS_2$ of 15 mm diameter and length 40 mm	Korczak et al.	1974	51
GF	Growth of $CuInSe_2$ after direct synthesis	Yu	1976	25
SG	Growth of II-V-VI_2 crystals by ACRT from Sn or Pb solution	Pamplin	1974, 1975	65, 66
VB	Growth of $AgGaS_2$ measuring 5 × 5 × 5 mm³	Matthes et al.	1975	37
SG	Growth of $CuAl_2$ from In solution	Yamamoto et al.	1975	55
CVT	Growth of $CuGaS_2$ using iodine reagent	Yamamoto et al.	1975	77
SG	Growth of $CuGaS_2$ from Sn solution	Plovnik et al.	1975	59
ZM	Growth of $CuInSe_2$ by the zone melting method	Tell et al.	1977	32
VB	Growth of $CuInSe_2$ of 10 mmϕ × 50 mm	Haupt et al.	1977	38
CVT	Growth of $CuInSe_2$ of 10 × 10 × 2 mm³ using iodine reagent	Haupt et al.	1977	38
GF	Growth of $CuInSe_2$ after direct synthesis	Hörig et al.	1977	26
LEC	Growth of $ZnGaP_2$ and $CuInSe_2$	Buehler	1978	52
GF	Growth of $CuInSe_2$ after direct synthesis	Irie et al.	1979	27
SG	Growth of $CuGaS_2$ from In solution	Höbler et al.	1980	56
VB	Growth of $CuInSe_2$ by the horizontal method.	Sobotta et al.	1981	39
SG	Growth of $CuGaS_2$ from Pb and Sn solutions	Höbler et al.	1981	60
flux	Growth of $CuGaS_2$ from $CdCl_2$ and CdI_2	Höbler et al.	1981	61
GF	Growth of $CuInSe_2$ from nearly stoichiometric melt	Bachmann et al.	1984	28
THM	Growth and characterization of $CuInSe_2$	Hsu et al.	1984	70
VB	Growth of $CdGa_2S_4$ of 28 mmϕ × 16 mm	Feigelson et al.	1985	40
SG	Growth of $CuGaS_2$ from Te, Te-Cu and Te-Cu-S solutions	Hirakawa et al.	1985	62
LEC	In situ synthesis and growth of $CuInSe_2$	Ciszek et al.	1985	53
GF	Growth of $CuInSe_2$ by the normal freezing	Endo et al.	1986	29
VB	Growth of $AgGaS_2$ of 28 mmϕ × 100 mm and $AgGaSe_2$ of 37 mmϕ × 100 mm	Feigelson et al.	1987	4
SG	Growth of $CuGaSe_2$ from In solution	Sugiyama et al.	1987	57
SSR	$CuInSe_2$ growth at 1050°C for 60 hrs and characterizaion	Miyauchi et al.	1988	82
CVT	Growth of $CuAl_xIn_{1-x}S_2$ using iodine reagent	Miyake et al.	1988	80
VB	Growth of $CdGa_2Se_4$ of 7 × 7 × 4 mm²	Okudo et al.	1989	41
THM	Growth of $CuGaSe_2$ at 890-980°C	Sugiyama et al.	1989	72
VB	Application of ACRT for the growth of $CuInSe_2$ Crystals of 50 mm² × 5 mm	Weng et al.	1989	42
CVT	Growth of $CuAlS_{1-x}Se_2$ using iodine reagent	Iwasaki et al.	1988	79
SSR	Growth of $CuInSe_2$ and $CuInTe_2$ crystals	Yamamoto et al.	1989, 1990	83, 84
THM	Growth of $CuInSe_2$ from In solution	Yamamoto et al.	1990	71
THM	Growth of $CuGaS_2$ from In solution	Miyake et al.	1990	73
VB	Tapering of crucibles and seeding for growth of $AgGaSe_2$ with square shaped crystals	Feigelson et al.	1990	43
SSR	Growth of various I-III-VI_2 crystals	Yamamoto et al.	1990	85
THM	Growth of various Cu-III-V_2 crystals	Miyake et al.	1993	74
others	Selenization of liquid Cu-In alloy	Nomura et al.	1993	87

Table 25.2 Crystal Growth of Chalchopyrite Crystals (continued)

Method	Contents	Authors	Year	Ref.
HB	Growth of $CuInSe_2$ with selenization by the 2T-HB method	Matsushita et al.	1993-1996	44, 45
HGF	Growth of $CuInSe_2$ with temperature gradient control	Arsene et al.	1996	33
SG	Growth of $CuGaS_2$ from In solution by the Cz method	Tanaka et al.	1997	64
SG	Growth of $AgCd_2GaS_4$ from $AgGaS_2$ solvent	Olekseyuk et al.	2005	63
VB	Growth of $LiMX_2$ (M = Al, In, Ga and X = S, Se, Te) and their optical properties	Iserenko et al.	2003-2005	46-48
VB	Growth of $CuInSe_2$ from stoichiometric and non-stoichiometric charges	Lachab et al.	2005	49
SG	Growth of P doped $CuInS_2$ from In solution	Yamaguchi et al.	2005	58
others	Preparation of $CuInS_2$ by microwave heating process	Susaki	2005	88
HGF	Growth of $AgGaGeS_4$ and $AgGaGe_5Se_{12}$	Schunemann et al.	2006	34
VB	Growth of $LiGaTe_2$ and their optical properties	Petroc et al.	2006	50
GF	Growth of $CuIn_5Se_8$	Bodnar et al.	2006	35
SG	Growth of $CuIn(S_{1-x}Se_x)_2$ from TEA	Chavhan et al.	2006	67
others	Preparation of $CuAlGa_{1-x}Se_2$ by direct fusion	Castro et al.	2006	89
CVT	Growth of $CuGaS_2$ using iodine reagent	Prabukanthan et. al.	2007	81
others	Growth of $AgGaSe_2$ by a hot-press method	Kinoshita et al.	2007	86

GF: Gradient Freezing, HB: Horizontal Bridgman, HGF: Horizontal Gradient Freezing, SG: Solution Growth, VB: Vertical Bridgman, CZ:Czochralski, LEC: Liquid Encapsulated Czochralski, ZL: Zone Leveling, ZM: Zone Melting, THM: Travelling Heater Method, CVT: Chemical Vapor Transport, SSR: Solid State Reaction

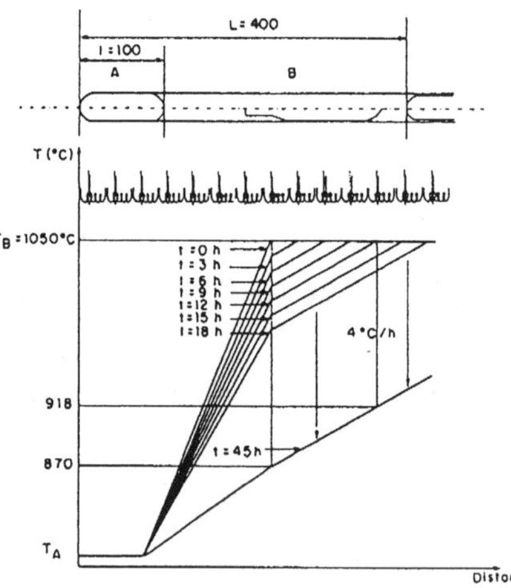

Fig. 25.6 Schematic diagram of the multiple zone furnace. The position of the ampoule inside the furnace and the variation of the temperature and gradient with time in zone B (reprinted from Ref. 33 with permission, copyright 1996 Elsevier).

(2) Bridgman method

Various chalcopyrite crystals have been grown by the Bridgman method.[4, 36-50] Matthes et al.[37] have grown $AgGaS_2$ single crystals by the vertical Bridgman method as shown in Fig. 25.7. $AgGaS_2$ polycrystals were synthesized from AgS_2 and GaS_2 prior to crystal growth, and the polycrystals were charged in a quartz crucible and sealed in a quartz ampoule. Since the thermal expansion coefficients along the a axis and the c axis are very different, the quartz cracks so that a double quartz structure is necessary. In order to prevent the quartz from cracking, it is important to grow crystals parallel to the c axis and to achieve this, seed crystals cut along the c axis are set in the bottom of the crucible.[4] The inner diameter of the crucible was also tapered from bottom to top to reduce the stress to the grown crystals.[4] $AuGaSe_2$ crystals were also grown by the same method.[43] Feigelson et al.[4] have grown twin- and crack-free single crystals of $AgGaSe_2$ and $AgGaS_2$ as shown in Fig. 25.8.

Weng et al.[42] have grown $CuInSe_2$ single crystals by the vertical Bridgman method, using an accelerated crucible-rotation technique (ACRT) for better mixing of the melt during growth. Crystals without voids or cracks as large as 50 mm^2 and of thickness greater than 5 mm could be obtained

Matsushita et al.[44,45] have grown $CuInSe_2$ single crystals by selenization using the horizontal Bridgman method with two temperature zones (2T-HB) as shown in Fig. 25.9 in order to control the Se vapor pressure. Hall mobility was maximum and the carrier concentration was minimum at about 20 Torr of Se vapor pressure.

$LiGaTe_2$ crystals, with the potential for frequency doubling of CO_2 lasers operating at 10.6 μm, have been grown by the vertical Bridgman method. Crystals were grown at a rate of 2-5 mm/day and typical axial temperature gradients of 2-4°C/mm. Maximum sizes of the grown crystals were about 0.5 cm^3.[50]

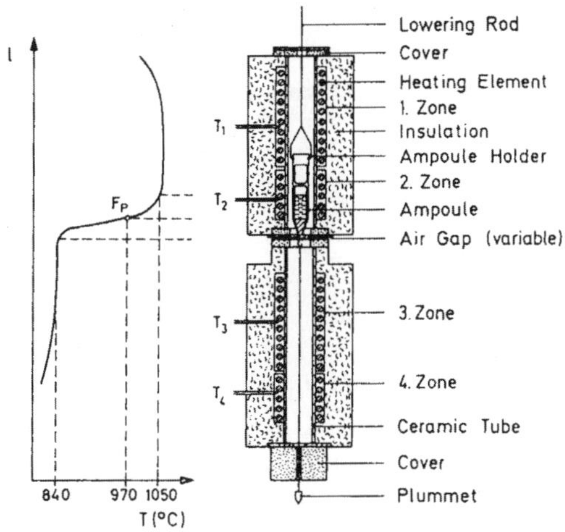

Fig. 25.7 Bridgman furnace for the crystal growth of $AgGaS_2$ and the temperature profile (reprinted from Ref. 37 with permission, copyright 1975 EDP Sciences).

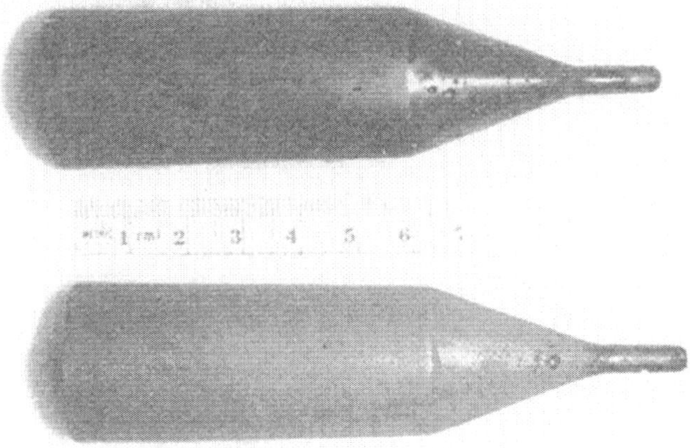

Fig. 25.8 28mm twin- and crack-free boules of $AgGaSe_2$ (top) and $AgGaS_2$ (bottom) (reprinted from Ref. 4 with permission, copyright 1987 International Society for Optics and Photonics).

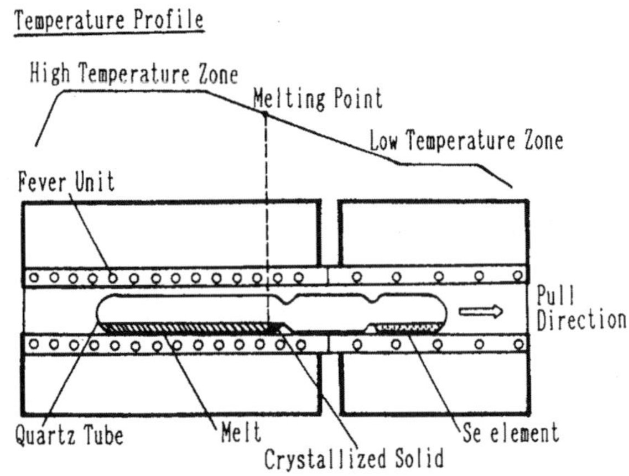

Fig. 25.9 Schematic diagram of electric furnace used for 2T-HB method (reprinted from Ref. 44 with permission, copyright 1993 Elsevier).

(3) Liquid Encapsulated Czochralski (LEC) method

Korczak and Staff[51] have grown $AgGaS_2$ crystals by the LEC method using the furnace configuration shown in Fig. 25.10. In order to overcome the difficulty in controlling the diameter due to the shallow radial temperature gradient in the encapsulated melt, the crucible base was cooled using a gas jet directing a stream of nitrogen onto it. Using this enhancement, crystals of uniform shape up to 40 mm in length and 15 mm in diameter were obtainable (Fig. 25.11).

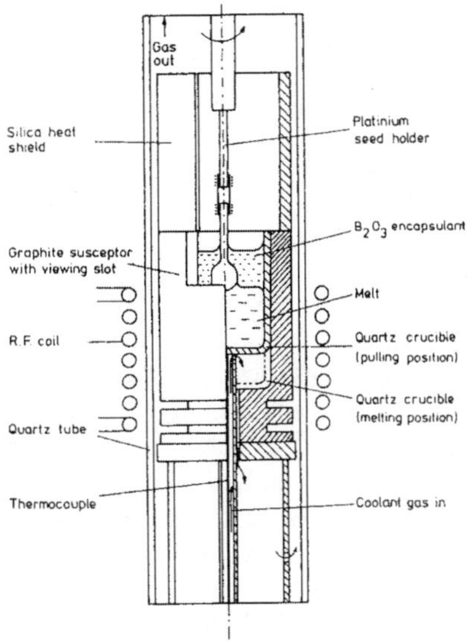

Fig. 25.10 LEC crystal growth system with crucible-base cooling (reprinted from Ref. 51 with permission, copyright 1974 Elsevier).

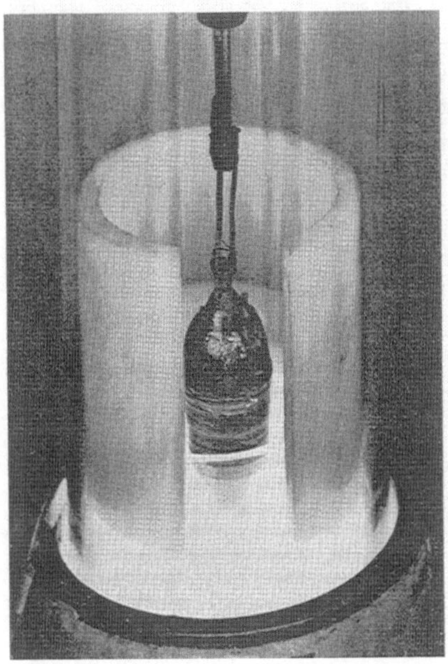

Fig. 25.11 Single crystal boule of $AgGaS_2$ in the LEC growth system (reprinted from Ref. 51 with permission, copyright 1974 Elsevier).

Buehler[52] has developed a small scale LEC high pressure furnace and has grown $ZnGeP_2$ and $CuInS_2$ single crystals. Ciszek[53] has performed the in-situ synthesis of $CuInSe_2$ in the crucible and has grown crystals by lowering the crucible by the Bridgman method or by the LEC method.

25.3.2 Solution Growth

(1) Solution growth method

Solution growth has been studied for $CuGaS_2$, using various solvents such as In [54-58], Sn [59,60], Pb [60], $CdCl_2$ [61], CdI_2 [61], Te [62], Te-Cu [62], Te-Cu-S [62] and $AgGaS_2$.[63] In Fig. 25.12, the solubility of $CuGaS_2$ in various solvents is shown.[61] The growth of $CuInS_2$[54] and $CuAlS_2$[55] is also reported using In as the solvent. Tanaka et al.[64] have grown $CuGaS_2$ crystals by the Czochralski (CZ) method using a GaAs seed crystal. Pamplin[65] has reviewed the growth of chalcopyrite crystals by accelarated crucible rotation technique (ACRT) solution growth. An ACRT solution growth arrangement is shown in Fig. 25.13. Many references to the growth of II-VI-V_2 crystals are cited in Ref. 65. $ZnCdGe_2P_4$ crystals have been grown by ACRT solution growth.[66] Chavhan and Sharma[67] have grown $CuIn(S_{1-x}Se_x)_2$ thin films from solutions including triethanolamine (TEA). Liquid phase epitaxial (LPE) growth of $CuGaSe_2$[68] and $CuGaS_2$[69] are also reported.

(2) Traveling Heater Method (THM)

The THM method using In as the solvent for the growth of I-III-VI_2 was first applied to $CuInS_2$.[70] Yamamoto et al.[71] have grown $CuInS_2$ crystals by this method. In is charged at the bottom of the quartz tube which is given a conical shape, and $CuInS_2$ polycrystals synthesized from the melt are set on In. The quartz ampoule is then lowered in the three

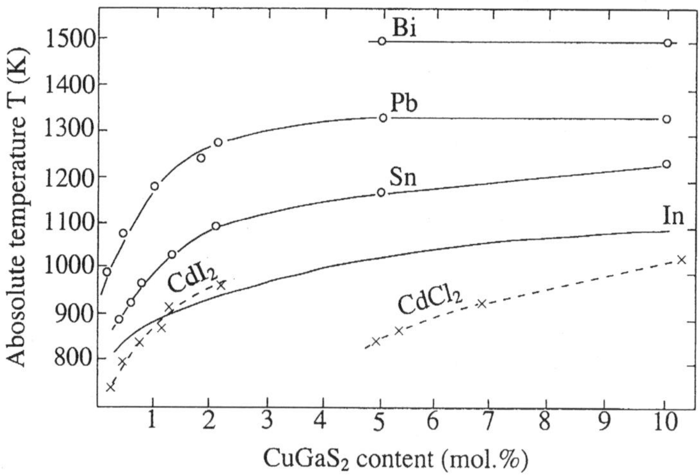

Fig. 25.12 Solubilities of $CuGaS_2$ in various solvents (reprinted from Ref. 61 with permission, copyright 1981 The Japan Society of Applied Physics).

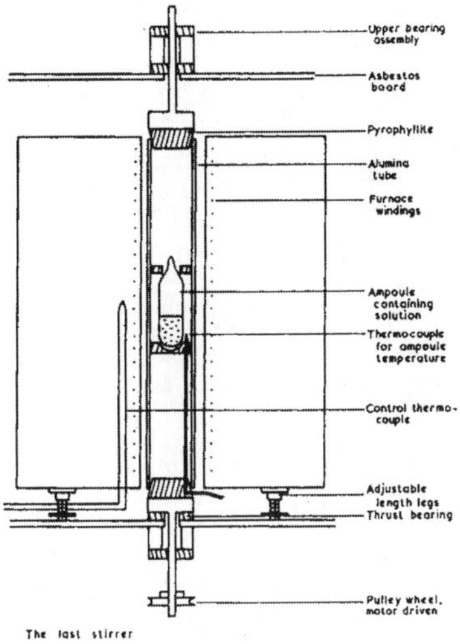

Fig. 25.13 Furnace for accelerated crucible rotation of molten metal soulution growth technique (reprinted from Ref. 65 with permission, copyright 1974 Elsevier).

zone furnace as shown in Fig. 25.14 at a rate of 5-10 mm/day. When the nucleation at the conical part is successful, single crystals can be obtained. In solvent is included in the grown $CuInS_2$ crystals but this inclusion is avoided when the growth rate is less than 10 mm/day.

$CuGaSe_2$ crystals have been grown by THM using In as the solvent.[72] In this case, In was included as the secondary phase or In replaced Ga atoms in the crystals making quaternary $CuGa_xIn_{1-x}Se_2$ crystals. When the growth rate was about 5 mm/day, the inclusion of In was eliminated but the replacement of Ga atoms by In atoms occurred with $1-x = 0.04-0.07$.

$CuGaS_2$ crystals were also grown by the THM using In as the solvent.[71,73] In order to start the nucleation from the center of the melt, it is desirable to make the solid-liquid interface concave upwards. Since In solvent is wetting to the quartz surface, the solid-liquid interface tends to be concave downwards. When carbon was coated on to the quartz surface the solid-liquid interface shape could be made concave upwards. Using this interface shape, the cracking of the quartz tube due to the anisotropy of the thermal expansion coefficient was avoided. Nucleation from the center of the melt could also be achieved by making the temperature of the center of the furnace lower than the periphery.[71]

The THM has been applied to grow various Cu-III-V_2 such as $CuGaS_2$, $CuGaSe_2$, $CuInS_2$ and $CuInSe_2$.[74] It was shown that using In as the solvent, quaternary alloys could been grown.

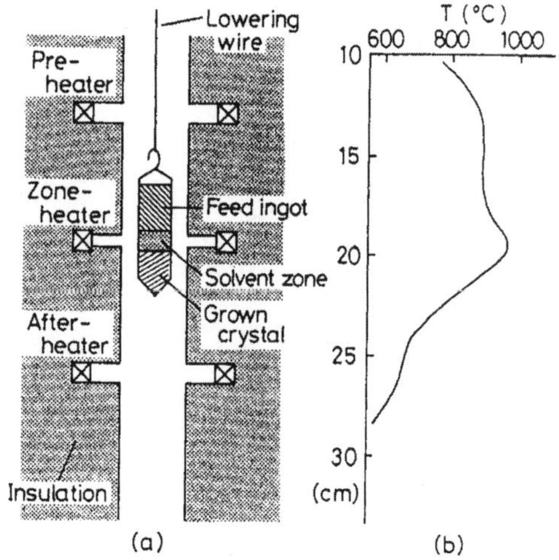

Fig. 25.14 Three zone furnace for THM growth method and the temperature profile (reprinted from Ref. 73 with permission, copyright 1993 The Japan Society of Applied Physics).

25.3.3 Vapor Phase Growth

Iodine is used as the agent in the chemical vapor transport method for growing I-III-VI_2 small single crystals.[38, 75-81] As starting materials, a mixture of three elements, or of binary compounds or the target compound synthesized by the solidification from a melt are used. In a quartz ampoule, the startingt material and iodine of 5-20 mg/cm^3 are sealed and set in a two zone furnace, with the starting material in a higher temperature zone (700-1000°C) kept for several days. Small single crystals of several mm could be grown as needlelike crystals or platelets.

25.3.4 Solid State Reaction (SSR)

The solid state reaction method was first applied for growing $CuInS_2$ crystals.[82-84] Cu_2Se and In_2Se_3 powders were crushed to 1 μm fine grains and were compacted by 1 ton/cm^2 pressing. These compacts were charged in graphite containers as shown in Fig. 25.15 and sintered in the δ phase temperature region. (112) oriented crystals consisting of 90% single crystals could be obtained. This method was applied also for $CuInTe_2$ and other I-III-VI_2 crystal growth.[83,85]

25.3.5 Others

Kinoshita et al.[86] have grown $AgGaSe_2$ crystals by the hot-press method. Mixed materials Ag_2Se and Ga_2Se_3 powders were compressed to 25 MPa at growth temperatures from 400°C to 700°C. $AgGaSe_2$ single phase was obtained at 700°C. All samples were polycrystalline and n-type conductivity .

Nomura and Takizawa[87] have selenized liquid Cu-In alloy for growing $CuInSe_2$ single crystals. Selenization temperatures above the melting point of $CuInSe_2$ were in-

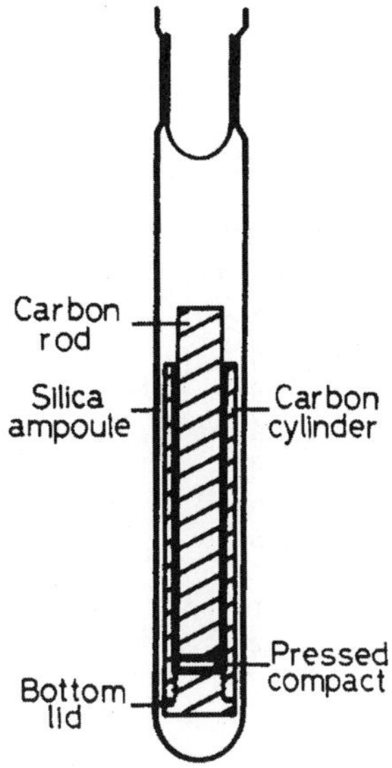

Fig. 25.15 Setting of two pressed compacts in closely fitted carbon cylinder with capped carbon rods (reprinted from Ref. 71 with permission, copyright 1988 The Japan Society of Applied Physics).

dispensable to achieve a single phase synthesis while preventing additional phases such as Cu_7Se_4 and In_2Se from developing.

Susaki[88] has grown $AgGaSe_2$ crystals by the microwave heating process. Castro et al.[89] have grown $CuAl_xGa_{1-x}Se_2$ alloys by direct fusion of the stoichiometric mixture of the elements.

25.4 CHARACTERIZATION

25.4.1 Defects

Point defects in chalcopyrite crystals have a serious effect on the electrical properties. Wasim[90] have shown various possible intrinsic defects in $CuInSe_2$ as shown in Table 25.3. Müller et al.[91] have calculated the formation energies of vacancies and antisite defects in $CuInSe_2$ crystals and discussed their effect on the electrical properties. Radautsan et al.[92] made a theoretical study of the point defects in $II\text{-}III_2VI_4$ compounds. Kerr et al.[93] have evaluated native defects in CIS ($CuInSe_2$) and CIGS($Cu(In, Ga)Se_2$) by deep level transient spectroscopy (DLTS) and found differences in defect properties. Employing the generalization of Van Vechten's cavity model, formation energies of neutral point defects in chalcopyrites have been estimated.[94] The obtained data can be used to calculate point defect concentrations and homogeneity ranges as a function of partial pressure and temperature. Lany and Zunger[95,96] have found a new local atomic reconstruction mode around anion vacancies resulting from the formation of metal dimers and predicted persistent hole photoconductivity caused by V_{Se} in $Cu\text{-}III\text{-}Se_2$ chalcopyrites.

Table 25.3 Various Possible Intrinsic Defects in $CuInSe_2$ and the Corresponding Formation Energies and Electrical Activites (reprinted from Ref. 90 with permission, copyright 1974 Elsevier)

Intrinsic defect	Formation energy (eV)	Electrical activity
In_{Cu}	1.4	donor
Cu_{In}	1.5	acceptor
V_{Se}	2.4	donor-acceptor
V_{Cu}	2.6	acceptor
V_{In}	2.8	acceptor
Cu_i	4.4	donor
In_{Se}	5.0	donor
Se_{In}	5.5	acceptor
Cu_{Se}	7.5	acceptor
Se_{Cu}	7.5	donor
In_i	9.1	donor
Se_i	22.4	acceptor

25.4.2 Electrical Properties

In most chalcopyrite crystals, the electrical properties can be controlled by controlling anion defect densities by annealing. In the case of $CuInSe_2$, donor and acceptor levels have been determined from the temperature dependence of Hall measurements as shown in Fig. 25.16.[97] It is known that when the crystal is In rich, it becomes n-type and when Se rich, it becomes p-type.[98] In fact, Tell et al.[99] have studied the effect of annealing on the electrical properties of various chalcopyrite crystals as shown in Table 25.4. Wasim[90] have reviewed the electrical properties of $CuInSe_2$ single crystals, based on defect chemistry considerations and the formation energy of the predominant electrically active intrinsic defects. Neumann and Tomlinson[100] have studied the relationship between electrical properties and the composition of $CuInSe_2$ single crystals.

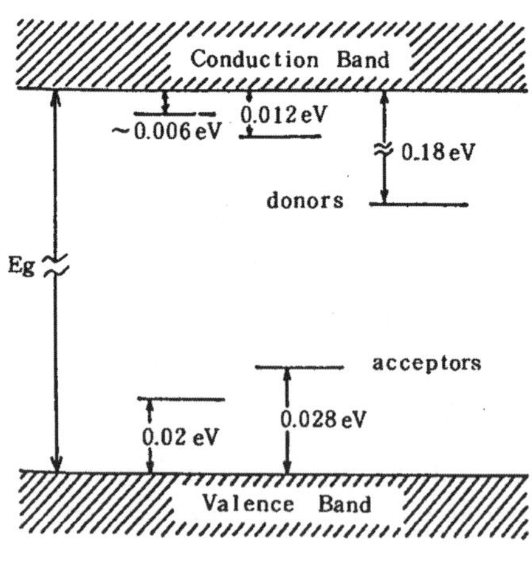

Fig. 25.16 Donor and acceptor levels in $CuInSe_2$ single crystals, calculated from the tempeature dependence of Hall coefficients (from Ref. 97 with permission).

Table 25.4 Room Temperature Electrical Properties of I-III-VI_2 Compounds (reprinted from Ref. 99 with permission, copyright 1972 American Institute of Physics).

Compound	Energy gap (eV)	Annealed under maximum S or Se pressure type	ρ (Ω•cm)	C/C cm^{-3}	μ(cm^2/V•sec	Annealed under minimum S or Se pressure type	ρ (Ω•cm)	C/C cm^{-3}	μ(cm^2/V•sec
$CuAlS_2$	3.5	p	10^2-10^3	$>3 \times 10^{15}$	<2		$>10^5$		
$CuAlSe_2$	2.7	p	10^2-10^3	$>1 \times 10^{16}$	~1		$>10^5$		
$CuGaSe_2$	2.5	p	1	4×10^{17}	15		$>10^5$		
$CuGaSe_2$	1.7	p	0.05	5×10^{18}	20		$>10^5$		
$CuInS_2$	1.5	p	5	1×10^{17}	15	n	1	3×10^{16}	200
$CuInSe_2$	0.8	p	0.5	1×10^{18}	10	n	0.05	4×10^{17}	320
$AgGaS_2$	2.7		$>10^5$				$>10^5$		
$AgInS_2$	2.0		$>10^5$			n	10	4×10^{15}	150
$AgGaSe_2$	1.8		$>10^5$				$>10^5$		
$AgInSe_2$	1.2	n	10^4	8×10^{11}	750	n	0.02	5×10^{17}	600

25.4.3 Optical Properties

Tell et al.[99] have measured of the various optical properties of $CuGaS_2$ and $CuInS_2$. Hara et al.[101] have studied the photoluminescence of chalcopyrites grown by metal organic chemical vapor deposition (MOCVD). Abou-Elfotouh et al.[102] and Dagan et al.[103] have systematically studied point defects in $CuInSe_2$ by photoluminescence and assigned the origin of various photoluminescence peaks. An example of photoluminescence (PL) spectra of n-type and p-type $CuInSe_2$ is shown in Fig. 25.17. Ozaki et al.[104] have proposed an energy-band scheme for the explanation of PL emission observed in $CdGa_2Te_4$. Wakita et al.[105] have examined the time-resolved PL of free and bound excitons in $CuInS_2$ single crystals grown by THM. Yakushev et al.[106] have examined the PL and reflectance in high quality $CuInS_2$ single crystals grown by THM and found a number of sharp peaks associated with excitons. Siebentritt et al.[107] have studied the reconciliation of luminescence on the $CuGaSe_2$.

25.5 APPLICATIONS

Potential applications of chalcopyrite crystals are solar cells, emitting devices, nonlinear optical devices and optical filters as reviewed in Refs. 1-8.

25.5.1 Solar Cells

$CuInSe_2$ which is also called CIS is a good material for solar cells since it has a large absorption coefficient compared with silicon and GaAs and can absorb most of the solar spectrum within about 1 μm thickness.[108] CIS applications have therefore been extensively studied[109-112] in thin film solar cells. Recently, $CuGaInSe_2$ mixed crystals known as CIGS have also been studied for improved high efficient solar cells.[113-117] Other mixed crystals such as $CuInAlSe_2$ have also been studied.[118]

25.5.2 Short Wavelength Emitting Devices

I-III-VI_2 compounds whose bandgap exceeds 2.4 eV are $CuAlS_2$, $AgGaS_2$, $CuAlSe_2$ and $CuGaS_2$. Among them, $AgGaS_2$ is not conductive and $CuAlSe_2$ is difficult to grow crys-

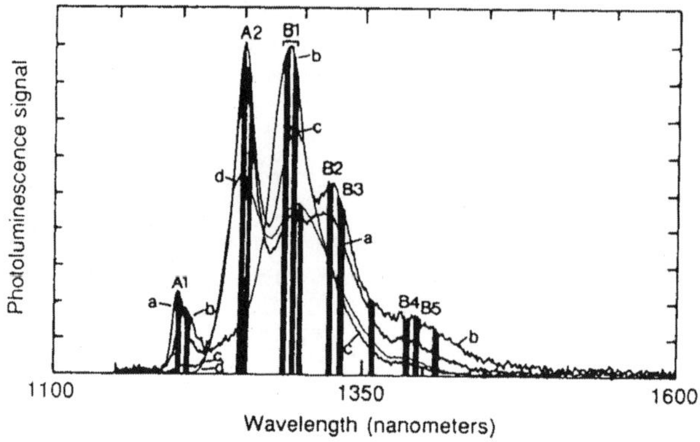

(a)

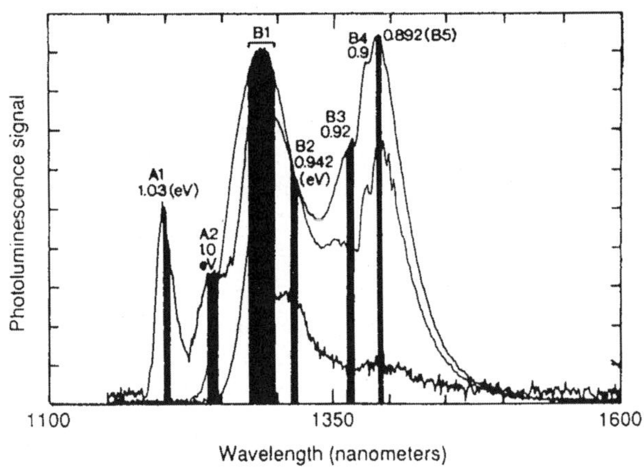

(b)

Fig. 25.17 Photoluminescence spectra of (a) n-type $CuInSe_2$ and (b) p-type $CuInSe_2$. The assignment of each peak is given in the reference (reprinted from Ref. 102 with permission, copyright 1989 Elsevier).

tals. $CuAlS_2$ whose bandgap is 3.6 eV and $CuGaS_2$ whose bandgap is 2.4 eV are promising as short wavelength emitting devices. These crystals have direct transition so that if p-n junctions can be achieved, emitting devices can be fabricated. It is however difficult to obtain n-type crystals.

25.5.3 Nonlinear Optical Devices

Conventional nonlinear optical crystals are $LiNbO_3$ or $LiIO_3$ but their transparent region is limited to 5 μm. I-III-VI_2 compounds which satisfy the phase matching conditions and are applicable as nonlinear optical devices are $AgGaSe_2$ and $AgGaS_2$. They are transparent up to 17 μm and 12 μm respectively and can be used in CO_2 lasers.[118,119]

$AgGaSe_2$ can be used in second harmonic generators (SHGs) for fundamental wavelengths from 3.1 to 13 μm[4], three wave mixing for wavelength as short as 1.2 μm[4] and optical parametric oscillators.[120]

Since $AuGaS_2$ has short wavelength transparency up to 0.5 μm, it is possible to make infrared-visible conversion by nonlinear mixing.[121,122]

25.5.4 Narrow Range Optical Filters

I-III-VI_2 compounds normally have birefringence and optical rotary power. $InCuAlS_2$, $AgGaS_2$, $CuAlSe_2$ $CuGaS_2$ and $AgGaSe_2$, the birefringence changes from positive to negative in their transparent region and the birefringence vanishes at room temperature at 370 nm, 497 nm, 536 nm, 642 nm and 811 nm, respectively.[123] By using this property, it is possible to fabricate selective transparent optical filters with the narrow bandpass width of 0.5-3 nm for these specific wavelength.[124]

REFERENCES

1. A. Rockett and R.W. Birkmire, *J. Appl. Phys.* **70** (1991) R81.
2. S. Nomura, S. Endo and T. Irie, *Jpn. Soc. Appl. Phys. (in japanese)* **55** (1986) 795.
3. A. Iida and A. Ooe, *Jpn. Soc. Appl. Phys. (in japanese)* **57** (1988) 871.
4. R.L. Feigelson and R.K. Route, *Opt. Engineer.* **26** (1987) 113.
5. A.L. Gentile, *Prog. Crystal Growth Chrater.* **10** (1985) 241.
6. R.S. Feigelson, *J. Phys.* **36** (1975), C3-57.
7. J.L. Shay and J.H. Wernick, *Ternary Chalcopyrite Semiconductors: Growth, Electronic Properties and Applications* (Pergamon Press, Oxford, 1975)
8. N. Yamamoto, *Towards New Functional Semiconductor Materials (in japanese)* (IPC, Tokyo, 1989)
9. S. Endo, *Physics (in japanese)* **8** (1987) 441.
10. *Landolt-Börnstein New Series. III-17h, Physics of Ternary Compounds*, ed. O. Madelung (Springer-Verlag, 1985).
11. B.R. Pamplin, T. Kiyosawa and K. Matsumoto, *Prog. Crystal Growth Charact.* **1** (1979) 331.
12. S. Isomura, *Jpn. Soc. Appl. Phys. (in japanese)* **43** (1974) 1184.
13. J.E. Jaffe and A. Zunger, *Phys. Rev. B* **29** (1984) 1882.
14. L.S. Palatnik and E.I. Rogacheva, *Sov. Phys. Dokl.* **12** (1967) 503.
15. L.S. Palatnik and E.K. Belova, *Sov. Phys. Inorg. Mater.* **3** (1967) 1914.
16. G. Brandt and V. Kramer, *Mater. Res. Bull.* **11** (1976) 1381.
17. J.J. Binsma, L.J. Giling and J. Bloem, *J. Crystal Growth* **50** (1980), 429.
18. J.C. Mikkelsen, Jr., *J. Electron. Mater.* **10** (1981) 541.
19. M.L. Fearheiley, *Solar Cells* **16** (1986) 91.
20. U.C. Boehnke and G. Kuhn, *J. Mater. Sci.* **22** (1987) 1635.
21. J.S. Shah, *Prog. Crystal Growth Charact.* **3** (1981) 333.
22. R.D. Tomlinson, *Solar Cells* **16** (1986) 17.
23. I.G. Austin, C.H.L. Goodman and A.E. Pengelly, *J. Electrochem. Soc.* **103** (1956) 609.
24. J.L. Shay, B. Tell, H.M. Kasper and L.M. Schiavone, *Phys. Rev. B* **8** (1973) 4485.
25. P.W. Yu, *J. Appl. Phys.* **47** (1976) 677.
26. W. Hörig, H. Neumann, H.J. Hobler and G. Kuhn, *phys. stat. sol. (b)* **80** (1977) K21.
27. T. Irie, S. Endo and S. Kimura, *Jpn. J. Appl. Phys.* **18** (1979) 1303.
28. K.J. Bachmann, M. Gearheley, Y.H. Shing and N. Tran, *Appl. Phys. Lett.* **4** (1984) 407.
29. S. Endo, T. Irie and H. Nakanishi, *Solar Cells* **16** (1986) 1.
30. P. Migliorato, J.L. Shay and H.M. Kasper, *J. Electron. Mater.* **4** (1975) 209.
31. J. Parkes, R.D. Tomlinson and M.J. Hampshire, *J. Crystal Growth* **20** (1973) 315.
32. B. Tell and P.M. Bridenbaugh, *J. Appl. Phys.* **48** (1977) 2477.
33. M.A. Arsene, A. Albacete, F. Voillot, J.P. Peyrade, A. Barra, J. Galiebrt, S.M. Wasim and E. Hernandez, *J. Crystal Growth* **158** (1996) 97.
34. P. G. Schunemann, K.T. Zawilski and T.M. Pllak, *J. Crystal Growth* **287** (2006) 248.
35. I.V. Bodbar, I.T. Bodnar, V.F. Gremenok, A.M. Kovalchuk and M. Leon, *J. Crystal Growth* **293** (2006) 324.

36. R.K. Route, R.S. Feigelson and R.J. Raymakers, *J. Crystal Growth* **24/25** (1974) 390.
37. H. Matthes, R. Viehmann, N. Marschall and P. Korczak, *J. Phys.* **C-3, Suppl.** (1975) 105.
38. H. Haupt and K. Hess, *Inst. Phys. Conf. Ser.* **35** (1977) 35.
39. H. Sobotta, H. Neumann, W. Kissinger, G. Kuhn and V. Riede, *phys. stat. sol. (b)* **103** (1981) K125.
40. R.S. Feigelson and R.K. Route, *Prog. Crystal Growth Charact.* **10** (1985) 189.
41. T. Okudo, A. Uemura, H. Horinaka and N. Yamamoto, *36th Spring Meeting, Jpn. Soc. Appl. Phys. and Related Socs.* **I** (1989) 349.
42. W.S. Weng, L.S. Yip, I. Shih and C.H. Champness, *Can. J. Phys.* **67** (1989) 294.
43. R.L. Feigelson and R.K. Route, *J. Crystal Growth* **104** (1990) 789.
44. H. Matsushita, S. Endo, T. Irie and H. Nakanishi, *J. Crystal Growth* **128** (1993) 655.
45. H. Matsushita and T. Takizawa, *J. Crystal Growth* **160** (1996) 71.
46. L. Isaenko, A. Yelisseyev, S. Lobanov, A. Titov, V. Petrov, J.-J. Zondy, P. Krinitsin, A. Merkulov, V. Vedenyapin and J. Smirnova, *Cryst. Res. Technol.* **38** (2003) 379.
47. L. Isaenko, I. Vasilyeva, A. Merkulov, A. Yelisseyev, S. Lobanov, *J. Crystal Growth* **275** (2005) 217.
48. L. Isaenko, P. Krinitsin, V. Vedenyapin, A. Yelisseyev, A. Merkulov, J.-J. Zondy, V. Petrov and V. Petrov, *Crystal Growth Des.* **5** (2005) 1325.
49. M. Lachab, A.A. Attia and C. Linares, *J. Crystal Growth* **280** (2005) 474.
50. V. Petrov, L. Isaenko, A. Yeliseyev, P. Krinitsin, V. Vedenyapin, A. Merkulov and J.-J. Zondy, *J. Non-Crystal. Solids* **352** (2006) 2434.
51. P. Korczak and C.B. Staff, *J. Crystal Growth* **24/25** (1974) 386.
52. E. Buehler, *J. Crystal Growth* **43** (1978) 584
53. T.F. Ciszek, *J. Electron. Mater.* **14** (1985) 451.
54. N. Yamamoto and T. Miyauchi, *Jpn. J. Appl. Phys.* **9** (1972) 1383.
55. N. Yamamoto, H. Kubo and T. Miyauchi, *Jpn. J. Appl. Phys.* **14** (1975) 299.
56. H.J. Höbler, G. Kühn and A. Tempel, *Krist. Tech.* **15** (1980) K45.
57. K. Sugiyama, A. Sawada, K. Ito, S. Iwasaki and T. Endo, *J. Crystal Growth* **84** (1987) 673.
58. H. Yamaguchi, T. Nakamura, Y. Nabetani, T. Matsumoto and T. Kato, *J. Jpn. Assoc. Crystal Growth (in japanese)* **32** (2005), 114.
59. R.H. Plovnik, *Mat. Res. Bull.* **10** (1975) 555.
60. H.J. Höbler, G. Kühn, A. Tempel and H. Neumann, *J. Appl. Phys.* **53** (1981), 451.
61. H.J. Höbler, G. Kühn and H. Neumann, *Jpn. J. Appl. Phys.* **20** (1981) 307.
62. K. Hirakawa, H. Nakamura and M. Aoki, *Jpn. J. Appl. Phys.* **24** (1985) 265.
63. I.D. Olekseyuk, O.V. Parasyuk, O.M. Yurchenko, V.Z. Pankevych, V.I. Zaremba, R. Valiente and Y.E. Romanyuk, *J. Crystal Growth* **279** (2005) 140.
64. A. Tanaka, K. Aikawa, H. Katsuno, M. Kimura and T. Sukegawa, *J. Crystal Growth* **178** (1997) 350.
65. B.R. Pamplin, *J. Crystal Growth* **26** (1974) 239.
66. B.R. Pamplin, *J. Phys.* **c3/9/36** (1975) C3-35.
67. S. Chavhan and R. Sharma, *J. Phys. Chem. Solids* **67** (2006) 767.
68. H. Asai and K. Sugiyama, *Jpn. J. Appl. Phys.* **20** (1981) 1401.
69. N. Yamamoto, T. Takayama and B.R. Pamplin, *Prog. Cryst. Growth Character.* **10** (1985) 235.
70. H.J. Hsu, M.H. Yang, R.S. Tang, T.M. Hsu and H.L. Hwang, *J. Crystal Growth* **70** (1984) 427.
71. H. Yamamoto, Y. Kimura and H. Horinaka, *Proc. 8th Int. Conf. Ternary Multinary Compounds (ICTMC-8)* (Kishinev, Moldova, 1990) p. 184.
72. K. Sugiyama, H. Kato and H. Miyake, *J. Crystal Growth* **98** (1989) 610.
73. H. Miyake and K. Sugiyama, *Jpn. J. Appl. Phys.* **29** (1990) L1859.
74. H. Miyake and K. Sugiyama, *Proc. 9th Int. Conf. Ternary Multinary Compounds* (Yokohama, Japan, 1993), *Jpn. J. Appl. Phys.* **Suppl., 32-3**(1993) 156.
75. W.N. Honeyman, *J. Phys. Chem. Solids* ***30*** (1969) 1935.
76. W.N. Honeyman and K.H. Wilkinson, *J. Phys. D; Appl. Phys.* **4** (1971) 1182.
77. N. Yamamoto, N. Tohge and T. Miyauchi, *Jpn. J. Appl. Phys.* **14** (1975) 192.
78. Y. Morita, T. Yokogawa and T. Narusawa, Ext. Abstr. *36th Spring Meeting, Jpn. Soc. Appl. Phys. and Related Socs.* **I** (1989) p. 349.
79. S. Iwasaki, A. Sawada, G. Sato, T. Endo and K. Sugiyama, Ext. Abstr. *35th Spring Meeting, Jpn. Soc. Appl. Phys. and Related Socs.* **I** (1988) p. 289.
80. H. Miyake, S. Iwasaki, T. Endo and K. Sugiyama, Ext. Abstr. 49th Autumn Meeting, Jpn. Soc. Appl. Phys. **I** (1988) p. 241.
81. P. Prabukanthan and R. Dhanasekaran, *Cryst. Growth Des.* **7** (2007) 618.
82. T. Miyauchi, N. Yamamoto and H. Higuchi, *Jpn. J. Appl. Phys.* **27** (1988) L1178.
83. N. Yamamoto, S. Ishida and H. Horinaka, *Jpn. J. Appl. Phys.* **28** (1989) 1780.
84. N. Yamamoto, J. Ogihara and H. Horinaka, *Jpn. J. Appl. Phys.* **29** (1990) 650.
85. N. Yamamoto, K. Yokota and H. Horinaka, *J. Crystal Growth* **99** (1990) 747.

86. A. Kinoshita, K. Nomoto, K. Yoshino and T. Ikari, *Bull. Fac. Engeer. Miyazaki Univ.* **36** (2007) 189.
87. S. Nomura and T. Takizawa, *J. Crystal Growth* **128** (1993) 659.
88. M. Susaki, *Jpn. J. Appl. Phys.* **44** (2005) L866.
89. J. Castro, J. Naranjo, L. Duran, L. Melendez, J.R. Fermin, J. Ceudros and C.A. D. Rincon, *Cryst. Res. Technol.* **41** (2006) 946.
90. S.M. Wasim, *Solar Cells* **16** (1986) 289.
91. H.J. Möller, *Solar Cells,* **31** (1991) 77.
92. S.I. Radautsan and I.M. Tiginyanu, *Jpn. J. Appl. Phys.* **32 Suppl. 32-3** (1993) p. 5.
93. L.L. Kerr, S.S. Li, S.W. Johnston, T.J. Anderson, O.D. Crisalle, W.K. Kim, J. Abushama and R.N. Noufi, *Solid-State Electron.* **48** (2004) 1579.
94. S. Fiechter, Solar Energy Mater. *Solar Cells* **83** (2004) 459.
95. S. Lany and A. Zunger, *Phys. Rev. Lett.* **93** (2004) 156404.
96. S. Lany and A. Zunger, *Phys. Rev. B* **72** (2005) 035215.
97. S. Endo, T. Irie and M. Nakanishi, in *Copper Indium Diselenide for Photovoltaic Applications*, eds. T.J. Coutts, L.L. Kazmerski and S. Wagner, (Elsevier, 1986) p. 1.
98. B. Tell, J.L. Shay and H.M. Kasper, *Phys. Rev. B* **4** (1971) 2463.
99. B. Tell, J.L. Shay and H.M. Kasper, *J. Appl. Phys.* **43** (1972) 2469.
100. H. Neumann and R.D. Tomlinson, *Solar Cells* **28** (1990) 301.
101. K. Hara, T. Shinozawa, J. Yohsino and H. Kukimoto, *J. Crystal Growth* **93** (1988) 771.
102. F. Abou-Elfotouh, D.J. Dunlavy and T.J. Coutts, *Solar Cells* **27** (1989), 237.
103. G. Dragan, F. Abou-Elfotouh, D,J. Dunlavy, R.J. Matson and D. Cahen, *Chem. Mater.* **2** (1990), 286.
104. S. Ozaki, K. Muto, H. Nagata and S. Adachi, *J. Appl. Phys.* **97** (2005) 043507.
105. K. Wakita, K. Nishi and H. Ohta, *Proc. SPIE* **6118** (2006) 61180G-1.
106. M.V. Yakushev, A.V. Mudryi, Y. Feofanov, A.V. Ivaniukovich and I.V. Vicorov, *Thin Solid Films* **511-512** (2006) 130.
107. S. Siebentritt, I. Beckers, T. Riemann, J. Christen, A. Hoffmann and M. Dworzak, *Appl Phys. Lett.* **86** (2005) 091909.
108. A. Herman, L. Febick, K. Zweibel and R. Hardy, *Proc. 16th IEEE PV Specialists Conference* (1982) p. 840.
109. T. Coutts, L.L. Kazmerski and S. Wagner, *Sollar Cells, Special Issue of $CuInSe_2$*, **Vol. 16** (Elsevier Sequoia, S.A., Lausanne, 1986).
110. S. Endo et al, T. Coutts, L.L. Kazmerski and S. Wagner, *Sollar Cells, Special Issue of $CuInSe_2$*, **Vol. 16** (Elsevier Sequoia, S.A., Lausanne, 1986) p. 1.
111. T. Coutts, L.L. Kazmerski and S. Wagner, *Sollar Cells, Special Issue of $CuInSe_2$*, **Vol. 27** (Elsevier Sequoia, S.A., Lausanne, 1989).
112. K. Kushiya, *Oyo Buturi (Monthly Publ. Jpn. Soc. Appl. Phys.) (in japanese)* **75** (2006) 870.
113. K.W. Mitchell, C. Eberspacher, J. Ermer and D. Pier, *Proc. 20th Inst. Electr. Electron. Enginer. Spec. Conf.* (Las Vegas, 1988) p. 1384.
114. T. Nakada, *Oyo Buturi (Monthly Publ. Jpn. Soc. Appl. Phys.) (in japanese)* **69** (2000) 1445.
115. M.A. Contreras, M.J. Roemro and R. Noufi, *Thin Solid Films* **511-512** (2006) 51.
116. S. Niki, K. Sakurai, S. Ishizuka, K. matsubara, A Yamada, M. Tanbo and T. Terada, *Solid Physics (in japanese)* **41** (2006) 295.
117. Y. Bharath, K. Ruddy, V.S. Raja, *Solar Energy Mater. Solar Cells* **90** (2006) 1656.
118. G.D. Boyd, H.M. Kasper and J.H. McFee, *IEEE J. Quantum Electron.* **QE-7** (1971) 563.
119. G.D. Boyd, H.M. Kasper, J.H. McFee and F.G. Storz, *IEEE J. Quantum Electron.* **QE-8** (1972) 900.
120. X.X. Fan, R.C. Eckardt, R.L. Byer, R.K. Route and R.S. Feigelson, *Appl. Phys. Lett.* **45** (1984) 313.
121. E.S. Voronin, v.S. Soloniatin, N.I. Cherepov, V.V. Shuvalov, V.V. Badikov and O.N. Pivovarov, *Sov. J. Quan. Electon.* **5** (1975) 597.
122. W. Jantz and P. Koidl, *Appl. Phys. Lett.* **31** (1977) 99.
123. H. Horinaka, T. Okubo and N. Yamamoto, *Jpn. J. Appl. Phys.* **28** (1989) 1328.
124. T. Yamamoto, H. Takehara, H. Horinaka, T. Miyauchi, *Jpn. J. Appl. Phys.* **25** (1986) 1397.

APPENDIX: RAW MATERIALS

1. INTRODUCTION

Since the electrical and optical properties of compound semiconductors are determined by their purity, it is very important to use high purity materials in their preparation. Compared with Si, methods of improving the purity of the constituents of compound semiconductors developed later. The improvement in purity has been however largely driven by the demand from the compound semiconductor material manufacturers. It is a matter of fact that the purest material is the best but it is always necessary to elucidate which impurity has what effect in what way in compound semiconductor materials to get the best improvement in purity. By clarifying the effect of each impurity, the improvement in purity is maximised.

2. PURIFICATION PROCESS

Most of the constituent elements of compound semiconductors are produced as by-products of main metal smelting and refining processes. For instance, Ga is a by-product of Al smelting, Te, Se, As of Cu smelting, and so on. In order to know which impurity is predominant in the said element, it is very helpful to know the refining and purification processes because several elements behave in a similar way during the purification process. The details of these purification processes can be seen in Refs. 1-11.

High purification is not achieved by one process alone. Each purification process has its specific purification characteristics and to remove various impurities from a given element, it is necessary to combine some of these processes in the knowledge of which impurities are contained in large amounts. The main purification processes are classified as follows.

2.1 Electrolysis[2]

Electrolysis is often used as the initial stage of high purification. In this process, the element whose purification is desired is electrolytically dissolved from a positive electrode in a solution and is deposited on the negative electrode. Impurities which are less electronegative than the element are not disssolved in the solution and are deposited at the bottom of the electrolysis cell. In this process, impurities which are more electronegative than the host material to be purified, are disssolved in the solution as ions and are deposited on the negative electrode with the element. These impurities can not therefore be removed. The order of electronegativity is determined by the following ionization tendency.

$$K > Ca > Na > Mg > Al > Zn > Cr^{III} > Fe^{II} > Cd > Co^{II} > Ni > Sn^{II} > Pb > Fe^{III} > H > Cu^{II} > Hg^{I} > Ag > Au \quad (A.1)$$

The drawback of electrolysis is that more electronegative impurities are incorporated into the host material to be purified. These impurities therefore must be removed by other processes. The other drawback is due to new impurities being incorporated from the solution and cell materials themselves. Because of contamination, the purity level that can be achieved by electrolysis is limited. Electrolysis is therefore to be used mainly in the initial stage of purification.

2.2 Solvent Extraction[3-5]

When host materials are in solution, it is possible to extract some impurities by using a solvent. The removal of impurities is determined by the distribution coefficient, K,

$$K = C_{solvent}/C_{solution} \quad (A.2)$$

where C is the concentration of the impurity. Because of the distribution coefficient, when the impurity level is high, the achievable purity level is limited by the distribution coefficient. This is also a method which is preferred to be used in the initial stage of purification.

2.3 Ion Exchange[6,7]

There are two types of ion resins used for ion exchange, cation exchange resins and anion exchange resins.

In the case of cation exchange, metallic impurity ions, M^+ are absorbed on cation resins (HR) according to the following equation.

$$HR + M^+ \rightarrow MR + H^+ \quad (A.3)$$

In the case of anion exchange, impurity complex ions, $[M_m^{\ +}Cl_n^{\ -}]_{m-n}$, formed in a hydrogen chloride medium are absorbed on anion resins when $m < n$.

In the case of ion exchange, impurities can be removed to the level of the exchange equilibrium coefficient so that ion exchange is very effective in removing impurities. Only when the resin is saturated by impurities does, the impurity concentration rise to the original level. In the ion exchange process, it is important to be able to detect when the resins are fully exchanged by impurities. Since the ion exchange resins do not function when they are saturated with impurities, it is important to reduce the impurity level to the lowest possible by other methods before applying the ion exchange process in order to prolong the life time of the resins as long as possible.

2.4 Vacuum Distillation[8,9]

Each element has a different vapor pressure as shown in Fig. A.1 so that vacuum distillation will achieve purification.. The material to be purified is heated in a crucible in order to vaporize and condense it to the condenser part. By this process, impurities with higher vapor pressures and lower vapor pressures than the host material can be removed. Most impurities can be removed by this process. It is however difficult to sepa-

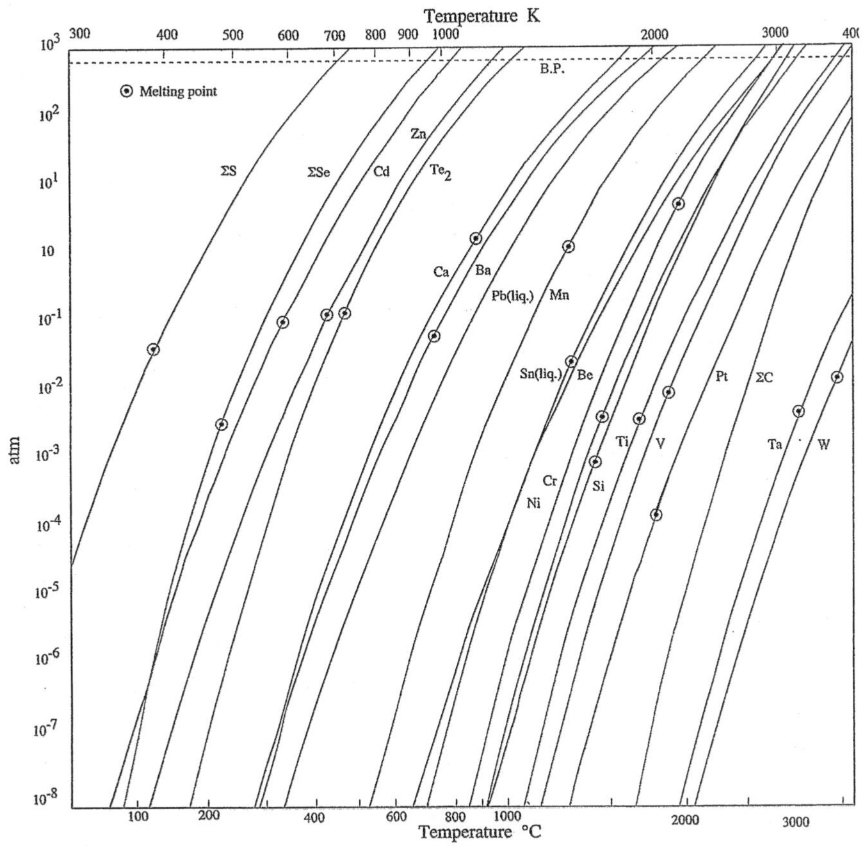

Fig. A.1 Vapor pressure curves of metals.

rate impurities whose vapor pressures are similar to that of the host material. In order to remove these impurities, other methods are required.

2.5 Zone Melting[10, 11]

Zone melting is a method that uses the differences in segregation coefficients of impurities. The host material to be purified is set in a crucible and it is heated from the top part and the melt zone is moved slowly from one end to the other as shown in Fig. A.2 Because of the differences in segregation coefficients, impurities with coefficients less than unity are swept into the melted zone and those with coefficients higher than unity are left. The theoretical impurity concentration is expressed as follows.

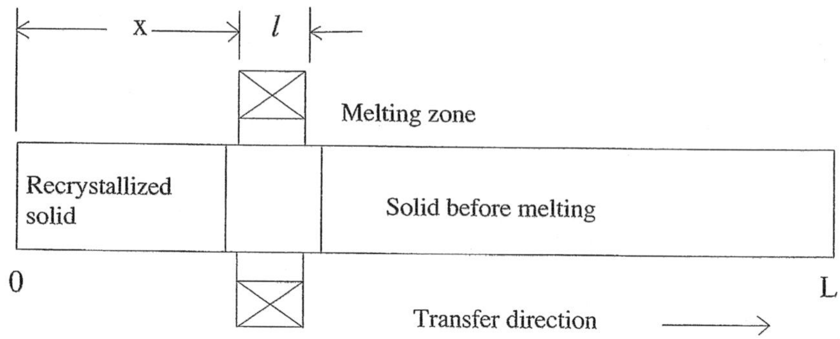

Fig. A.2 Purification by zone refining.

$$C^S = C_0^S\left[1-(1-k)e^{-k(x/l)}\right] \tag{A.4}$$

Here, C_0^S is the initial impurity concentration in the rod, l is the width of the zone, x is the position of the zone and k is the segregation coefficient.

For effective purification, multi-pass zone melting is applied in which purification by segregation is multiply performed. In this case, the limit of the purification is expressed as follows.

$$C_\infty = Ae^{Bl} \tag{A.5}$$

Here the constants A and B are expressed as follows.

$$A = \frac{C_0^S BL}{e^{BL} - 1} \tag{A.6}$$

$$k = \frac{Bl}{e^{Bl} - 1} \tag{A.7}$$

3. ZINC

The purification of zinc is reviewed by Dubois.[12] Zinc is obtained from zinc blende minerals (ZnS) by the dry process or the wet process.

In the dry process, ZnS is oxidation-roasted to ZnO, and ZnO is reduced to metallic zinc by reduction distillation with carbon as the reducing agent.

In the wet process, ZnS is roasted in a roasting furnace to ZnO, and ZnO is dissolved in sulfuric acid and $ZnSO_4$ solution is obtained. Metallic zinc is then extracted by elec-

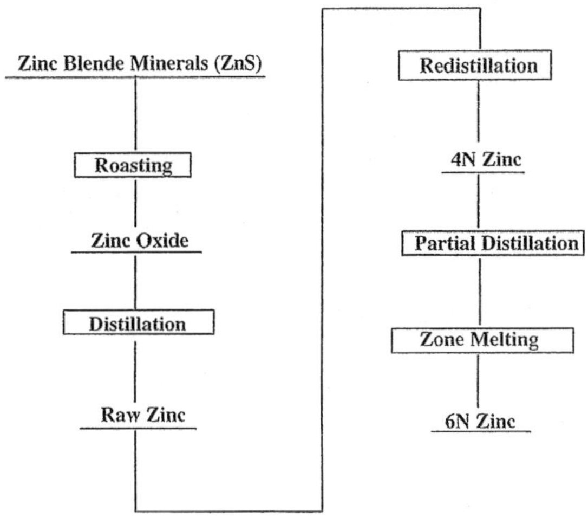

Fig. A.3 Purification process for zinc.

Table A.1 Typical Purity of High-Purity Zinc (from Ref. 18 with permission, The Mining and Materials Processing Institute of Japan)

(ppmw)

Fe	Ni	Cu	Cd	Tl	Pb	Cr	Sn
<0.05	<0.02	<0.04	<0.01	<0.1	<0.02	<0.04	<0.05

Analysis: wet flameless atomic adsorption

trolysis. Prior to the electrolysis process, impurities such as Fe, As, Sb are removed by precipitation by adding MnO_2 powders and impurities such as Cu, Ni, Co and Cd are removed by adding Zn or arsenious acid powders.

Raw zinc thus obtained is refined by distillation[13,14] and/or zone-melting[15-17] and high purity zinc is obtained. A typical refining process is shown in Fig. A.3. Typical purity of high-purity zinc is shown in Table A.1.[18]

For further purification, Isshiki et al.[19] have performed vacuum distillation and multi-zone melting and a residual resistivity ratio (RRR) of 25,600 was obtained as shown in Fig. A.4.

4. CADMIUM

Purification of cadmium has been reviewed by several authors.[20-22] Cadmium is included in zinc minerals. Cadmium is therefore obtained as a by-product of zinc smelting. Raw cadmium is produced by the wet method or the dry method as in the case of zinc smelting. Because of its higher volatility and oxygen affinity, cadmium can be separated from zinc. Raw cadmium is purified by electrolysis[23,24], vacuum distillation[24], and zone melting.[25,26] In Fig. A.5, a typical refining process is shown. For further purifica-

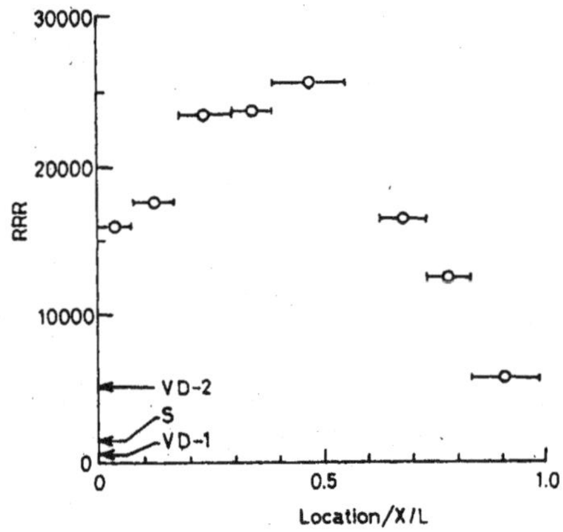

Fig. A.4 RRR of multiple zone refining purification for zinc after vacuum distillation. S; starting material, VD-1 and VD-2; the first and the second vacuum distillation (reprinted from Ref. 19 with permission, copyright 1984 The Japan Institute of Metals).

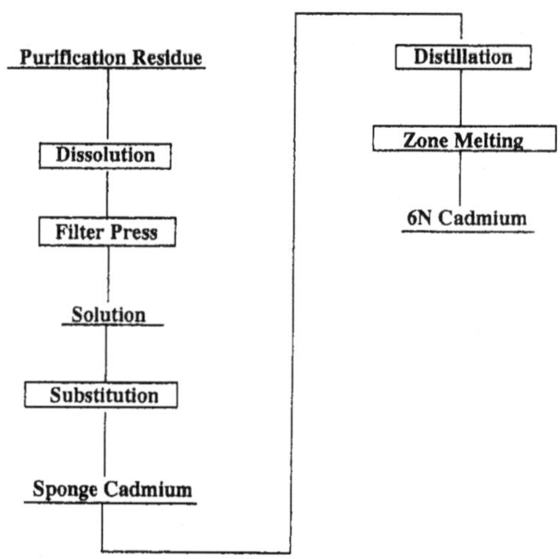

Fig. A.5 A typical refining process of cadmium.

tion, Isikawa et al.[27] performed vacuum distillation and multi-zone melting and an RRR of 21,000 was obtained as shown in Fig. A.6. Typical purity of high-purity cadmium is shown in Table A.2.[21]

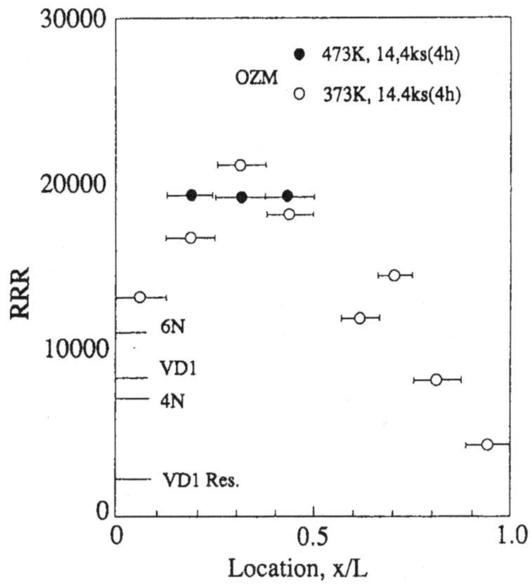

Fig. A.6 RRR of multiple zone refining purification for cadmium. 4N and 6N; reference mterials, VD1; after vacuum distillation, VD res.; residue of the first vaccum distillation. (reprinted from Ref. 27 with permission, copyright 1994 The Mining and Materials Processing Institute of Japan).

Table A.2 Typical Purity of Vacuum Distilled Cadmium (reprinted from Ref. 21 with permission, copyright 1981 Elsevier)

Impurities in ppm

	Pb	Cu	Ni	Zn	Al	Tl	Fe	Co	Mg	As
Feed	35	15	10	3	3	1	2	<1	<1	<1
Single distillation	0.5	0.5	ND	<1	0.1	ND	0.1	ND	0.2	<1
Double distillation	0.1	<0.1	ND	ND	0.1	ND	0.1	ND	0.2	<1

5. MERCURY

Purification of mercury is reviewed in Refs. 21 and 28. Raw mercury is obtained by condensing mercury vapor which comes from the reduction in air of HgS (red sulfide cinnabar) in a roasting furnace. Air is blown into raw mercury to make oxides of various impurities such as copper, lead, zinc, tin, and bismuth. Dilute nitric acid is then added to dissolve soluble constituents and filtration is performed. To remove noble metal impurities, low pressure distillation is performed. For further purification, multiple distillation is performed.

A typical refining process is shown in Fig. A.7 and a typical purity of high-purity mercury is shown in Table A.3.[21]

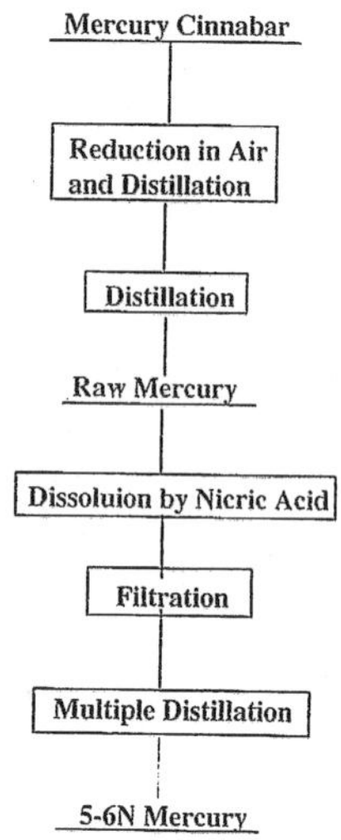

Fig. A.7 Purification process of mercury.

Table A.3 Typical Analysis of Vacuum Distilled Mercury (reprinted from Ref. 21 with permission, copyright 1981 Elsevier)

(ppmw)

Al	Bi	Cd	Ca	Cu	Cr	In	Fe	K	Mg
0.001	(0.01)[a]	(0.01)[a]	0.001	<0.001	(0.01)[a]	(0.01)[a]	<0.01	0.04	<0.001
Mn	**Ni**	**Pb**	**Si**	**Ag**	**Na**	**Sn**	**Ti**	**Cl**	**S**
(0.01)[a]	(0.01)[a]	(0.1)[a]	<0.001	<0.001	0.02	(0.01)[a]	(0.01)[a]	0.04	0.06

[a] Not detected wih detection limits given in parenthesis

6. GALLIUM

Gallium purification is summarized in Refs 29 and 30. Gallium is obtained as a by-product of smelting of Al or zinc. The former example is the Alcoa method and the latter examples are the Eagle Picher method and the Dowa mining method. The details of these three refining methods are as follows.

(i) Alcoa method

Sodium aluminate solution is obtained from bauxite minerals by the Byer method, and carbonic acid gas is blown into condensed gallium. Filtered aluminium-gallium hydroxide is dissolved in strong caustic soda (sodium hydroxide) and is refined by electrolysis to obtain metallic gallium.

(ii) Eagle-Picher method

Aluminum-gallium hydroxide is obtained by caustic soda oxidation from precipitates including gallium obtained from wet zinc smelting. Gallium trichloride is then extracted by ether and metallic gallium is obtained by electrolysis from the caustic soda solution.

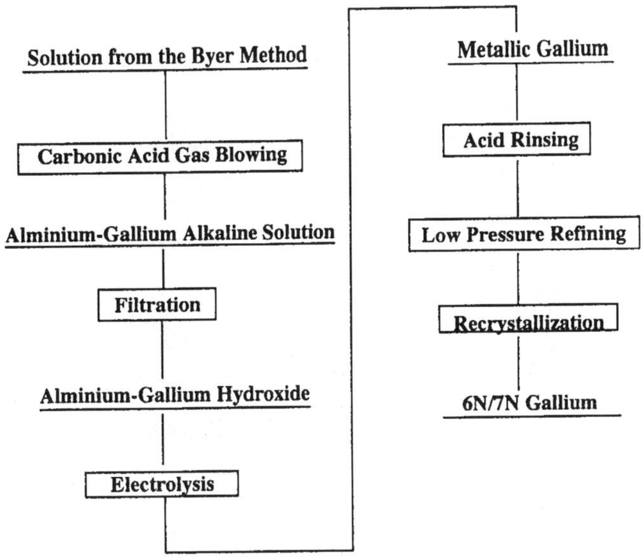

Fig. A.8 Purification process of gallium by the Alcoa method.

Table A.4 Typical Purity of High-Purity Gallium after Recrystallization (reprinted from Ref. 29 with permission, copyright 1972 Editions DUNOD)

(ppm)

Ag	Al	Ca	Cr	Cu	Fe	Hg	In	Mg	Mn	Na
ND	0.04	0.05	ND	ND	ND	ND	ND	0.03	ND	ND
Ni	Pb	Si	Sn	Ti	V	Zn				
ND	ND	<0.3	ND	ND	ND	ND				

(iii) Dowa Mining method
Gypsum is obtained by a neutralization process of the residue after zinc extraction from black ore. After ether extraction and removal of iron, metallic gallium is obtained by electrolysis.

The metallic gallium which is obtained from one of these methods is then rinsed by acid and purified by recrystallization. The typical purification process is shown in Fig. A.8. The final purity of gallium is improved up to 7N as shown in Table A.4.[29]

7. INDIUM

Purification of indium is reviewed in Refs. 31 and 32. Indium is obtained as a by-product from Zn refining or Pb refining. Fumes containing indium are treated by hot sulfuric acid and water-extracted as indium sulfate. Zinc sulfide or sodium sulfide is added to the solution of indium sulfate to remove Bi, Cu, Sb and As as deposits. When reducing metals such as Al or Zn are immersed in the remained solution, sponge indium is deposited by substitution. This sponge indium is then redissolved and elec-

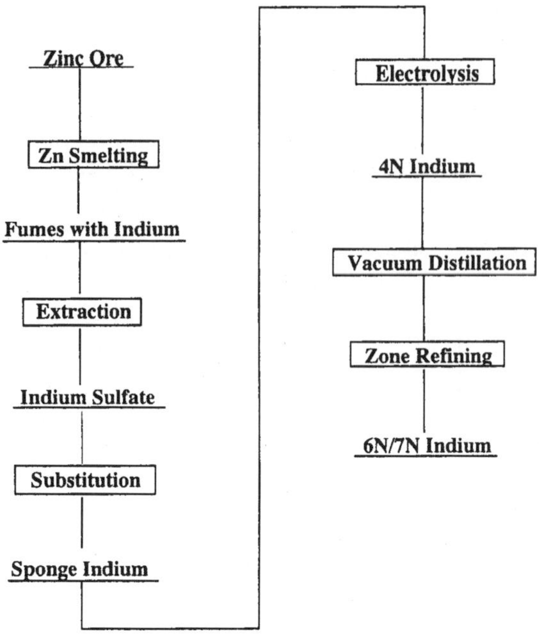

Fig. A.9 Purification process for indium.

trolysis is repeated to obtain 4-5N Indium. This indium is then refined by vacuum distillation and/or by zone-melting to 6N Indium. The indium purification process is shown in Fig. A.9. In Table A.5, the purity development of high-purity indium is shown.[33,35]

8. PHOSPHORUS

The purification of phophorus is reviewed by Takanoha et al.[34] Phosphorus minerals such as apatite fluoride and apatite are mixed with silicate as flux and coke as reducing agent and are reduced to yellow phosphorus at 1400-1500 °C in an electric furnace. Phosphorus vapor in CO gas is introduced to water to be condensed and recovered. Raw phosphorus is then liquefied with hot water and filtered to make it 99%. Yellow phosphorus is then filtered, washed with water, distilled in water vapor and dry-distilled. This raw yellow phosphorus is reacted with chlorine gas to form phosphorus trichloride (PCl_3) which is then refined by fractional distillation. Purified phosphorus trichloride is then reduced by hydrogen to high purity yellow phosphorus.

Since yellow phosphorus is combustible, it is converted to red phosphorus by heating in a quartz ampoule at 240-300 °C for several days. Unreacted yellow phosphorus is removed by a sodium hydride solution and red phosphorus is dried in nitrogen. The purification process is shown in Fig. A.10.

Typical purity values of high-purity phosphorus from various manufacturers are shown in Table A.6.[35] Of various impurities, it is known that arsenic which behaves

Table A.5 Purity Analysis of High Purity Indium (reprinted from Ref. 33 with permission, copyright 1990 Institute of Physics)

Impurity	Detection Limit	Year 1983	1986	1989
B	0.0003	0.006	0.001	<
Mg	0.001	ND	ND	ND
Al	0.001	0.005	0.001	<
Si	0.003	0.01	0.01	<
S	0.04	ND	ND	ND
Ca	0.002	0.03	0.008	<
V	0.004	ND	ND	ND
Cr	0.004	<	<	<
Mn	0.004	ND	ND	ND
Fe*	0.05	<	<	<
Co	0.005	ND	ND	ND
Ni	0.02	<	<	ND
Cu	0.008	ND	ND	ND
Zn	0.02	ND	ND	ND
As	0.007	ND	ND	ND
Se	0.02	ND	ND	ND
Br	0.02	ND	ND	ND
Ag	0.02	ND	ND	ND
Cd*	0.01	<	<	<
Sn	0.06	ND	ND	ND
Sb	0.03	ND	ND	ND
Tl*	0.03	<	<	<
Pb*	0.05	<	<	<
Bi*	0.02	<	<	<

ND; not detected
<; less than the detection limit
*; atomic absroption method

Table A.6 Purities of Phosphorus from Various Manufactureres (reprinted from Ref. 35 with permission, copyright 1990 The Japanese Association for Crystal Growth)

Impurity	Detection Limit	Manufacturer A			B			C	
		Lot 1	Lot 2	Lot 3	Lot 1	Lot 2	Lot 3	Lot 1	Lot 2
Cu	0.01	ND	ND	ND	ND	0.01	0.01	ND	ND
Pb	0.01	ND	ND	ND	0.04	0.04	0.03	ND	ND
Fe	0.1	ND	ND	ND	0.28	0.30	0.21	ND	ND
Cd	0.01	ND	ND	ND	ND	ND	ND	ND	ND
Zn	0.1	ND	ND	ND	ND	ND	ND	ND	ND
S*	0.2	ND	ND	ND	ND	0.3	ND	ND	ND
As	0.01	1.8	0.28	0.51	7.4	5.7	5.6	ND	ND

similarly to phosphorus is an important impurity. High-purity phosphorus with arsenic below the detection limit is now available.

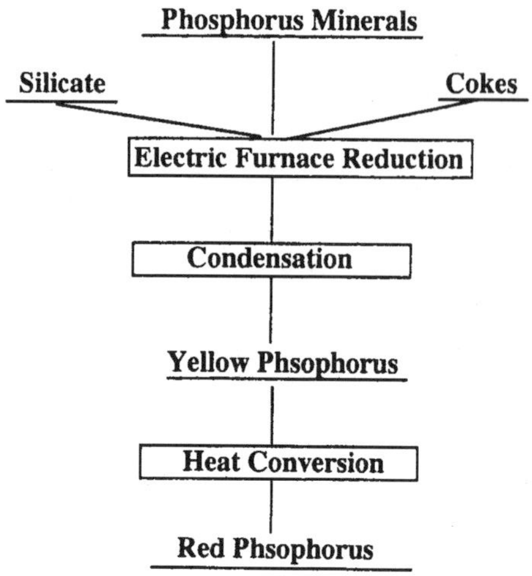

Fig. A.10 Purification process for phosphorus.

9. ARSENIC

Purification of arsenic is reviewed by Toyabe et al.[36-38] Source minerals such as FeAsS, $FeAs_2$ and As_2S_3 are roasted at 600-700°C and the resulting As_2O_3 gas is cooled to obtain fine arsenious acid crystals with a purity of about 95 %. When As_2O_3 crystals are resublimed, white As_2O_3 with a purity of about 99 % is obtained. As_2O_3 is then converted to arsenic chloride (III) by dissolving in hydrogen chloride and by dehydration in sulfuric acid. Arsenic chloride (III) is distilled at temperatures lower than 100 °C in order to remove not only heavy metal impurities but also alkali metal and alkaline-earth metal impurities. Purified arsenic chloride (III) is then converted to purer As_2O_3 by hydrolysis. As_2O_3 can be easily reduced to metallic arsenic by reduction with hydrogen or carbon. Arsenic chloride (III) itself can also be reduced to metallic arsenic by hydrogen reduction.[39]

High purity metallic arsenic can be obtained from raw metallic arsenic or from purified arsenious acid by various methods as follows.

(i) Sublimation of metallic arsenic

Metallic arsenic can be sublimed at 615°C. Heavy metal impurities can be removed by sublimation in inert gas or in vacuum. When metallic arsenic is heated at around 200-300 °C, As_2O_3 at the surface of metallic arsenic can also be removed.

(ii) Sublimation in hydrogen[40,41]

When metallic arsenic is sublimed in hydrogen, S, Se, Te can be removed as hydrides. Heavy metal impurities can be removed by this process.

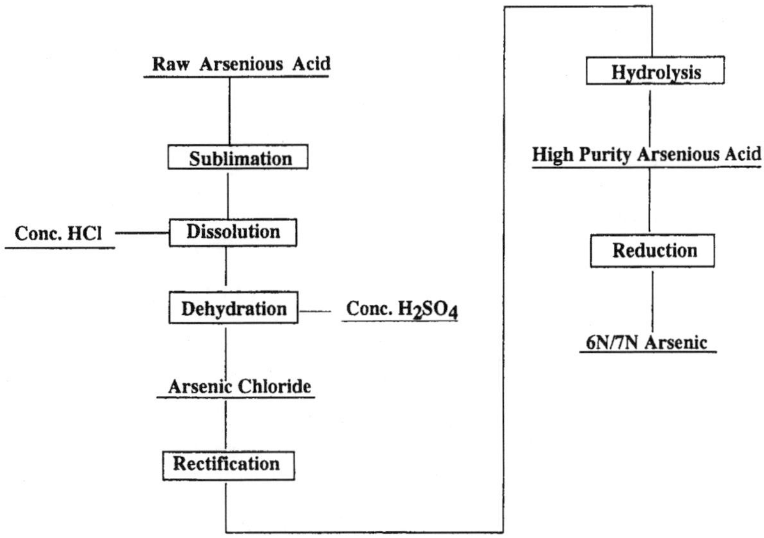

Fig. A.11 Purification process for arsenic

Table. A.7 Typical Analysis for 99.9999+% As (reprinted from Ref. 44 with permission, copyright 1997 The Mining and Materials Processing Institute of Japan) (ppm)

	Fe	Cu	Ca	Mg	Si	Sb	S	Se
Spectroscopic Analysys	ND	ND	ND	ND or trace	ND	ND	ND	ND
Chemical Analysys	ND	ND	ND	<0.05	ND	ND	<0.1	<0.1

(iii) Arsenic-lead alloy method[42]

Arsenic is alloyed with lead and heated to remove S, Se, Te as PbS, PbSe, and PbTe. Arsenic is then deposited at lower temperatures and Pb is separated. Arsenic is resublimed to remove the residual lead.

(iv) Bridgman method[43]

Arsenic is sealed in vacuum in a quartz ampoule and single crystals are grown from the bottom by gradient freezing. It is known that S, Se, Ni, Pb, Mn, Sb, Ge, Zn, Bi and Fe can be removed while Bi and Mg are difficult to remove by this process.[44]

A typical purification process is shown in Fig. A.11 and a typical purity of high-purity arsenic is shown in Table A.7.[44] Purification up to 7N is also reported as high purity material for MBE applications.[45]

10. ANTIMONY

Purification of antimony is described in Refs. 46 and 47. Source materials such as antimony oxide or sulfide minerals are roasted and volatile Sb_2O_3 is gathered. This Sb_2O_3 is reduced by charcoal. High grade sulfide minerals can be directly reduced to raw antimony by Fe. Sulfide minerals can be reduced by roasting and by interaction to raw

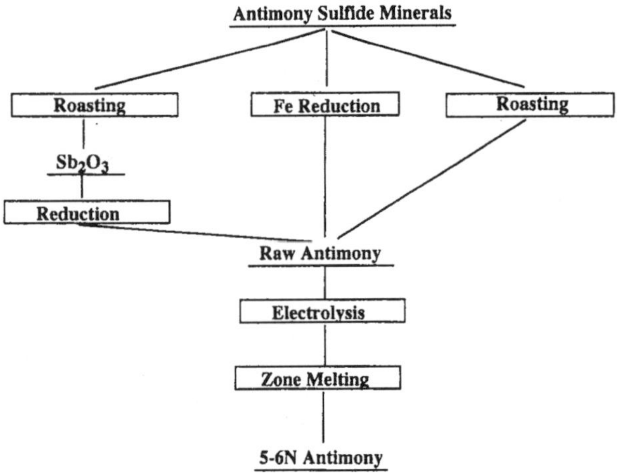

Fig. A.12 Purification process for antimony.

Table A.8 Typical Purity of High-Purity Antimony (from Ref. 68 with permission,The Mining and Materials Processing Institute of Japan)

(ppm)

As	Ag	Bi	Cu	Fe	Ni	Pb	Sn
<5	<0.5	<1	~1	<1	<5	<2	<2

antimony. Raw antimony is then melted with sulfuric sodium, sodium carbonate, sodium chloride and pure antimony oxide minerals in a reflection furnace and 97-99% Sb can be obtained. This antimony is then refined by electrolytic refining. The metallic antimony is then purified by distillation or fractional distillation of antimony chloride and by reduction. The material is then purified by zone melting, distillation and reelectrolysis to 5-6 N Sb.

Commercial antimony is purified by distillation of $SbCl_3$ dissolved in 6M HCl[48,49], fractional distillation [50,51], vacuum distillation[52-55] and by electrorefining.[56-65] Purified $SbCl_3$ thus obtained is then converted to metal by reduction in hydrogen [49,50], conversion to oxide followed by reduction in hydrogen[55], and electrowinning.[49,59,61] Further purification is performed by zone refining.[53,66,67]

A typical purification process is shown in Fig. A.12 and the typical purity of high-purity antimony is shown in Table A.8.[68]

11. BISMUTH

Purification of bismuth is described in Refs. 47 and 69-71. Anode slimes of lead and copper smelting or lead dross (Kroll-Betterson debismuth process) are used as the starting material. Impurities in these starting materials are removed by dissolution with soda oxidation or with chlorination. High purification is then performed by vacuum distillation and zone melting. The typical purification process is shown in Fig. A.13 and a typical purity of high-purity bismuth is shown in Table A.9.[68]

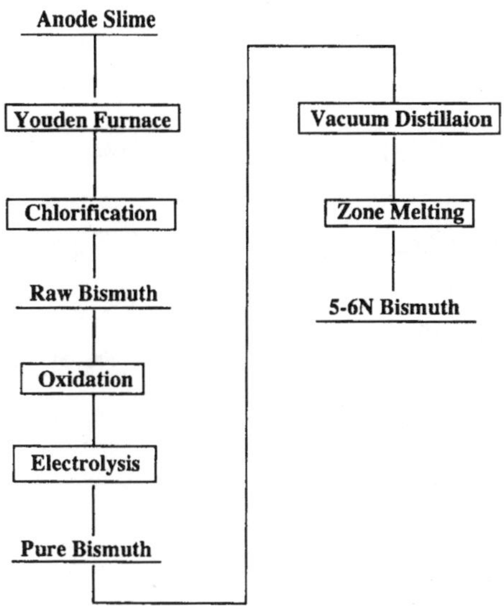

Fig. A.13. Purification process for bismuth.

Table A.9 Typical Purity of Bismuth (from Ref. 68 with permission, The Mining and Materials Processing Institute of Japan)

(ppm)

Cu	Ag	Zn	Pb	Fe
0-2	~0	0-3	~0	0-2

12. SELENIUM

Purification of selenium is described in Refs. 72-77. Selenium is obtained as a by-product during copper smelting. When copper is obtained as a metal electrode by electrolysis, metals less electronegative than copper remain as electrolysis deposits in the electrolysis bath. These deposits are burned in air and selenium dioxide which has high vapor pressure is separated from other metals whose oxide vapor pressure is lower. In this process, arsenic, the vapor pressure of whose oxide is relatively high is difficult to eliminate. Selenium dioxide is then dissolved in water as selenious acid and is reduced by sulphur dioxide gas blowing. Ion exchange is used to purify the selenious acid before reduction. The selenium obtained is then dried and purified by vacuum distillation. The final selenium product is in the form of water shots of diameter several millimeters. It should be noted that selenium is not easily crystallized and is obtained in the amorphous state.

In Fig. A.14, a typical purification process is shown and typical purity of high-purity selenium is shown in Table A.10.

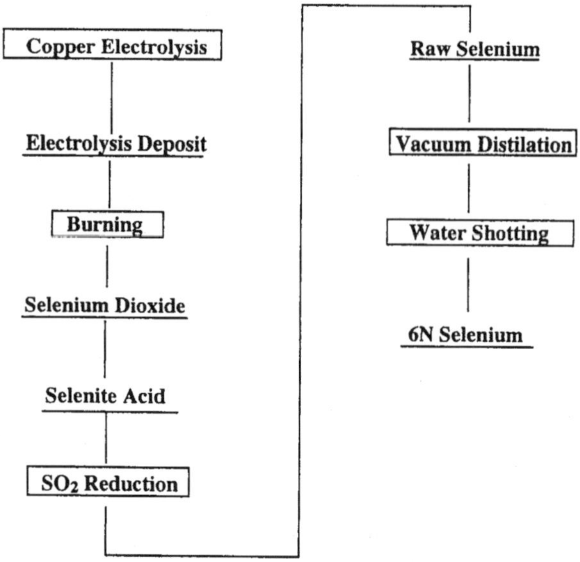

Fig. A.14. Purification process for selenium.

Table A.10 Typical Purity of High-Purity Selenium

(ppmw)

Te	As	Sb	Cu	Pb	Zn	Fe	Ni	Mg	Sn
≤0.01	0.03	≤0.01	≤0.01	≤0.01	0.03	0.05	~0.25	≤0.01	≤0.01

13. TELLURIUM

Purification of tellurium is described in Refs. 78-88. Tellurium is obtained as a by-product during copper smelting as anode slimes as in the case of selenium. When copper is obtained as a metal electrode by electrolysis, metals less electronegative than copper remain as a deposit in the electrolysis bath. These deposits are roasted in air and selenium dioxide which has a higher vapor pressure than that of tellurium oxide is removed from the residue. Tellurium remains in the residue as tellurium oxide which is soluble in acid. The residue is immersed in the copper electrolysis solution based on sulfuric acid and copper powders are added to the solution to deposit tellurium with silver and selenium. The deposit is treated in a silver separating furnace with Chili niter and soda ash and soda slag containing selenium and tellurium is obtained. Hot water is added to the slag to immerse selenious sodium and tellurous sodium. These acids are neutralized to obtain tellurium dioxide. Tellurium dioxide is dissolved in sodium hydroxide and lead is precipitated out by adding sodium sulfide. The residue is neutralized by sulfuric acid to obtain 96 % tellurium dioxide. Tellurium dioxide is then reduced to metallic tellurium by the following three methods.

(i) TeO_2 is dissolved in sodium hydroxide and reduced by electrolysis.

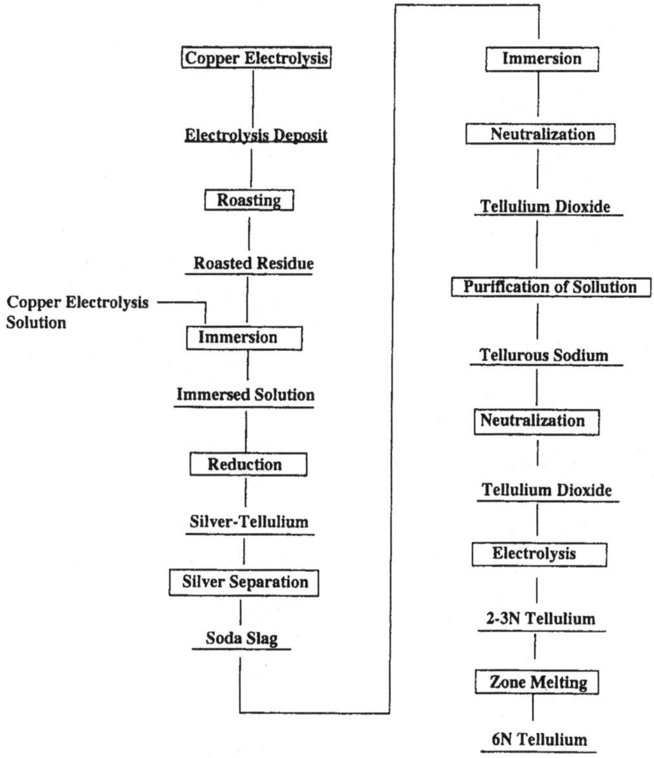

Fig. A.15 Purification process for tellurium

Table A.11 Mass Spectrograph Anaylses of Quaruple Zone Refined Tellurium (reprinted from Ref. 21 with permission, copyright 1981 Elsevier)

(ppmw)

C	N_2	O_2	H_2	Na	Mg	Al	Si	P
0.8	0.02	3	X[a]	0.02	0.03	0.08	0.02	Int.[b]
Cl	K	Cr	Fe	Ni	Cu	Ga	As	Se
0.02	In.b	002	0.1	0.01	Int.[b,c]	ND[d]	ND[d]	0.1

[a] X, not looked for.

[b] Int. = line interference.

[c] Determined at less that 0.1 ppmw by emission spectrograph

[d] ND = not detected

(ii) TeO_2 is reduced by carbon with covering borax.
(iii) TeO_2 is dissolved by hydrogen chloride or sulfuric acid and is reduced by SO_2 gas after the addition of HCl.

Tellurium thus obtained is 2-3 N and is purified by vacuum distillation with or without hydrogen[78-80] up to 6 N and is ultra-purified by zone refining.[78-80, 83, 85] Zone refining is very effective for purification since most of the impurities have favorable segregation coefficients. For ultra-high purification, a multiple pass zone refining technology has been developed.[86-88]

The typical purification process is shown in Fig. A.15. Typical purity of high-purity tellurium is shown in Table A.11.[80]

14. BORIC OXIDE (B_2O_3)

Boric oxide is used as a liquid encapsulant for liquid encapsulated Czochralski (LEC) crystal growth (Vol. 1, Sec. 2.2.3) and is an important material for the purification of compound crystals. Boric oxide is very effective in purifying melt compound materials by the solvent extraction effect but if the purity is not satisfactory or if it contains an impurity whose partition coefficient is small for B_2O_3, the purity of the compound semiconductor melt is compromised.

From minerals containing boron, boron is extracted as $Na_2B_4O_7$ solution and boric acid (H_3BO_3) is deposited and is refined by recrystallization. Boric acid is then refined by repeated recrystallization. Dewaterizing is performed by heating in a high frequency furnace. B_2O_3 is then heated in a Maffle furnace with nitrogen gas babbling to reduce the water content.

The water content in B_2O_3 is measured by infrared (IR) spectroscopy by measuring the peak at 1.41 μm and the accuracy is ± 5 ppm.[89] Since the water content in B_2O_3 has a large influence on carbon concentration in GaAs crystal growth, its controllability is very important.

15. PYROLYTIC BORON NITRIDE (pBN)

Quartz was mainly used in the past to make crucibles for crystal growth, but because of Si contamination, high-purity pBN crucibles are now more often used.[90]

pBN is a BN material which is produced from BCl_3 and NH_3[91] by the thermal CVD method at high temperatures under low pressure. Layers of high quality pBN which is dense and stable in air can be obtained at high temperatures (>1800°C) under low pressure (<10 Torr). These pBN layers are formed in a graphite mould up to sufficient thickness and removed from the mould to obtain a pBN crucible with the desired size and shape.

pBN is nontoxic and has a white yellow color and a turbostratic hexagonal BN structure.[92] It is very anisotropic between a axis and c axis, in its mechanical strength and thermal conductivity, and electrical resistivity. This is because BN hexagonal crystals are layered towards the plane parallel to the a axis.[93, 94] pBN has very high purity, high temperature tensile strength and high insulation compared with sintered BN. pBN also has a very high temperature stability. It can be used up to a maximum temperature of 3000°C.

pBN has higher thermal conductivity towards a axis compared with c axis. This property is an advantage for crucible applications since the temperature gradient can be reduced for vertical and radial directions. The thermal expansion coefficient is much greater in the c axis than the a axis so that when pBN is used with isotropic graphite material, it is necessary to keep in mind this difference in designing the inner structure of the growth furnace.

The chemical stability of pBN is excellent, resistant against acids, alkalis and organic solvents at room temperature and is inactive against various metals, compounds and glasses.

Graphite material can be coated with pBN and this coating allows different shapes of crucible materials to be made. This possibility helps various crystal growth methods since new arrangements for crystal growth can be designed without contamination.

Typical purity of pBN is shown in Table A.12[90] and various crucibles are shown in Fig. A.16.

Table A.12 Typical Purity of pBN by the Inducd Coupled Plasma (ICP) Method (from Ref. 90 with permission)

(ppm)

Ca	Cu	Co	Fe	Mn	Ni	Zn	Na*	Si**
<1	<1	<1	<1	<1	<1	<1	<1	<1

* Flameless atomic absorption analysis

** DC ark method

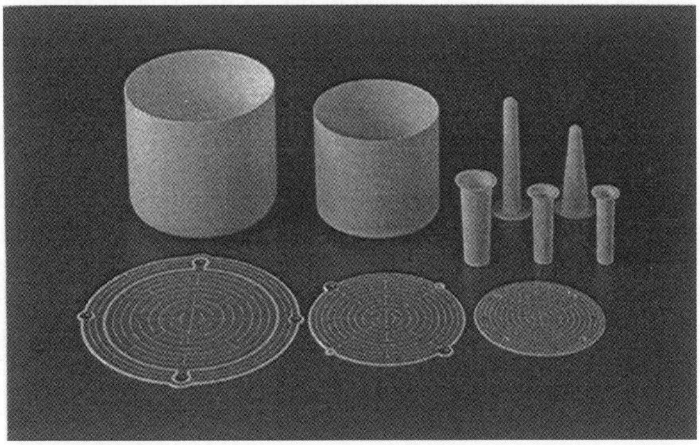

Fig. A.16 pBN crucibles (courtesy of Shin-Etsu Chemicals).

16. GRAPHITE[95-101]

Graphite is an important material for crystal growth of compound semiconductors since it is used in crucible succeptors, heaters, heat insulating materials and others. For these applications, high grade isotropic graphite is mainly used. Graphite itself has a hexagonal system and a layer lattice structure, so that it is essentially anisotropic. Isotropic graphite is produced by the cold isostatic press (CIP) method, by which graphite powders are pressed homogeneously and then sintered so that it appears macroscopically homogeneous even though graphite itself is inhomogeneous microscopically.

Fig. A.17 Various graphite parts including heaters and susceptors (Courtesy of Covalent Materials Co.).

Isotropic graphite has characteristics such as heat shock resistance, high purity, rapid degassing, low contamination by carbon dust. It is also important in the preparation of fine structure and high mechanical strength materials.

Graphite suppliers have been developing it in various directions to improve its properties. Alternatives to carbon material are being considered, for example metaphase small sphalerite graphite. The raw material size is optimized and high density graphite material is also being developed. Coating by amorphous carbon is being developed to cover the surface pore and internal cavity by amorphous carbon. SiC coating is also being developed, with a view to avoiding contamination from graphite.

It is important to select the appropriate material for the particular crystal growth required, from the viewpoint of strength, purity, avoidance of carbon contamination and homogeneity. When it is used for heaters (Fig. A.17), it is important to control the homogeneity more precisely since inhomogeneity influences the temperature distribution. Inhomogeneity makes for differences in resistivity from heater to heater and this variance causes instability of crystal growth from heater to heater.

REFERENCES

1. G. Chaudron, in *Monographies sur les Métaux de Haute Pureté*, ed. G. Chaudron (Masson et Cie, Paris, 1972) Vol. 1, p. 5.
2. A.T. Kuhn and P.M. Wright, *Industrial Electrochemical Processes* (Elsevier, Amsterdam, 1971).
3 *Handbook of Solvent Extraction*, eds. T.C. Lo, M.H.I. Baired, C. Hanson, (Wiley, New York, 1983)
4. G.M. Ritce and A.W. Ashbrook, *Solvent Extractin: Principles and Applications to Process Metallurgy* (Elsevier, Amsterdam, 1979).
5. T. Sekine and Y. Hasegawa, *Solvent Extraction Chemistry: Fundamentals and Applications* (M. Dekker, New York, 1977).
6. R.W. Grimshaw and C.E. Harland, *Ion-Exchange: Introduction to Theory and Practice* (Chemical Society, London, 1975).
7. *Ion Exchange: Theory and Application*, ed. F.C. Nachod (Academic Press, New York, 1949).
8. O. Winkler and R. Bakish, *Vacuum Metallurgy* (Elsevier, 1971).
9. *Vacuum Metallurgy*, ed. by R.F. Bunshah, (Reinhold Publ., New York, 1958).
10. W.G. Pfann, *Zone Melting* (John Wilwy & Sons Inc., New York, 1964).
11. J. Hernacz and A. Pardo, *Rev. Metal. Madrid* **21** (1985) 159.
12. B. Dubois, in *Monographies sur les Métaux de Haute Pureté Vol. 1*, ed. G. Chaudron (Masson et Cie, Paris, 1972) p. 412.

13. M. Kyoya, *J. Nippon Metal Society* **20** (1956) 631.
14. B. V. Aleksandorov and V. I. Udovikov, *Izv. Akad. Nauk SSSR, Metal* **21** (1973) 241.
15. M. Kyoya, *J. Nippon Metal Society* **21** (1957) 241.
16. A. Desalvo, P. Gondi, F. A. Leviand and F. Zigani, *Met. Ital.* **55** (1963) 511.
17. J. H. Wernick and E. E. Thomas, *Trans. Met. Soc. AIME* **213** (1960) 645.
18. Y. Awakura and T. Ohishi, *Proc. Mining and Mineral Processing. Inst. Japan (Fall meeting, 1996)* B-4-1, p. 100.
19. M. Isshiki, T. Tomizono, T. Yoshida, T. Ohkawa and K. Igaki, *J. Japan Inst. Metals* **48** (1984) 1176.
20. B. Dubois, in *Monographies sur les Métaux de Haute Pureté Vol. 1*, ed. G. Chaudron (Masson et Cie, Paris, 1972) p. 430.
21. H. E. Hirsch, S. C. Liang and A. G. White, *Semicond. Semimet.* **18** (1981) 21.
22. R. E. Lund and R. E. Sheppard, *J. Met.* **16** (1964) 724.
23. D. M. Chizhikov, *Cadmium* (translated by D. E. Hayler, Pergamon, Oxford, 1966).
24. K. Zanio, *Semicond. Semimet.* **13** (1978) 39.
25. B. N. Aleksandrov, *Phys. Met.* **11** (1961) 99.
26. J. H. Wernick and E. E. Thomas, *Trans. Metall. Soc. TIME* **218** (1960) 763.
27. Y. Ishikawa, Y. Bailing, K. Mimura, T. Tomizono and M. Isshiki, *Shigen To Sozai, J. Mineral. Mater. Proc. Inst. Jpn.(in japanese)* **110** (1994), 1175.
28. J. Rzepski, in *Monographies sur les Métaux de Haute Pureté Vol. 1*, ed. G. Chaudron (Masson et Cie, Paris, 1972) p. 444.
29. P. de la Bretèque, in *Monographies sur les Métaux de Haute Pureté Vol. 1*, ed. G. Chaudron (Masson et Cie, Paris, 1972) p. 508.
30. *Rare Earch Metal Data Book: Gallium (in japanese)* (Metal Mining Agency of Japan, 1983).
31. M. Cunha Belo, in *Monographies sur les Métaux de Haute Pureté Vol. 1*, ed. G. Chaudron (Masson et Cie, Paris, 1972) p. 525.
32. T. Yoshimi, *III-Vs Review* **5** (1992) 56.
33. O. Oda, K. Kainosho, K. Kohiro, H. Shimakura, T. Fukui, R. Hirano and S. Katsura, *Semi-insulating III-V Materials* (IOP Publishing, 1990) p. 189.
34. A. Takanoha, H. Yanai and S. Ishii, *Phosphorus and Potassium* **153** (1988), 29.
35. O. Oda, *J. Jpn. Assoc. Crystal Growth (in japanese)* **17** (1990) 168.
36. K. Toyabe, C. Segawa, H. Kadoya and H. Ohwa, *Arsenic Metallurgy Fundamentals and Applica tions: Symposium, Auunal Meeting* (1988) p. 351.
37. T. Terayama, *Bullutin of Shigen, Sozai Society* **105** (1989) 585.
38. H.E. Hirsch, *Lead-Zinc-Tin '80, Proc. World Symp. Metal. Environ. Control* (1980) 360.
39. R. H. Harada and A. J. Strauss, *J. Appl. Phys.* **30** (1959) 121.
40. J. M. Whelen J. D. Struthers and J. Ditzenberger, *J. Electrochem. Soc.* **107** (1960) 982.
41. G. A. Silvey, V. J. Lyons and V. J. Silvestri, *J. Electrochem. Soc.* **108** (1961) 653.
42. H. Gatos, *Properties of Elemental and Compound Semiconductors* (Inter. Publ., 1960)
43. L. R. Weisberg and P. R. Celmer, *J. Electrochem. Soc.* **110** (1963) 56.
44. M. Asano, K. Murata. S. Ishiguro, *J. Nippon Mining Soc.* **97** (1981), 841.
45. R.G.L. Barnes, *Crystal Prop. Prep.* **32-34** (1991) 351.
46. K. Iyer and S.G. Beshpande, *J. Appl. Eectrochem.* **17** (1987), 936.
47. A. Fuwa and Y. Fukunaka, *Proc. Mining and Mineral Processing. Inst. Japan (Fall meeting, 1996)* B-4-7, p. 123.
48. A. Müller, S.H. Fox and R.I. Day, *Anal. Chem.* **22** (1950) 1570.
49. M. S. Brooks and J. F. Kenedy (ed.), in *Ultrapurification of Semiconductor Materials* (The Macmillan Company, New York, 1962) p.182.
50. V. A. Milkhailov, L. I. Shchukin, S. I. Sokolova and I. G. Yudelevich, *Chem. Abstr.* **97** (1982) 96050t.
51. L. A. Niel'son, Z. N. Orshanskaya, K. V. Tret'yakeva, A. D. Miklyaev, V. P. Borisova and V. N. Kazimirov, *Izv. Akad. Nauk SSSR, Metal* **4** (1982) 51.
52. F. Deml and V. Gaislova, *Coll. Czech. Chem. Commun.* **25** (1960) 2213.
53. G. N. Nikolaenko, *Chem. Abstr.* **55** (1961) 1349a.
54. A. J. Singh, P. Suryanarayana and B. S. Mathur, *Indian J. Technol.* **5** (1967) 162.
55. R. K. Iyer and T. R. Bhat, *Trans. Ind. Inst. Metals* **21** (1968) 49.
56. M. F. W. Heberlein, *Trans. Electrochem. Soc.* **77** (1940) 119.
57. D. Schlain, *J. Electrochem. Soc.* **95** (1949) 145.
58. R. Beau, US 3 108 934 29, 29 Oct. 1963.
59. G. S. Rao, R. Krishna Iyer and S. G. Deshpande, *Indian J. Technol.* **10** (1972) 275.
60. P. P. Baiborodov, A. V. Uvarova, I. K. Gerasimov, A. B. Ezhkov and N. A. Kolbein, *Ger. Offen.* 2 063 307 (Cl. C22b), 6 July 1972.

61. E. V. Kantslev, V. M. Rudoi, A. I. Zevin and M. S. Khadyev, *Chem Abstr.* **81** (1974) 144623a.
62 A. G. Preussag, *Metall, Belg.* 861 488(C1, C25C) 31 March 1978.
63. A. Kisza, K. Gatner, J. Urbanowicz and M. Kusowski, Pol. PL 122 590(C1, C25C1/22), 5 April 1984.
64. L. E. Vaaler, *Electrochem. Technol.* **5** (1967) 170.
65. R. L. Murray and M. S. Kirchev, *Trans Electronchem. Soc.* **86** (1944) 83.
66. I. Miura, *Oyo Buturi (Monthly Publ. Jpn. Soc. Appl. Phys.) (in japanese)* **24** (1955) 203.
67. A. J. Singh, B. S. Mathur, P. Suryanarayana and S. N. Tripathi, *Proc. Symp Non-ferrous Metals Vol. III* (Jamshedpur, India, 1968) p. 82.
68. A. Fuwa and Y. Fukunaka, *Proc. Mining and Mineral Processing. Inst. Japan (Fall meeting, 1996)* B-4-7, p. 123.
69. I. L. Barker, *J. Met.* **8** (1956), 1058.
70. W. C. Smith and P. J. Hickey, *Trans. Met. Soc. AIME* **224** (1962), 224.
71. M. Maone and R. Tarrico, *Metall. Ital.* **62** (1970), 482.
72. R. A. Zingaro and W. C. Cooper, *Selenium* (Van Nostrand, New York, 1957).
73. J. E. Hoffmann, *J. Met.* **42** (1989) 33.
74. J. E. Hoffmann, *J. Met.* **42** (1989) 50.
75. Von. T. H. Renner, *Z. Anorg. Allg. Chem.* **298** (1959) 30.
76. D. Belforti, S. Blum and B. Bovarnick, *Nature* **4779** (1961) 901.
77. O. J. Guentert, R. L. Mozzi, *Nature* **4815** (1962) 570.
78. D. M. Chizikov and V. P. Shchastilivyi, *Tellurium and Tellurides* (Collet's Publ., 1970).
79. H. E. Hirsch, S. C. Liang and A. G. White, *Semicon. Semimet.* **18** (1981) 21.
80. A. B. Bollong and R. P. Bult, *JOM* **July** (1989) 39.
81. P. H. Jennings, in *Tellurium*, ed. by W. C. Cooper, (Van Hostrand Reinhold. Co., New York, 1971) p. 14.
82. K. Zanio, *Semicon. Semimet. Vol. 13*, ed. Willardson and Beer, (Academic Press, 1978)
83 J. E. Hoffman, *J. Met.* **41** (1989) 32.
84. J. E. Hoffman, *J. Met.* **42** (1990) 50.
85. B. Schaub and C. Potard, *Proc. Int. Symp. Cadmium Telluride Mater. Gamma Ray Detectors* (Strasbourg, 1971) p. 29.
86. L. Kuchar, J. Drapala and J. Linacek, *J. Crystal Growth* **161** (1996) 94.
87. J. Drapala, L. Kuchar and G.S. *Burkhanov* **34** (1998) 165.
88. N.R. Munirathnam, D.S. Prasad, Ch. Sudheer and T.L. Prakash, *J. Crystal Growth* **254** (2003) 262.
89. G.A. Pasteur, *J. Amer. Ceram. Soc.* **56** (1973) 548.
90. Y. Kubota, *Denshi Zairyo (Electronic Materials) (in japanese)* **July** (1985) 34.
91. Von. T. H. Renner, *Z. Anorg. Allg. Chem.* **298** (1959) 30.
92. D. Belforti, S. Blum and B. Bovarnick, *Nature* **4779** (1961) 901.
93. O. J. Guentert, R. L. Mozzi, *Nature* **4815** (1962) 570.
94. M. Basche and D. Schiff, *Materials in Des. Engeer.* **2** (1964) 78.
95. E. Shibutami, *Nikkęi New Materials (in japanese)* **May** (1988) 36.
96. S. Nishikawa, *Metals (in japanese)* **58** (1988) 30.
97. M. Isshiki, *Metals (in japanese)* **58** (1988) 21.
98. U. Kumashiro, *Metals (in japanese)* **58** (1988) 46.
99. A. Katsui, *Metals (in japanese)* **58** (1988) 29.
100. M. Nanjo, *Kinozairyo (in japanese)* **58** (1988) 5.
101. S. Goto, *Metals (in japanese)* **58** (1988) 34.

INDEX

$I-III-VI_2$ 360
$I-III_5-VI_8$ 352
$II-III_2VI_4$ 361
15R-SiC 214
3C-SiC 211, 219, 229, 242
4H-SiC 216
6H-4H transformation 244
6H-SiC 214

A

α-HgI_2 153
a-plane 58, 226
Absorption edge 339
Accelerated crucible rotation technique (ACRT) 157, 161, 211, 355
Acheson method 209
Acidic mineralizer 54
Activation analysis 252
$AgGaS_2$ 349, 355
$AgInTe_2$ 349
Air-bridge lateral epitaxial growth (ABLEG) 69
Alcoa method 376
$AlGa_{1-x}As_x$ 20
Aluminum nitride (AlN) 30, 73
Aluminum antimonide (AlSb) 73
Ammonia nitridation 43
Ammonothermal method 54
Anion exchange 370
Annealing 188
Anti-phase-boundary (APB) 232
Antimony 381
Antisite defects 82, 263
Arsenic 380
Atomic force microscopy (AFM) 253
Avalanche photodiode (APD) 276, 280

B

β-HgI_2 153
Back-reflection X-ray topography 257
BaF_2 324
Barikov Process 237
Basal plane dislocation (BPD) 242, 246, 256, 275
Basic mineralizer 54
Bipolar junction transistor (BJT) 289
Birefringence 365
Bismuth 382
Blue LD 97
Blue LED 94, 279
Boron nitride (BN) 19, 29, 73, 85
Boric oxide (B_2O_3) 11, 385
Boron phosphide (BP) 73
Bound exciton (BE) 270
Breakdown field 282
Breakdown voltage 208, 276
Bridgman method 154, 355, 317, 381
Burgers vector 242, 256
Byer method 376

C

Cadmium 373
Carbon inclusion 257
Carbon vacancy 260, 263
Carrier concentration 139
Carrot defect 276
Cathodoluminescence (CL) 89, 141, 253, 275
Cation exchange 370
$Cd_{1-x}Ze_xTe$ (CZT) 194
Charge injection device (CID) 194
Charged coupled device (CCD) 194
Chemical vapor deposition (CVD) 232
Chemical vapor transport (CVT) 7, 136, 176, 331
Chemomechanical polishing 142
CIGS($Cu(In, Ga)Se_2$) 361, 363
CIS ($CuInSe_2$) 361, 363
Closed space vapor transport (CSVT) 41, 86, 235, 365
Cold isostatic press (CIP) 387
Cold traveling heater method (CTHM) 322
Compensation ratio 267
Complex defect 263
Constitutional supercooling 12
Contactless electroreflectance spectroscopy (CER) 58
Continuous feed physical vapor transport (CF-PVT) 229
Copper smelting 383
Couette flow 157
Cr doping 85
$Cu-III-V_2$ 359
Cubic BN (c-BN) 29
$CuGaSe_2$ 349
$CuInSe_2$ 349

Czochralski (CZ) method 10, 14, 15, 358

D

D-A pair emission 270
Deep band emission (DBE) 141
Deep level 82, 261, 272
Deep level optical spectroscopy (DLOS) 84
Deep level transient spectroscopy (DLTS) 84, 184, 260
Digital versatile disk (DVD) 76
Dimethylsulfoxide (DMSO) 168
Directional freezing 352
Directional solidification 21
Dislocation 75, 138, 179, 246, 255, 275, 333
Distillation 373, 382
Divacancy 82, 260, 263
Donor-acceptor (DA) pair 91, 272
Double positioning boundary (DPB) 229
Dowa Mining method 377
Drift mobility 87
Drift velocity 208

E

E etchant 76
e-h recombination 277
Eagle-Picher method 376
Edge dislocation 242
Edge emission 141, 270
Ekman flow 157
Electroluminescence (EL) 275
Electrolysis 369, 373, 382, 384
Electron affinity 100
Electron back scattering diffraction (EBSD) 229
Electron energy loss spectroscopy (EELS) 257, 260
Electron paramagnetic resonance (EPR) 260
Electron spin resonance (ESR) 79, 84, 261
Electron-beam-induced current (EBIC) 256
Electronegativity 205
Electrorefining 382
Epitaxial lateral overgrowth (ELO, ELOG) 68, 76, 97, 98
Energy dispersive X-ray spectroscopy (EDXA) 91, 165, 253
Etch pit density (EPD) 15, 222, 244, 334
Etch back process 247
Exciton emission 270

F

Facet controlled epitaxial lateral overgrowth (FACELO) 69
Facet initiated epitaxial overgrowth (FIELO) 68
Facet 240
Far infrared photodetector 343
Fe doping 85
Fermi level 270
Fermi level pinning 333
Field emitter 100
Filtration 375
Flat panel X-ray imaging sensor 196
Flux method 50, 130
Forced flux method (FFM) 172
Fourier transform infrared (FTIR) spectroscopy 178, 263
Frank screw dislocation 331
Free exciton (FE) 141, 270
Freestanding 38, 72

G

Ga-Na flux 50
Gallium oxide (Ga_2O, Ga_2O_3) 66
Gallium arsenide (GaAs) 71
$GaAs_{1-x}P_x$ 19
Gallium chloride ($GaCl_3$) 62
Gallium hydride (GaH_x) 67
Gallium 376
Gamma-ray detector 195
Gas fed top-seeded solution growth (GF-TSSG) 212
Gate turn-off thyristor (GTO) 289
Germanium telluride (GeTe) 328
Glow discharge mass spectroscopy (GDMS) 252
Gradient freezing 45, 352
Graphene 92
Graphite 387
Green LD 93, 97
Gunn oscillator 22

H

Halide CVD (HCVD) 235
Halide vapor transport epitaxy (HVTE) 43
Hall measurement 264
Hetero bipolar transistor (HBT) 100
Hexagonal BN (h-BN) 29, 386
$Hg(Br_xI_{1-x})_2$ 167
$Hg_{1-x}Cd_xI_2$ 167
$Hg_{1-x}Cd_xTe$ 151, 153
$Hg_{1-x}In_xTe$ 165
$Hg_{1-x}Zn_xTe$ (MZT) 165
HgI_2 152
High electron mobility transistor (HEMT) 99
High pressure solution growth (HP-SG) 45
High purity semi-insulating (HP-SI) 227
High resolution X-ray diffraction (HRXRD) 77, 143, 253
High temperature CVD (HT-CVD) 43, 234

High-resolution photoinduced transient spectroscopy 261
Hydride vapor phase epitaxy (HVPE) 42, 67
Hydrothermal method 131

I

In-process etching 227
$In_{1-x}Ga_xAs$ 10
$In_{1-x}Ga_xP$ 8
$In_{1-x}Ga_xSb$ 14
$In_{1-x}Px_xAs$ 18
Inclusion 240, 254, 257
Indium 377
Indium nitride (InN) 72
Infrared focal plane array (IRFPA) 194
Infrared laser 340
$InSb_{1-x}Bi_x$ 20
Insulated gate bipolar transistor (IGBT) 251, 289
Interstitial 82, 260
Ion exchange 370
Ion implantation 261
Ionicity 205
Ionization energy 81, 87, 88, 187, 190, 250
IR laser diode (IR-LD) 341
Isothermal current transient spectroscopy (ICTS) 84, 261
Isotropic graphite 388

J

Junction field effect transistor (JFT) 288
Junction termination extension (JTE) 285

K

KCl 62
KI 58
KNH_2 56
KOH etchant 255

L

Laser diode (LD) 22, 97, 144, 278
Lely method 208, 213
Li_3N 62
$LiAlO_2$ (LAO) 62
LiCl 62
$LiGaO_2$ (LGO) 62, 72
$LiGaTe_2$ 355
Light emission imaging (LEI) 275
Light emitting diode (LED) 21, 92, 93, 144, 278
Linear muffin-tin orbital (LMTO) method 184
$LiNH_2$ 56
Liquid crystal display (LCD) 94
Liquid Encapsulated Czochralski (LEC) 7, 10, 18, 19, 73, 316, 386, 356
Liquid phase electroepitaxy (LPEE) 8, 10, 14, 20
Liquid phase epitaxy (LPE) 61, 194, 245, 343, 358
Local vibration mode (LVM) 260
Long-wave infrared (VLWIR) 193
Low angle grain boundary (LAGB) 247, 257, 249
Low temperature photoluminescence (LT-PL) 252

M

M etchant 76
m-plane 58, 226
Macrodefects 254
Macrostep formation 259
Markov-Devydov method 328
Maximum oscillation frequency 290
Mechanical stress 334
Melt/crystal interface 158
Mercury 375
Mercury cadmium telluride (MCT) 151, 153
Mercury halide 176
Mercury indium telluride (MIT) 165
Mercury iodide (HgI_2) 152
Merged PiN Schottky (MPS) diode 278
Metal insulating (oxide)-silicon carbide (MISiC/MOSiC) 281
Metal organic chemical vapor deposition (MOCVD) 42, 194
Metal-insulator-semiconductor (MIS) 144, 194
Metal-oxide-semiconductor field effect transistor (MOSFET) 100, 275, 286, 289
Metal-semiconductor field effect transistor (MESFET) 276
Metal-semiconductor-metal (MSM) photodiode 98
Methyl trichlorosilane (MTCS) 237
Micro electromechanical systems (MEMS) 281
Microcathodoluminescence (MCL) 82
Microgravity 322
Micropipe 209, 220, 229, 241, 275
Micropipe closing 245
Micropipe density 241, 229
Micropipe-free 221, 228, 243
Microplasma 276
Mid-wave infrared (MWIR) 193
Mineralizer 54
Minority carrier lifetime 188
Minority carrier spectroscopy (MCTS) 261
Miscibility gap 3
Mixed dislocation 242
Mobility 139, 264
Mobility-lifetime measurement 184

Modified Lely method 216, 240, 243
Modified PVT (MPVT) 213
Molecular beam epitaxy (MBE) 72, 194
Monte Carlo simulation 139
Multiple bound exciton complex (MBEC) 272
Multiquantum well (MQW) 97

N

NaN_3 50
Nanocrystal-assembled (NAB) 72
NaOH etchant 255
Native defects 138, 141, 337
$NdGaO_3$ (NGO) 68, 71
Neutral mineralizer 54
Nitridation 43, 72
Nonlinear optical device 364
Nonstoichiometry 138, 181, 334, 336

O

Off-cut angle seed 247
On-resistance 282
Optical admittance spectroscopy (OAS) 261, 263
Optical beam induced current (OBIC) 276
Optical DLTS (ODLTS) 84
Optical emission microscopy (OEM) 275
Optical filter 365
Optical parametric oscillator 365
Optically detected magnetic resonance (ODMR) 260

P

$Pb_{1-x}Ge_xTe$ 328
$Pb_{1-x}Sn_xSe$ 321
$Pb_{1-x}Snx_xTe$ 316
$PbS_{1-x}Se_x$ 315
PbSe 315
PbTe 315
PENDEO 69
Periodic oscillation of the crystal temperature (POST) 172
Periodic oscillation of the source temperature (POCT) 172
Persistent photoconductivity (PPC) 84, 263, 333
Phase diagram 3, 153, 205
Phase separation 4
Phosphorus 378
Phosphorus doping 250
Photocapacitance spectroscopy (PCS) 84
Photoconductive detector 193
Photodetector 22
Photoinduced current transient spectroscopy (PICTS) 84, 184, 263
Photoluminescence (PL) 141, 250, 257, 260, 261 263, 270, 273, 275, 363
Photomagentoelectric effect (PME) 188
Photon-stimulated current (PSC) 184
Physical vapor transport (PVT) 34, 63, 168, 323
Pin diode 276, 280
Pinhole 249
Piper-Polish method 328
Pn diode 281, 285
Point defect 253, 260
Polar surface 143
Polishing 142
Polyethylene 176
Polygonization 334
Polytype 207, 239
Polytype inclusion 249
Positron annihilation spectroscopy (PAS) 80, 260, 263
Power added efficiency (PAE) 252
Power transistor 285
Pressure-controlled solution growth (PC-SG) 48
Pulling method 170
Purification 369
Purity control 226
Pyrolytic boron nitride (pBN) 386

Q

Quasi equilibrium condition 247
Quaternary III-V mixed crystal 20

R

Radiation detector 195, 281
Raman scattering 90, 257
Raman spectroscopy 183, 229, 270
Reactive ion etching (RIE) 71
Recombination-enhanced dislocation glide (REDG) 277
Recrystallization 332, 377, 386
Red luminescence (RL) 92
Reduction of growth rate 175
Repeated a-face (RAF) 224, 247
Repeated temperature modulation (RTM) 140
Rotational Bridgman method 11, 21

S

Saturated velocity 282
$ScAlMgO_4$ 144
Scanning electron microscopy (SEM) 253
Schottky barrier diode (SBD) 60, 275, 281, 284
Scintillator 144
Screw dislocation 242, 276

Second harmonic generator (SHG) 365
Secondary ion mass spectroscopy (SIMS) 74, 252
Seebeck coefficient 343
Seed polarity 240
Segregation 157, 160, 372
Selective etching 257, 258
Selenium 383
Self-buffer layer technique 140
Self-seeded PVT 36
Self-selecting vapor growth (SSVG) 328, 329
Semi-insulating (SI) 58, 84, 88, 251
Sensor 280
Short wavelength emitting devices 363
Signal processing in the element (SPRITE) 193
Silane (SiH_4) 235
Silicon inclusion 257
Silicon on insulator (SOI) 238
Silicon vacancy 260, 263
Simulation 230
Small angle grain boundaries (SAGB) 227, 334
Smart Cut process 238
SnTe 315
Solid sate reaction (SSR) 162, 360
Solid source mass spectroscopy (SSMS) 252, 333
Solid/liquid interface 20, 334, 359
Solubility 54
Solute redistribution 158
Solute solution diffusion (SSD) 8, 10
Solute-feeding Czochralski (SF-CZ) 15
Solution growth 33, 45, 167, 209, 322, 358
Solvent extraction 370
Space growth 175
Spinodal decomposition 4
Spontaneous nucleation 29
Spontaneous seeding 52
Spontaneous separation 72
Stacking fault (SF) 219, 242, 248, 257, 275
Star defect 257, 258
Static evaporation 168
Static induction accumulation field-effect transistor (SIAFET) 289
Static induction transistor (SIT) 288
Step bunching 259
Stoichiometry 183, 336
Striation 335
Structural defect 138, 254
Subgrain boundary 257, 258
Sublimation method 34, 213, 216, 380
Sublimation sandwich method (SSM) 65, 66
Submerged heater method (SHM) **15**
Supercell approach 247
Supercritical 54
Synchrotron white beam X-ray topography (SWBXT) 77, 253, 257, 275

T

TaC crucible 40
Tantalum container 231
Tellurium 384
Temperature difference method under controlled vapor pressure (TDM-CVP) 8, 253, 342
Temperature oscillation method (TOM) 171
Terahertz detector 341, 343
Thermal conductivity 208, 386
Thermal CVD method 386
Thermal stress 335
Thermally stimulated current (TSC) 184, 261, 263
Thermally stimulated luminescence (TSL) 84
Thermoelastic stress 247
Thermoelectric effect spectroscopy (TEES) 84
Thermoelectric generator 343
Thermogravimetry 65
Thermopower 343
Threading edge dislocation (TED) 256
Threading screw dislocation (TSD) 247
Time resolved four wave mixing 263
Tl doping 339
Top-seeded solution growth (TSSG) 131, 209
Transient charge technique (TCT) 184
Transmission electron microscopy (TEM) 77, 275
Transport properties 264
Traveling heater method (THM) 8, 19, 73, 160, 322, 358
Traveling liquidus zone (TLZ) 8, 12
Travelling-solvent floating zone (TSFZ) 131
Trimethylsilane (TMS) 229
Twin boundary (TB) 233
Two temperature horizontal Bridgman (2T-HB) 355

U

Ultraviolet (UV) photodetector 280
Ultraviolet (UV) LD 97
Ultraviolet (UV) LED 92
Undoped semi-insulating 251
Unidirectional freezing 322

V

Vacancy 81, 260, 263
Vacuum distillation 370, 373, 382
Van Arkel process 237
Van der Waals bond 153
Van Vechten's cavity model 361

Vanadium doping 222, 251
Vapor reaction 135
Vapor-liquid-solid (VLS) 8, 211, 329
Vaporization 175
Vertical cavity surface emitting laser (VSEL) 342
Violet LD 97
Violet LED 94

W

Water content 386
WC crucible 38

X

X-ray detector 196
X-ray radiography 253

Y

Yellow luminescence (YL) 91

Z

Zeeman spectroscopy 263
Zinc 372
$Zn_{1-x}Mg_xO$ 135
Zone leveling (ZL) 11, 158
Zone melting (ZM) 11, 158, 322, 352, 373, 382
Zone refining 372, 386